Managing Soil Quality

Challenges in Modern Agriculture

_____

Edited by

P. Schjønning, S. Elmholt and B.T. Christensen

Danish Institute of Agricultural Sciences
Research Centre Foulum
Tjele
Denmark

CABI Publishing

CABI Publishing is a division of CAB International

CABI Publishing
CAB International
Wallingford
Oxon OX10 8DE
UK

CABI Publishing
875 Massachusetts Avenue
7th Floor
Cambridge, MA 02139
USA

Tel: +44 (0)1491 832111
Fax: +44 (0)1491 833508
E-mail: cabi@cabi.org
Website: www.cabi-publishing.org

Tel: +1 617 395 4056
Fax: +1 617 354 6875
E-mail: cabi-nao@cabi.org

A catalogue record for this book is available from the British Library, London, UK.

Library of Congress Cataloging-in-Publication Data
Managing soil quality : challenges in modern agriculture / edited by P. Schjonning, S. Elmholt, and B.T. Christensen.
     p. cm.
Includes bibliographical references and index.
  ISBN 0-85199-671-X (alk paper)
  1. Soils--Quality. 2. Soil management. I. Schjonning, P. (Per) II. Elmholt, S. (Susanne) III. Christensen, B. T. (Bent Tolstrup)

  S591.M323 2004
  631.4--dc21   2003008927

ISBN 0 85199 671 X

Typeset by AMA DataSet, UK
Printed and bound in the UK by Biddles Ltd, King's Lynn

Contents

Contributors

Alabouvette, C., *CMSE-INRA, U.M.R. Biochimie, Biologie Cellulaire et Ecologie des Interactions Plantes Microorganismes, F 21065 Dijon, France.*

Andrews, S.S., *USDA-NRCS, Soil Quality Institute, 2150 Pammel Drive, Ames, Iowa 50011, USA.*

Askegaard, M., *Danish Institute of Agricultural Sciences, Department of Agroecology, PO Box 50, DK-8830 Tjele, Denmark.*

Backhouse, D., *University of New England, School of Environmental Sciences and Natural Resources Management, Armidale, New South Wales 2351, Australia.*

Bloem, J., *Wageningen University and Research Centre, Alterra Green World Research, PO Box 47, NL-6700 AA Wageningen, The Netherlands.*

Bouma, J., *Wageningen University and Research Center, Environmental Sciences Group, PO Box 47, NL-6700 AA Wageningen, The Netherlands.*

Brussaard, L., *Wageningen University and Research Centre, Sub-department of Soil Quality, PO Box 8006, NL-6700 EC Wageningen, The Netherlands.*

Burgess, L.W., *University of Sydney, School of Land, Water and Crop Sciences, New South Wales 2006, Australia.*

Carter, M.R., *Agriculture and Agri-Food Canada, Crops and Livestock Research Centre, 440 University Avenue, Charlottetown, Prince Edward Island C1A 4N6, Canada.*

Christensen, B.T., *Danish Institute of Agricultural Sciences, Department of Agroecology, PO Box 50, DK-8830 Tjele, Denmark.*

Condron, L.M., *Soil, Plant & Ecological Sciences Division, PO Box 84, Lincoln University, Canterbury 8150, New Zealand.*

Dick, W.A., *The Ohio State University, School of Natural Resources, 1680 Madison Avenue, Wooster, Ohio 44691, USA.*

Didden, W.A.M., *Wageningen University and Research Centre, Sub-department of Soil Quality, PO Box 8006, NL-6700 EC Wageningen, The Netherlands.*

Donovan, N.J., *New South Wales Agriculture, Elizabeth Macarthur Agricultural Institute, Private Mail Bag 8, Camden, New South Wales 2570, Australia.*

Drinkwater, L.E., *Cornell University, Department of Horticulture, Plant Science Building, Ithaca, New York 14853, USA.*

Edel-Hermann, V., *CMSE-INRA, U.M.R. Biochimie, Biologie Cellulaire et Ecologie des Interactions Plantes Microorganismes, F 21065 Dijon, France.*

Elmholt, S., *Danish Institute of Agricultural Sciences, Department of Agroecology, PO Box 50, DK-8830 Tjele, Denmark.*

Eriksen, J., *Danish Institute of Agricultural Sciences, Department of Agroecology, PO Box 50, DK-8830 Tjele, Denmark.*

de Goede, R.G.M., *Wageningen University and Research Centre, Sub-department of Soil Quality, PO Box 8006, NL-6700 EC Wageningen, The Netherlands.*

Goossens, D., *Catholic University of Leuven, Laboratory for Experimental Geomorphology, Redingenstraat 16, B-3000 Leuven, Belgium.*

Govers, G., *Catholic University of Leuven, Laboratory for Experimental Geomorphology, Redingenstraat 16, B-3000 Leuven, Belgium.*

Gregorich, E.G., *Agriculture and Agri-Food Canada, Central Experimental Farm, Ottawa K1A 0C6, Canada.*

Johnston, A.E., *Rothamsted Research, Harpenden, Herts AL5 2JQ, UK.*

Karlen, D.L., *USDA-ARS, National Soil Tilth Laboratory, 2150 Pammel Drive, Ames, Iowa 50011, USA.*

Kay, B.D., *University of Guelph, Department of Land Resource Science, Guelph, Ontario N1G 2W1, Canada.*

Kuyper, T.W., *Wageningen University and Research Centre, Sub-department of Soil Quality, PO Box 8006, NL-6700 EC Wageningen, The Netherlands.*

Lal, R., *The Ohio State University, Carbon Management and Sequestration Center, Columbus, Ohio 43210, USA.*

Locke, M.A., *USDA-ARS, Southern Weed Science Research Unit, PO Box 350, Stoneville, Mississippi 38776, USA.* Present address: *National Sedimentation Laboratory, Water Quality and Ecological Processes Research Unit, PO Box 1157, Oxford, Mississippi 38455, USA.*

Megharaj, M., *University of South Australia, Centre for Environmental Risk Assessment and Remediation, Mawson Lakes Boulevard, Mawson Lakes, South Australia 5095, Australia.*

Munkholm, L.J., *Danish Institute of Agricultural Sciences, Department of Agroecology, PO Box 50, DK-8830 Tjele, Denmark.*

Naidu, R., *University of South Australia, Centre for Environmental Risk Assessment and Remediation, Mawson Lakes Boulevard, Mawson Lakes, South Australia 5095, Australia.*

Owens, G., *University of South Australia, Centre for Environmental Risk Assessment and Remediation, Mawson Lakes Boulevard, Mawson Lakes, South Australia 5095, Australia.*

Poesen, J., *Catholic University of Leuven, Laboratory for Experimental Geomorphology, Redingenstraat 16, B-3000 Leuven, Belgium.*

Schjønning, P., *Danish Institute of Agricultural Sciences, Department of Agroecology, PO Box 50, DK-8830 Tjele, Denmark.*

Steinberg, C., *CMSE-INRA, U.M.R. Biochimie, Biologie Cellulaire et Ecologie des Interactions Plantes Microorganismes, F 21065 Dijon, France.*

Van den Akker, J.J.H., *Wageningen University and Research Centre, Alterra Green World Research, PO Box 47, NL-6700 AA Wageningen, The Netherlands.*

Wienhold, B.J., *USDA-ARS, Soil and Water Conservation Research Unit, University of Nebraska, East Campus, Lincoln, Nebraska 68583, USA.*

Zablotowicz, R.M., *USDA-ARS, Southern Weed Science Research Unit, PO Box 350, Stoneville, Mississippi 38776, USA.*

Preface

'*Soil quality is how well soil does what we want it to do.*' The statement, extracted from the website of the USDA Soil Quality Institute, represents the very essence of the soil quality concept. At first glance, one might be tempted to leave the subject, overwhelmed by the enormous complexity embedded in this statement. Alternatively, one might analyse the two aspects of soil quality separately: 'how well' relates to grading soils, while 'what we want' relates to priority of soil functions. Most previous books on soil quality have emphasized the descriptive grading of soils or management effects, often by focusing on soil-quality indicators. In the editorial group, however, we were more concerned with soil quality as it relates to what we want the soil to do. Clearly, we must define what we want before we can consider how well this service is delivered. What glues the two aspects together is soil management. Our ambition was therefore to switch from a more passive attitude to a more active and management oriented attitude. The readers will judge how successful we have been, but if this book can promote a revitalized discussion on the soil quality conceptual framework, we certainly consider our effort worthwhile.

Science is a human activity, and science and society interact. The focus of science will inevitably reflect the priorities of society. In societies with a shortage of food supply, the focus will be on soil productivity, while in developed countries with an abundant supply of affordable food, the focus will switch from sheer productivity to the overall sustainability of the food production activities. Sustainable agriculture involves a sustained productivity but also the protection of natural resources. The concept of soil quality is deeply rooted in considerations on sustainable production, but since the priorities of society change over time and differ from one society to another, soil quality cannot be aligned with the universal laws of nature. The concept of soil quality is a human construct allowing specific soil functions to be evaluated against specific purposes.

In this book, we have identified a number of specific challenges in modern agriculture. All contributors were encouraged to identify the thresholds in terms of management, which are necessary to secure soil quality. This does not mean that soil-quality indicators are not discussed, but the focus of this book is on management and the identification of research needs and implementation of existing knowledge. Although most contributions are concerned with challenges facing industrialized countries, the book also includes a chapter considering soil quality in developing countries.

The editorial work was financially supported by the Danish Ministry of Food, Agriculture and Fisheries and based on a decision by the Parliament to create an overview of the influence of modern agriculture on soil quality. We wish to thank the Danish Institute of Agricultural Sciences for hosting the project. Thanks to Jesper Waagepetersen, Head of The Department of

Agroecology, who chaired the project executive committee, and to committee members for fruitful suggestions during the initial phases of the work.

The book links to the efforts of the European Society for Soil Conservation (ESSC). In 1998, the ESSC decided to accentuate a number of soil protection issues and the senior editor was appointed to lead a task force on the soil quality concept. This was a main incentive to give the present work a truly international perspective. We sincerely hope that the book will serve to facilitate communication among scientists and between scientists and decision makers in society. Although still developing, the soil quality concept may be useful in the creation of 'codes of conduct' by governments and inter-governmental organizations.

We wish to thank all contributors to the book and we most gratefully acknowledge their efforts. Although busy with numerous other serious commitments, these distinguished experts took the time to prepare high quality manuscripts. Thanks also to all anonymous referees for reviewing the contributions.

We thank Dr Hugo Fjelsted Alrøe at the Danish Research Center for Organic Farming (DARCOF) for fruitful discussions on the role of values in science. We acknowledge the very positive and constructive cooperation with Tim Hardwick and colleagues at CABI Publishing. Last but not least, we thank Ms Anne Sehested for carrying out the tedious work of bringing the contributions into full accordance with the requests of the publisher. Part of the editorial work was linked to ongoing projects financed by DARCOF (ROMAPAC, PREMYTOX and NIMAB).

Per Schjønning, Susanne Elmholt, Bent T. Christensen
Research Centre Foulum
March 2003

Chapter 1
Soil Quality Management – Concepts and Terms

P. Schjønning, S. Elmholt and B.T. Christensen

Danish Institute of Agricultural Sciences, Department of Agroecology, PO Box 50, DK-8830 Tjele, Denmark

---

Summary

The industrialization of agriculture and the concurrent increase in societal concerns on environmental protection and food quality have put the focus on agricultural management and its impact on soil quality. Soil quality involves the ability of the soil to maintain an appropriate productivity, while simultaneously reducing the effect on the environment and contributing to human health. This development has changed society's expectations of science and there is an urgent need to improve the communication among researchers from different scientific disciplines. The interaction of scientists with decision makers is a topic of utmost relevance for future developments in agriculture. *Reflexive objectivity* denotes the exercise of raising one's consciousness of the *cognitive context*, i.e. societal priorities, and the values and goals of the researcher. The term *sustainability* comprehends the priorities in the cognitive context and thus constitutes a valuable tool for expressing the basis of scientific work. Soil quality evaluations should include awareness of the *stability* of any given quality attribute to disturbance and stress. This implies addressing *resistance* and *resilience* of the soil functions and/or the physical form in question. Most existing literature on soil quality focuses on assessment of soil quality rather than the management tools available to influence soil quality. Identification of *management thresholds* rather than soil-quality *indicator thresholds* is suggested as an important means of implementing the soil quality concept. The major challenges facing modern agriculture include proper nutrient cycling, maintained functions and diversity of soil, protection of an appropriate physical form and avoidance of chemical contamination. It is suggested that these challenges and problems as related to the soil quality concept are discussed in the framework expounded above.

©CAB International 2004. *Managing Soil Quality: Challenges in Modern Agriculture* (eds P. Schjønning, S. Elmholt and B.T. Christensen)

Agricultural Research in a Changing World

The foundation of modern agriculture was laid more than 150 years ago. At that time an awareness of the role of plant nutrients in crop production emerged, supported by experiments showing the beneficial effects of adding mineral fertilizers to the soil. However, the most rapid development has occurred since the early 1950s. This development has been driven not only by scientific achievements, but also by access to affordable energy, traction power and other technological achievements that reduced the time and manpower required for agricultural production. Mineral fertilizers, pesticides and cultivars that respond effectively to increased nutrient levels were important requisites in the dramatic increase in productivity. The development of modern agriculture was supported by government policies introducing systems of production and commodity subsidies with the overall aim to secure adequate and reliable sources of food of good quality and at affordable prices. The side effects were structural changes towards larger and more specialized production units and a massive movement of labour force from agriculture to the industry and service sectors. Government policies also involved a substantial increase in the research supporting agricultural production.

In the developed and industrialized countries, modern agriculture achieved these primary goals, and even more so, as demonstrated by surplus production and subsidized export of agricultural products. This has contributed to a switch in societal concerns from sheer productivity to sustainability of agriculture, including the effects of production methods on the environment, the diversity of the natural flora and fauna, the welfare of domestic animals, and the soil resource itself. The quality of air, water and – as yet to a minor extent – soil has come more into focus.

Almost every aspect of modern agriculture is now under scrutiny from concerned producers, environmentalists and consumers, from researchers and government as well as non-governmental organizations, and

agricultural sustainability is on the agenda of most political movements and parties. Concerns, attitudes and opinions about agricultural production are effectively communicated and amplified by news media. At the same time, the number of economic subsidies devoted to agriculture is being questioned. The demands for economic and ecological sustainability are bound to introduce changes in the production concepts of modern agriculture. This development has increased the demand for scientifically based solutions that incorporate a wider range of aspects. Scientists have been involved in problem solving and development in society for centuries, but the pressure from society for a proactive role of science is much greater than previously.

Another aspect is the increased interaction between *descriptive* and *prescriptive* branches of science (Ellert *et al.*, 1997). Typically, scientists in ecology, geography and other classical scientific disciplines perceive soil as an ecosystem component, and their approach is descriptive and observational in nature. Agricultural researchers, on the other hand, are concerned primarily with the production of food and fibre, and perceive soils mainly as media to support plant growth. Fertility trials, crop rotation studies, tillage experiments, etc. have provided the basis for an increasing productivity. Thus, researchers involved in agricultural sciences are accustomed to producing prescriptions with the clear aim of increasing yields. Ellert *et al.* (1997) advocated a combination of the conceptual/descriptive approaches of ecologists and the quantitative/prescriptive approaches of agronomists.

However, the vast amount of scientific literature concerned with ecosystem health, sustainable farming, soil fertility and soil quality reveals problems in communication. As an example, Doran *et al.* (1996) reported on communication failures due to different opinions on the use of values in science. In the section below we discuss some basic issues regarding the role of science in society, which we believe may facilitate communication. The philosophical deductions should be regarded as a layman's view, not as a professional contribution to the theory of science.

Science and Society – the Need for Reflexive Objectivity

Agricultural research is an applied science with the main objective of improving production methods and developing production systems. In consequence, agricultural science influences its own subject area, agriculture, in important ways (Lockeretz and Anderson, 1993). In general, science that influences its own subject area is defined as *systemic science* (Alrøe and Kristensen, 2002). This characteristic is also true for health, environmental and engineering sciences. The fact that science plays a proactive role in the world that it studies makes the criterion of objectivity as a general scientific ideal less straightforward. The general understanding of objectivity is derived from the *positivistic* criterion of verifiability of knowledge. Freeman and Skolimowski (1974) defined 'object' as 'the totality of external phenomena constituting the not-self' and hence 'objective' as 'something that is external to the mind'. That is, objectivity is defined as opposite to the subjective. However, when the 'subject' (the scientist in systemic sciences) is part of the 'object' (the system studied), an extra dimension is added to his/her role as a scientist. It is, therefore, important that the scientist is able to view her- or himself as part of the system (self-reflection). As an example, the researcher involved in the optimization of crop yields by management strategies should be able to recognize the consequences of his/her prescriptions on other aspects than just yield. This ability to take an 'objective' stance but at the same time being aware of the intentional and value-laden aspects of science is denoted *reflexive objectivity*, and the framework in which these reflections take place is labelled the *cognitive context* (Alrøe and Kristensen, 2002). The cognitive context may be divided into three dimensions: the observational, the societal and the intentional (Fig. 1.1). The observational context includes the actual methodological aspects of the research, the societal context is the group or segment for which the research is relevant, and the intentional context is the goals and values employed.

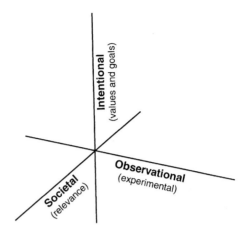

Fig. 1.1. The three dimensions of the cognitive context in science. Cognition (perception) is dependent on observational aspects (experimental set-up), societal priorities, and intentions and goals for the scientist or scientific group performing the scientific studies. Please consult the text for details. Based on Alrøe and Kristensen (2002).

The *observational context* comprises the characteristics of a scientific work, which are evaluated by the procedure of peer review (such as the experimental set-up, statistical treatment of data and discussion of results in relation to other relevant studies). The selection of research topics and the choice of methods will frame the outcome of the work (Dumanski *et al.*, 1998), and the methodological aspects of a work are more important to the results and conclusions than often realized. For example, a study of phosphorus availability in soil might reach quite different conclusions depending on the analytical method. Extraction by sulphuric acid would yield much more P than a resin (anion exchange membrane based) methodology. Obviously, you would say. The point is, however, that when judging the results, one uses present-day knowledge of the lability of different P-pools in soil. There may well be plant–soil interactions of importance for P-uptake by plants that we have yet to realize. And such knowledge might induce new methodologies. Our cognition regarding P availability in soil is thus highly dependent on how we establish our analyses.

The relevance of the scientific work depends on the *societal context* pervading at the time of the study. There is no 'universal' science that is independent of social context. When pesticides became available to farmers in the mid 20th century, the most relevant task for agricultural researchers was to optimize their use for maximum production and minimum costs. When – on the other hand – an agricultural scientist is engaged in the development of organic farming, completely different topics dominate. The paradigm associated with organic farming gives priority to quality aspects of crops, soil and the environment. Concerning pesticides, today's scientists in industrialized countries are engaged in studies of the detrimental rather than the beneficial effects of pesticides (e.g. groundwater pollution, bioaccumulation, side effects on non-target organisms). These examples serve to illustrate that the societal context has changed dramatically during the period discussed here.

The *intentional context* in science is perhaps the most controversial. It has to do with values and goals for the specific research group or scientist. Sojka and Upchurch (1999) gave a critical review on the concept of soil quality. Some of their concerns were abstracted as 'we are . . . reluctant to endorse redefining the soil science paradigm away from the value-neutral tradition of edaphology and specific problem solving to a paradigm based on variable, and often subjective societal perceptions of environmental holism'. That is, the authors support the classical understanding of objectivity in science. In their paper, however, they draw attention to articles dealing with different aspects of soil quality and raise the query of whether a high biodiversity in soil is more valuable than animals at the other end of the food chain. We interpret their statement as giving a high production of foods (for higher animals) a higher priority than a high biodiversity in the soil. This is of course a legitimate standpoint, but the point is that this opinion also reflects an 'intention' or a 'value/goal'. Awareness of these values is what reflexive objectivity is all about. And the example clearly illustrates that reflexive objectivity in the 'room' of the cognitive context would facilitate or even be a

prerequisite for communication. We concur with the statement by Jamieson (1992) and Ellert *et al.* (1997) that frank discussions about the values involved in concepts like soil quality may be equally or more important than the technical development and use of indicators to manage ecosystems.

Scientific work cannot be fully understood when detached from the societal and intentional contexts. Campbell *et al.* (1995) stated that the classification of sustainability and 'health' of an agroecosystem require the establishment of specific judgement criteria, and concluded that such judgement criteria must be established from a viewpoint that is ecologically, politically, socially and economically acceptable. As stated by Munasinghe and Shearer (1995) there is bound to be conflict among such interests. The task of scientists is thus to provide information that enables decision makers to choose among conflicting objectives by assessing the trade-offs among these objectives and the consequences of their application.

The Soil Quality Concept

The term *quality* implies value judgement (degree of excellence). Thus *soil quality* is concerned with some measure of a property or function of soil (good/bad, low/high, etc.). Fundamentally, classification of data and information about soil seems to be a basic human need, and the concept of 'soil capability' ('good' or 'bad' for a specific purpose) is as old as civilization itself (Carter *et al.*, 1997). Patzel *et al.* (2000) stated that 'soil quality encompasses an indefinite (open) set of tangible or dispositional attributes of the soil'. Thus the concept of soil quality may be regarded as a 'vessel' for various attributes of interest in any given situation. As an example, soil quality in the context of highway constructions is concerned with the bearing capacity of the soil medium but does not consider soil functions for plant growth. Although some people may regard this open (indefinite) concept as truly 'academic' and of little use, we think it facilitates reflections on the value-laden character of the soil quality concept. Any decision on quality attributes enclosed by the concept of soil quality will necessarily be based on viewpoints, values and goals from the societal and intentional contexts.

Blum and Santelises (1994) and Blum (1998) considered the functions and services of soil as related to human activity and grouped them into six categories. Three ecological uses are: (i) the production of biomass; (ii) the use of soils for filtering, buffering and transforming actions; and (iii) the provision of a gene reserve for plant and animal organisms. Three other functions relate to non-agricultural human activities; (iv) a physical medium for technical and industrial structures; (v) a source of raw materials (gravel, minerals, etc.); and (vi) a cultural heritage. This classification of human interest in and interaction with soil may facilitate an operational definition of soil quality.

Several definitions of soil quality have been advanced (see Karlen *et al.*, Chapter 2, this volume). Most definitions relate soil functions to: (i) biological productivity; (ii) the environment; and (iii) different expressions of plant, animal and/or human health (e.g. Doran and Parkin, 1994; Doran *et al.*, 1996). A committee appointed by the Soil Science Society of America (SSSA) offered the following definition (Fig. 1.2): *Soil quality is the capacity of a specific kind of soil to function, within natural or managed ecosystem boundaries, to sustain plant and animal productivity, maintain or*

enhance water and air quality, and support human health and habitation (Allan *et al.*, 1995; Karlen *et al.*, 1997). Although this definition creates a framework for considering soil quality, it does not eliminate the value-laden character of the concept. To determine soil quality, the functions or services expected of the system must be defined and delineated (Ellert *et al.*, 1997). Judgement of what is good or bad is influenced by subjective and/or societal priorities and decisions. Accordingly, Pankhurst *et al.* (1997) noted that most authors contributing to their book on biological indicators of soil health emphasized the holistic nature of the soil health concept and accepted subjective assessments of what is healthy. The same holds for the soil quality concept.

Early papers on soil quality emphasized terms like *fitness for use* in regard to agricultural use of soil (Larson and Pierce, 1991, 1994). Letey *et al.* (2003) preferred the term *use* to *function* because 'use' highlights the management aspect of the term. However, a function like carbon sequestration in soil and its interaction with greenhouse gases occurs irrespective of agricultural use; only the magnitude of this soil function can be manipulated by agricultural management.

As opposed to other definitions of soil quality, the SSSA definition mentions humans only in the 'health' part of the text. Concerns regarding plants and animals are associated with the 'productivity' part. We find this noteworthy because the expression 'promotion of plant and animal health' (Doran and Parkin, 1994; Doran *et al.*, 1996) in its extended interpretation is very ambitious (animals include nematodes and collembola, for example). We agree that the activity and the diversity of the soil community are important, and that a large biomass and a high biodiversity in soil may link to the degree of soil quality. However, our attitude emphasizes that agriculture by definition is a human activity designed for the production of food and fibres.

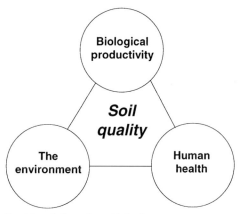

Fig. 1.2. Soil quality with its three concerns: biological productivity, the environment and human health. Based on Allan *et al.* (1995).

Sustainability

The term 'sustainability' is frequently used in scientific papers dealing with agricultural systems and is closely linked to societal and individual priorities. The term may be regarded as a manifestation of priorities, values and goals of researchers and society. A link between soil quality and sustainability is important because soil quality should not remain an abstract concept but rather something to be strived for by management (Bouma *et al.*, 1998).

Sustainability entered public debate following the work of the World Commission on Environment and Development, labelled the 'Brundtland Report' (WCED, 1987). To *sustain* means to 'keep up, maintain' (Oxford Advanced Learner's Dictionary of Current English, 1974). If applied only in this sense, sustainability does not make much sense for the constantly changing human society. Originally, sustainability more accurately translates into sustainable *development* (Bossel, 1999). Accordingly, the concept of sustainable development was proposed by the Brundt-land Commission as 'economic development that meets the needs of the present generation without compromising the ability of future generations to meet their own needs' (WCED, 1987).

When applying the concept of sustain-ability to agriculture, a somewhat more tangi-ble definition has to be constructed, although Swift (1994) noted that the concept would still be complex. It would embody issues of economic viability, the quality of life and human welfare, and ecological stability and resilience over time. Several other papers and documents have discussed the issue of sustainability in greater depth, all emphasiz-ing the combination of biophysical and social aspects of the concept (e.g. Stewart *et al.*, 1991; Smyth and Dumanski, 1993; Lal, 1994, 1998; Herdt and Steiner, 1995; Munasinghe and Shearer, 1995).

Smyth and Dumanski (1993) stated that

> Sustainable land management combines technologies, policies and activities aimed at integrating socio-economic principles with environmental concerns so as to simulta-neously: (i) maintain or enhance production and services; (ii) reduce the level of produc-tion risk; (iii) protect the potential of natural resources and prevent degradation of soil and water quality; (iv) be economically viable; and (v) socially acceptable.

Bouma *et al.* (1998) underscored the five criteria for sustainability in this definition, i.e. productivity, security, protection, viability and acceptability, and suggested that these criteria should also be used for judging soil quality. We have adopted this suggestion for framing the soil quality discussion in this book. We further endorse the viewpoint of Stewart *et al.* (1991) and Pankhurst (1994) that sustainability should be considered dynamic because, ultimately, it will reflect the changing needs of an increasing global population.

Stability in Terms of Resistance and Resilience

Evaluation of systems requires estimates of their stability when stressed or disturbed. Stability may express: (i) the *resistance* to change in function or form during a stress event, or (ii) the capacity to recover functional and structural integrity (*resilience*) after a disturbance. It is important to distinguish between resistance and resilience. In popu-lation ecology, resistance is defined as 'the capacity to resist displacement from an equilibrium condition', whereas resilience is defined as 'the capacity of a population (or system) to return to an equilibrium following displacement in response to a perturbation' (Swift, 1994). We tend to follow Seybold *et al.* (1999) by using the term *resistance* instead of *stability*, which occasionally has been used to express the capacity of resisting disturbance (e.g. Kay, 1990). We find that stability is more appropriate as a common denominator for resistance and resilience.

Eswaran (1994) emphasized that soil res-ilience relates to either '*performance*' or '*state or structure*' of the system. The same applies to resistance. According to Eswaran, 'perfor-mance' refers to functions and processes in the soil while 'state or structure' refers to the pedological composition of the material. The latter is analogous to the *structural form* (Kay, 1990), although Eswaran had a larger time span in mind than Kay. Thus, resilience relates to the ability of recovering functions as well as to physical form. Figure 1.3 illustrates the relationship between the terms discussed.

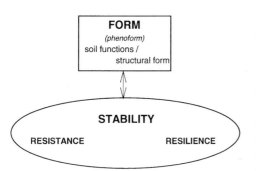

Fig. 1.3. Soil stability in terms of resistance and resilience as related to the suggested term *form* comprising soil functions as well as structural form. A stable form may be due to a high resistance and/or a high resilience. The arrow indicates that a given stability is assigned to a given form, but also that the stability may change with a change in (pheno)form. Based on Kay (1990) and Droogers and Bouma (1997).

A soil may exhibit a high resistance but a poor resilience with respect to some specific property. This would be the case if subjecting a dry clay soil to heavy mechanical loads. The soil strength and thus its resistance to compaction is high. If, however, the 'structural form' collapses, which would happen at a very high load, it would probably be associated with a compaction along the 'virgin compression line' (Larson *et al.*, 1980). And the resilience – the ability to recover – from such compaction effects is poor (e.g. Håkansson and Reeder, 1994). Alternatively, a soil may exhibit a poor resistance but a high resilience for some attribute. A number of microbial soil functions show examples of this when subjected to, for example, pesticide applications. Pesticides may cause response deficits of more than 90% and yet the soil function may return to its original level so quickly that the ecotoxicological effect can be regarded as insignificant when compared to natural stress effects (Domsch *et al.*, 1983).

Although the stability of soil systems should be assessed both in terms of resistance and resilience, the latter property particularly deserves attention when evaluating soil quality in managed ecosystems. As any form of agriculture disturbs the original equilibrium of the native ecosystem, it is evident that resilience is a key parameter when judging

the sustainability of agricultural systems. The concept of resilience was originally coined by Holling (1973) with emphasis on the persistence of *relationships* within a system. Resilient systems may show the capacity to occupy more than one state of equilibrium (Swift, 1994). Each state of equilibrium may maintain a qualitative structural and functional integrity but the quantitative properties may differ among equilibria. This dimension of the resilience concept is crucial when dealing with managed ecosystems. Any form of agricultural activity disturbs the original equilibrium of the native ecosystem, and soil resilience can be invoked to connote the ability of management to maintain the performance of the soil (Eswaran, 1994). This interpretation may be controversial, but is logical when dealing with managed ecosystems. Management is an integrated part of the agroecosystem, and resilience should be related to equilibria in the *managed* system, not the performance or state that would prevail in the *original, native* ecosystem (Blum, 1998).

Resilience has been defined from various points of view for various purposes (Szabolcs, 1994). One important aspect is the time scale. The rate of soil formation from the parent rock is extremely low compared with the potential rate of soil loss in unsustainable agricultural systems (Lal, 1994; Pennock, 1997). Lal (1994) reviewed the estimates of rates of soil formation for a number of soil types and concluded that most soils can be considered a non-renewable resource within the human life span. However, a soil subjected to severe gully erosion may be judged resilient also to this disturbance if regarded in the context of geological time spans of hundreds or thousands of years. Thus, the time factor has to be considered when discussing soil resilience.

It should be emphasized that the expression of resilience has no meaning without an explicit statement of the agents, forces or effects (disturbance) facing the soil (Szabolcs, 1994). Blum (1998) discussed the potential 'disturbances' and classified the corresponding 'type' of resilience into three groups: (i) resilience to physical disturbances; (ii) chemical resilience; and (iii) resilience to biological disturbances.

Soil-quality Indicators

Soil quality assessment typically includes the quantification of *indicators* of soil quality. Such indicators may be derived from reductionistic studies, i.e. specific soil parameters obtained from different disciplines of soil science (e.g. Larson and Pierce, 1991). However, descriptive indicators, which are inherently qualitative, can also be used in assessing soil quality (Seybold *et al.*, 1998; Munkholm, 2000). Soil-quality indicators condense an enormous complexity in the soil. They are measurable surrogates for processes or end points such as plant productivity, soil pollution and soil degradation (Pankhurst *et al.*, 1997). Herdt and Steiner (1995) and Carter *et al.* (1997) drew attention to situations where individual indicators show opposite or different trends. Larson and Pierce (1994) and later Doran and Parkin (1996) realized the weaknesses in expressing soil quality information in single numbers, at least in comparative studies of soil management. As stated by Doran and Parkin, such indicators may provide little information about the processes creating the measured condition or performance factors associated with respective management systems. Thus, the interpretation of soil-quality indicators requires the experience and 'skill' of the researcher and/or soil manager. Doran (2002) realized that several soil-quality indicators would be too complex to be used by land managers or policy makers. Hence, he suggested concentrating on simple indicators, which have meaning to farmers. The use of indicators like topsoil depth and soil protective cover in a given management system were hypothesized to be the most fruitful means of linking science with practice in assessing the sustainability of management practices (Doran, 2002). Schjønning *et al.* (2000) showed that quantitative soil mechanical properties derived by analytical procedures in the laboratory correlated well with qualitative behaviour of soil in the field. It seems important to evaluate such links when considering the use of soil-quality indicators obtained by reductionistic studies in controlled environments.

Larson and Pierce (1991) suggested a *minimum data set* to describe the quality of a soil. This data set should consist of a number of indicators describing the quality/health of the soil. Using an analogue to human medicine, reference values for each indicator would set the limit for a healthy soil (Larson and Pierce, 1991). The use of indicators has been widely discussed in the literature on soil quality (e.g. Doran and Jones, 1996). Lilburne *et al.* (2002) and Sparling and Schipper (2002) presented achievements obtained in a New Zealand soil quality project. In contrast to most other soil quality assessments, their focus was on a regional rather than on a farm or field scale. Management was similarly addressed in terms of distinct land uses (e.g. arable cropping, dairy farms, pine plantations). Much effort was allocated to identify the most adequate indicators, and seven key parameters were chosen: soil pH, total C and N, mineralizable N, Olsen P, bulk density and macroporosity (Sparling and Schipper, 2002). Lilburne *et al.* (2002) identified the difficult task of isolating the relevant target/threshold values of indicators. Sparling and Schipper (2002) acknowledged the problem in addressing satisfactorily all combinations of soil types and land uses. Generally, however, they found the approach useful to raise an awareness of soil quality issues among regional council staff, scientists and the general public.

We agree that indicators *per se* as well as their thresholds may be important in order to make the soil quality concept operational. The authors of the individual chapters of this book have been encouraged to identify indicators and thresholds whenever it was possible to establish generally applicable limits. However, we realized that this endeavour would be difficult due to the vast number of soil types and agroecosystems addressed. The human species is well defined compared with soils and a body temperature of 37°C is an established threshold for a healthy person, at least regarding infectious diseases. Seybold *et al.* (1998) and Sojka and Upchurch (1999) stressed the difficulty in dealing with the 18,000–20,000 soil series occurring in the USA. Considering the diverse agricultural uses of soils (e.g. growing different crops with dissimilar soil requirements) and the different optima associated with each specific use, Sojka and Upchurch (1999) emphasized *understanding* rather than *rating* of the soil resource. However, within a well-defined scenario, for example research in agricultural management at one specific site or region, the quantification of soil attributes and the use of these as indicators of soil quality may be quite useful (e.g. Campbell *et al.*, 1997).

Indicator Threshold and Management Threshold

Threshold was defined by Smyth and Dumanski (1993) as 'levels beyond which a system undergoes significant change; points at which stimuli provoke response'. Thus threshold links to resilience. As an example, Smyth and Dumanski mentioned the threshold for erosion as the level (extent of erosion) beyond which erosion is no longer tolerable (in order to maintain sustainability). Gomez *et al.* (1996) adopted this definition and used the term threshold to denote the boundary between sustainable and unsustainable indicator values. Thus, thresholds are values of a variable beyond which rapid, often exponential, negative changes occur (Pieri *et al.*, 1995). Because of their intimate association with resilience, we encourage that focus is on *thresholds* rather than on *references, baselines* or *benchmarks*, often employed in the literature on soil-quality indicators.

A main issue when considering the quality of agricultural soil is how to identify sustainable management. One major aim of this book is to promote a shift from *assessing* soil quality to *managing* soil quality. Of course management cannot be addressed without evaluating soil attributes (i.e. indicators), but by focusing on the effects of management we intended to establish a more relevant foundation for the soil quality concept. Our ambition was to concentrate on the challenges facing agriculture in the context of maintaining soil quality. When the common knowledge on soil functions and properties (including indicator thresholds) is combined with that derived from studies on the effects of specific management tools, the potential outcome can be

management thresholds, i.e. the most severe disturbance any management may accomplish without inducing significant changes towards unsustainable conditions. Regarding soil acidity, soil pH is a soil-quality indicator for which a threshold can be established, whereas the rate of liming (e.g. kg $CaCO_3$/ha/year) required to maintain the pH at some prescribed level represents the management threshold.

The *management threshold* approach may seem less ambitious than the *indicator threshold* approach, which includes the identification of a universal minimum dataset. However, the former may be more successful in solving key management problems in agriculture. Exerting all efforts in coping with the problem of non-universality in indicator thresholds implies the risk of never approaching the management problems. The management approach, however, also needs to consider differences among soil types and agroecosystems, and should be based on a thorough understanding of the reaction of individual soils to management. Figure 1.4 illustrates the differences in the two approaches discussed.

Challenges in Modern Agriculture

Modern agriculture faces a number of challenges, which are subject to intense research, but they are seldom defined and discussed in the context of all three aspects of the soil quality concept (Fig. 1.2). As an example, farmers are challenged to manage plant nutrients in order to maintain production volumes, minimize losses of nutrients to the environment and create a high quality in plant products for animal and human consumption.

When addressing the challenges of modern agriculture, a main issue is the identification of management procedures that are sustainable, that is, simultaneously meet societal concerns and recognize the vulnerability of the soil system to degradation. The authors of all chapters have been encouraged to explain their judgement of *sustainability*. Ideally, management options are considered in relation to the three 'concerns' of the SSSA

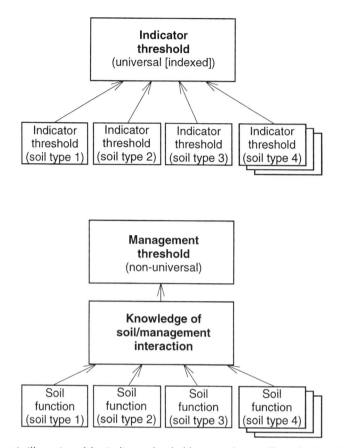

Fig. 1.4. Schematic illustration of the 'indicator threshold' approach typically applied in soil quality studies (top) and the 'management threshold' approach suggested for this volume (bottom). In the 'indicator threshold' approach, the focus is on identifying (universal) thresholds for specific soil-quality indicators, whereas for the 'management threshold' approach, the focus will be put on identifying thresholds (probably non-universal) for specific management tools.

definition of soil quality (Fig. 1.2); that is, how will different soil management affect biological *productivity*, the *environment* framing the managed soil system and *human health*. The latter relates primarily to the quality of products for human consumption. We have further asked for a consideration of soil stability to a given management practice, applying the concept suggested above (Fig. 1.3). This implies identifying *resistance* as well as *resilience* of the soil to the influence from the specific management applied. Finally, a goal-directed approach includes discussion and, if possible, identification of *soil indicator thresholds* as well as *management thresholds* for the soil characteristics and the management

procedures discussed in each chapter (Fig. 1.4).

Figure 1.5 summarizes the approach used in this book for discussing soil management as related to soil quality. In the centre stands the major challenges and management tools, which will be discussed in relation to: (i) the three aspects of soil quality; (ii) the stability of 'form' (physical form or soil functions); and (iii) the potential of identifying soil-quality indicator thresholds as well as management thresholds. Figure 1.5 also illustrates how these considerations are framed by the understanding of sustainability and further by societal priorities, and the values and goals of the scientist (the cognitive context).

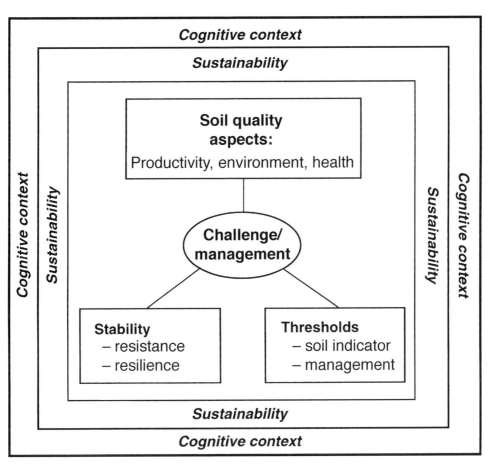

Fig. 1.5. Illustration of the approach used in focusing major challenges in modern agriculture as related to scientifically based terms and as framed by societal values and priorities. Note that only the societal and intentional dimensions of the cognitive context (cf. Fig. 1.1) are active in defining sustainability.

Outlining the Book Content

One major concern in agriculture is an adequate supply of nutrients to the crops. Chapters 3, 4, 5 and 6 of this volume address aspects crucial to basic soil processes and plant nutrition. Soil acidity influences most soil functions. Nitrogen, phosphorus and potassium are three important macro-nutrients, i.e. nutrients taken up by crops in amounts of kilograms per hectare. More general aspects of the soil ecosystem are topics of Chapters 7, 8 and 9, which deal with soil diversity, including carbon dynamics and biodiversity. The physical form of soils is treated in Chapters 10, 11 and 12, with emphasis on physical degradation of

agricultural soils. Chemical contaminants are major threats to soil quality. Chapters 13 and 14 evaluate the potential hazards from the use of organic waste materials and pesticides.

The contributions addressing specific management problems are framed by four conceptual chapters. Chapter 2 reviews the history of and advances in soil quality research. Chapter 15 is an important reminder that systems research may reveal mechanisms not perceived in analytical research. Finally, any work on soil quality should reflect on how the knowledge gained can be implemented. Hence, Chapters 16 and 17 discuss how to put soil quality knowledge to work for industrialized and developing countries, respectively. Figure 1.6 gives an outline of the book content.

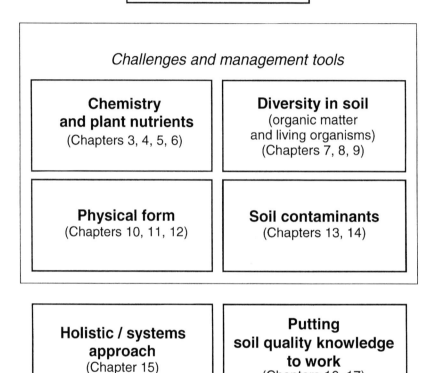

Fig. 1.6. An outline of the book chapters indicating the four groups of challenges addressed in specific chapters.

References

Allan, D.L., Adriano, D.C., Bezdicek, D.F., Cline, R.G., Coleman, D.C., Doran, J.W., Haberen, J., Harris, R.G., Juo, A.S.R., Mausbach, M.J., Peterson, G.A., Schuman, G.E., Singer, M.J. and Karlen, D.L. (1995) SSSA statement on soil quality. In: *Agronomy News*, June 1995, ASA, Madison, Wisconsin, p.7.

Alrøe, H.F. and Kristensen, E.S. (2002) Towards a systemic research methodology in agriculture. Rethinking the role of values in science. *Agriculture and Human Values* 19, 3–23.

Blum, W.E.H. (1998) Basic concepts: degradation, resilience, and rehabilitation. In: Lal, R., Blum, W.E.H., Valentine, C. and Stewart, B.A. (eds) *Methods for Assessment of Soil Degradation*. CRC Press, Boca Raton, Florida, pp. 1–16.

Blum, W.E.H. and Santelises, A.A. (1994) A concept of sustainability and resilience based on soil functions: the role of ISSS in promoting sustainable land use. In: Greenland, D.J. and Szabolcs, I. (eds) *Soil Resilience and Sustainable Land Use*. CAB International, Wallingford, UK, pp. 535–542.

Bossel, H. (1999) *Indicators for Sustainable Development: Theory, Method, Applications. A Report to the Balaton Group*. International Institute for Sustainable Development, Manitoba, Canada, 124 pp.

Bouma, J., Finke, P.A., Hoosbeek, M.R. and Breeuwsma, A. (1998) Soil and water quality at different scales: concepts, challenges, conclusions and recommendations. *Nutrient Cycling in Agroecosystems* 50, 5–11.

Campbell, C.A., Janzen, H.H. and Juma, N.G. (1997) Case-studies of soil quality in the Canadian prairies: long-term field experiments. In: Gregorich, E.G. and Carter, M.R. (eds) *Soil Quality for Crop Production and Ecosystem Health*. Developments in Soil Science 25, Elsevier, Amsterdam, pp. 351–397.

Campbell, C.L., Heck, W.W., Neher, D.A., Munster, M.J. and Hoag, D.L. (1995) Biophysical measurement of the sustainability of temperate agriculture. In: Munasinghe, M. and Shearer, W. (eds) *Defining and Measuring Sustainability. The Biogeophysical Foundations*. The World Bank, Washington, DC, pp. 251–273.

Carter, M.R., Gregorich, E.G., Anderson, D.W., Doran, J.W., Janzen, H.H. and Pierce, F.J. (1997) Concepts of soil quality and their significance. In: Gregorich, E.G. and Carter, M.R. (eds) *Soil Quality for Crop Production and Ecosystem Health*. Developments in Soil Science 25, Elsevier, Amsterdam, pp. 1–19.

Domsch, K.H., Jagnow, G. and Anderson, T.H. (1983) An ecological concept for the assessment of side-effects of agrochemicals on soil microorganisms. *Residue Reviews* 86, 65–105.

Doran, J.W. (2002) Soil health and global sustainability: translating science into practice. *Agriculture, Ecosystems and Environment* 88, 119–127.

Doran, J.W. and Jones, A.J. (eds) (1996) *Methods for Assessing Soil Quality*. Soil Science Society of America Special Publication Number 49, 410 pp.

Doran, J.W. and Parkin, T.B. (1994) Defining and assessing soil quality. In: Doran, J.W., Coleman, D.C., Bezdicek, D.F. and Stewart, B.A. (eds) *Defining Soil Quality for a Sustainable Environment*. Soil Science Society of America Special Publication Number 35, pp. 3–21.

Doran, J.W. and Parkin, T.B. (1996) Quantitative indicators of soil quality: a minimum data set. In: Doran, J.W. and Jones, A.J. (eds) *Methods for Assessing Soil Quality*. Soil Science Society of America Special Publication Number 49, pp. 25–37.

Doran, J.W., Sarrantonio, M. and Liebig, M.A. (1996) Soil health and sustainability. In: Sparks, D.L. (ed.) *Advances in Agronomy*. Academic Press, San Diego, California, pp. 1–54.

Droogers, P. and Bouma, J. (1997) Soil survey input in explanatory modeling of sustainable soil management practices. *Soil Science Society of America Journal* 61, 1704–1710.

Dumanski, J., Pettapiece, W.W. and McGregor, R.J. (1998) Relevance of scale dependent approaches for integrating biophysical and socio-economic information and development of agroecological indicators. *Nutrient Cycling in Agroecosystems* 50, 13–22.

Ellert, B.H., Clapperton, M.J. and Anderson, D.W. (1997) An ecosystem perspective of soil quality. In: Gregorich, E.G. and Carter, M.R. (eds) *Soil Quality for Crop Production and Ecosystem Health*. Developments in Soil Science 25, Elsevier, Amsterdam, pp. 115–141.

Eswaran, H. (1994) Soil resilience and sustainable land management in the context of AGENDA 21. In: Greenland, D.J. and Szabolcs, I. (eds) *Soil Resilience and Sustainable Land Use*. CAB International, Wallingford, UK, pp. 21–32.

Freeman, E. and Skolimowski, H. (1974) The search for objectivity in Peirce and Popper. In: Schilpp, P.A. (ed.) *The Philosophy of Karl R. Popper*. The Open Court Publishing Co., La Salle, Illinois, pp. 464–519.

Gomez, A.A., Kelly, D.E.S., Syers, J.K. and Coughlan, K.J. (1996) Measuring sustainability of agricultural systems at the farm level. In: Doran, J.W. and Jones, A.J. (eds) *Methods for Assessing Soil Quality*. Soil Science Society of America Special Publication Number 49, pp. 401–409.

Herdt, R.W. and Steiner, R.A. (1995) Agricultural sustainability: concepts and conundrums. In: Barnett, V., Payne, R. and Steiner, R. (eds) *Agricultural Sustainability: Economic, Environmental and Statistical Considerations*. John Wiley & Sons, Chichester, UK, pp. 1–13.

Holling, C.S. (1973) Resilience and stability of ecological systems. *Annual Review of Ecology and Systematics* 4, 1–23.

Håkansson, I. and Reeder, R.C. (1994) Subsoil compaction by vehicles with high axle load – extent, persistence and crop response. *Soil and Tillage Research* 29, 277–304.

Jamieson, D. (1992) Ethics, public policy, and global warming. *Science Technology & Human Values* 17, 139–153.

Karlen, D.L., Mausbach, M.J., Doran, J.W., Cline, R.G., Harris, R.F. and Schuman, G.E. (1997) Soil quality: a concept, definition, and framework for evaluation. *Soil Science Society of America Journal* 61, 4–10.

Kay, B.D. (1990) Rates of change of soil structure under different cropping systems. *Advances in Soil Science* 12, 1–52.

Lal, R. (1994) Sustainable land use systems and soil resilience. In: Greenland, D.J. and Szabolcs, I. (eds) *Soil Resilience and Sustainable Land Use*. CAB International, Wallingford, UK, pp. 41–67.

Lal, R. (1998) Soil quality and agricultural sustainability. In: Lal, R. (ed.) *Soil Quality and Agricultural Sustainability.* Ann Arbor Press, Chelsea, Michigan, pp. 3–12.

Larson, W.E. and Pierce, F.J. (1991) Conservation and enhancement of soil quality. In: *Evaluation for Sustainable Land Management in the Developing World.* International Board for Soil Research and Management, Bangkok, Thailand, pp. 175–203.

Larson, W.E. and Pierce, F.J. (1994) The dynamics of soil quality as a measure of sustainable management. In: Doran, J.W., Coleman, D.C., Bezdicek, D.F. and Stewart, B.A. (eds) *Defining Soil Quality for a Sustainable Environment.* Soil Science Society of America Special Publication Number 35, pp. 37–51.

Larson, W.E., Gupta, S.C. and Useche, R.A. (1980) Compression of agricultural soils from eight soil orders. *Soil Science Society of America Journal* 44, 450–457.

Letey, J., Sojka, R.E., Upchurch, D.R., Cassel, D.K., Olson, K., Payne, B., Petrie, S., Price, G., Reginato, R.J., Scott, H.D., Smethurst, P. and Triplett, G. (2003) Deficiencies in the soil quality concept and its application. *Journal of Soil and Water Conservation* 58, 180–187.

Lilburne, L.R., Hewitt, A.E., Sparling, G.P. and Selvarajah, N. (2002) Soil quality in New Zealand: policy and the science response. *Journal of Environmental Quality* 31, 1768–1773.

Lockeretz, W. and Anderson, M.D. (1993) *Agricultural Research Alternatives.* University of Nebraska Press, Lincoln, Nebraska, 239 pp.

Munasinghe, M. and Shearer, W. (1995) An introduction to the definition and measurement of biogeophysical sustainability. In: Munasinghe, M. and Shearer, W. (eds) *Defining and Measuring Sustainability. The Biogeophysical Foundations.* The World Bank, Washington, DC, pp. xvii–xxxii.

Munkholm, L.J. (2000) *The Spade Analysis – a Modification of the Qualitative Spade Diagnosis for Scientific Use. DIAS report, Plant Production* No. 28. The Danish Institute of Agricultural Sciences, Tjele, Denmark, 73 pp.

Oxford Advanced Learner's Dictionary of Current English (1974) Oxford University Press, Oxford.

Pankhurst, C.E. (1994) Biological indicators of soil health and sustainable productivity. In: Greenland, D.J. and Szabolcs, I. (eds) *Soil Resilience and Sustainable Land Use.* CAB International, Wallingford, UK, pp. 331–351.

Pankhurst, C.E., Doube, B.M. and Gupta, V.V.S.R. (1997) Biological indicators of soil health: synthesis. In: Pankhurst, C.E., Doube, B.M. and Gupta, V.V.S.R. (eds) *Biological Indicators of Soil Health.* CAB International, Wallingford, UK, pp. 419–435.

Patzel, N., Sticher, H. and Karlen, D.L. (2000) Soil fertility – phenomenon and concept. *Journal of Plant Nutrition and Soil Science* 163, 129–142.

Pennock, D.J. (1997) Effects of soil redistribution on soil quality: pedon, landscape, and regional scales. In: Gregorich, E.G. and Carter, M.R. (eds) *Soil Quality for Crop Production and Ecosystem Health.* Developments in Soil Science 25, Elsevier, Amsterdam, pp. 167–185.

Pieri, C., Dumanski, J., Hamblin, A. and Young, A. (1995) *Land Quality Indicators.* World Bank Discussion Papers 315, The World Bank, Washington, DC, 63 pp.

Schjønning, P., Munkholm, L.J., Debosz, K. and Elmholt, S. (2000) Multi-level assessment of soil quality – linking reductionist and holistic methodologies. In: Elmholt, S., Stenberg, B., Grønlund, A. and Nuutinen, V. (eds) *Soil Stresses, Quality and Care. Proceedings from NJF Seminar 310, 10–12 April 2000, Ås, Norway. DIAS report* 38, Danish Institute of Agricultural Sciences, Tjele, Denmark, pp. 43–52.

Seybold, C.A., Mausbach, M.J., Karlen, D.J. and Rogers, H.H. (1998) Quantification of soil quality. In: Lal, R., Kimble, J.M., Follett, R.F. and Stewart, B.A. (eds) *Advances in Soil Science.* CRC Press, Boca Raton, Florida, pp. 387–404.

Seybold, C.A., Herrick, J.E. and Brejda, J.J. (1999) Soil resilience: a fundamental component of soil quality. *Soil Science* 164, 224–234.

Smyth, A.J. and Dumanski, J. (1993) *FESLM: an International Framework for Evaluating Sustainable Land Management.* World Resources Reports 73, Land and Water Development Division, FAO, Rome, 77 pp.

Sojka, R.E. and Upchurch, D.R. (1999) Reservations regarding the soil quality concept. *Soil Science Society of America Journal* 63, 1039–1054.

Sparling, G.P. and Schipper, L.A. (2002) Soil quality at a national scale in New Zealand. *Journal of Environmental Quality* 31, 1848–1857.

Stewart, B.A., Lal, R. and El-Swaify, S.A. (1991) Sustaining the resource base of an expanding world agriculture. In: Lal, R. and Pierce, F.J. (eds) *Soil Management for Sustainability.* Soil and Water Conservation Society, Ankeny, Iowa, pp. 125–144.

Swift, M.J. (1994) Maintaining the biological status of soil: a key to sustainable land management? In: Greenland, D.J. and Szabolcs, I. (eds) *Soil Resilience and Sustainable Land Use.* CAB International, Wallingford, UK, pp. 235–247.

Szabolcs, I. (1994) The concept of soil resilience. In: Greenland, D.J. and Szabolcs, I. (eds) *Soil Resilience and Sustainable Land Use.* CAB International, Wallingford, UK, pp. 33–39.

WCED (1987) *Our Common Future: the Brundtland Report.* Report from the World Commission on Environment and Development (WCED). Oxford University Press, Oxford.

Chapter 2

Soil Quality, Fertility and Health – Historical Context, Status and Perspectives

D.L. Karlen,[1] S.S. Andrews[2] and B.J. Wienhold[3]

[1]USDA-ARS, National Soil Tilth Laboratory, 2150 Pammel Drive, Ames, Iowa 50011, USA; [2]USDA- NRCS, Soil Quality Institute, 2150 Pammel Drive, Ames, Iowa 50011, USA; [3]USDA-ARS, Soil and Water Conservation Research Unit, University of Nebraska, East Campus, Lincoln, Nebraska 68583, USA

Summary

Evolution of the soil quality concept and its relationship to soil fertility and health are discussed. A framework for evaluating soil quality is also presented. Soil quality assessment begins by selecting the management goal(s) (e.g. productivity, waste management, carbon sequestration) for which the evaluation is being made. Critical soil functions (e.g. nutrient cycling, water infiltration and retention, filtering and buffering) associated with each goal are identified. Finally, appropriate physical (e.g. aggregate stability, bulk density), chemical (e.g. pH, organic carbon, total N, EC, phosphorus) and biological (e.g. potentially mineralizable N, microbial biomass, soil enzymes) indicators are selected to measure how well each function is being performed. Scoring algorithms help to interpret the indicator data, with each having soil- and site-specific threshold and optimum values to accommodate differences in soil properties, climate or management practices (e.g. tillage, fertilization, crop rotation or water management). The framework can be modified to evaluate construction sites, athletic fields, forests or other land uses by selecting different critical functions, indicators and scoring algorithms. We conclude by stressing that soil quality is not 'an end in itself' but rather a science-based soil management tool for modern agriculture and other land uses.

Evolution of the Soil Quality Concept

Introduction

Alexander (1971) suggested developing 'soil quality criteria' in reference to agriculture's role in environmental improvement. Warkentin and Fletcher (1977) subsequently introduced the soil quality concept *per se* at an international seminar on soil environment and fertility management for intensive agriculture. They stressed that the concept was needed to facilitate better land-use planning because of the increasing number of functions (e.g. food and fibre production, recreation, and recycling or assimilation of wastes or other by-products) that soil resources must either provide or accommodate.

During its evolution, the soil quality concept has been examined using several approaches. These range from simple scorecard and test-kit monitoring for educational purposes to comprehensive laboratory-based assessments and indexing as a tool to evaluate the sustainability of various soil management practices (Larson and Pierce, 1991; Doran and Parkin, 1994; Doran *et al.*, 1996; Gregorich, 1996; Karlen *et al.*, 1997, 2001). None of the approaches, however, ever implied that the soil quality concept was expected to replace modern soil survey programmes or diminish the importance of technology and scientifically-based soil management. Rather, as the intensity of modern agriculture increases (i.e. profit margins narrow, land and other natural resources become more scarce, world population increases, and public concern for off-site impacts of agriculture grows), the need to understand soil quality not as 'an end in itself' but as a science-based tool that can be used to help guide soil management decisions will only increase. It is within this context that this book emphasizes soil quality *management* in relation to agricultural practices rather than simply *monitoring* soil-quality indicators, and

thus appropriately brings the soil quality concept full-circle from when it was introduced.

Warkentin and Fletcher (1977) suggested several ways to manage soil quality. One was to determine the critical soil function(s) within an ecosystem and then to evaluate how adequately those functions were being performed. This approach was suggested for intensive agricultural operations, assuming the information could be used to improve future land-use decisions. A second approach was to determine the number of feasible options for which a specific soil resource could be used. High quality soils would be those capable of supporting a larger number of different land uses. Soil quality could also be determined based on the absence of pollutants, analogous to some water and air quality determinations. For this approach, soil quality would be quantified by determining the suitability of a soil for several different uses. For example, if soil resources were used for disposal of toxic waste materials and this created irreversible changes that would make those soils unsuitable for other functions (e.g. crop production, recreation or urban development), the overall soil quality rating would be lower (less desirable) than if applying waste materials did not impair the resource for other potential uses. This consideration of irrevocable system changes anticipated the state and transition models currently being developed for range management (Friedel, 1991). Warkentin and Fletcher (1977) concluded by stating that a soil quality concept was needed to complement soil science research by making our understanding of soils more complete and to help guide labour, fiscal and input allocation as agriculture intensifies and expands to meet increasing world demands.

As we begin the 21st century, the demands and expectations for our soil resources continue to increase. Worldwide need for food, feed and fibre production, recreational areas, reforestation and carbon sequestration, urban development and remediation of wastes are just some of the sustainable land use and development issues for which different stakeholder groups have expressed concern. Wynen (2002) stated that desertification, salinization and waterlogging of soils are clear indications that the modern ways of managing soils are not sustainable. She listed other examples including contamination of soil, ground and surface waters with nitrate and pesticides; eutrophication of inland and near-shore waters; and continuing soil erosion in all parts of the world.

The importance of soil protection was also recognized at the Rio Summit in 1992 with the result being the adoption of conventions on climate change, biological diversity and desertification in some countries. The European Commission has outlined a strategy to protect soil resources by including a thematic strategy on soil protection within the 6th Environment Action Programme. Attention is being given to preventing erosion, deterioration, contamination and desertification of soil resources with emphasis on the decline in soil organic matter (OM) and prevention of pollution (COM, 2002). Recommended activities include describing the multiple functions of soils, identifying the main threats to soil resources and outlining soil characteristics relevant to policy development. To meet those needs, we suggest evaluating both inherent (intrinsic) and dynamic (use-dependent) soil properties and processes to ensure that the immediate and long-term effects of modern agriculture are sustainable.

Defining soil quality

Prior to the mid-1980s, controlling soil erosion and minimizing its effect on crop productivity were major foci for North American soil management research. Gradually attention broadened to include sustainable agriculture, environmental health and prevention of further soil resource degradation. An important outcome of this expansion was the Canadian Soil Quality Evaluation Program (SQEP) and its assessment of soil health (Acton and Gregorich, 1995).

During this same period, Larson and Pierce (1991) defined soil quality as the capacity of soils to function within the ecosystem boundaries and to interact positively with the environment external to that ecosystem. They were among the first to propose a quantitative formula for assessing soil quality and relating the changes to soil management practices. As a result, soil quality was recognized and interpreted as a more sensitive and dynamic way to measure soil condition, response to management changes and resilience to stresses imposed by natural forces or human uses. The emerging soil quality concept also provided the focus for an International Workshop entitled 'Assessment and Monitoring of Soil Quality' at the Rodale Institute Research Center in Emmaus, Pennsylvania. One outcome was agreement that the soil quality concept should not be limited to soil productivity, but should encompass environmental quality, human and animal health, and food safety and quality.

Soil quality became more prevalent in the vocabulary of policy makers, natural resource conservationists, scientists and farmers after the US National Academy of Sciences published the book *Soil and Water Quality: an Agenda for Agriculture* (National Research Council, 1993) and specifically identified the need for more holistic soil quality research. The increasing interest in the concept resulted in several symposia and publications. Each, however, seemed to provide its own definition, list of critical soil functions and applications for which soil quality should be assessed.

Some equated soil quality with soil fertility and in response, Patzel *et al.* (2000) attempted to differentiate the two terms within the German-language literature by extensively examining soil science, agronomic and ethnic studies. They concluded that none of the existing soil fertility terminology was synonymous with the soil quality concept and that both terms were appropriate. Others considered soil quality to be synonymous with soil productivity (Sojka and Upchurch, 1999), an effort that was prematurely institutionalized, and a diversion of resources from efforts aimed directly at developing improved soil management practices (Sojka *et al.*, 2003).

Linkages between soil productivity and soil quality are logical because productivity is a critical function for agricultural sustainability. Productivity can be reduced through wind or water erosion, nutrient mining, salinization, acidification, waterlogging or compaction, all of which are conditions reflecting reduced soil quality. Furthermore, the effects of soil management on productivity can be assessed using soil quality attributes. We acknowledge the partial overlap, but argue that the soil quality concept is even more encompassing than soil productivity because of its emphasis on environmental externalities, chemical, physical, and biological properties, multiple land uses and human health.

In 1994, Dr Larry Wilding, president of the Soil Science Society of America (SSSA) appointed a 14-person committee to define the soil quality concept, examine its rationale and justification, and identify the soil and plant attributes that would be useful for describing and evaluating it. The Committee presented its first report in the June 1995 issue of *Agronomy News*, stating that the simplest definition for soil quality is 'the capacity (of soil) to function'. An expanded version (Karlen *et al.*, 1997) defined soil quality as 'the capacity of a specific kind of soil to function, within natural or managed ecosystem boundaries, to sustain plant and animal productivity, maintain or enhance water and air quality, and support human health and habitation'. This definition was adopted for the present volume (Chapter 1).

Assessment tools and indicator selection

Soil quality research and education activities have resulted in the development of teaching materials, assessment tools and the evaluation of many different soil biological, chemical and physical indicators. In the USA, educational materials were developed primarily in partnership with the NRCS-Soil Quality Institute (SQI). These include information sheets, the Soil Biology Primer, Guidelines for Soil Quality Assessment in Conservation Planning and an evolving website with user-friendly information and a framework for soil quality evaluation. Assessment tools include farmer-based scorecards, soil quality test kits and a spreadsheet designed to help interpret indicator data and compute soil quality indices.

Patterned after the Wisconsin soil health card (Romig *et al.*, 1995), the scorecards are intended to provide a qualitative self-assessment of a farmer's current soil and crop management practices. The visual soil assessment (VSA) protocol (Shepherd, 2000) and a soil quality monitoring system (SQMS) (Beare *et al.*, 1999) are two other tools developed to help improve the sustainability of land-management decisions. For all of these visual assessment tools, scoring is relatively simple (e.g. poor, fair, good) and based on general observations of tilth, earthworms, runoff, ponding, plant vigour, yield, ease of tillage, soil colour, aroma, structure, cloddiness or similar indicators.

Soil quality test kits were developed to provide semiquantitative indicator data primarily at the soil surface (0–7.5-cm depth). Bulk density, infiltration rate, water-holding capacity, electrical conductivity, soil pH, soil nitrate and soil respiration were identified as a reasonable minimum data set for evaluating soil quality at points within a field. Preliminary studies in several locations showed that test-kit data compared favourably with laboratory analyses (Liebig *et al.*, 1996). The results demonstrated the potential to use these tools for screening agricultural soil quality (Sarrantonio *et al.*, 1996) and for evaluating non-agricultural soil conditions (e.g. New York City's Central Park; L. Norfleet, NRCS-SQI, Iowa, 2002, personal communication).

For assessments using laboratory data, many different indicators have been evaluated to identify those most appropriate for making assessments at points within a single field, across entire fields, farms, watersheds or Major Land Resource Areas. Some (e.g. Larson and Pierce, 1991) have suggested that identifying a minimum data set (MDS) could provide sensitive, reliable and meaningful information for soil quality assessment. Realistically a single MDS will probably remain undefined because of the inherent variability among soils, but it may be feasible to identify a suite of biological, chemical and physical indicators that are useful for evaluating site-specific, temporal trends in soil quality. Wide variation in magnitude and importance of various indicators, failure to clearly define soil quality or soil health, and disagreement among soil scientists, conservationists and other land managers regarding which indicators should be measured, as well as when and how, are unresolved challenges.

As the soil quality concept continues to evolve into a tool for modern agriculture, there are several issues associated with indicator selection that need to be resolved. Two are spatial and temporal scale (Halvorson *et al.*, 1997; Wander and Drinkwater, 2000). Another is the need to demonstrate causal relationships between soil quality and ecosystem functions (Herrick, 2000). The accuracy, precision and cost of making the MDS measurements are also questions that have not been resolved. However, as people become aware that soil is a vital and largely non-renewable resource, we anticipate that the use of soil quality assessment will increase. We are confident that those assessments will help quantify resistance to degradation (defined as the capacity of a system to continue functioning without change when disturbed (Pimm, 1984)) and the resilience of a soil resource to recover following disturbance. Ultimately, rules for indicator selection will be created to identify general trends in soil quality, if not specific index values. We suggest that an important role for soil scientists is to help determine those rules so that the assessments will be understood by and useful to land managers, who are the ultimate stewards of soil quality and soil health (Doran and Zeiss, 2000).

regard to sustainability. It is also important to understand that inherent and dynamic indicators are observations in time along a continuum and that the values will be variable because soils are living and dynamic systems. The philosophy that both inherent and dynamic soil properties and processes influence soil quality has been discussed previously (e.g. Seybold *et al.*, 1998; Karlen *et al.*, 2001) but without reference to the genoform/phenoform concept of Droogers and Bouma (1997). We suggest that the two concepts are closely related.

Evaluating inherent soil properties and interpreting how they affect land use have provided the foundation for soil survey, classification and land-use recommendations for more than a century (Kellogg, 1955). The modern survey has focused on identifying and grouping soils with similar morphology, properties or functional characteristics and emphasized the suitability or limitations of each soil for various uses (e.g. crop production, recreation, forestry, wetlands, drainage fields, roads or building sites). In contrast, dynamic soil quality assessment has focused on the surface 20–30 cm and attempts to describe the status or condition of a specific soil due to relatively recent (i.e. < 2–10 years) land-use or management decisions. Therefore, traditional soil survey, classification and interpretation (i.e. inherent soil quality evaluation) and dynamic soil quality assessment are not competing concepts, but complementary. Furthermore, as stated by Herrick (2000), true calibration of soil quality requires more than merely comparing values across soils or management systems. Soil quality must be viewed in a landscape context since most ecosystem functions depend on multiple connections through time and space.

Inherent and dynamic soil-quality indicators

Inherent soil characteristics are those determined by the soil-forming factors of climate, parent material, time, topography and biota (Jenny, 1941), whereas the dynamic or use-dependent soil properties are those influenced by the management practices imposed by humankind. Both are very important with

Implementation of the soil quality concept

Institutional as well as research and education activities have contributed to the worldwide evolution and implementation of the soil quality concept. In the USA, the reorganization of the USDA-Soil Conservation Service (SCS) into the Natural Resources

Conservation Service (NRCS) was just one of these institutional changes. One outcome of that 1994 reorganization was the creation of the SQI, whose mission is to '*cooperate with partners in the development, acquisition and dissemination of soil quality information and technology to help people conserve and sustain our natural resources and the environment*'. By pursuing this mission, the SQI has been able to effectively promote the soil quality concept within the US and around the world. For additional information regarding the SQI, the reader is encouraged to view their website (http://soils.usda.gov/SQI).

Another national soil science effort is the European Soil Bureau (ESB). Created in 1996, the ESB is an integral part of the European Commission (EC). It is located at the Joint Research Centre (JRC), the European Union's scientific and technical research laboratory, and is managed through a permanent Secretariat at Ispra, Italy (http://esb.aris.sai.jrc.it/). A primary objective for the ESB is to provide compatible and coherent information on European soils to both policy makers and users of soil data.

Institutionalization of the soil quality concept in Europe has been different from in the USA for several reasons. One is that the scale for soil quality evaluation is often larger and may be better described as land quality (e.g. Bouma, 2000, 2002) or soil condition. Another, reported by Singer and Ewing (2000), is that in Canada and Europe, contaminant levels and their effects on soil resources have been more central to the soil quality debate than in the USA. They also stated that the reasons for assessing soil quality within managed ecosystems are often different from those for monitoring natural ecosystems.

In Germany, the Federal Soil Protection Act (BbodSchG, 1998) institutionalized the soil quality concept by recognizing soil as: (i) a basis for life and habitat for animals, plants and soil organisms; (ii) part of natural systems, especially water and nutrient cycles; and (iii) a filter and buffer, especially for water protection. The focus for the Act is on protecting or restoring critical soil functions (Höper, 2000) and using good agricultural practices (i.e. appropriate tillage, conserving or improving soil structure, avoiding compaction,

reducing erosion or conserving soil organic matter (OM)). In England, the Department of the Environment, Transport and the Regions (DETR) attempted to draft a soil strategy, but concluded that existing soil indicators were not sufficient for its needs (Wynen, 2002). A new project was therefore initiated to develop key indicators '... to derive soil targets and to help communicate and evaluate the effects of our soil policies and programmes'.

In New Zealand, the VSA (Shepherd, 2000) and SQMS (Beare *et al.*, 1999) were developed as monitoring and educational tools. The effectiveness of the VSA was evaluated through a series of workshop presentations to more than 300 farmers, regional council staff, agribusiness representatives, researchers and teachers on both the North and South Islands (Shepherd *et al.*, 2001). A review of the workshop evaluations showed good agreement between soil quality assessments made by 'expert' and 'non-expert' groups, and that approximately 90% of the participants indicated they would use VSA again. Similar utility and acceptance have been reported for the SQMS. As with many NRCS-SQI programmes, both the VSA and SQMS emphasize education regarding land management and its impact on soil quality.

The European Society for Soil Conservation (ESSC), Danish Ministry of Food, Agriculture and Fisheries (DMFAF), Commission of the European Communities (COM), and many other governmental and non-governmental (NGO) groups have contributed substantially towards the institutionalization and implementation of the soil quality concept. In a communication to the Council of the European Union, European Parliament, Economic and Social Committee, and Committee of the Regions, the COM outlined a thematic strategy for soil protection (COM, 2002) that includes proposals for a series of environmental measures designed to prevent soil contamination. The proposed legislation will address mining waste, sewage sludge and compost, and pursue integration of soil protection concerns into other major European Union (EU) policies. A proposal for soil monitoring legislation will be developed by 2004 with an emphasis on soil erosion, soil contamination and the decline in soil OM.

In a report prepared for the DMFAF, Wynen (2002) concluded that public and political leaders do not recognize the severity of soil management problems throughout the world. Desertification, salinization and waterlogging of soils were given as clear indications that the modern ways of managing soils are not sustainable. The project was conducted to consider the feasibility and advisability of working towards a convention within the United Nations to protect soil health. Specific goals for the proposed convention were to: (i) create a focus on soil problems for public debate; (ii) provide a focus on available information for alleviating the problems; and (iii) make protection of soil health an obligation of governments and, by doing so, put pressure on politicians to take the problems seriously and to take action. Through her analysis, Wynen (2002) concluded that although deterioration in soil quality may be obvious to many soil scientists, it is not obvious to the public, which needs to be the driving force behind any change in national or international policy. Furthermore, scientists currently do not even agree upon the effects of different management systems or on indicators that could be expected to show the effects.

Wynen (2002) concluded that it would be nearly impossible to develop an international convention addressing soil health. As an alternative, she suggested developing a Code of Conduct or guidelines that would not be legally binding, but might provide for increased flow of information concerning sustainable land-use and management practices. This Code of Conduct should identify: (i) the best practices, technologies and thresholds for action to improve soil management in each country; (ii) gaps in research; and (iii) institutional problems within countries that result in continued degradation of soil resources.

An ESSC Task Force report also shows the soil quality concept being institutionalized throughout the world. It stated that the soil quality concept is a valuable tool for getting scientists involved with managed soil systems and may help focus their research efforts. Schjønning (1998) also outlined soil quality criteria and suggested that the concept might be regarded as a means of differentiating and quantifying our understanding of soil behaviour, especially in relation to soil conservation.

Worldwide implementation of the soil quality concept has been facilitated by numerous research projects that, for brevity, can only be briefly summarized. Studies have been used to: (i) establish land values; (ii) monitor degradation; and (iii) address challenges affecting food security. The latter is especially important in developing countries where soil quality is exacerbated by rapidly decreasing per capita land area and water resources (Lal, 1999).

Eswaran et al. (1999) stated that worldwide about 2 billion people are malnourished and an equal number live below the poverty level. These two groups include both socially and economically disadvantaged people who often eke out a living from a plot of land that does not belong to them. As competition for land increases, the landless groups move to more fragile ecosystems, often permanently destroying them simply to survive. This emphasizes that in developing countries, it is even more important to identify thresholds and to link soil quality directly to management because there are fewer resources for monitoring and the rate of soil degradation is faster than in most developed countries. It also demonstrates that for soil quality to be a useful indicator of sustainable land management, the early-warning system must include not only biophysical and socio-economic conditions and trends, but also models that incorporate feedback between soil quality and those trends (Herrick, 2000). Soil quality thresholds are needed to help people make fundamental decisions about whether it is better to keep working a plot of land or to move on and cut forest in another area.

Relationships Among Soil Fertility, Health, Productivity and Management

The concept of soil quality focuses on the interactions among physical, chemical and biological properties and processes occurring within the soil and thus overlaps with issues associated with soil fertility, soil health, soil productivity and response to soil management. In this section, we attempt to point out the subtle differences among the five concepts and review their relationships to the critical services (Blum, 1998) that humankind expects from their soil resources.

Soil fertility and soil quality

Patzel *et al.* (2000) attempted to differentiate soil fertility and soil quality within the German-language literature by conducting a thorough review of soil science, agronomy and ethnic studies literature. Their analysis showed that 30–40% of the soil fertility literature focuses on 'providing yield', but of more importance it revealed how different linguistics, literature background, research momentum and scientific discussion have resulted in very different and sometimes conflicting definitions for what most would consider a 'well-defined' term. They further stated that to understand the subtle differences among the various concepts of soil fertility, it was important not only to examine *what* was said, but also *how* it was said. This illustrates the human impact associated with conceptual terminology.

After having identified several problems associated with the conceptual approaches to soil fertility, Patzel *et al.* (2000) proposed that both 'soil fertility' and 'soil quality' should be incorporated into the German-language literature. To facilitate this recommendation, they suggested that 'soil quality' encompasses an indefinite or open set of tangible attributes that can be substituted or added to without changing the term itself, whereas 'soil fertility' is a definite feature of the soil. Their rationale was that the well-established term 'soil fertility' should not be expected to encompass all of the characteristics associated with an ideal soil, because that would result in an expansion of the definition to where the term no longer had any true meaning. Conversely, because of its indefinite nature, the relatively new term 'soil quality' could easily and logically be used to encompass all of the attributes that are valued as being important for measuring the performance of a soil against a standard. This approach would enable land managers,

researchers and others to determine the relative capacity of a specific soil to do whatever is expected with regard to sustainability, productivity, environmental buffering, biodiversity or other anthropogenic goals. Indeed, this recommendation verifies that soil quality is defined by value judgements, an approach that is of grave concern to some scientists (Sojka and Upchurch, 1999; Sojka *et al.*, 2003). We acknowledge their concerns, but disagree because our values highlight the need to prevent further degradation of soil resources.

Soil services, soil health and soil quality

As might be expected from the linguistic analysis of soil fertility, a definitive, specific feature of soils (Patzel *et al.*, 2000), the conceptual and indefinite characteristics of soil quality have resulted in a plethora of definitions and overlapping terminology. Furthermore, the clarity with which the soil quality concept is articulated may be even more complicated because of its close association with sustainability and sustainable agriculture, two other conceptual terms that are often described using words that convey different messages depending on 'not only what is said, but how it is said'.

From its inception, the soil quality concept has been closely associated with the critical functions that soil resources perform within the biosphere (Doran *et al.*, 1996). Therefore, we reiterate that the simplest definition for the concept is 'the capacity [of soil] to function' (Karlen *et al.*, 1997) or stated in another way 'how a soil is functioning' for a specific goal or use. This definition closely parallels many others (i.e. 'suitability for chosen uses' or 'range of possible uses') that have been used (Doran *et al.*, 1996). The close association between soil function and the soil quality concept also helps illustrate the concept of soil services used to describe the concepts of sustainability and soil resilience (Blum, 1998). These services (also outlined by Schjønning *et al.*, Chapter 1, this volume) have been grouped in two categories, ecological uses and non-agricultural human activities. The former include biomass production (food, fibre and energy), soil as a reactor (filtering, buffering and transforming actions), or soil as a biological habitat and genetic reserve. The latter include soil as a physical medium, as a source of raw materials and as a repository for cultural heritage that helps preserve the history of earth and humankind (Doran *et al.*, 1996).

Soil health and dynamic soil quality are often used interchangeably, and that is our preference. However, as noted through the linguistic analysis of 'soil fertility', there can be subtle differences depending on the exact choice of words and how they are used. For an in-depth discussion of soil health and its relationship to sustainability, readers should see Doran *et al.* (1996). However, regardless of the terminology, we fully agree that any definition of soil health should: (i) reflect the soil as a living system; (ii) address all essential functions of soil in the landscape; (iii) compare the condition of a given soil against its own unique potential within climatic, landscape and vegetation patterns; and (iv) somehow enable meaningful assessment of trends. We also agree with the definition proposed for soil health, which is: 'the continued capacity of soil to function as a vital living system, within ecosystem and land-use boundaries, to sustain biological productivity, maintain the quality of air and water environments, and promote plant, animal, and human health' (Doran *et al.*, 1996).

Soil productivity and soil quality

Having evolved from common research programmes (e.g. Pierce *et al.*, 1984; Larson and Pierce, 1991) some may argue that soil productivity and soil quality are synonymous. Certainly, productivity is a critical soil function for sustainable agriculture systems, however, we argue that the soil quality concept is much broader because of its emphasis on environmental impacts, multiple land uses and human health (Warkentin and Fletcher, 1977).

Soil productivity can certainly be lost through erosion, nutrient mining or other processes such as salinization, acidification,

waterlogging and compaction (Oldeman, 1994). The linkage between soil productivity and soil quality is apparent when changes in soil attributes used to assess soil quality are linked to causes of productivity loss. The effects of management practices on productivity can also be assessed using soil quality attributes. These assessments assume that a high-quality soil will have greater yield stability and be more productive because climatic stresses such as drought will have a reduced impact due to the greater resistance and resilience of those soils.

Soil management effects on soil quality

Tillage, fertilization, crop rotation, water management, liming and cover crops are soil management practices that can significantly affect soil quality. Tillage is used to incorporate residues, prepare a seedbed, control weeds, and incorporate lime, fertilizer and other chemicals, and by doing so will often enhance plant growth and thus improve soil quality. Negative effects associated with tillage include erosion caused by the physical downhill movement of soil (i.e. tillage erosion), exposure of the soil surface to wind and water erosion, and loss of soil OM through oxidation. To balance these factors, no-tillage or conservation tillage practices are being developed and recommended as management strategies to improve soil quality throughout the world.

Fertilizer applications can have either positive or negative effects on soil quality. Identifying yield-limiting nutrients and using fertilizers to correct the deficiencies often increases crop yield and organic inputs (above and below ground). However, repeated application of ammoniacal fertilizers and leaching of excess nitrate nitrogen (Barak *et al.*, 1997) can degrade soil quality through acidification. Crop rotations can be used to improve soil quality by altering the quantity and quality (i.e. C:N ratio and lignin content) of residue added to the soil, varying the soil space utilized for nutrient and water uptake by using crops with different rooting patterns, and providing cover to protect soil from erosion.

Water management affects soil quality primarily through its effects on plant growth. In regions where precipitation is sufficient to support adapted crops, the primary soil quality concerns are to minimize runoff and leaching by achieving good infiltration and storage within the soil profile. If soil water levels are consistently high (i.e. hydric soils), plants must be adapted to the saturated conditions (e.g. lowland rice (*Oryza sativa* L.)) or drainage must be installed. Drainage generally improves aeration and allows the production of a wider range of crops, but can degrade soil quality by enhancing soil OM decomposition. In regions requiring irrigation for crop production, irrigation water quality, irrigation scheduling, method of irrigation and drainage potential (for leaching of salts from the soil profile and prevention of waterlogging) are critical management concerns.

Soil quality management strategies

When European immigrants settled the North American Great Plains during the late 1800s and early 1900s, they immediately adapted familiar temperate crop-production systems to the semiarid region. Cereal grains were grown in a crop–fallow sequence using intensive tillage. Soil water accumulation and increased nitrogen availability through mineralization of soil OM during fallow periods greatly reduced the risk of crop failure when compared to annual cropping. However, wind and water erosion combined with the continued decomposition of soil OM resulted in a major decline in soil fertility and soil quality. By the late 1900s, soils in the northern Great Plains region contained 20–30% less soil OM than did their uncultivated counterparts (e.g. Janzen *et al.*, 1998).

Traditionally, farms were small integrated crop and livestock operations. Nutrient cycles were localized and balanced with crops being grown, harvested and fed to livestock on-site, and the manure was redistributed on the land that produced the crops. Over the last several decades there has been a steady trend for farms to become larger and more specialized. Crops are transported from where they

are grown to other areas to support animal production. Then manure is inefficiently applied to land that did not grow the crop. As a result of this enterprise separation, both manure management and soil quality problems (Fig. 2.1) have emerged (Karlen *et al.*, 1998). Overapplication of poultry litter increases nitrate contamination of groundwater, phosphate contamination of surface waters, and heavy metal contamination of soils and the plants produced on them. Conversely, in areas not receiving periodic applications of manure and bedding materials (i.e. input of more carbon than is returned through crop residue alone), soil OM often decreases more rapidly than if these materials are applied as part of the overall nutrient management programme. The lower carbon input often affects soil quality by decreasing the water stability of soil aggregates, microbial activity, earthworm populations and water retention. The net result is a less favourable soil biological, chemical and physical environment for crop growth.

In the 1960s, research was initiated on developing conservation tillage practices and by the 1980s those systems were widely adopted throughout the region. Conservation tillage and increased use of fertilizers allowed for more intensive cropping. These more intensive systems resulted in abatement of soil OM losses in many soils and accumulation of OM in others (Halvorson *et al.*, 2002). The increased organic carbon content was estimated to have an on-site value of \$1–4/ ton/year (Smith *et al.*, 2000). Several soil biological properties were also improved through the more intensive cropping. These included increased microbial activity, microbial biomass and N-mineralization rates.

Interpreting Soil Quality Data

Interpretation and integration of indicator data are two of the more difficult and controversial issues associated with the soil quality concept. This was recognized and has been an integral part of our soil quality research programme since our first attempt to combine a variety of soil indicator information into an

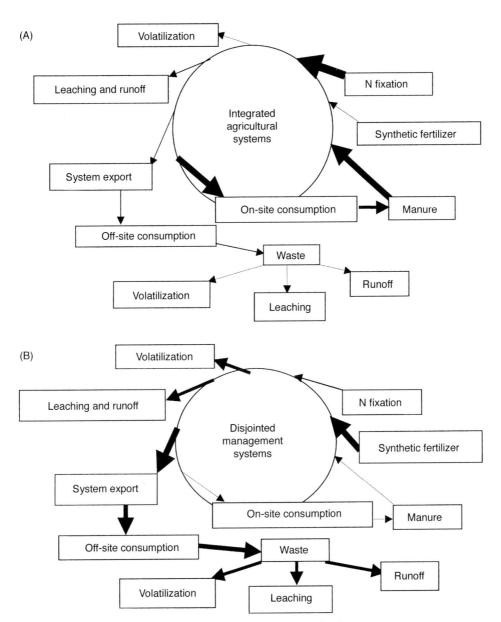

Fig. 2.1. Material flow within integrated (A) and disjointed (B) agricultural systems.

index (Karlen *et al.*, 1994). There are several indexing techniques, but a common theme since Larson and Pierce (1991) suggested indexing soil quality is that the indicators and indices should be selected based on the critical soil functions needed to achieve the 'management goals' (Andrews *et al.*, 2002) for which the assessment is being made. By starting with goal identification (Fig. 2.2), soil quality can be viewed as one component within a nested hierarchy describing agroecosystem sustainability (Fig. 2.3). The nested approach also provides a framework for assessing management practices other than tillage and crop production (e.g. bioremediation, or forest, rangeland and pasture management).

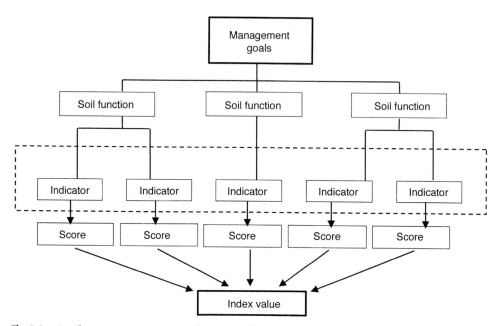

Fig. 2.2. A soil management assessment framework being developed to evaluate soil quality in response to various land uses or practices.

Once the purpose for the assessment or 'management goals' are identified, soil quality indexing involves three main steps: (i) choosing appropriate indicators; (ii) transforming indicator data to scores; and (iii) combining the indicator scores into an index (Fig. 2.2). The selection of indicators for evaluation is often based on expert opinion (e.g. Doran and Parkin, 1994) or accomplished with statistical procedures such as principal components or factor analysis (e.g. Andrews and Carroll, 2001). Expert opinion is often easier but carries the possibility of disciplinary biases; statistical approaches require large existing data sets. Both approaches gave similar results for a vegetable production study on irrigated soils in northern California, USA (Andrews *et al.*, 2002).

Scoring and combining the indicators into indices can be done in a variety of ways (Andrews *et al.*, 2002). Our preference is for non-linear scoring because it accommodates threshold and optimum values as well as transition areas where small changes in indicator values can have a large effect on the score. Linear scoring can be used and may be desirable for indicators that change gradually along a continuum. Step-functions (i.e. good or bad, yes or no) may be appropriate for indicators that measure 'contaminated versus non-contaminated' situations. Andrews *et al.* (2002) found non-linear scoring more accurately reflected soil function when compared to a linear method. Non-linear approaches are not unique, since they are commonly used for utility functions in economics (Norgaard, 1994), multi-objective decision making (Yakowitz *et al.*, 1993) and systems engineering (Wymore, 1993).

Development of non-linear scoring functions does require in-depth knowledge of each indicator's behaviour and function within the system, but it accommodates spatial and temporal issues (Halvorson *et al.*, 1997) that are based on inherent soil and/or climatic factors. For each indicator, baseline and threshold, levels are defined based on inherent soil properties and with sensitivity to scale and anticipated rates of change. Several methods, including the use of benchmark sites, have been suggested for establishing baseline values. We prefer to use measurements for a specific soil at T_0 and to then determine the net change (i.e. aggrading, degrading or stable) at

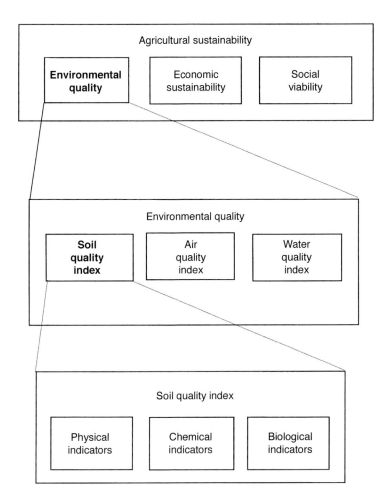

Fig. 2.3. Hierarchy of agricultural indices showing soil quality as one of the critical foundations for assessing sustainable land management (Andrews *et al.*, 2002).

future times (T_N) that are appropriate for each indicator included in the MDS.

An assessment tool that accommodates different indicator thresholds for multiple soil, climate and crop combinations has been developed (S.S. Andrews, D.L. Karlen, C.A. Seybold and J.P. Mitchell, 2002, unpublished results). Others within the USDA-ARS and around the world are testing the tool in both spreadsheet and web-based formats. The tool uses scoring curves for each indicator (*sensu* Karlen and Stott, 1994) but alters the inflection points of each curve based on site-specific information. This database approach requires knowledge of the appropriate curve shape

(based on indicator performance of ecosystem function) and direction of change in curve inflections based on soil properties such as OM or texture, climate factors like annual precipitation and temperature, or crop requirements. For example, the scoring curve for pH is a Gaussian function (bell-shape) based on crop productivity. The ideal or threshold range is crop dependent, i.e. the crop is the determining factor for pH score. Most crops have a threshold between 5.5 and 7.0. For soil test P, the left side of the curve is based on crop yield and is very well supported by research. The right side is based on environmental risk but is currently less well defined (Fig. 2.4). The

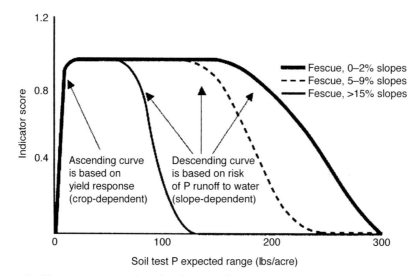

Fig. 2.4. Flexible scoring functions based on factors such as average annual precipitation, temperature, inherent soil OM levels or crop nutrient requirements (S.S. Andrews, D.L. Karlen, C.A. Seybold and J.P. Mitchell, 2002, unpublished results).

primary factor determining the inflection point of the left side is crop P requirement, whereas the right-side inflection point is the percentage slope of the land. Both parts of the curve are further moderated by inherent soil OM, soil texture and soil test P method. By identifying trends in function due to determining factors, this approach yields site-specific soil indicator interpretations without the need to construct formal thresholds for every possible soil, climate and crop combination.

To date, most soil quality assessments have focused more on past land-use effects than on the future management practices. Our goal for developing a more dynamic assessment tool is to provide land managers with multiple strategies triggered by a sustainability assessment based on soil quality. These might include implementation of reduced or no-tillage practices, use of cover crops or more diverse rotations, or the application of manure or compost to increase annual carbon input. One of the goals for this book is to provide strategies that can be used within the context of modern agriculture to improve soil quality. We believe this is consistent with the original intent of the soil quality concept (Warkentin and Fletcher, 1977), and that it demonstrates that soil quality is not an end unto itself, but rather a tool to help land managers and

decision makers develop more sustainable soil management practices.

Perspectives for Using the Soil Quality Concept

The soil quality concept evolved rapidly throughout the past decade with two distinct areas of emphasis – education and assessment – both based soundly on principles of science. Certainly, there are a number of constraints to adoption including the issues of scale (Halvorson *et al.*, 1997; Wander and Drinkwater, 2000) and the need to demonstrate causal relationships between soil quality and ecosystem functions (Herrick, 2000). We also agree with Wynen (2002) who stated that 'for many interested in soils it seems that unless drastic action is taken in the near future, soil degrading processes will take their irreversible toll. Immediate and binding action is therefore seen as a must' and that 'without international cooperation in policies benefiting the environment, a "race-to-the-bottom" as far as the environment is concerned does not seem to be that far-fetched.'

One concern regarding the soil quality concept (Sojka and Upchurch, 1999) was that tillage and crop production issues had received most of the early US soil quality emphasis. This occurred primarily because of the disciplinary focus of the early adopters, but can be corrected by selecting different goals and identifying the critical soil functions, indicators and scoring algorithms (Fig. 2.2) in the assessment frameworks. Providing guidelines for bioremediation is one non-agricultural land use for which a better understanding of soil science is needed (Sims et al., 1997). This is already an important application of soil quality assessment in Europe (BbodSchG, 1998). For restoration, uncontaminated natural sites with similar inherent soil properties could provide the critical reference values needed for scoring and the critical soil-quality indicators.

Other land uses for which soil quality assessment and the identification of management baselines and threshold values could be developed include forests, rangelands or pastures, and urban soils. It is beyond the scope of this chapter to discuss these activities, but based on the literature, recent conferences and activities coordinated by the SQI, ESB, ESSC, LandCare, AgResearch and many other groups around the world, we are confident that these needs and applications for soils information will be met through the overall effort to implement the soil quality concept.

References

Acton, D.F. and Gregorich, L.J. (eds) (1995) *The Health of our Soils: Toward Sustainable Agriculture in Canada*. Pub. No. 1906/E, Centre for Land and Biological Resources Research, Agriculture and Agri-Food Canada, Ottawa, 138 pp.

Alexander, M. (1971) Agriculture's responsibility in establishing soil quality criteria. In: *Environmental Improvement – Agriculture's Challenge in the Seventies*. National Academy of Sciences, Washington, DC, pp. 66–71.

Andrews, S.S. and Carroll, C.R. (2001) Designing a decision tool for sustainable agroecosystem management: soil quality assessment of a poultry litter management case study. *Ecological Applications* 11, 1573–1585.

Andrews, S.S., Karlen, D.L. and Mitchell, J.P. (2002) A comparison of soil quality indexing methods for vegetable production systems in northern California. *Agriculture, Ecosystems and Environment* 90, 25–45.

Barak, P., Jobe, B.O., Krueger, A.R., Peterson, L.A. and Laird, D.A. (1997) Effects of long-term soil acidification due to nitrogen fertilizer inputs in Wisconsin. *Plant and Soil* 197, 61–69.

BbodSchG (1998) *Federal Soil Protection Act* (Bundes-Bodenschutzgesetz) of the 17.03.1998. Federal Law Gazette I, 1998, p. 502.

Beare, M.H., Williams, P.H. and Cameron, K.C. (1999) On-farm monitoring of soil quality for sustainable crop production. In: Currie, L.D., Hedley, M.J., Horne, D.J. and Loganathan, P. (eds) *Best Soil Management Practices for Production*. Proceedings of the 1999 Fertilizer and Lime Research Centre Conference, Occasional Report No. 12, Massey University, Palmerston North, New Zealand, pp. 81–90.

Blum, W.E.H. (1998) Basic concepts: degradation, resilience, and rehabilitation. In: Lal, R., Blum, W.E.H., Valentine, C. and Stewart, B.A. (eds) *Methods for Assessment of Soil Degradation*. CRC Press, Boca Raton, Florida, pp. 1–16.

Bouma, J. (2000) The land quality concept as a means to improve communications about soils. In: Elmholt, S., Stenberg, B., Gronlund, A. and Nuutinen, V. (eds) *Soil Stresses, Quality and Care*. Proceedings from NJF Seminar 310, 10–12 April 2000, Ås, Norway. DIAS report 38, Danish Institute of Agricultural Sciences, Tjele, Denmark, pp. 1–14.

Bouma, J. (2002) Land quality indicators of sustainable land management across scales. *Agriculture Ecosystems and Environment* 88, 129–136.

Commission of the European Communities (COM) (2002) Towards a thematic strategy for soil protection. Availble at: http://europa.eu.int/comm/environment/agriculture/soil_protection.htm

Doran, J.W. and Parkin, T.B. (1994) Defining and assessing soil quality. In: Doran, J.W., Coleman, D.C., Bezdicek, D.F. and Stewart, B.A. (eds) *Defining Soil Quality for a Sustainable Environment*. Soil Science Society of America Special Publication No. 35, pp. 3–21.

Doran, J.W. and Zeiss, M.R. (2000) Soil health and sustainability: managing the biotic component of soil quality. *Applied Soil Ecology* 15, 3–11.

Doran, J.W., Sarrantonio, M. and Liebig, M.A. (1996) Soil health and sustainability. In: Sparks, D.L. (ed.) *Advances in Agronomy* 56, Academic Press, San Diego, California, pp. 1–54.

Droogers, P. and Bouma, J. (1997) Soil survey input in explanatory modeling of sustainable soil management practices. *Soil Science Society of America Journal* 61, 1704–1710.

Eswaran, H., Beinroth, F. and Reich, P. (1999) Global land resources and population-supporting capacity. *American Journal of Alternative Agriculture* 14, 136.

Friedel, M.H. (1991) Range condition assessment and the concept of thresholds: a viewpoint. *Journal of Range Management* 44, 422–426.

Gregorich, E.G. (1996) Soil quality: a Canadian perspective. In: Cameron, K.C., Cornforth, I.S., McLaren, R.G., Beare, M.H., Basher, L.R., Metherell, A.K. and Kerr, L.E. (eds) *Soil Quality Indicators for Sustainable Agriculture in New Zealand. Proceedings of a Workshop, Lincoln Soil Quality Research Centre*. Lincoln University, Christchurch, New Zealand, pp. 40–52.

Halvorson, A.D., Wienhold, B.J. and Black, A.L. (2002) Tillage, nitrogen, and cropping system effects on soil carbon sequestration. *Soil Science Society of America Journal* 66, 906–912.

Halvorson, J.J., Smith, J.L. and Papendick, R.I. (1997) Issues of scale for evaluating soil quality. *Journal of Soil and Water Conservation* 52, 26–30.

Herrick, J.E. (2000) Soil quality: an indicator of sustainable land management? *Applied Soil Ecology* 15, 75–83.

Höper, H. (2000) The German Federal soil protection act. In: *Southern Cooperative Soil Survey Conference Proceedings, Auburn, AL. 18–22 June, 2000*. Available at: http://www.ga.nrcs.usda.gov/mlra15/soilconf/hoeper2.pdf

Janzen, H.H., Campbell, C.A., Izaurralde, R.C., Ellert, B.H., Juma, N., McGill, W.B. and Zentner, R.P. (1998) Management effects on soil C storage on the Canadian prairies. *Soil and Tillage Research* 47, 181–195.

Jenny, H. (1941) *Factors of Soil Formation, a System of Quantitative Pedology*. McGraw Hill, New York, 281 pp.

Karlen, D.L. and Stott, D.E. (1994) A framework for evaluating physical and chemical indicators of soil quality. In: Doran, J.W., Coleman, D.C., Bezdicek, D.F. and Stewart, B.A. (eds) *Defining Soil Quality for a Sustainable Environment. Soil Science Society of America Special Publication* No. 35, pp. 53–72.

Karlen, D.L., Wollenhaupt, N.C., Erbach, D.C., Berry, E.C., Swan, J.B., Eash, N.S. and Jordahl, J.L. (1994) Crop residue effects on soil quality following 10-years of no-till corn. *Soil and Tillage Research* 31, 149–167.

Karlen, D.L., Mausbach, M.J., Doran, J.W., Cline, R.G., Harris, R.F. and Schuman, G.E. (1997) Soil quality: a concept, definition, and framework for evaluation. *Soil Science Society of America Journal* 61, 4–10.

Karlen, D.L., Russell, J.R. and Mallarino, A.P. (1998) A systems engineering approach for utilizing animal manure. In: Hatfield, J.L. and Stewart, B.A. (eds) *Animal Waste Utilization: Effective Use of Manure as a Soil Resource*. Ann Arbor Press, Chelsea, Michigan, pp. 283–315.

Karlen, D.L., Andrews, S.S. and Doran, J.W. (2001) Soil quality: current concepts and applications. *Advances in Agronomy* 74, 1–40.

Kellogg, C.E. (1955) Soil surveys in modern farming. *Journal of Soil and Water Conservation* 10, 271–277.

Lal, R. (1999) Soil management in the developing countries. *Soil Science* 165, 57–72.

Larson, W.E. and Pierce, F.J. (1991) Conservation and enhancement of soil quality. In: Dumanski, J. *et al.* (eds) *Evaluation for Sustainable Land Management in the Developing World. Vol. 2: Technical Papers*. Proceedings of the International Workshop, Chiang Mai, Thailand, 15–21 September 1991. International Board for Soil Research and Management, Bangkok, Thailand, pp. 175–203.

Liebig, M.A., Doran, J.W. and Gardner, J.C. (1996) Evaluation of a field test kit for measuring selected soil quality indicators. *Agronomy Journal* 88, 683–686.

National Research Council (1993) *Soil and Water Quality: an Agenda for Agriculture*. National Academic Press, Washington, DC, 516 pp.

Norgaard, R.B. (1994) Ecology, politics, and economics: finding the common ground for decision making in conservation. In: Meffe, G.K. and Carroll, C.R. (eds) *Principles of Conservation Biology*. Sinauer Associates, Sunderland, Massachusetts, pp. 439–465.

Oldeman, L.R. (1994) The global extent of soil degradation. In: Greenland, D.J. and Szabolcs, I. (eds) *Soil Resilience and Sustainable Land Use*. CAB International, Wallingford, UK, pp. 99–118.

Patzel, N., Sticher, H. and Karlen, D. (2000) Soil fertility – phenomenon and concept. *Journal of Plant Nutrition and Soil Science* 163, 129–142.

Pierce, F.J., Larson, W.E. and Dowdy, R.H. (1984) Soil loss tolerance: Maintenance of long-term soil productivity. *Journal of Soil and Water Conservation* 39, 136–138.

Pimm, S.L. (1984) The complexity and stability of ecosystems. *Nature* 307, 321–326.

Romig, D.E., Garlynd, M.J., Harris, R.F. and McSweeney, K. (1995) How farmers assess soil health and quality. *Journal of Soil and Water Conservation* 50, 229–236.

Sarrantonio, M., Doran, J.W., Liebig, M.A. and Halvorson, J.J. (1996) On-farm assessment of soil quality and health. In: Doran, J.W. and

Jones, A.J. (eds) *Methods for Assessing Soil Quality. Soil Science Society of America Special Publication* No. 49, pp. 83–105.

Schjønning, P. (1998) *Soil Quality Criteria*. European Society for Soil Conservation (ESSC) website. Available at: http://www.zalf.de/essc/task2.htm

Seybold, C.A., Mausbach, M.J., Karlen, D.L. and Rogers, H.H. (1998) Quantification of soil quality. In: Lal, R., Kimble, J.M., Follett, R.F. and Stewart, B.A. (eds) *Soil Processes and the Carbon Cycle*. CRC Press, Boca Raton, Florida, pp. 387–404.

Shepherd, T.G. (2000) *Visual Soil Assessment*, Vol. 1, *Field Guide for Cropping and Pastoral Grazing on Flat to Rolling Country*. Landcare Research, Palmerston North, New Zealand, 84 pp.

Shepherd, T.G., Bird, L.J., Jessen, M.R., Bloomer, D.J., Cameron, D.J., Park, S.C. and Stephens, P.R. (2001) Visual soil assessment of soil quality – trial by workshops. In: Currie, L.D. and Loganathan, P. (eds) *Precision Tools for Improving Land Management*. Occasional report No. 14. Fertilizer and Lime Research Centre, Massey University, Palmerston North, New Zealand, pp. 119–126.

Sims, J.T., Cunningham, S.D. and Sumner, M.E. (1997) Assessing soil quality for environmental purposes: roles and challenges for soil scientists. *Journal of Environmental Quality* 26, 20–25.

Singer, M.J. and Ewing, S. (2000) Soil quality. In: Sumner, M.E. (ed.) *Handbook of Soil Science*. CRC Press, Baca Raton, Florida, pp. G-271–G-298.

Smith, E.G., Lerohl, M., Messele, T. and Janzen, H.H. (2000) Soil quality attribute time paths: optimum levels and values. *Journal of Agricultural and Resource Economics* 25, 307–324.

Sojka, R.E. and Upchurch, D.R. (1999) Reservations regarding the soil quality concept. *Soil Science Society of America Journal* 63, 1039–1054.

Sojka, R.E., Upchurch, D.R. and Borlaug, N.E. (2003) Quality soil management or soil quality management: performance vs semantics. *Advances in Agronomy* 79, 1–48.

Wander, M.M. and Drinkwater, L.E. (2000) Fostering soil stewardship through soil quality assessment. *Applied Soil Ecology* 15, 61–73.

Warkentin, B.P. and Fletcher, H.F. (1977) Soil quality for intensive agriculture. Proceedings of the International Seminar on Soil Environment and Fertility Management. In: *Intensive Agriculture Society Science Soil and Manure*, National Institute of Agriculture Science, Tokyo, pp. 594–598.

Wymore, A.W. (1993) *Model-Based Systems Engineering: an Introduction to the Mathematical Theory of Discrete Systems and to the Tricotyledon Theory of System Design*. CRC Press, Boca Raton, Florida, 710 pp.

Wynen, E. (2002) A UN convention on soil health or what are the alternatives? Available at: http://okologiens-hus.dk/PDFs/Muldrap.doc

Yakowitz, D.S., Stone, J.J., Lane, L.J., Heilman, P., Masterson, J., Abolt, J. and Imam, B. (1993) A decision support system for evaluating the effect of alternative farm management systems on water quality and economics. *Water Science Technology* 28(3–5), 47–54.

Chapter 3
Soil Acidity – Resilience and Thresholds

A.E. Johnston

Rothamsted Research, Harpenden, Herts AL5 2JQ, UK

Summary

The maintenance of soil acidity (pH) at a value appropriate for optimizing crop production is an important first step in soil management because soil pH affects soil properties, nutrient availability and plant nutrition. Soil pH is easily determined and, in regions where calcitic or magnesian limestones are readily available, pH is easily adjusted and maintained at the required value. This value is usually a compromise because all those factors affected by soil acidity are not optimum at the same pH. In temperate regions, the target pH is broadly similar for all mineral soils with average soil organic matter (OM) levels when growing arable crops and is slightly lower for grassland. The exceptions are organic mineral soils, and peat soils when the recommended pH is often less. The amount of liming material required to maintain the recommended soil pH varies with soil type. This chapter discusses the effect of soil acidity on soil properties and nutrient availability. It focuses on the need to maintain soil pH, because long-term acidification adversely affects the sustainability of agricultural systems, and how maintaining soil pH varies with soil type.

Introduction

The adjustment and maintenance of soil acidity (pH) to a value suitable for crop production is an important first step in soil management, and the history and management of soil acidity is a long and interesting one (Wild, 1988). Soil pH can have no precise value or unambiguous meaning, but a knowledge of soil pH is useful in comparing soils in terms of both plant nutrition and understanding soil properties (Rowell,

1988; Thomas, 1996). Differences in management required for soils and crops in the temperate and tropical regions are frequently related to soil pH (Sanchez, 1976). Recently, soil acidity has become an environmental issue with concerns about the acidifying effects of 'acid rain' and possible adverse effects on the soil and its inhabitants as a result of replacing grassland and deciduous woodland with coniferous forests.

Soil acidity can affect various factors that contribute to the biological, chemical and physical properties of soil that make it best able to fulfil its desired purpose. Fulfilling a 'desired purpose' is an important concept in soil quality (see Schjønning et al., Chapter 1, this volume), thus maintaining the optimum pH is a key aspect of soil quality. Soil acidity, which is relatively easy to determine, shows whether a soil is acidic (pH < 7) or basic (pH > 7). Managing soil quality in respect of soil acidity requires the removal of excess hydrogen ions (H^+), produced by various processes in soil, by liming – the application of any material (liming material) containing base cations able to replace excess H^+.

Maintaining soils within an appropriate range of acidity has been fully justified scientifically and is essential for economically viable, environmentally benign farming systems. Soil pH is maintained by appropriate inputs of liming materials, preferably by adding calcium (Ca) in a calcitic limestone ($CaCO_3$). Although magnesian limestone can be used, its application may lead to problems with excess magnesium (Mg) displacing potassium (K) on cation exchange sites. Naturally occurring deposits of calcitic and magnesian limestones are a non-renewable resource, but there are vast deposits of both although they are not uniformly distributed throughout the world. Even though these global reserves are very large, energy is required to mine, prepare, transport and apply them, so that both energy and transport costs are involved. These costs may be a significant proportion of the cost of producing crops where deposits are far removed from soils used for agriculture.

Measuring Soil Acidity

Measurements of soil acidity usually determine actual soil acidity – the concentration of H^+ in the soil solution. This is a 'snapshot' at the time of sampling/determination and will vary with time because cation exchange in soils is a dynamic process. The H^+ concentrations are usually small and are expressed in terms of pH, defined as the logarithm of the reciprocal of the H^+ concentration, in moles per litre. The pH scale ranges from 0 to 14, with 7 being neutral; the Danish scientist Sørensen (1909) first used the term pH. Actual soil acidity should be independent of soil type.

Besides the actual acidity, there is the potential acidity that includes the H^+ adsorbed to soil colloids. Determining the potential acidity requires exchanging these H^+ with other cations prior to measuring the pH. This potential acidity will depend on the extent to which the cation exchange sites are occupied by base cations and the number of exchange sites depends mostly on the clay and organic matter (OM) content of the soil, and thus varies with soil type.

Equipment for the laboratory determination of soil pH is now readily available at reasonable cost and in-field determinations can be made with hand-held pH meters or by using liquid reagents or papers whose colour changes according to pH. Generally, soil pH is determined in water or 0.1 M KCl or 0.01 M $CaCl_2$ solution using a glass electrode. The pH of a soil/water or soil/salt solution depends, in part, on the soil/solution ratio; the most frequently used is 1:2.5. To assess changes in pH over time the same ratio should always be used. Soil pH is lowered as the concentration of soluble salts, particularly chlorides, sulphates and nitrates, increases. The amount of salts present depends on cropping, fertilization and weather, and can vary throughout the year and from year to year (Thomas, 1996 and references therein). To follow changes in pH on individual fields, the soil should be sampled to the same depth at the same time of year. On soils where there is sufficient through drainage during the winter to remove excess salts, sampling is best done in spring. To overcome the effect of salts Schofield and Taylor (1955) suggested that pH should be determined in 0.01 M $CaCl_2$. For many soils the pH measured in $CaCl_2$ or KCl is often about 0.5 pH unit lower than that in water. However, it is more important to interpret the data and take an appropriate course of action than be concerned with what solution to use. Data given here are pH values in water.

Soil Acidity and Soil Properties

Soil acidity derives from a number of sources that include: (i) dissociated and un-dissociated H^+ associated with layer silicate clays; (ii) ionizable H^+ in acidic functional groups in soil OM; (iii) various forms of soil aluminium (Al) that have a positive charge; (iv) H^+ resulting from acidic precipitation, soluble organic acids and acid-forming reactions in soil, such as the oxidation of ammonium-based fertilizers; and (v) greater biological activity, which results in the formation of carbonic acid when carbon dioxide produced by soil microorganisms dissolves in the soil solution.

The nature of the parent material from which the soil was formed, soil weathering (breakdown) and the interaction between these two and the length of time of weathering greatly influence the degree of acidification and the forms of acidity (McLean, 1982). The removal of base cations, especially Ca and Mg, by leaching and erosion results in their replacement by acidic cations, H, Al and iron (Fe), on cation exchange sites and in the soil solution. Soils developed from parent materials with small amounts of base cations tend to acidify more rapidly than those with larger amounts.

Besides the actual soil acidity, the buffer capacity (resilience) of a soil to respond to changes in H^+ concentrations is important and will vary with soil type. Hydrogen ions, produced by various processes in soil, can exchange with other cations adsorbed to soil colloids and these cations are released into the soil solution. This exchange has a number of implications. First, soils rich in organic and inorganic colloids have a larger cation exchange capacity (CEC), and therefore greater resilience to a decline in actual pH than soils poor in colloids. This buffer capacity depends both on the size of the CEC and on the extent to which it is saturated with cations other than H^+. Thus soil texture and soil OM are important because much of the CEC resides in the clay and soil OM fractions. Secondly, maintaining a large proportion of Ca on the CEC is facilitated by the presence of free $CaCO_3$, as in calcareous soils, or by regular additions of $CaCO_3$. Calcium is the dominant cation on the CEC in temperate soils but even in strongly leached soils in warm and wet climates, like New Zealand and Malaya, what little CEC they contain is frequently dominated by Ca (Cooke, 1967). Thirdly, as excess rainfall drains through soil it removes both anions (sulphate, nitrate, chloride, bicarbonate, etc.) and cations (calcium, magnesium, hydrogen, etc.) in the soil solution, but the sum of the anion and cation charges must be equal to maintain the electrical neutrality of the drainage water. Drainage water analyses show that Ca is frequently lost as the accompanying cation.

Long-term Changes in Soil Acidity

Testing the effects of acidity and liming requires long-term experiments because only in exceptional cases does soil pH change rapidly in temperate agriculture. For the same acidifying input, the pH of sandy soils will change more quickly than that of heavier textured, clayey soils and the decline in pH affects crop yields. This is illustrated by changes in pH in the topsoil, 0–23 cm, and the yields of winter wheat grown continuously in two long-term experiments, one at Rothamsted and the other at Woburn. Both tested nitrogen (N) applied at 96 kg/ha as ammonium sulphate ((NH_4)$_2SO_4$) with phosphorus (P) and K fertilizers applied annually. At Woburn in 1877 the sandy loam topsoil (~10% clay) had pH 6.3 and acidification with ((NH_4)$_2SO_4$) decreased pH to 5.0 by 1927. Average annual grain yield in the first decade (1877–1886) was 2.04 t/ha and this declined to 0.64 t/ha in 1917–1926 (Johnston, 1975a;

Johnston and Chater, 1975). By comparison, in 1843 the silty clay loam soil (~25% clay) on Broadbalk, Rothamsted, had pH 7.8 and 40 years later it was still 8.0. With 96 kg N/ha, the average annual grain yield did not decline; it was 2.42 t/ha in the first decade and 2.46 t/ha in the fourth decade. The pH had been buffered by a decline in free $CaCO_3$, from about 4.7% $CaCO_3$ to about 3.2% in 1881 (Bolton, 1972). By 1944, 100 years after the start, soil pH was still 7.5 but there was only 0.8% $CaCO_3$ (Johnston, 1969). An ample reserve of free $CaCO_3$ prevented a decline in pH, thus maintaining soil quality over a long period of time.

In another experiment from 1942 to 1972 on the sandy loam at Woburn, soil pH at the start was 5.3. NPK fertilizers were applied every year and lime about every second year. Over the first 16 years a total of 23 t/ha $CaCO_3$ were applied and the pH increased to 6.8. Interestingly, over the next 12 years it required a total of about 18 t/ha $CaCO_3$ to maintain pH 7.1 (Johnston, 1975b). This need to apply quite large amounts of $CaCO_3$ to maintain a neutral pH is a financial cost to the farmer but the lost Ca has no adverse environmental effect. The only way to minimize Ca losses would be to minimize anion losses.

Evidence suggests that when no lime is applied, an equilibrium pH is reached which depends on the acidifying inputs, parent material, climate, topography, biological activity, management and time. The Park Grass experiment at Rothamsted was started in 1856 on a site that had been in permanent grass for at least 200 years (Warren and Johnston, 1964). Unmanured and ammonium sulphate treatments have remained unchanged since the start and one-quarter of every plot has never received any lime. It has been estimated that in 1856 soil pH was 5.7–5.8. On the unmanured plot, pH declined a little in both the 0–23 and 23–46 cm soil horizons between 1923 and 1959, and by 1959 it was about the same value in both (Table 3.1). By 2002, 43 years later, pH was still 5.3 in both horizons, i.e. under grass vegetation and with natural acidifying inputs, the pH had stabilized at about 5.3. Where 470 kg/ha $(NH_4)_2SO_4$ was applied annually, the pH of the top 23 cm had declined to 3.8 by 1923 and has remained at this value since. Also by 1923 the soil in the next 23 cm layer was much more acid than on the unmanured plot, and the pH had declined further, to 3.7, by 2002 (Table 3.1). Johnston (1972) showed how difficult it was to change the pH of these acid soils under permanent grassland. Preventing the development of subsoil acidity is even more important because correcting it is very time consuming unless lime can be incorporated into the subsoil. Calcium leached from the topsoil will eventually neutralize subsoil acidity, but excessive amounts of lime have to be added to the topsoil and this can lead to problems with nutrient availability.

In the Continuous Wheat and Barley experiments on the sandy loam at Woburn,

Table 3.1. Effect of both natural and additional fertilizer acidifying inputs on soil pH at different depths under grassland and woodland at Rothamsted (adapted from Johnston et al.,1986, with additional data).

Input and horizon (cm)	Year and experiment										
	Park Grass, Grassland						Geescroft Wilderness, Woodland				
	1876	1923	1959	1984	1991	2002	1883	1904	1965	1983	1991
Natural											
0–23	5.4	5.7	5.2	5.0	4.8	5.3	7.1	6.1	4.5	4.2	4.3
23–46	6.3	6.2	5.3	5.7	5.4	5.4	7.1	6.9	5.5	4.6	5.1
46–69	6.5	–	–	–	5.7	5.9	7.1	7.1	6.2	5.7	6.0
Fertilizer[a]											
0–23	5.4	3.8	3.7	3.6	3.6	3.6					
23–46	6.3	4.4	4.1	3.7	4.0	3.7					
46–69	6.5	–	–	4.0	4.4	4.1					

[a]Annual application of ammonium sulphate at 470 kg/ha.

the pH in the top 23 cm in 1876 was 6.2 and with natural acidifying inputs, it had declined to 5.4 by 1927, similar to the equilibrium value in the unmanured soil on Park Grass. With annual applications of 225 kg/ha $(NH_4)_2SO_4$ between 1877 and 1927, pH declined to 5.2 by 1954 (Johnston and Chater, 1975). This decline in pH to 5.2 is in proportion to the decline in pH on Park Grass for the time and amount of N applied. That the equilibrium pH is the same on these two contrasted soil types with identical inputs suggests that soil type has less effect on equilibrium pH than it has on resilience to pH change. In Denmark a survey of farmland soils appeared to show a difference in equilibrium pH with soil type (Madsen and Munk, 1987). However, this apparent relationship could have been due to differences in acidifying inputs from the differing farming systems and farmland soils. What is important is that, at these low equilibrium pH values, most temperate agricultural crops either will not grow or will produce only small yields. Maintaining soil quality is very dependent on maintaining an adequate pH throughout the soil profile.

Natural acidifying inputs interact with aboveground vegetation to affect soil pH as illustrated by two adjacent Rothamsted experiments, Park Grass and Geescroft Wilderness (Table 3.1). At present the semimature oak trees on Geescroft represent the slowly changing vegetation since the growing of arable crops ceased in 1886. The 0–23 and 23–46 cm soil horizons on Park Grass had reached pH 5.3 by 1959 and there has been little change since (Table 3.1). Under the developing deciduous tree canopy on Geescroft, the pH in the top 23 cm had declined from 7.1 in 1883 to just over 4 in 1983 and has changed little since. This soil is now almost as acid as that receiving 470 kg/ha $(NH_4)_2SO_4$ annually on Park Grass. Besides this intense acidification of the topsoil on Geescroft, the pH has declined in both the 23–46 and 46–69 cm horizons (Table 3.1) and the decline started before the topsoil had reached its lowest value. The two sites are exposed to the same potentially acidifying aerial inputs and presumably the much greater acidification on Geescroft is related

to trees being more efficient than grass at trapping them.

Soil pH and Nutrient Availability

The effects of soil acidity on plant growth are partly through the effect of pH as such on root function and partly through its effect on soil properties, especially the indirect effect of an excess of different ionic species of Al and manganese (Mn) (Rowell, 1988). The effect of Al toxicity is mainly on root growth and the release of Al in various forms from clay minerals is largely pH dependent, the dominant species being Al^{3+} and $AlOH^{2+}$ (Bache, 1985). $AlOH^{2+}$ is much more toxic to plants than is Al^{3+} and the concentration of $AlOH^{2+}$ is larger at pH 4.5 than at 4.0, thus the toxicity of Al is greatest at pH 4.5 (Moore, 1974). Figure 3.1 shows that in the pH range 5–6 most nutrients have maximum availability and below pH 5 the availability of most nutrients declines. An important exception is Mn, which, like some other elements, becomes more available with increasing acidity and can reach toxic levels under very acid conditions. Also, some elements become toxic to plants at levels they can normally tolerate when reducing conditions develop in soil due to poor aeration, for example when ferric iron is reduced to toxic ferrous compounds. Above pH 6 the availability of P, boron (B) and Mn decreases. In the case of P, plants principally take up $H_2PO_4^-$ and to a lesser extent HPO_4^{2-} and pH regulates the ratio of these two ions in the soil solution, $H_2PO_4^-$ being favoured below pH 7. Rubaek et al. (1998) discussed the interaction of lime and P on a very acid soil in Denmark. The availability of K, Mg, copper (Cu) and zinc (Zn) declines above pH 7 but molybdenum (Mo) becomes more available under alkaline conditions.

Nutrient uptake rate is also pH dependent. Generally anions, like nitrate and phosphate, are taken up at a faster rate under weakly acid conditions whereas the cation uptake rate seems to be faster in the more neutral pH range. Rorison (1980) discussed relationships between soil pH, nutrient uptake and the physiology of plants.

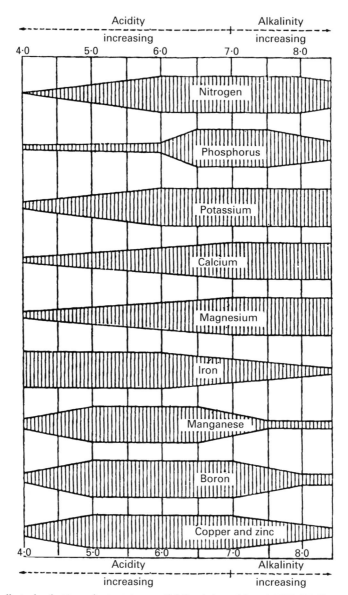

Fig. 3.1. The effect of soil pH on plant nutrient availability (adapted from MAFF, 1976).

There are innumerable examples of the direct effects of acidity on crop yields (see for example, Walker, 1952; Gardner and Garner, 1957; Adams, 1984). The optimum pH for maximum growth differs between crops. Some prefer more acid conditions, for example, potatoes 5.5–6.0, winter wheat 6.0, but sugarbeet, spring barley and lucerne are less tolerant of acidity and prefer pH 6.5–7.0. Of interest is the interaction between P, K and soil acidity shown by the yields of spring barley and potatoes in two long-term experiments (Table 3.2). The interaction of pH and nutrient availability over many decades on the yield and species composition of permanent grassland has been discussed by Thurston *et al.* (1976) and Johnston *et al.* (2001 and references therein).

Soil pH influences the occurrence and activity of soil microorganisms. In general

Table 3.2. Yields of spring barley and potatoes given NPK fertilizers at various soil pH values and the yield when P or K was omitted. Rothamsted and Woburn, 1970–1974 (adapted from Bolton, 1971, 1977).

	Rothamsted				Woburn		
Barley, grain, t/ha, 1970–1973							
pH$_{water}$	4.6	5.3	6.2	7.4	5.8	6.7	7.2
Yield with NPK	–[a]	4.86	4.96	5.15	4.89	4.92	5.14
Yield with NK	–	2.36	3.92	4.31	3.82	4.13	5.02
Yield with NP	–	3.32	4.53	4.75	4.20	4.60	4.61
Potatoes, tubers, t/ha, 1974							
pH$_{water}$	4.4	5.0	5.6	6.5	5.4	6.3	6.7
Yield with NPK	53.2	56.7	60.5	56.3	57.9	53.2	55.7
Yield with NK	11.0	32.6	39.3	40.4	32.4	41.0	45.3
Yield with NP	35.7	22.0	23.3	10.2	16.0	17.5	15.0

[a]Barley failed due to acidity.

fungi dominate at soil pH < 5.5 and in the rhizosphere whereas at a higher pH bacteria are more abundant (Trolldenier, 1971). Mycorrhizal fungi play an important role in phosphate uptake in acid, P-deficient soils. *Nitrosomonas* and *Nitrobacter* that nitrify NH_4-N and NO_2-N, respectively, prefer more neutral soil conditions, as do free-living and symbiotic bacteria that fix N_2. As bioremediation of polluted soils becomes ever more important, the optimum pH for appropriate microbial activity will need to be studied.

The assessment of the optimum pH has to consider, therefore, the direct effects of acidity on nutrient availability, the uptake rate of available nutrients by roots, the effects that soil microorganisms have on modulating nutrient availability and the crops to be grown. Although crops have an optimum pH, most can cope to a varying degree with differences in bulk soil pH by actively modifying the acidity of the rhizosphere, the narrow cylinder of soil around the roots. As roots excrete H^+, rhizosphere soil pH changes quite quickly, often with beneficial effects for that crop. Any change in pH will be quite small relative to the pH of the total soil mass and soil mixing during cultivation after harvest will quickly correct the very localized acidity.

As for other aspects of soil management, compromises must be made in deciding the recommended pH because not all the factors are optimum at the same pH. In temperate climates, for mineral soils ranging from sands to clays with average amounts of soil OM, the recommended value for arable soils would be

pH 6.5 and for grassland soils pH 6.0. For organic mineral soils (10–25% soil OM) and peat soils the values would be pH 6.2 and 5.8, respectively, for arable crops, and 5.7 and 5.3, respectively, for grassland. These lower values are justified because there is sufficient soil OM to strongly complex Al.

Indirect Effects of Soil Acidification

Soil acidification may have indirect effects that can take many years to develop. Data from the Broadbalk experiment discussed above showed that reserves of free $CaCO_3$ buffered soil pH for more than 100 years. In the Agdell experiment started in 1848 on a soil with some free $CaCO_3$ and pH ~8, arable crops were grown in rotation. There were three treatments, unmanured, PK and NPK with the fertilizers applied only once every 4 years. The $(NH_4)_2SO_4$ used slowly acidified the soil and after some 80 years, yields of turnips given NPK on soils now with pH 5.6, were little better than those on plots always without fertilizers but with pH 8.2. Although the acidity could be, and later was, corrected, the acid conditions encouraged the development of the fungus *Plasmodiophora brassica*, the causal agent of 'finger-and-toe', which decreased yield and is very difficult to eradicate unless brassicas are not grown for a long period. The yields of winter wheat were not affected (Johnston, 1994). The ability to grow the widest possible range of crops within the constraints of climate and topography must be a primary aim of good soil management. Optimizing soil pH to minimize the risk of establishing soilborne pests and diseases must be a part of that management strategy.

Effects on crop quality are less well documented than those on yields. Increasing soil pH decreased the concentration of cadmium in herbage (Johnston and Jones, 1995). Increased crop growth, as a result of liming soils that are below optimum pH, can return more OM to the soil resulting in increased soil microbial and faunal activity (Haynes and Naidu, 1998). Soil pH also affects the survival of many soilborne fungi; for example, mention has already been made of the effects of *Plasmodiophora brassica*, whereas *Streptomyces scabies*, which causes a skin blemish on potatoes, is better adapted to survive in neutral rather than acid soils. Occasionally, within-field differences in weed and pest control by soil-applied pesticides have been related to differences in soil pH (Eagle, 1978). More recently, soil acidity has been shown to affect glyphosate sorption by soil (de Jonge *et al.*, 2001).

It is often considered that liming accelerates the decomposition of soil OM. In the top 23 cm of soil in the Park Grass experiment this has only happened on the $(NH_4)_2SO_4$ plots (Table 3.3). In contrast, on the plots without N increasing the pH has increased soil OM, whereas with sodium nitrate ($NaNO_3$) increasing pH has not changed the level of soil OM (Table 3.3) despite yields being larger with $NaNO_3$. Presumably there was no difference in root turnover in the two soils. In this grassland experiment these differences in soil OM in relation to soil properties are difficult to assess. Effects might be expected in arable soils but Haynes and Naidu (1998) could find no data in the literature to show the effects of liming on soil OM in arable soils. There is a scarcity of data on the effects of liming on soil physical conditions (Schjønning *et al.*, 1994). It is generally thought that soil structure benefits from liming but the mechanisms are unclear (Davies and Payne, 1988), possibly because it takes many years to establish soil structural differences in field experiments. Long-term improvements probably arise from Ca flocculating soil colloids into stable aggregates but in the short term liming has

been shown to disperse soil colloids (Haynes and Naidu, 1998 and references therein). Liming light-textured soils lessened their tendency to cap under heavy rainfall (Gardner and Garner, 1957) and improved tractor-ploughing speeds and decreased drawbar pull on heavy-textured soils (Russell and Keen, 1921).

Liming and Lime Requirements

Little Ca is removed from soil in harvested crops, often not more than 100 kg/ha, and only very modest applications of $CaCO_3$ would be required to replace such losses. The annual loss of Ca in drainage increases as soil pH increases (Table 3.4). Gasser (1973) used data from some earlier experiments and

Table 3.3. Carbon and nitrogen concentrations in the 0–23 cm horizon of permanent grassland soils of different pH and given different forms of nitrogen (adapted from Warren and Johnston, 1964).

			Nitrogen treatment			
	None		Ammonium sulphate		Sodium nitrate	
pH	5.2	7.1	3.7	5.4	5.7	7.3
%C	3.3	3.9	4.1	3.7	3.1	3.7
%N	0.26	0.33	0.31	0.29	0.25	0.31
C:N ratio	12.7	11.8	13.2	12.8	12.4	11.9

Table 3.4. Postulated annual amounts of calcium carbonate lost by leaching from soils of pH_{water} 5 to 8 (adapted from Johnston *et al.*, 2001).

	$CaCO_3$ lost, kg/ha each year		
	Gasser (1973)		Chambers and
Soil pH_{water}	Mean	Range	Garwood (1998)[a]
5.0	118	79–157	235
5.5	168	112–224	438
6.0	235	156–313	640
6.5	336	224–448	842
7.0	471	314–628	1045
7.5	672	448–896	1248
8.0	942	628–1256	1450

[a]Losses derived from their equation: Annual loss $CaCO_3$ (kg/ha) = (pH value $\times$ 405) – 1790.

recognized that there could be a range of Ca losses depending on soil type and fertilizer addition. Chambers and Garwood (1998) derived an equation (Table 3.4) for Ca loss from data for additions of CaCO₃ required to maintain the initial pH in a number of field experiments in the UK in the 1980s and 1990s. These authors concluded that over-winter drainage volume and soil texture did not appear to be important variables in determining losses and they calculated only a single value for Ca loss at each pH. Their estimated losses are larger than Gasser's. This is presumably because in the interval between the two sets of experiments there had been an increase in the quantity of anions lost with increasing N applications, larger crop residue returns and more acidifying aerial inputs. For a soil at pH 7, Chambers and Garwood's estimated Ca loss, ~1000 kg CaCO₃, is similar to that reported for the Market Garden experiment at Woburn and the general 'rule-of-thumb' value used in Denmark (P. Schjønning, Foulum, 2002, personal communication).

Table 3.4 shows that as pH declines, Ca losses decrease. The reason for this is not readily obvious because H⁺ concentration increases tenfold for every one-unit decrease in pH. Declining subsoil pH noted above in a number of experiments would suggest that as the number of Ca ions on the CEC declines then H⁺ is lost from the surface soil, as well as Ca²⁺, causing development of acidity in the subsoil.

Lime requirement is defined as the amount of liming material required to raise soil pH to the desired value. Goulding and Annis (1998) discussed both the calculation and measurement of lime requirement. These authors showed that calculation can be done in two ways: (i) by calculating the Ca lost in drainage, for example, using the equation of Chambers and Garwood (1998) that requires only knowing the pH when lime is to be applied (see Table 3.4); or (ii) calculating the lime needed to neutralize acidity. The second approach requires calculating the acidity produced by each input every year since the last lime application. Goulding and Annis (1998) showed that this method, although more time consuming, gave lime requirements

much closer to those required as shown by field experiments.

The measurement of lime requirement is based on laboratory soil analysis (see, for example, Woodruff, 1948; Shoemaker *et al.*, 1961; Peech *et al.*, 1965; McLean, 1982; van Lierop, 1990). These methods often give different lime requirements for the same soil, perhaps because different amounts of H⁺ on the cation exchange sites are replaced. A satisfactory method must ensure that the amount of lime recommended will increase the existing pH to the required value for the cropping and environmental conditions that exist in the field. Because liming is usually done periodically rather than annually, the amount of lime applied often exceeds that which is determined analytically to give a small reserve. To avoid the need for analysis, currently the Agricultural Development and Advisory Service in England and Wales uses 'look up' tables (see Table 3.5) for lime requirement based on soil pH (ADAS, 1986). The tables, derived from many years' experimental data, are based on a linear relationship between soil pH and lime requirement. The lime requirement is calculated as (target pH *minus* measured pH) times the 'lime factor'. The target pH is a little higher than the recommended pH to allow for the pH declining below the recommended value before the next application is given. The lime factor, i.e. the amount of lime needed to raise soil pH by one unit, increases a little as soil texture becomes heavier for mineral soils with average soil OM. The lime factors are much larger for organic mineral soils and peat soils because they have large numbers of cation exchange sites. Currently a lime requirement model is being developed at Rothamsted (Goulding and Annis, 1998).

Conclusions

Soil acidity affects the biological, chemical and physical properties of soil that individually or through interactions affect the sustainability of biomass production in both managed and natural ecosystems. Although crops vary considerably in their tolerance of acidity, one aspect of sustainability is the

Table 3.5.   Lime recommendation tables based on current and target pH to achieve a recommended pH.[a]

Soil type	Recommended pH (water)	Target pH (water)	Lime factor for 1 pH unit change CaCO₃ t/ha
Arable soil to a depth of 20 cm			
Sands, loamy sands	6.5	6.7	6
Sandy loams, sandy silt loams, silt loams	6.5	6.7	7
All clay loams and clays	6.5	6.7	8
Organic mineral soils	6.2	6.4	10
Peaty soils and peats	5.8	6.0	16
Grassland soil to a depth of 15 cm			
Sands, loamy sands	6.0	6.2	4.5
Sandy loams, sandy silt loams, silt loams	6.0	6.2	5.0
All clay loams and clays	6.0	6.2	6.0
Organic mineral soils	5.7	5.9	7.5
Peaty soils and peats	5.3	5.5	12.0

[a]For details see text. Organic mineral soils 10–25% soil OM, peaty soils and peats > 25% soil OM (adapted from ADAS, 1986).

opportunity to grow a wide range of crops for food or industrial use and to do this soil must not be allowed to become very acid.

For unamended soils, their equilibrium pH is a function of their buffering capacity and the magnitude of the acidifying inputs. In agricultural soils, acidity can be corrected but it is preferable to maintain pH by regular liming because the sustainability of crop production is jeopardized when surface and subsoils become acid and subsoil acidity is difficult to correct. In temperate regions, a general recommendation for mineral soils with average amounts of soil OM is to maintain a threshold pH ~6.5 for arable soils and ~6.0 for grassland soils. These values are somewhat lower for organic mineral and peat soils. To achieve the ameliorative effects of liming, large inputs are required and the amount may vary with soil type if H⁺ on cation exchange sites has to be replaced to achieve and maintain the recommended pH. Globally, liming materials are not a scarce resource but they are not readily available everywhere. Besides their direct cost there is an environmental cost in transport and energy use. Therefore excessive liming, with increased Ca losses from soil, should be avoided in temperate regions. By contrast, liming highly weathered soils in tropical regions is costly and usually aims to alleviate Al toxicity and

Ca deficiency by maintaining pH in the range 5.3–5.6 (Pearson, 1975; Kamprath, 1984). Where lime is not readily available it may be necessary to breed cultivars of crop plants that, while maintaining their yield potential, are tolerant of acidity. Soils that are not actively managed but provide landscape amenity often have a low buffering capacity, i.e. low resilience to change, and maintaining the appropriate vegetation will depend on lowering aerial acidifying inputs.

References

Adams, F. (ed.) (1984) *Soil Acidity and Liming*, 2nd edn. Agronomy Monograph No.12. ASA, Madison, Wisconsin, 380 pp.

ADAS (1986) *Changes in ADAS Lime Recommendations*. TFS 731, SS/86/2. Ministry of Agriculture, Fisheries and Food, London.

Bache, B.W. (1985) Soil acidification and aluminium mobility. *Soil Use and Management* 1, 10–14.

Bolton, J. (1971) Long-term liming experiments at Rothamsted and Woburn. *Rothamsted Experimental Station Report for 1970* Part 2, 98–112.

Bolton, J. (1972) Changes in magnesium and calcium in soils of the Broadbalk wheat experiment at Rothamsted from 1856 to 1966. *Journal of Agricultural Science, Cambridge* 79, 217–223.

Bolton, J. (1977) Changes in soil pH and exchangeable calcium in two liming experiments on contrasting soils over 12 years. *Journal of Agricultural Science, Cambridge* 89, 81–86.

Chambers, B.J. and Garwood, T.W.D. (1998) Lime loss rates from arable and grassland soils. *Journal of Agricultural Science, Cambridge* 131, 455–464.

Cooke, G.W. (1967) *The Control of Soil Fertility.* Crosby Lockwood & Sons Ltd, London, 526 pp.

Davies, D.B. and Payne, D. (1988) Management of soil physical properties. In: Wild, A. (ed.) *Russell's Soil Conditions and Plant Growth,* 11th edn. Longman Scientific and Technical, Harlow, UK, pp. 412–448.

de Jonge, H., de Jonge, L.W., Jacobsen, O.H., Yamaguchi, T. and Moldrup, P. (2001) Glyphosate sorption in soils of different pH and phosphorus content. *Soil Science* 166, 230–238.

Eagle, D.J. (1978) *Effect of Lime on the Effectiveness of Pesticides.* SS/PU/8 and SS/PU/17. Ministry of Agriculture, Fisheries and Food, London, 2 pp.

Gardner, H.W. and Garner, H.V. (1957) *The Use of Lime in British Agriculture.* E. & F.N. Spon, London, 216 pp.

Gasser, J.K.R. (1973) An assessment of the importance of some factors causing losses of lime from agricultural soils. *Experimental Husbandry* 25, 86–95.

Goulding, K.W.T. and Annis, B.J. (1998) *Lime, Liming and the Management of Soil Acidity.* Proceedings No. 410. The International Fertiliser Society, York, UK, 36 pp.

Haynes, R.J. and Naidu, R. (1998) Influence of lime, fertilizer and manure applications on soil organic matter content and soil physical conditions: a review. *Nutrient Cycling in Agroecosystems* 51, 123–137.

Johnston, A.E. (1969) Plant nutrients in Broadbalk soils. *Rothamsted Experimental Station Report for 1968* Part 2, 93–112.

Johnston, A.E. (1972) Changes in soil properties caused by the new liming scheme on Park Grass. *Rothamsted Experimental Station Report for 1971* Part 2, 177–180.

Johnston, A.E. (1975a) Experiments made on Stackyard Field, Woburn, 1876–1974. I. History of the field, details of cropping and manuring and the yields in the Continuous Wheat and Barley experiments. *Rothamsted Experimental Station Report for1974* Part 2, 29–44.

Johnston, A.E. (1975b) The Woburn Market Garden Experiment, 1942–1969. II. The effects of the treatments on soil pH, soil carbon, nitrogen, phosphorus and potassium. *Rothamsted Experimental Station Report for 1974* Part 2, 102–130.

Johnston, A.E. (1994) The Rothamsted classical experiments. In: Leigh, R.A. and Johnston, A.E.

(eds) *Long-term Experiments in Agricultural and Ecological Sciences.* CAB International, Wallingford, UK, pp. 9–37.

Johnston, A.E. and Chater, M. (1975) Experiments made on Stackyard Field, Woburn, 1876–1974. II. Effects of the treatments on soil pH, P and K in the Continuous Wheat and Barley experiments. *Rothamsted Experimental Station Report for 1974* Part 2, 45–60.

Johnston, A.E. and Jones, K.C. (1995) *The Origin and Fate of Cadmium in Soil.* Proceedings No. 366, The International Fertiliser Society, York, UK, 39 pp.

Johnston, A.E., Goulding, K.W.T. and Poulton, P.R. (1986) Soil acidification during more than 100 years under permanent grassland and woodland at Rothamsted. *Soil Use and Management* 2, 3–10.

Johnston, A.E., Poulton, P.R., Dawson, C.J. and Crawley, M.J. (2001) *Inputs of Nutrients and Lime for the Maintenance of Fertility of Grassland Soils.* Proceedings No. 486, The International Fertiliser Society, York, UK, 40 pp.

Kamprath, E.J. (1984) Crop response to lime on soils in the tropics. In: Adams, F. (ed.) *Soil Acidity and Liming,* 2nd edn. Agronomy Monograph No. 12. ASA, Madison, Wisconsin, pp. 349–368.

Madsen, H.B. and Munk, I. (1987) The influence of texture, soil depth and geology on pH in farmland soils: a case study from southern Denmark. *Acta Agriculturae Scandinavica* 37, 407–418.

MAFF (1976) *Lime and Liming.* Bulletin 35. Ministry of Agriculture, Fisheries and Food, Her Majesty's Stationery Office, London, 44 pp.

McLean, E.O. (1982) Soil pH and lime requirement. In: Page, A.L. and Miller, R.H. (eds) *Methods of Soil Analysis Part 2,* 2nd edn. Agronomy Monograph No. 9. ASA and SSSA, Madison, Wisconsin, pp. 199–224.

Moore, D.P. (1974) Physiological effects of pH on roots. In: Carson, E.W. (ed.) *The Plant Root and its Environment.* University Press of Virginia, Charlottesville, Virginia, pp. 135–151.

Pearson, R.W. (1975) *Soil Acidity and Liming in the Humid Tropics.* Cornell International Agricultural Bulletin No. 30, Cornell University, Ithaca, New York, 65 pp.

Peech, M., Cowan, R.L. and Baker, J.H. (1965) A critical study of the $BaCl_2$-triethanolamine and the ammonium acetate methods for determining the exchangeable hydrogen content of soils. *Soil Science Society of America Proceedings* 26, 37–40.

Rorison, I.H. (1980) The effects of soil acidity on nutrient availability and crop response.

In: Hutchinson, T.C. and Havas, M. (eds) *Effects of Acid Precipitation on Terrestrial Ecosystems.* Plenum Press, New York, pp. 283–304.

Rowell, D.L. (1988) Soil acidity and alkalinity. In: Wild, A. (ed.) *Russell's Soil Conditions and Plant Growth*, 11th edn. Longman Scientific and Technical, Harlow, UK, pp. 844–898.

Rubaek, G.H., Sinaj, S., Frossard, E., Sibbesen, E. and Borggaard, K.O. (1998) Long-term effects of liming and P fertilisation on P state and P exchange kinetics in an acid, sandy soil. *Symposium no. 12, ISSS World Congress, Montpellier, France.*

Russell, E.J. and Keen, B.A. (1921) The effect of chalk on the cultivation of heavy land. *Journal of the Ministry of Agriculture* 28, 419–422.

Sanchez, P.A. (1976) *The Properties and Management of Soil in the Tropics.* John Wiley & Sons, Chichester, UK, 618 pp.

Schjønning, P., Christensen, B.T. and Carstensen, B. (1994) Physical and chemical properties of a sandy loam receiving animal manure, mineral fertilizers or no fertilizer for 90 years. *European Journal of Soil Science* 45, 257–268.

Schofield, R.K. and Taylor, A.W. (1955) The measurement of soil pH. *Soil Science Society of America Proceedings* 19, 164–167.

Shoemaker, H.E., McLean, E.O. and Pratt, P.F. (1961) Buffer methods for determination of lime requirements of soils with appreciable amounts of extractable aluminium. *Soil Science Society of America Proceedings* 25, 274–277.

Sørensen, S.P.L. (1909) Enzymstudier: II. Om maalingen og betydningen af brintion-koncentrationen ved enzymatiske processer. *Meddelelser fra Carlsberg Laboratoriet* 8, 1–153.

Thomas, G.W. (1996) Soil pH and soil acidity. In: Sparks, D.L., Page, A.L., Helmke, P.A., Loeppert, R.H., Soltanpour, P.N., Tabatabai, M.A., Johnston, C.T. and Sumer, M.E. (eds) *Methods of Soil Analysis. Part 3. Chemical Methods.* Book Series No. 5. SSSA, Madison, Wisconsin, pp. 475–490.

Thurston, J.M., Williams, E.D. and Johnston, A.E. (1976) Modern developments in an experiment on permanent grassland started in 1856; effects of fertilizers and lime on botanical composition and crop and soil analysis. *Annales Agronomique* 27, 1043–1082.

Trolldenier, G. (1971) *Soil Biology, the Soil Organisms in the Economy of Nature.* Franckh'sche Verlagshandlung, Stuttgart, 116 pp.

van Lierop, W. (1990) Soil pH and lime requirement. In: Westerman, R.L. (ed.) *Soil Testing and Plant Analysis*, 2nd edn. SSSA, Madison, Wisconsin, pp. 73–126.

Walker, T.W. (1952) The estimation of the lime requirement of soils. *Journal of Soil Science* 3, 261–276.

Warren, R.G. and Johnston, A.E. (1964) The Park Grass Experiment. *Rothamsted Experimental Station Report for 1963*, 240–262.

Wild, A. (ed.) (1988) *Russell's Soil Conditions and Plant Growth*, 11th edn. Longman Scientific and Technical, Harlow, UK, 991 pp.

Woodruff, C.M. (1948) Soil tests for lime requirement by means of a buffer solution and the glass electrode. *Soil Science* 66, 53–63.

Chapter 4
Tightening the Nitrogen Cycle

B.T. Christensen

Danish Institute of Agricultural Sciences, Department of Agroecology,
PO Box 50, DK-8830 Tjele, Denmark

Summary

The availability of nitrogen to crop plants is a universally important aspect of soil quality, and often nitrogen represents the immediate limitation to crop productivity in modern agriculture. Nitrogen is decisive for the nutritive value of plant products and plays a key role in the environmental impact of agricultural production. The fundamental doctrine of nitrogen management is to optimize the nitrogen use efficiency of both introduced and native soil nitrogen by increasing the temporal and the spatial coincidence between availability and root uptake of mineral nitrogen. Natural ecosystems have evolved to produce a high degree of coincidence and maintain a relatively tight nitrogen cycle. The management needed to assure the vitality of crop cultivars with a high production potential causes agroecosystems to be relatively open with respect to nitrogen. By shifting the focus from sheer productivity to a balance between productivity, product quality and environmental impact, the management of agroecosystems can be redesigned to allow

©CAB International 2004. *Managing Soil Quality: Challenges in Modern Agriculture*
(eds P. Schjønning, S. Elmholt and B.T. Christensen)

for a higher nitrogen use efficiency. The important management measures to improve the nitrogen aspect of soil quality are crop sequences that incorporate cover crops, judicious use of soil tillage, improved timing and use of animal manures, crop residues and mineral fertilizers, and a sustainable balance between plant production potential and animal stocking density. This chapter addresses the characteristics of nitrogen cycling in plant production and the premises upon which an improved nitrogen use has to be based, and discusses management principles that can improve nitrogen use in cropping systems.

Introduction

Nitrogen and soil quality

One universally important attribute of soil quality is the availability of plant nutrients, whether these are derived from residues of previous nutrient dressings, atmospheric deposition or the soil parent material, or are provided directly to the crop by mineral fertilizers and animal manures. An adequate and balanced availability of a wide range of macro- and micronutrients is required to optimize plant growth, but, in most developed agroecosystems, the immediate limitation to productivity is the availability of mineral nitrogen. Nitrogen is also a major element in terms of the environmental impacts of agricultural activity, and nitrogen containing compounds of biosynthesis (amino acids and proteins) are crucial for the nutritive value of plant products and thus for animal and human health. Developing a sustainable nitrogen management policy is a particular challenge in maintaining soil quality.

Characteristics of the nitrogen cycle

The nitrogen cycle in agriculture and various aspects of nitrogen turnover in arable soils have been thoroughly reviewed (e.g. Stevenson, 1982; Jarvis *et al.*, 1996; Silgram and Shepherd, 1999). This section briefly outlines aspects of particular relevance to nitrogen management. In the soil, nitrogen undergoes a variety of largely microbially mediated transformations, and almost any transformation of nitrogen is associated with organic matter (OM) turnover. Agricultural soils contain a large pool of organically bound nitrogen. Figure 4.1 shows a schematic outline of the nitrogen inputs, outputs and pools in the rooting zone of a 'typical' arable soil, the size of the boxes indicating their relative importance. The soil layers exploited by plant roots typically hold 5000–15,000 kg N/ha of which, however, only 1–2% may become mineralized and available to crop uptake within a growth period. The pools of nitrogen that dominate the shorter-term nitrogen turnover are the decomposer biomass and labile OM pools. These pools of nitrogen are relatively dynamic and respond more readily to inputs of plant residues and animal manure, and to soil disturbances, e.g. by tillage.

Due to its chemical nature and the large microbial potential in soil, nitrogen can occur in forms with widely different characteristics in terms of availability to plants and susceptibility to loss to the environment. In contrast to other plant nutrients, nitrogen can be considered a renewable resource in crop production since lost nitrogen can be replaced through biological or industrial fixation of atmospheric N_2. This capacity of course does not offset concerns regarding excessive nitrogen losses to the environment, and the energy needed to manufacture and distribute mineral fertilizers. Loss of nitrate from the soil by leaching contributes to the eutrophication of freshwater bodies and coastal areas. Nitrification and denitrification cause gaseous losses of nitrous oxide (N_2O), which is a most potent agent in global warming. Volatilization of ammonia may deteriorate nutrient-restrained natural ecosystems and cause acidification when returned from the atmosphere through wet or dry deposition. Whereas nitrate is mobile in water and subject to denitrification, nitrogen in ammonium form is retained in the soil through adsorption to soil colloids or fixation in clay minerals. By far the largest fraction of organically bound nitrogen will be retained

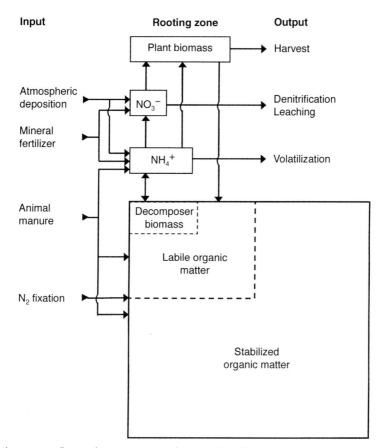

Fig. 4.1. The nitrogen flow in the rooting zone of a 'typical' arable soil showing major inputs, outputs and pools of nitrogen.

in the soil, but under some circumstances losses of organic nitrogen dissolved by percolating water may be significant (Murphy et al., 2000).

Sustainable nitrogen management

Losses of nitrogen from the soil–plant system not only relate to the environmental component of soil quality. If lost nitrogen is not replaced by management, the soil pool of mineralizable and thus potentially plant-available nitrogen will gradually be depleted. This will reduce not only the productivity of the agroecosystem, but will also affect the nutritive quality of the products. Although the protein component of diets may be derived from leguminous crops with symbiotic N_2-fixing capabilities, a balanced diet requires sources of dietary energy from crops relying on mineral nitrogen.

The accumulation of organic nitrogen in the soil is a characteristic feature of terrestrial ecosystems, and understanding the regulation of soil nitrogen turnover in natural ecosystems can enable us to develop management that better sustains soil quality. In unmanaged (natural) ecosystems, nitrogen is returned to the soil either directly in plant litters or in residues from the herbivore-based food chain. External sources of nitrogen are atmospheric deposition and N_2-fixing components of the vegetation. Through the mineralization–immobilization turnover, mineral nitrogen becomes available for root uptake but also for loss to the environment.

The rate and the seasonal and spatial distribution of nitrogen release in forms available to plants influence the composition and productivity of the vegetation. Thereby mineralization provides a feedback on the amount and quality of plant and animal residues that eventually enter the soil. In unmanaged ecosystems with mixed plant communities, the spatio-temporal pattern of the aggregated nitrogen uptake by annual and perennial plant species has evolved in response to the nitrogen release profile of the soil supporting the vegetation. Natural temperate ecosystems often exhibit a high degree of temporal and spatial coincidence (termed synchrony and synlocation, respectively) between nitrogen release and root uptake potentials within the soil, causing the transfer of nitrogen across system borders to be small compared to the amount of nitrogen cycled within the ecosystem (e.g. Likens *et al.*, 1977). In such systems, the nitrogen cycle can be considered as relatively tight, as substantiated by the small nitrate leaching losses from temperate forest soils (e.g. Dise and Wright, 1995).

In contrast, most agroecosystems are relatively open with respect to nitrogen. Due to the very purpose of producing biomass suitable for consumption outside the system, external inputs are needed to compensate for nitrogen removed in exported products. The inclusion of N_2-fixing plants in the cropping system may only alleviate to some extent the effects of nitrogen removal at harvest. Other major differences between natural and agricultural systems are of significance to the nitrogen turnover in soil. These include the replacement of vegetation of mixed native annual and perennial species (herbaceous and woody) with monocultures of introduced cultivars relevant to the nutrition of animals and humans. Crop cultivars used in intensive agriculture generally differ markedly from native species and exhibit an intensive nitrogen uptake in the active, but often relatively short, growth phase. Management also introduces a massive physical disturbance of the soil structure through tillage and traffic and affects the hydrology through drainage and irrigation. Moreover, agriculture may cause soil compaction, alleviate soil acidity by liming, and modify topography and wind exposure.

The management of the soil–plant system therefore causes the nitrogen dynamics of agroecosystems to differ from those of natural ones. Although the various processes in the nitrogen cycle are the same as in natural systems, their quantitative significance differs. Reduced synlocation and synchrony in the nitrogen turnover of agroecosystems and a reduced return of OM to the soil affects the nitrogen use efficiency.

Various management measures to improve synchrony and synlocation in nitrogen turnover and increase the nitrogen use efficiency of both added and native soil nitrogen are discussed below.

Improving the Nitrogen Use Efficiency

Premises and measures

Results of long-term field experiments clearly demonstrate a significant year-to-year variation in crop nitrogen demands. Obviously, the weather conditions during the growth period, but also crop protection intensity and timing, the choice of crop cultivars and other management parameters will affect the nitrogen use efficiency realized on a given site. Thus for irrigated crops and crops fed nitrogen through fertigation, higher and less variable nitrogen use efficiencies are possible.

The mineralization–immobilization turnover of nitrogen in the soil plays an important role in the nitrogen supply to the growing plant even in cropping systems receiving recommended levels of nitrogen in mineral fertilizers. This has been well illustrated in studies tracing the fate of introduced ^{15}N-labelled nitrogen. Generally, 30–70% of the ^{15}N added in mineral fertilizers is in the harvestable fractions (grain, straw, herbage) of the crop, most of the remaining ^{15}N being retained in the soil–root system (Kjellerup and Kofoed, 1983; Powlson *et al.*, 1986a; Glendining *et al.*, 1997). It appears that the recovery of ^{15}N is similar across a range of fertilizer additions, annual crop species and soil types, whereas the climatic conditions and the characteristics of introduced organic amendments significantly influence the recovery of added N. Pilbeam (1996) assembled data from experiments with ^{15}N-labelled mineral fertilizer applied to wheat grown in different parts of the world and observed that more ^{15}N was recovered in the crop in humid than in dry environments. The retention of residual ^{15}N in the soil, however, increased with increasing climate dryness.

Postharvest losses of residual fertilizer nitrogen by nitrate leaching are small, however. Lysimeter studies under northwest European climate conditions have shown that less than 5% of ^{15}N added in mineral fertilizer is lost by nitrate leaching over the following 3–4 years (e.g. Kjellerup and Kofoed, 1983; Barraclough et al., 1984), indicating that nitrate susceptible to leaching during autumn and winter originates mainly from mineralization of organic nitrogen accumulated in the soil.

Crop recoveries are smaller and the retention in the soil larger for ^{15}N added in crop residues and animal manure for which a large part of the nitrogen is in organic forms at the time of application (e.g. Thomsen and Jensen, 1994; Thomsen and Christensen, 1996; Jensen et al., 1999; Thomsen et al., 2001). Higher crop ^{15}N-recoveries and reduced nitrogen losses are found for spring than for autumn applications of crop residues and manures.

An improved nitrogen use efficiency can therefore only be achieved by considering both the turnover of nitrogen added in fertilizers, manure and crop residues, and the dynamics of the nitrogen already in the soil. Important management measures to enhance the nitrogen aspect of soil quality are to deliberately manipulate the nitrogen turnover in soil, to determine the nitrogen mineralization capacity of the soil and, based on these measures, to calculate the additional nitrogen input that provides the best compromise between production and environment.

The art of nitrogen demand prognoses

Traditionally, the mission of forecasting crop nitrogen demands has been to establish the optimum rates of nitrogen fertilizers with respect to productivity and economic returns. The environmental concerns have more recently been introduced into fertilizer recommendations, most often driven by legislative actions to restrict fertilizer use. Nevertheless, any prognosis system has to be based on the biologically founded interactions between added nitrogen, soil nitrogen turnover and nitrogen uptake in the crop.

The simplest way to predict the nitrogen demand of a given crop is to establish its response to nitrogen added in increasing increments. Replicated experiments over a number of years will provide good guidance for the average nitrogen demand of a crop subject to a specific management. Indeed, nitrogen response curves obtained from a number of different sites constitute the backbone of most standard fertilizer recommendations provided by agroindustrial companies, advisory services and public authorities concerned with agricultural production and environmental protection.

The black-box approach described above provides little information on the mechanisms that govern the nitrogen turnover in soil. Such information is needed to improve our ability to explain and predict the variability in the needs for nitrogen inputs and, if mechanistic simulation models are to be used, to predict the changes in plant-available nitrogen in the soil before, within and outside the active growth phase. However, the use of more complex simulation models of nitrogen turnover in the agroecosystems is still limited (e.g. Jensen et al., 1996; Smith et al., 1996). More simple and empirical decision support tools rely on nitrogen balance sheets, supported by a traditional bookkeeping of easily measured and estimated field-scale inputs and outputs of nitrogen. The use of nitrogen response curves from field experiments and of various other more refined decision-support tools has improved the nitrogen use efficiency considerably compared to sheer trial and error. There is, however, plenty of room for improvement. Most attempts to determine more accurately the variations in the needs for external nitrogen inputs rest on measurement and prediction of nitrogen mineralization in soil.

Methods to assess and predict mineralization potentials of relevance for crop uptake have been subject to intensive research for many years. The approaches taken in various studies can be divided into: chemical extractions of mineral and labile organic nitrogen, laboratory and field incubations of soil, application of ^{15}N tracers, plant analyses by chemical methods and remote sensing (analyses of spectral reflectance), and computer simulation models (e.g. Jarvis *et al.*, 1996; Jensen *et al.*, 1996; Smith *et al.*, 1996; Schröder *et al.*, 2000; Delgado *et al.*, 2001).

Whereas the aim of chemical extractions of mineral nitrogen is to establish the pool of soil nitrogen in plant-available forms at a given point in time, the aims of extractions including labile organic nitrogen and of incubation methods are to establish the pools of nitrogen that are potentially available to the crop within the entire growth period or fractions thereof. Whether this potential becomes fully or only partly realized depends on weather conditions and management parameters, which affect both plant performance and the turnover of nitrogen in the soil. Application of ^{15}N-labelled nitrogen is often used to establish pathways and rates of nitrogen turnover in soil and eventually to determine the contribution of soil nitrogen to crop uptakes. Thus the use of ^{15}N tracers provides information related both to plant uptakes and to pools of readily and potentially available soil nitrogen as outlined in the section 'Premises and measures'.

Plant analysis by chemical or spectral methods represents an integrated but retrospective approach. These methods provide information on the nutritive state of the plants at the time of sampling, which in turn reflects previous growth conditions including nitrogen supply. Previous development and present performance of the crop plant in relation to nitrogen supply link more closely to the concepts of synchrony and synlocation. However, the nutritive status of plants is influenced by fluctuations in plant physiological processes (including water fluxes and photosynthetic capacity and rate). The superiority of plant-based methods over methods considering only the capacity of soil to mineralize nitrogen may warrant their wider use as diagnostic tools for establishing needs for additional nitrogen inputs. It remains to be demonstrated, however, that the use of methods based on plant analyses increases the efficiency by which subsequent nitrogen inputs can be exploited. This would provide an overall increase in the nitrogen use efficiency of the cropping system.

The use of mechanistic computer models to simulate nitrogen cycling in agroecosystems appears attractive when considering the shortcomings associated with most other diagnostic tools. However, even strongly simplified and condensed versions of research models, designed for use in nitrogen prognosing, require substantial inputs of soil, climate, crop, management and other site-specific data, many of which are not readily available. Mechanistic simulation models of nitrogen cycling integrate the synchrony and synlocation concepts and may incorporate relevant knowledge on processes and their controls as it becomes available in experimental research.

Sources of nitrogen in plant production

The nitrogen recovered by a crop may originate from a range of different sources. External inputs of nitrogen include mineral fertilizers, animal manures, urban and industrial waste products, seeds and atmospheric deposition. Symbiotic fixation of N_2 from the atmosphere is a most significant but variable source of nitrogen in leguminous crops (e.g. Ledgard and Steele, 1992; Evans *et al.*, 2001), whereas the source of nitrogen derived by free-living N_2-fixing soil microorganisms and blue-green algae is generally considered to be of limited agricultural significance in soils dominated by aerobic conditions (e.g. Stevenson, 1982).

Besides nitrogen mineralized from soil organic pools, sources of nitrogen inherent to the cropping system are above- and belowground residues of previous crops, including cover crops and main crops with or without N_2-fixing capability. The significance of nitrogen in these sources is related to the type and amount of plant biomass returned to the soil.

Crop recoveries of nitrogen given in mineral fertilizers are reasonably well established and can be used as reference when reporting crop recoveries of nitrogen derived from other sources. For instance, the availability of nitrogen in ^{15}N-labelled straw can be related to that of nitrogen in ^{15}N-labelled mineral fertilizer added to the same test crop by calculating the mineral fertilizer equivalent (MFE). The MFE is (% uptake of ^{15}N added in straw) × 100 divided by (% uptake of ^{15}N given in fertilizer). The use of MFE facilitates comparison of the fertilizer value of different nitrogen sources applied in different years, on different sites and to different crops. The MFE indicates how many kilograms of nitrogen in mineral fertilizer it takes to replace 100 kg of nitrogen given in some other source.

The value of crop residues, animal manure and other organic amendments as a source of nitrogen to crops can be highly variable. Much effort has been put into research trying to identify parameters that control their net release of inorganic nitrogen. Attempts have included chemical analyses of elemental contents and of more or less well-defined organic fractions (lignin, cellulose, acid detergent fibres, fatty acids, water- or ether-soluble compounds, etc.), but incubation methods combined with modelling have also been assessed (e.g. Chescheir et al., 1986; Jensen, 1996a; Gordillo and Cabrera, 1997; Henriksen and Breland, 1999). However, the turnover of the added material is determined not only by its chemical composition but also by soil properties and management of both the soil and the added nitrogen source, including storage, particle size and location of the material in the soil, and interactions with other amendments (Christensen, 1986; Bernal and Kirchmann, 1992; Mary et al., 1996; Ambus and Jensen, 1997; Angers and Recous, 1997). It appears that the C : N ratio may provide some indication of the first-year net mineralization potential of nitrogen added in crop residues and animal manure (Jarvis et al., 1996; Kyvsgaard et al., 2000). However, the concept of critical C : N ratios probably carries little weight in determining the exact breaking point between net immobilization and net mineralization of nitrogen during residue turnover (Fog, 1988).

To evaluate the efficiency with which nitrogen from different sources is used in the cropping system, experiments under field conditions with ^{15}N-labelled mineral fertilizer, crop residues and animal manure have been particularly useful. This approach allows for quantification of nitrogen exploited by one or more following crops, lost from the soil by nitrate leaching or retained in the soil organic nitrogen pools and later remineralized. Designed properly, such experiments provide information on both short- and longer-term dynamics of the added nitrogen under in situ conditions and thus constitute a truly integrated measure of the overall nitrogen use efficiency in the agroecosystems. For N_2-fixing leguminous crops, the use of changes in the abundance of ^{15}N with reference to ^{15}N in the soil or in a simultaneously grown non-fixing crop provides efficient ways of determining the amount of nitrogen derived from the atmosphere and that derived from the soil (Shearer and Kohl, 1989; Ledgard and Steele, 1992).

For good reasons, experiments based on ^{15}N have increased in popularity in research on nitrogen use efficiencies under field conditions. To avoid serious pitfalls (Jenkinson et al., 1985; Hart et al., 1986), these studies have to be carefully planned and conducted. In combination with the associated costs, these reservations limit the applicability of ^{15}N-labelling as a practical tool in farm-level nitrogen prognosis.

Towards Improved Nitrogen Management

A number of management factors can be tuned to improve the nitrogen use efficiency in plant production. One obvious means is to adjust external nitrogen inputs as closely as possible to crop demands and thereby eliminate excessive dressings of fertilizers (see the section 'The art of nitrogen demand prognoses'). To obtain the most efficient use of nitrogen in the soil–plant system, the vitality of the crop must be assured as far as possible by proper plant densities, best choice of cultivars, sufficient availability of other plant nutrients and water (e.g. by irrigation), and optimal protection against weeds, pests and plant pathogens. The following sections address additional management aspects that can be involved to improve the nitrogen use efficiency in intensive cropping systems under climatic conditions resembling those that prevail in northwest Europe.

Time of nitrogen application

Under temperate humid conditions, more nitrogen is recovered by the crop and retained in the soil when nitrogen in mineral fertilizers is applied in the spring when active growth commences (see the section 'Premises and measures'). Autumn applications of nitrogen can lead to highly variable and often significantly reduced recoveries of added nitrogen. Thus, Powlson et al. (1986b) found that 11–42% of the ^{15}N-labelled mineral fertilizer applied in the autumn was recovered by winter wheat at harvest with total crop + soil (0–50 cm) recovery being only 22–61% of the added ^{15}N. For springtime additions of nitrogen, crop + soil recoveries are usually in the range of 70–90%, leaving a 20% loss of the added nitrogen (Powlson et al., 1986a; Glendining et al., 1997), but reduced recoveries and larger losses are occasionally reported for nitrogen applied in the spring or within the growth period (e.g. Limaux et al., 1999). The advantage of split-applications within the growth season is variable and depends on climate and crop growth activity. Reduced nitrogen use efficiencies are obtained when nitrogen is given too late in the season and is followed by adverse climatic conditions. This is also true for nitrogen added in animal manures (e.g. Beckwith et al., 1998). For animal manure, storage facilities must allow manure to be kept over winter without significant losses of nitrogen. Applications in the spring are a prerequisite for maximizing the use of nitrogen in this source.

The handling of nitrogen in crop residues is less flexible than that of fertilizer and manure nitrogen, the period of residue application being determined by the time of crop harvest and the planting of a subsequent crop. Incorporation in the autumn of physiologically mature crop residues with a wide C : N

ratio can reduce leaching losses of nitrate by net immobilization of nitrogen during residue turnover (Christensen, 1986; Thomsen and Christensen, 1998), whereas incorporation of residues high in nitrogen (e.g. sugarbeet tops and pea residues) can lead to elevated losses of nitrate (Thomsen and Christensen, 1996; Thomsen et al., 2001). Management strategies to allow for a more efficient use of nitrogen in residues from green manures, nitrate catch crops and grasslands are to postpone incorporation of residues until late winter or early spring (Cuttle and Scholefield, 1995; Francis, 1995; Hansen and Djurhuus, 1997a). Although the nitrogen released from green residues ploughed under in the spring has been shown to contribute significantly to the nitrogen recovered by a subsequent crop (e.g. Thomsen, 1993; Hansen et al., 2000a; Eriksen, 2001a), the effect of postponing residue incorporation from autumn to spring appears to have a greater impact on nitrate leaching than on the nitrogen recovery by the subsequent crop (Hansen and Djurhuus, 1997b; Baggs et al., 2000), suggesting that the nitrate saved from leaching is incorporated into organic soil pools from which it is remobilized at a slower rate than it was immobilized. However, if residues are incorporated too late in the spring, have too slow a turnover or contain large quantities of nitrogen, the leaching loss of nitrate may be increased following crop harvest.

Crop residue disposal

Crop residues returned to the soil may represent a significant input of nitrogen but also of easily available carbon to the decomposer populations. The disposal of crop residues therefore has a significant influence on the nitrogen turnover in soil. Mature residues with a wide C : N ratio immobilize mineral nitrogen after their incorporation. When incorporated in the autumn, well-managed soils usually maintain levels of mineral nitrogen that are sufficiently large to meet the needs of both the decomposing residues and an autumn-sown crop. Shorter-term residue turnover rates can be affected by manipulating the particle size (Jensen, 1994; Angers and

Recous, 1997), whereas particle size appears less important to longer-term nitrogen turnover (Ambus and Jensen, 1997). Residues placed above or on the soil surface immobilize less nitrogen than residues incorporated into the soil, but the depth of placement within the plough layer seems less important (e.g. Christensen, 1986).

The fate of nitrogen added in ^{15}N-labelled residues has been examined under field conditions using enclosures that allow for normal crop development (Table 4.1). Mature crop residues applied in August–October produce MFE values of 8 to 13. Early incorporation of green residues reduced the recovery of residue nitrogen, whereas more nitrogen was recovered when residues were applied in October–November (MFE = 17–26). Ryegrass residues incorporated in March produced the greatest recovery of added nitrogen. Typically, the recovery in the second year was 50% of that found in the first year after application.

Residues from ploughed temporary grasslands, in particular grazed leguminous grass swards, contain large amounts of nitrogen which mineralize and contribute to crop nitrogen uptakes during the following years. The mineralization potential of swards ploughed under in the spring will contribute

to the nitrate leaching in subsequent winter periods unless nitrate catch crops are included in the rotation that follows termination of the grasslands (Francis, 1995; Eriksen, 2001a).

Efficient use of animal manure

Before environmental concerns were truly integrated into crop production, the abundance of industrially manufactured and affordable nitrogen fertilizers gradually caused farmers to consider animal manure to be a waste product of animal husbandry rather than a valuable source of plant nutrients. One of the main challenges in modern agriculture is to change this attitude and to demonstrate a more efficient use of the nitrogen contained in animal manure.

The value of nitrogen in manure can be significantly improved provided that it is applied in adequate quantities in the spring by methods that prevent losses of ammonia by volatilization. Direct injection of slurry into the soil and trail-hose application onto the soil surface beneath a well developed crop cover are very efficient ways of improving its nitrogen use efficiency (e.g. Petersen, 1996;

Table 4.1. Crop recovery of ^{15}N added in plant residues (month of application in parentheses) in the first and second growing seasons after application. MFE is mineral fertilizer equivalent (see text for explanation).

[15]N-labelled amendment			1st growing season		2nd growing season		
Residue type	kg N/ha	Test crop	% of ^{15}N recovered	MFE	Test crop	% of ^{15}N recovered	Ref.
Barley straw (Sep)	16	Spring barley	4.5	13	Spring barley	2.6	a
Oilseed rape, stubbles (Sep)	14	Winter wheat	4.9	8			b
Oilseed rape, pods (Sep)	63	Winter wheat	5.7	10			b
Oilseed rape, pods (Sep)	188	Winter wheat	6.1	10			b
Oilseed rape, straw (Sep)	123	Winter wheat	5.2	9			b
Ryegrass (Nov)	21	Spring barley	9.0	26	Spring barley	2.3	a
Ryegrass (Mar)	39	Spring barley	35.0	73	Spring barley	12.8	c
Beet tops (Oct)	71	Spring barley	6.8	17			b
Beet tops (Oct)	71	Winter rye	8.6	19			b
Green field pea (Jul)	160	Winter wheat	3.7		Green field pea	1.6	d
Green field pea (Jul)	160	Spring barley	3.3		Green field pea	1.5	d
Mature field pea (Aug)	77	Winter wheat	6.2		Mature field pea	3.4	d
Mature field pea (Aug)	77	Spring barley	4.7		Mature field pea	2.9	d

[a]Thomsen and Jensen (1994); [b]Thomsen and Christensen (1996); [c]Thomsen (1993); [d]Thomsen et al. (2001).

Schröder *et al.*, 1997a). For animal manure to be a reliable source of plant-available nitrogen, the ammonia loss associated with its application should be low, and the distribution of manure across the field should be predictable and as uniform as can be obtained with mineral fertilizers. To meet these demands under farm conditions, more technologically advanced (and also more costly) implements have to be employed in manure application. Furthermore, there is a need to develop rapid and inexpensive methods for on-farm determination of the potential availability to plants of manure nitrogen. The content of ammoniacal nitrogen in manures can be determined rapidly and accurately with, for example, the Agros Nitrogen Meter (Kjellerup, 1986; van Kessel *et al.*, 1999), and calculations of nitrogen availability based on the composition of the animal diet appear promising (e.g. Paul *et al.*, 1998; Kyvsgaard *et al.*, 2000). Recent studies suggest that NIR (near infrared reflectance) spectroscopy performed on manure samples may be a valuable tool to estimate the content of ammoniacal nitrogen and other relevant characteristics of the manure (Millmier *et al.*, 2000; Reeves and van Kessel, 2000). Due to variability in nitrogen transformations and ammonia volatilization losses during manure handling and storage, the available nitrogen in the manure has to be assessed just before application in the field and rapid on-farm analyses are pivotal for improving the efficiency by which nitrogen in manure can be used in plant production.

In recent research, the availability of manure nitrogen has been examined in confined plots under field conditions using ^{15}N-labelled urine and faeces obtained from animals fed ^{15}N-labelled and similar but unlabelled diets. By combining labelled and unlabelled manure components, the contribution of nitrogen from urine, faeces and bedding materials to subsequent crop uptakes can be determined separately. Table 4.2 summarizes results obtained in this way in a number of Danish studies. Crop recovery of nitrogen added in different manure components differs widely, nitrogen in urine being a dominant contributor to first year crop uptakes. Nitrogen recovery peaks for manure applied in the spring, with MFE values of 18–31 for faeces and 61–88 for urine and poultry excreta. Interactions between manure components were of importance to plant uptakes of nitrogen in the first growing season after manure application. The recovery of manure nitrogen in the second growing season was comparatively low (2.3–5.9%) and almost independent of the source in which nitrogen was originally introduced into the soil.

When animal manure is a dominant source of plant nitrogen, more manure nitrogen generally has to be applied to reach crop yields comparable to those obtained by mineral fertilizer nitrogen (i.e. the MFE value is less than 100). Provided that the application technique ensures little loss of ammonia by volatilization, most of the excess nitrogen will be incorporated into the soil organic pool from where it is mineralized only slowly. In cropping systems where animal manure is the dominating source of plant nitrogen, nitrate leaching during the autumn/winter period can be significantly higher than for systems based solely on mineral fertilizer nitrogen (Thomsen *et al.*, 1993; Thomsen and Christensen, 1999). There are good indications, however, that the nitrogen in animal manures can be used much more efficiently than previously anticipated, but only if the storage and application methods and the cropping sequences allow for proper handling of manure.

Cover crops

Cover crops are defined as crops grown in breaks between main crops and serve purposes related to several aspects of soil quality maintenance. With respect to nitrogen management, cover crops can be divided into green manure crops and nitrate catch crops according to their principal role in the cropping system. Green manure crops are grown to provide external nitrogen inputs by fixation of atmospheric N_2, whereas the main function of nitrate catch crops is to retain nitrogen already in the soil–plant system by reducing the nitrate leaching losses during the autumn/winter period. To some extent green manures may also reduce nitrate

Table 4.2. Crop recovery of ^{15}N added in animal manures (month of application in parentheses) in the first and second growing seasons after application. MFE is mineral fertilizer equivalent (see text for explanation).

		1st growing season			2nd growing season		
^{15}N-labelled component	Manure type	Test crop	% of ^{15}N recovered	MFE	Test crop	% of ^{15}N recovered	Ref.
Bedding	Fresh solid (Apr)	Spring barley[j]	13.1	27	Ryegrass	3.3	a
straw	Composted solid (Aug)	Winter wheat	8.8		Spring barley	2.9	b
	Stored solid (Aug)	Winter wheat	5.6		Spring barley	3.1	b
Ruminant	Fresh solid (Apr)	Spring barley[j]	8.8	18	Ryegrass	4.1	a
faeces	Faeces (Apr)	Spring barley	16.5	30	Ryegrass	5.9	c
	Slurry (May)	Spring barley	13.0	31			d
	Composted solid (Aug)	Winter wheat	7.2		Spring barley	2.3	b
	Stored solid (Aug)	Winter wheat	7.6		Spring barley	2.4	b
Ruminant	Fresh solid (Apr)	Spring barley[j]	30.3	61	Ryegrass	3.6	a
urine	Stored urine (May)	Ryegrass	56.5	88			e
	Fresh urine (Jun)	Ryegrass	61.0	83			e
	Slurry (May)	Ryegrass	56.0	72			e
	Slurry (May)	Spring barley	34.0	81			d
	Composted solid (Aug)	Winter wheat	8.5		Spring barley	2.4	b
	Stored solid (Aug)	Winter wheat	10.1		Spring barley	3.0	b
Poultry	Fresh excreta (Dec)	Spring barley[j]	15.9	33	Ryegrass	3.3	f
excreta	Fresh excreta (Mar)	Spring barley[j]	40.2	84	Ryegrass	5.1	f
Pig faeces	Slurry (May)	Spring barley	32.9	59			g
Pig urine	Slurry (May)	Spring barley	46.9	84			g
Ammonium	Pig slurry (Apr)	Spring barley	39.0	86	Spring barley[j]	3.5	h
	Cattle slurry (Apr)	Spring barley	37.0	79	Spring barley[j]	3.5	i

[a]Jensen et al. (1999); [b]Thomsen (2001); [c]Sørensen et al. (1994); [d]Thomsen et al. (1997); [e]Sørensen and Jensen (1996); [f]I.K.Thomsen, Foulum, 2002, personal communication; [g]P. Sørensen and I.K. Thomsen, Foulum, 2002, personal communication; [h]Sørensen and Amato (2002); [i]P. Sørensen, Foulum, 2002, personal communication; [j]Indicates that spring barley was undersown with ryegrass and that grass recovery of ^{15}N is included.

leaching losses since legume uptake of soil nitrogen is proportional to the amount of mineral nitrogen in the soil.

Nitrate catch crops are mainly species of Poaceae and Brassicaceae with a substantial nitrogen uptake potential. Grasses used as nitrate catch crops are undersown in the spring or established after harvest of the main crop, whereas species of Brassicaceae are sown post-harvest in late summer or early autumn. Similarly, green manure crops (mainly Fabaceae eguminosae) can be undersown in the main crops or established postharvest. The effectiveness of cover crops depends on the choice of species, the time of their establishment and of main crop harvest, and on the weather conditions during the autumn and winter period. Generally, undersown cover crops accumulate more nitrogen than cover crops established after harvest of main crops. The development of postharvest cover crops relies heavily on an early planting date and on favourable weather conditions during the initial growth phases. Due to the risk of unsuccessful development, postharvest establishment of cover crops is considered to be the least feasible option.

When cover crops are incorporated into the soil in the autumn or are severely affected by early frost periods, a significant proportion of the nitrogen accumulated in above- and belowground plant parts may become mineralized and lost by leaching over winter. Whenever possible, winter annuals should be used and incorporation of cover crops should be postponed until the spring. However, even for winter annuals some loss of nitrogen will occur over winter in response to the adverse growth conditions (low temperature, low light intensity, waterlogged soil). Assessment of nitrogen in cover crops in late autumn therefore does not provide a reliable estimate of their potential to supply nitrogen to the succeeding main crop.

As for grass crops, cover crops may contribute significantly to the nitrogen supply of succeeding crops when incorporated in the spring (Thomsen, 1993; Hansen and Djurhuus, 1997b; Thomsen and Christensen, 1999; Hansen et al., 2000a). The efficiency of cover crops varies among different species (Wallgren and Lindén, 1991; Thorup-Kristensen, 1994; Schröder et al., 1997b; Baggs et al., 2000; N'Dayegamiye and Tran, 2001), and occasionally spring incorporation of cover crops reduces the nitrogen uptake of the following crop. When incorporated in the spring, 15–20% of the nitrogen held in cover crops may appear in the succeeding crop, while another 50–75% is retained in the soil–root system at crop harvest. When incorporated in the autumn, 5–15% of the nitrogen in cover crops can be lost over winter by nitrate leaching, with another 5–15% being taken up by the next main crop.

Leguminous green manure crops usually contribute more nitrogen than non-leguminous nitrate catch crops and incorporation of green manure crops may substitute up to 100 kg N/ha in mineral fertilizer (e.g. Schröder et al., 1997b). For nitrate catch crops such as perennial ryegrass, incorporation in the spring may substitute up to 25 kg N/ha in mineral fertilizer (e.g. Hansen and Djurhuus, 1997b), although much smaller and even negative effects have been recorded (Wallgren and Lindén, 1991; Thorup-Kristensen, 1994; Garwood et al., 1999; Baggs et al., 2000). A negative effect may be ascribed to a combination of a relatively wide C : N ratio in the biomass and late-spring incorporation, which induces immobilization of fertilizer and mineral soil nitrogen during the growth of main crops.

The potential benefit of using cover crops to improve the nitrogen use efficiency in plant production is intimately linked to the cropping sequence and other management parameters (e.g. disposal of main crop residues, mechanical weeding, tillage system). The discontinuation of cover crop use in continuous spring cereal growing may increase nitrate

leaching losses (Thomsen and Christensen, 1999; Hansen et al., 2000b), and consequently the cropping sequence should be adapted to recover the extra mineralization of nitrogen.

Effects of soil tillage

Soil tillage maintains multiple functions in crop production. Tillage is used to prepare seedbeds, to control weeds, to incorporate crop residues, fertilizers and animal manure, to alleviate soil compaction, and to improve water drainage and soil aeration. Tillage implements vary considerably in their construction and mode of operation and therefore have different impacts on soil tilth. Traditionally, tillage practice has mainly been evaluated by the ability to fulfil these primary functions in the most cost-effective way. However, whether appreciated or not, tillage also accomplishes secondary functions of importance to soil quality, including effects on soil microorganisms (e.g. pathogens), soil fauna (e.g. earthworms), and on macropore transport of noxious substances (e.g. pesticides) and nutrients (e.g. phosphorus) to the subsoil.

The effect of tillage on OM dynamics in arable soil now attracts more attention. Silgram and Shepherd (1999) have reviewed the effects of tillage on nitrogen turnover in arable soils with particular emphasis on nitrogen mineralization. The fate of nitrogen in grass swards when ploughed has for some time been a major research area (e.g. Cuttle and Scholefield, 1995; Francis, 1995; Cuttle et al., 1998; Davies et al., 2001; Eriksen, 2001a).

The disturbance of soil structure brought about by tillage influences nitrogen turnover by modifying aeration and soil moisture, which in turn affect the activity of plant roots and soil organisms. Also, the spatial distribution and the particle size of soil amendments are affected by tillage operations, promoting a more intimate mixing of soil and nitrogen containing substrates. Tillage may disrupt macro-aggregates, whereby particulate OM protected within aggregates is released and exposed to decomposition. Depending on the nature of the newly exposed substrates, tillage may result in a temporary increase in nitrogen net mineralization or net immobilization.

Tillage in the autumn generally enhances losses of nitrate from ploughed grasslands as well as from soil in arable rotation (Silgram and Shepherd, 1999), the nitrate saving effects of eliminating or postponing autumn tillage being related to the nitrogen mineralization potentials in the soil. Following cereal crops, the over-winter leaching losses of nitrogen can be reduced by more than 25% when ploughing is omitted or postponed from autumn to spring (Goss *et al.*, 1993; Hansen and Djurhuus, 1997a; Stenberg *et al.*, 1999). Hansen and Djurhuus (1997a) found that leaching of nitrate in experiments with continuous spring barley was significantly reduced when stubble cultivation after harvest was omitted (weeds and volunteer plants were eliminated by paraquat). In general, moving soil tillage from early autumn to spring seems to reduce annual losses of nitrogen from nitrate leaching by 0–30 kg N/ha depending on soil type, climate and date of tillage. However, the nitrogen effect on the subsequent crop is not equivalent to the amount of nitrogen saved from leaching (e.g. Hansen and Djurhuus, 1997b). For grazed pastures, especially leguminous types, reductions in nitrate losses obtained by postponing their termination until early spring may exceed 100 kg N/ha (e.g. Francis, 1995).

Much less is known about the effect of tillage on nitrogen turnover within the growth period (Jarvis *et al.*, 1996; Silgram and Shepherd, 1999). The primary functions of tillage in the spring and summer period are associated with seedbed preparations, incorporation of animal manure and fertilizers, hoeing, potato ridging and mechanical weeding. The disturbance of soil structure derived from these tillage operations is envisaged to accelerate the nitrogen mineralization–immobilization turnover in soil, but research reports are not consistent regarding the quantitative importance of tillage effects on nitrogen turnover and nitrogen supply to the crop (Silgram and Shepherd, 1999). It has been demonstrated that the spatial distribution of fertilizers and animal manure in the soil affects the nitrogen use efficiency (Petersen, 2001; Sørensen and Amato, 2002), banding providing a higher crop recovery of added nitrogen than intimate mixing of soil and the added nitrogen source.

Principles of Nitrogen Management in Agroecosystems

Cropping system characteristics

Traditionally, the term 'crop rotation' has been used to describe the succession of main crops grown in a field, implying that a fixed suite of crops is repeated systematically. However, the term 'crop sequence' may provide a more adequate concept of the succession of main and cover crops (grown in monoculture or intercropped) involved in modern cropping systems. These systems are designed not only to provide suitable and sufficient plant biomass, but also to accomplish an efficient nitrogen use and a reduced environmental load.

The crop sequence affects plant productivity by modifying soil properties as well as the prevalence of weeds and plant pathogens. Effects of the nitrogen management factors addressed in the section 'Towards Improved Nitrogen Management' and interactions among soil management, OM amendments and individual crops in the sequence have both short- and long-term impacts on the nitrogen status of the soil. For example, continuous application of animal manure and perennial grass–clover leys with grazing animals will lead to accumulation of nitrogen in the soil and make such systems more prone to losses of nitrogen by ammonia volatilization, denitrification and nitrate leaching. The challenge is to design the crop sequence and the associated management to allow for maximum retention of nitrogen in the soil–plant system. For grassland systems this might include avoiding late season grazing, high protein supplementary feeds and autumn ploughing of grasslands. For cash cropping this could mean reduced autumn tillage, nitrate catch crops and spring application of manure without significant losses of ammonia by volatilization. Cropping systems that allow for subsurface injection of animal slurries have the potential to significantly improve the nitrogen use efficiency.

Intercropping is the simultaneous growing of mixtures of two or more crops in the same field. Agricultural productions based on ruminants often adopt intercropping of non-legumes (e.g. cereals and grass) and N_2-fixing crops (e.g. peas and clovers) to produce forage for grazing and silage. Wholecrop harvest of cereal–pea–grass intercrops is used to produce silage, the grass subsequently being employed for grazing and later in the season as a nitrate catch crop. Similarly, grass–clover intercrops may be cut for silage and later used for grazing animals. The use of intercropping has been very restricted in cash-crop production where crops are grown to physiological maturity. However, undersown cover crops are increasingly being introduced to fulfil postharvest purposes as nitrate catch crops, green manures and grazing area. Intercropping with N_2-fixing plants appears attractive in terms of reducing the need for mineral fertilizer nitrogen, but in-season transfer of nitrogen between intercropped plants appears to be insignificant (e.g. Jensen, 1996b). The main advantage of intercropped N_2-fixing plants is the extra nitrogen made available postharvest for late season uptake in non-legume companion crops and for uptake in subsequently planted crops.

The crop sequence is one of the most important tools in designing agroecosystems that allow for a high nitrogen use efficiency. Options are multiple, but the selection of crops in the sequence has to consider both short-term and long-term aspects of the specific agricultural production and the associated farm nitrogen management. Two prototypes of crop production systems are briefly discussed below.

Cereal-dominated plant production

Crop sequences dominated by cereals are typical for farms engaged in cash-crop production and in pig and poultry production. The autumn- or spring-sown cereals alternate with cash crops such as oilseed rape, grain legumes (e.g. field peas), root crops (e.g. potatoes or sugarbeet) and grass-seed production. The cereals can be sold for malting or breadmaking purposes or fed to pigs or poultry.

In systems dominated by spring-sown cereals, cover crops can be established as an underseed and ploughed under in the spring before planting of main crops. This scenario and its relationship to the nitrogen turnover synchrony is outlined in Fig. 4.2 for typical northwest European growth conditions. Root-zone percolation dominates from mid-October until mid-March, causing losses of nitrogen by nitrate leaching. The availability of mineral nitrogen is relatively high during spring and summer due to mineralization of nitrogen in the soil and addition of fertilizers and manure. Mineralization of nitrogen in the soil continues in the autumn and drops to relatively low levels in the winter period due to low soil temperatures. Postharvest mineralization may accumulate significant amounts of nitrate susceptible to leaching losses in the subsequent percolation period. For spring-sown cereals, crop nitrogen uptakes are intense in the late spring and early summer periods, but drop to low levels by the end of June. However, the undersown grass catch crop provides a significant sink for nitrogen mineralized in the late summer and autumn period, narrowing the period susceptible to elevated nitrate accumulation and leaching loss. For autumn-sown crops, nitrogen losses can be reduced by reducing preplant tillage and by incorporation of crop residues with a wide $C:N$ ratio. However, the nitrogen uptake potential of autumn-sown cereals is smaller than that of a cover crop established as an underseed in the spring.

Nitrogen should be added in the spring and incorporated prior to planting. To meet grain quality standards (e.g. breadwheat), in-season and foliar applied nitrogen can be required to raise grain protein contents. For autumn-sown cereals, split applications of nitrogen are often practised. When animal slurry is available, early spring additions can be provided in mineral fertilizers. Later in the spring when the soil is more suited for heavy traffic, the remaining nitrogen requirement can be provided in animal slurry whether this is applied by trail-hose technique on the soil surface or injected directly into the soil. The latter option for slurry application provides

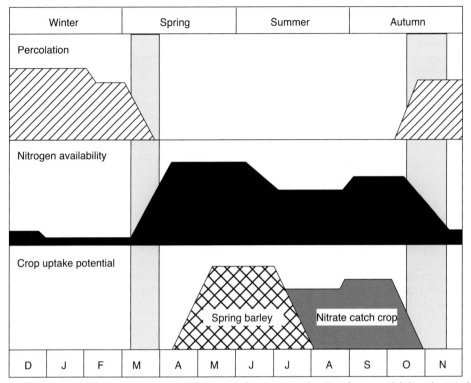

Fig. 4.2. The seasonal dynamics of potentials for percolation (nitrate leaching loss), availability of mineral nitrogen (mineralization + external inputs) and crop uptake in a 'standard' year under northwest European conditions. The spring barley is undersown with ryegrass acting as a nitrate catch crop. The vertical grey zones indicate periods susceptible to elevated nitrate leaching losses.

a higher and more reliable nitrogen use efficiency, but may result in crop stand damage. The likely impact of selected managements on the nitrogen use efficiency of the cropping system is illustrated in Fig. 4.3.

Whereas cereal crops allow for undersown cover crops, oilseed rape, field peas and root crops are usually grown in monoculture. Postharvest cover crops can be practised for oilseed rape and field peas, but usually these crops will be followed by autumn-sown cereals. Sugarbeets are harvested in late autumn and normally do not allow for autumn-sown cereals or cover crops. The thorough working of the soil associated with sugarbeet harvest and sugarbeet tops left in the field after harvest add to nitrate leaching losses and sugarbeet cropping constitutes a particular challenge in the process of tightening the nitrogen cycle in cereal-dominated cropping systems. Similarly, potato growing, especially

that of potatoes intended for starch production and harvested at a later date, leaves a vegetation-free and thoroughly reworked soil that may accommodate elevated nitrate leaching losses.

Outdoor pig production has gained interest due to animal welfare aspects and reduced costs of housing. However, pigs tend to demolish the vegetation and cause soil rooting whereby mineralization of nitrogen is increased and plant uptake potentials decreased. This will increase the asynchrony of the nitrogen cycling. Combined with the heterogeneity of excreted nutrients, outdoor pig production can lead to significant but highly variable nitrogen losses through nitrate leaching, ammonia volatilization and denitrification (Eriksen, 2001b; Eriksen *et al.*, 2002). Compared with pigs raised indoors with collection of excreta and spring-time application of slurry to an actively growing

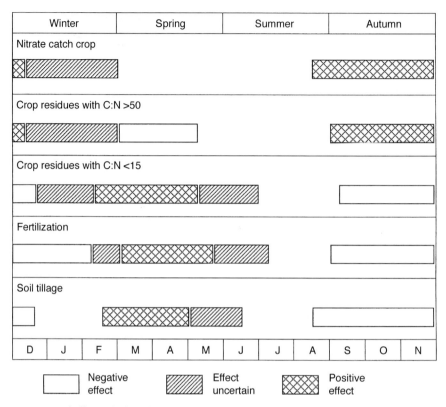

Fig. 4.3. Expected effects of selected managements on the nitrogen use efficiency in a cereal-dominated cropping system under northwest European climate. Open, hatched and cross-hatched bars indicate a negative, an uncertain or a positive effect on the nitrogen use efficiency, respectively.

crop, outdoor pig production may cause a reduced nitrogen use efficiency.

Forage cropping systems

Cropping systems for ruminant husbandry are designed to provide plant biomass for grazing purposes and for feeding of housed animals. Animal grazing has traditionally relied on perennial leys of grass mixtures and grass–clover. These are used for combined cutting for silage and grazing for a number of years before being ploughed and either reseeded or followed by cereals. Other grazing opportunities are provided by cover crops undersown in cereals that are harvested at physiological maturity. Following wholecrop harvest of cereal–pea–grass

intercrops at yellow ripening, the grass regrowth may be used for grazing.

Perennial grass-based leys potentially have a high nitrogen use efficiency, but, when used for grazing, the nitrogen losses may become substantial (Francis, 1995). The ruminants excrete a large fraction of the ingested nitrogen during grazing, including the nitrogen derived from imported feed concentrates. The surface deposition of excreta and their heterogeneous distribution across the field provide opportunities for gaseous and leaching losses of nitrogen, with areas affected by urine and faeces receiving nitrogen at rates that far exceed the crop uptake potentials. The nitrogen cycling in grazed systems has been subject to intensive research, and recommendations to increase the nitrogen use efficiency of grazed systems are available (e.g. Cuttle and Scholefield, 1995; Francis, 1995; Jarvis and

Aarts, 2000). Most measures to improve the nitrogen economy of grazed fields relate to the animal component and the feeding strategy of the production system. These aspects are beyond the scope of the present text. However, one main strategy for the cropping system is to achieve the right balance between grazing and cutting plant biomass for silage (see, e.g. Jarvis and Aarts, 2000).

In production systems where animals are housed all the year round, an improved nitrogen use efficiency can be obtained provided that nitrogen losses from animal housings and manure storage facilities can be controlled and kept low. Compared to deep litter and other solid manures rich in bedding materials, housings designed to collect excreta in liquid slurry will provide opportunities for an efficient use of the nitrogen in the manure, especially when slurries are stored and applied just before or early in the growing season, e.g. by subsurface injection.

Besides crops intended for grazing, forage cropping systems include crops intended for silage and other winter-feeding rations. The combination of cutting and grazing (provided that best practice grazing is observed) improves the nitrogen use of the grass and legume–grass crops, especially when these are employed as winter cover crops and terminated in the spring. When followed by cereal crops, these should be undersown with a nitrate catch crop because perennial leguminous leys may mineralize more nitrogen than can be adsorbed during the growth of the cereal crop. Maize is used to provide energy-rich silage. Although maize requires only moderate nitrogen inputs to produce high forage yields, a late harvest date may prevent establishment of postharvest nitrate catch crops. Maize management should allow for undersown winter-annual cover crops in order to minimize nitrate leaching losses. The relatively late planting date and the wide row spacing may allow for tillage to control weeds and to stimulate mineralization of soil nitrogen early in the growth period.

In the past, fodder beets were widely used to provide winter forage for housed ruminants. Both the roots and the tops were recovered, the tops being conserved as silage. Although labour-intensive, this crop is also very efficient at recovering nitrogen during the autumn period. However, the late harvest date excludes postharvest establishment of nitrate catch crops due to reduced light intensity and low temperatures.

Regardless of strategy, ruminant production accomplishes a lower nitrogen use efficiency than production based on cash crops. When plant biomass passes the livestock, 80–90% of the plant nitrogen is recycled on-farm. The handling and subsequent use of this nitrogen will unavoidably be associated with nitrogen losses. To reduce these losses as much as possible is a main challenge in tightening the nitrogen cycle of livestock production.

Conclusions

The fundamental doctrine of sustainable nitrogen management is to optimize the nitrogen use efficiency of both introduced and native soil nitrogen by increasing the temporal and the spatial coincidence between availability and crop demand of mineral nitrogen. The main challenge is to tighten the nitrogen cycle while balancing the productivity, environment and health components of the production system.

The nitrogen cycling in modern agriculture links to the aspect of soil quality that is dominated by management. Provided that knowledge and technological and economic means are available, nitrogen deficiency is reversible and can be readily restored, but due to the complex nature of the nitrogen turnover, the time-scale differs widely for the productivity, the environment and the health component.

Whether nitrogen is added in mineral fertilizers or in crop residues and animal manures, the mineralization–immobilization turnover in the soil has to be accounted for quantitatively. It appears that leaching losses of nitrate originate predominantly from mineralization of soil organic nitrogen rather than from residues of nitrogen applied immediately before or early in the active growth phase, unless excessive doses of nitrogen are given or the crop fails due to mismanagement.

A very significant proportion of the nitrogen eventually recovered by the plants will be due to mineralization of soil nitrogen.

The plant recovery of nitrogen from the various sources varies according to their proportions of inorganic and organic nitrogen, and the chemical and physical properties of the organic fraction. The time of application and the distribution of the nitrogen above or in the soil, the time and intensity of soil tillage, and the length of the period when plant growth is limited by nitrogen significantly affect the efficiency by which the nitrogen is used in the cropping system.

There are good reasons to consider more closely the secondary functions of tillage on nitrogen turnover in soil. The reductions in nitrate leaching losses accomplished by postponing tillage operations from autumn to spring suggest that tillage can have a quantitatively significant impact on the nitrogen turnover. However, more needs to be known about the effect of tillage on nitrogen turnover in the spring and during the growth period. Of particular importance is the question of how tillage practice can be optimized to fulfil both the primary and the secondary functions. The timing, intensity and frequency of tillage operations may be invoked to increase coincidence between nitrogen availability and crop nitrogen demands.

In intensive animal production based on protein supplements, the overall nitrogen balance may show very large surpluses. Efficient nitrogen management requires a sustainable balance between plant production potential (farm acreage) and animal stocking density (animal manure production). Alternatively, animal manure should be exported and used in plant production elsewhere. The nitrogen management in cash crop productions is simpler and the nitrogen use may potentially be more efficient than in animal husbandry.

Much remains to be done before the prediction of nitrogen demands turns from an art to a scientific discipline. Most likely, no single method is able to provide the information needed to establish a reliable nitrogen prognosis. The challenge is to assemble a minimum set of soil and crop analyses that combined with weather forecasts and management performance predicts the most likely nitrogen demand for the given crop. Current knowledge on the cycling of nitrogen in natural and arable ecosystems does, however, provide a good basis for improving the nitrogen use efficiency by adopting management practices that are known to increase the retention of nitrogen in the soil–plant system and, thereby, also promote a tighter nitrogen cycle. However, the loss of integrity in agro-ecosystems cannot be fully alleviated, as illustrated by larger leaching losses of nitrate and reduced levels of nitrogen in arable soils. Inevitably, losses of nitrogen from productive agroecosystems will be higher than those from pristine ecosystems.

The objectives of soil quality management are, in principle, site related and address the sustainability of a given agroecosystem. Some of the measures invoked in soil quality management are widely applicable, whereas others apply only to specific agroecosystems. The kind of production established on a given farm is determined by site characteristics and farmer preferences, but ultimately the viability of the production is governed by external factors, including price relationships, subsidies and other regulations related to farm products and environmental standards.

References

Ambus, P. and Jensen, E.S. (1997) Nitrogen mineralization and denitrification as influenced by crop residue particle size. *Plant and Soil* 197, 261–270.

Angers, D.A. and Recous, S. (1997) Decomposition of wheat straw and rye residues as affected by particle size. *Plant and Soil* 189, 197–203.

Baggs, E.M., Watson, C.A. and Rees, R.M. (2000) The fate of nitrogen from incorporated cover crop and green manure residues. *Nutrient Cycling in Agroecosystems* 56, 153–163.

Barraclough, D., Geens, E.L. and Maggs, J.M. (1984) Fate of fertilizer nitrogen applied to grassland. II. Nitrogen-15 leaching results. *Journal of Soil Science* 35, 191–199.

Beckwith, C.P., Cooper, J., Smith, K.A. and Shepherd, M.A. (1998) Nitrate leaching loss following application of organic manures to sandy soils in arable cropping. I. Effects of application time, manure type, overwinter crop cover and nitrification inhibition. *Soil Use and Management* 14, 123–130.

Bernal, M.P. and Kirchmann, H. (1992) Carbon and nitrogen mineralization and ammonia volatilization from fresh, aerobically and anaerobically treated pig manure during incubation with soil. *Biology and Fertility of Soils* 13, 135–141.

Chescheir, G.M., Westerman, P.W. and Safley, L.M. (1986) Laboratory methods for estimating available nitrogen in manures and sludges. *Agricultural Wastes* 18, 175–195.

Christensen, B.T. (1986) Barley straw decomposition under field conditions: effect of placement and initial nitrogen content on weight loss and nitrogen dynamics. *Soil Biology and Biochemistry* 18, 523–529.

Cuttle, S.P. and Scholefield, D. (1995) Management options to limit nitrate leaching from grassland. *Journal of Contaminant Hydrology* 20, 299–312.

Cuttle, S.P., Scurlock, R.V. and Davies, B.M.S. (1998) A 6-year comparison of nitrate leaching from grass/clover and N-fertilized grass pastures grazed by sheep. *Journal of Agricultural Science, Cambridge* 131, 39–50.

Davies, M.G., Smith, K.A. and Vinten, A.J.A. (2001) The mineralization and fate of nitrogen following ploughing of grass and grass-clover swards. *Biology and Fertility of Soils* 33, 423–434.

Delgado, J.A., Ristau, R.J., Dillon, M.A., Duke, H.R., Stuebe, A., Follett, R.F., Shaffer, M.J., Riggenbach, R.R., Sparks, R.T., Thompson, A., Kawanabe, L.M., Kunugi, A. and Thompson, K. (2001) Use of innovative tools to increase nitrogen use efficiency and protect environmental quality in crop rotations. *Communications in Soil Science and Plant Analyses* 32, 1321–1354.

Dise, N.B. and Wright, R.F. (1995) Nitrogen leaching from European forests in relation to nitrogen deposition. *Forest Ecology and Management* 71, 153–161.

Eriksen, J. (2001a) Nitrate leaching and growth of cereal crops following cultivation of contrasting temporary grasslands. *Journal of Agricultural Science, Cambridge* 136, 271–281.

Eriksen, J. (2001b) Implications of grazing by sows for nitrate leaching from grassland and the succeeding cereal crop. *Grass and Forage Science* 56, 317–322.

Eriksen, J., Petersen, S.O. and Sommer, S.G. (2002) The fate of nitrogen in outdoor pig production. *Agronomie* 22, 863–867.

Evans, J., McNeill, A.M., Unkovich, M.J., Fettell, N.A. and Heenan, D.P. (2001) Net nitrogen balances for cool-season grain legume crops and contributions to wheat nitrogen uptake: a review. *Australian Journal of Experimental Agriculture* 41, 347–359.

Fog, K. (1988) The effect of added nitrogen on the rate of decomposition of organic matter. *Biological Reviews* 63, 433–462.

Francis, G.S. (1995) Management practices for minimising nitrate leaching after ploughing temporary leguminous pastures in Canterbury, New Zealand. *Journal of Contaminant Hydrology* 20, 313–327.

Garwood, T.W.D., Davies, D.B. and Hartley, A.R. (1999) The effect of winter cover crops on yield of the following spring crops and nitrogen balance in a calcareous loam. *Journal of Agricultural Science, Cambridge* 132, 1–11.

Glendining, M.J., Poulton, P.R., Powlson, D.S. and Jenkinson, D.S. (1997) Fate of ^{15}N-labelled fertilizer applied to spring barley grown on soils of contrasting nutrient status. *Plant and Soil* 195, 83–98.

Gordillo, R.M. and Cabrera, M.L. (1997) Mineralizable nitrogen in broiler litter: I. Effect of selected litter chemical characteristics. *Journal of Environmental Quality* 26, 1672–1679.

Goss, M.J., Howse, K.R., Lane, P.W., Christian, D.G. and Harris, G.L. (1993) Losses of nitrate nitrogen in water draining from under autumn-sown crops established by direct drilling or mouldboard ploughing. *Journal of Soil Science* 44, 35–48.

Hansen, E.M. and Djurhuus, J. (1997a) Nitrate leaching as influenced by soil tillage and catch crop. *Soil and Tillage Research* 41, 203–219.

Hansen, E.M. and Djurhuus, J. (1997b) Yield and N uptake as affected by soil tillage and catch crop. *Soil and Tillage Research* 42, 241–252.

Hansen, E.M., Kristensen, K. and Djurhuus, J. (2000a) Yield parameters as affected by introduction or discontinuation of catch crop use. *Agronomy Journal* 92, 909–914.

Hansen, E.M., Djurhuus, J. and Kristensen, K. (2000b) Nitrate leaching as affected by introduction or discontinuation of cover crop use. *Journal of Environmental Quality* 29, 1110–1116.

Hart, P.B.S., Rayner, J.H. and Jenkinson, D.S. (1986) Influence of pool substitution on the interpretation of fertilizer experiments with ^{15}N. *Journal of Soil Science* 37, 389–403.

Henriksen, T.M. and Breland, T.A. (1999) Evaluation of criteria for describing crop residue degradability in a model of carbon and nitrogen turnover in soil. *Soil Biology and Biochemistry* 31, 1135–1149.

Jarvis, S.C. and Aarts, H.F.M. (2000) Nutrient management from a farming systems perspective. *Grassland Science in Europe* 5, 363–373.

Jarvis, S.C., Stockdale, E.A., Shepherd, M.A. and Powlson, D.S. (1996) Nitrogen mineralization in temperate agricultural soils: processes and measurement. *Advances in Agronomy* 57, 187–235.

Jenkinson, D.S., Fox, R.H. and Rayner, J.H. (1985) Interactions between fertilizer nitrogen and soil nitrogen – the so-called 'primary' effect. *Journal of Soil Science* 36, 425–444.

Jensen, B., Sørensen, P., Thomsen, I.K., Jensen, E.S. and Christensen, B.T. (1999) Availability of nitrogen in ^{15}N-labeled ruminant manure components to successively grown crops. *Soil Science Society of America Journal* 63, 416–423.

Jensen, C., Stougaard, B. and Østergaard, H.S. (1996) The performance of the Danish simulation model *DAISY* in prediction of N_{min} at spring. *Fertilizer Research* 44, 79–85.

Jensen, E.S. (1994) Mineralization–immobilization of nitrogen in soil amended with low C:N ratio plant residues with different particle sizes. *Soil Biology and Biochemistry* 26, 519–521.

Jensen, E.S. (1996a) Compared cycling in a soil-plant system of pea and barley residue nitrogen. *Plant and Soil* 182, 13–23.

Jensen, E.S. (1996b) Grain yield, symbiotic N_2 fixation and interspecific competition for inorganic N in pea-barley intercrops. *Plant and Soil* 182, 25–38.

Kjellerup, V. (1986) Agros Nitrogen Meter for estimation of ammonium nitrogen in slurry and liquid manure. In: Kofoed, A.D. *et al.* (eds) *Efficient Land Use of Sludge and Manure.* Elsevier Applied Science Publishers, London, pp. 216–223.

Kjellerup, V. and Kofoed, A.D. (1983) Nitrogen fertilization in relation to leaching of plant nutrients from soil. Lysimeter experiments with ^{15}N [in Danish with English summary]. *Tidsskrift for Planteavl* 87, 1–22.

Kyvsgaard, P., Sørensen, P., Møller, E. and Magid, J. (2000) Nitrogen mineralization from sheep faeces can be predicted from the apparent digestibility of the feed. *Nutrient Cycling in Agroecosystems* 57, 207–214.

Ledgard, S.F. and Steele, K.W. (1992) Biological nitrogen fixation in mixed legume/grass pastures. *Plant and Soil* 141, 137–153.

Likens, G.E., Bormann, F.H., Pierce, R.S., Eaton, J.S. and Johnson, N.M. (1977) *Biogeochemistry of a Forested Ecosystem.* Springer-Verlag, New York, 146 pp.

Limaux, F., Recous, S., Meynard, J.M. and Guckert, A. (1999) Relationship between rate of crop growth at date of fertilizer N application and fate of fertilizer N applied to winter wheat. *Plant and Soil* 214, 49–59.

Mary, B., Recous, S., Darwis, D. and Robin, D. (1996) Interactions between decomposition of plant residues and nitrogen cycling in soil. *Plant and Soil* 181, 71–82.

Millmier, A., Lorimor, J., Hurburgh, C., Fulhage, C., Hattey, J. and Zhang, H. (2000) Near-infrared sensing of manure nutrients. *Transactions of the ASAE* 43, 903–908.

Murphy, D.V., MacDonald, A.J., Stockdale, E.A., Goulding, K.W.T., Fortune, S., Gaunt, J.L., Poulton, P.R., Wakefield, J.A., Webster, C.P. and Wilmer, W.S. (2000) Soluble organic nitrogen in agricultural soils. *Biology and Fertility of Soils* 30, 374–387.

N'Dayegamiye, A. and Tran, T.S. (2001) Effects of green manures on soil organic matter and wheat yields and N nutrition. *Canadian Journal of Soil Science* 81, 371–382.

Paul, J.W., Dinn, N.E., Kannangara, T. and Fisher, L.J. (1998) Protein content in dairy cattle diets affects ammonia losses and fertilizer nitrogen value. *Journal of Environmental Quality* 27, 528–534.

Petersen, J. (1996) Fertilization of spring barley by combination of pig slurry and mineral nitrogen fertilizer. *Journal of Agricultural Science, Cambridge* 127, 151–159.

Petersen, J. (2001) Recovery of ^{15}N-ammonium–^{15}N-nitrate in spring wheat as affected by placement geometry of the fertilizer band. *Nutrient Cycling in Agroecosystems* 61, 215–221.

Pilbeam, C.J. (1996) Effect of climate on the recovery in crop and soil of ^{15}N-labelled fertilizer applied to wheat. *Fertilizer Research* 45, 209–215.

Powlson, D.S., Pruden, G., Johnston, A.E. and Jenkinson, D.S. (1986a) The nitrogen cycle in the Broadbalk Wheat Experiment: recovery and losses of ^{15}N-labelled fertilizer applied in spring and inputs of nitrogen from the atmosphere. *Journal of Agricultural Science, Cambridge* 107, 591–609.

Powlson, D.S., Hart, P.B.S., Pruden, G. and Jenkinson, D.S. (1986b) Recovery of ^{15}N-labelled fertilizer applied in autumn to winter wheat at four sites in eastern England. *Journal of Agricultural Science, Cambridge* 107, 611–620.

Reeves, J.B. III and van Kessel, J.S. (2000) Near-infrared spectroscopic determination of carbon, total nitrogen, and ammonium-N in dairy manures. *Journal of Dairy Science* 83, 1829–1836.

Schröder, J.J., ten Holte, L. and Brouwer, G. (1997a) Response of silage maize to placement of cattle slurry. *Netherlands Journal of Agricultural Science* 45, 249–261.

Schröder, J.J., ten Holte, L. and Janssen, B.H. (1997b) Non-overwintering cover crops: a significant source of N. *Netherlands Journal of Agricultural Science* 45, 231–248.

Schröder, J.J., Neeteson, J.J., Oenema, O. and Struik, P.C. (2000) Does the crop or the soil indicate how to save nitrogen in maize production? Reviewing the state of the art. *Field Crops Research* 66, 151–164.

Shearer, G. and Kohl, D.H. (1989) Estimates of N_2 fixation in ecosystems: the need for and basis of the ^{15}N natural abundance method. In: Rundel, P.W. *et al.* (eds) *Stable Isotopes in Ecological Research*. Springer-Verlag, New York, pp. 342–374.

Silgram, M. and Shepherd, M.A. (1999) The effects of cultivation on soil nitrogen mineralization. *Advances in Agronomy* 65, 267–311.

Smith, J.U., Bradbury, N.J. and Addiscott, T.M. (1996) SUNDIAL: a PC-based system for simulating nitrogen dynamics in arable land. *Agronomy Journal* 88, 38–43.

Sørensen, P. and Amato, M. (2002) Remineralisation and residual effects of N after application of pig slurry to soil. *European Journal of Agronomy* 16, 81–95.

Sørensen, P. and Jensen, E.S. (1996) The fate of fresh and stored ^{15}N-labelled sheep urine and urea applied to a sandy and a sandy loam soil using different application strategies. *Plant and Soil* 183, 213–220.

Sørensen, P., Jensen, E.S. and Nielsen, N.E. (1994) The fate of ^{15}N-labelled organic nitrogen in sheep manure applied to soils of different texture under field conditions. *Plant and Soil* 162, 39–47.

Stenberg, M., Aronsson, H., Linden, B., Rydberg, T. and Gustafson, A. (1999) Soil mineral nitrogen and nitrate leaching losses in soil tillage systems combined with a catch crop. *Soil and Tillage Research* 50, 115–125.

Stevenson, F.J. (1982) *Nitrogen in Agricultural Soils*. American Society of Agronomy, Madison, Wisconsin, 940 pp.

Thomsen, I.K. (1993) Nitrogen uptake in barley after spring incorporation of ^{15}N-labelled Italian ryegrass into sandy soils. *Plant and Soil* 150, 193–201.

Thomsen, I.K. (2001) Recovery of nitrogen from composted and anaerobically stored manure

labelled with ^{15}N. *European Journal of Agronomy* 15, 31–41.

Thomsen, I.K. and Christensen, B.T. (1996) Availability to subsequent crops and leaching of nitrogen in ^{15}N-labelled sugarbeet tops and oilseed rape residues. *Journal of Agricultural Sciences, Cambridge* 126, 191–199.

Thomsen, I.K. and Christensen, B.T. (1998) Cropping system and residue management effects on nitrate leaching and crop yields. *Agriculture, Ecosystems and Environment* 68, 73–84.

Thomsen, I.K. and Christensen, B.T. (1999) Nitrogen conserving potential of successive ryegrass catch crops in continuous spring barley. *Soil Use and Management* 15, 195–200.

Thomsen, I.K. and Jensen, E.S. (1994) Recovery of nitrogen by spring barley following incorporation of ^{15}N-labelled straw and catch crop material. *Agriculture, Ecosystems and Environment* 49, 115–122.

Thomsen, I.K., Hansen, J.F., Kjellerup, V. and Christensen, B.T. (1993) Effects of cropping system and rates of nitrogen in animal slurry and mineral fertilizer on nitrate leaching from a sandy loam. *Soil Use and Management* 9, 53–58.

Thomsen, I.K., Kjellerup, V. and Jensen, B. (1997) Crop uptake and leaching of ^{15}N applied in ruminant slurry with selectively labelled faeces and urine fractions. *Plant and Soil* 197, 233–239.

Thomsen, I.K., Kjellerup, V. and Christensen, B.T. (2001) Leaching and plant uptake of N in field pea/cereal cropping sequences with incorporation of ^{15}N-labelled pea harvest residues. *Soil Use and Management* 17, 209–216.

Thorup-Kristensen, K. (1994) The effect of nitrogen catch crop species on the nitrogen nutrition of succeeding crops. *Fertilizer Research* 37, 227–234.

van Kessel, J.S., Thompson, R.B. and Reeves, J.B. III (1999) Rapid on-farm analysis of manure nutrients using quick tests. *Journal of Production Agriculture* 12, 215–224.

Wallgren, B. and Lindén, B. (1991) Residual nitrogen effects of green manure crops and fallow. *Swedish Journal of Agricultural Research* 21, 67–77.

Chapter 5

Phosphorus – Surplus and Deficiency

L.M. Condron

Soil, Plant and Ecological Sciences Division, PO Box 84, Lincoln University, Canterbury 8150, New Zealand

---

Summary

A combination of biological, chemical and physical properties and processes, together with the history and intensity of land use and management determine the forms, dynamics and mobility of phosphorus (P) in the soil–plant system. Amounts of total and available soil P in intensively managed agroecosystems have increased steadily since the early 1950s due to continued inputs of P, often in excess of crop requirements. The elevated P status of many agricultural soils has helped to sustain high levels of crop and animal production, although there is also an increased risk of diffuse P transfer in overland and subsurface flow. Rates of diffuse P transfer are generally small (1–6 kg P/ha/year), but this P loss can have a significant impact on water quality and health through accelerated eutrophication. Extensive research has been carried out to examine P mobility in soil and establish relationships between P inputs, soil P status and the amounts and forms of P loss to water. This in turn has resulted in the development of strategies for environmental management of P at the field and catchment scale. Phosphorus deficiency is not common in modern agriculture, although it could become an issue in reduced-input systems such as organic farming. Organic farming is expanding rapidly in many countries due to increased consumer demand for food produced with minimum inputs of synthetic soluble P fertilizers. Continued inputs of P in the form of imported feed, manure and sparingly soluble reactive phosphate rock should enable soil P fertility and productivity to be sustained in organic farming systems. Achieving the correct balance between maintaining productivity and minimizing P transfer will be a vital component of future strategies for effective soil quality management in low- and high-intensity agroecosystems. This in turn will require continued research to improve our understanding of the key properties and processes that determine the availability and mobility of P in the soil–plant system.

©CAB International 2004. *Managing Soil Quality: Challenges in Modern Agriculture*
(eds P. Schjønning, S. Elmholt and B.T. Christensen)

Sustainability in Phosphorus Management

Phosphorus (P) is a non-renewable resource and continued inputs of P are necessary to sustain the productivity of agroecosystems. These P inputs are primarily designed to increase and maintain soil P fertility and replace P removed in off-farm produce. In essence, sustainable P management involves the provision of appropriate P inputs to maintain production at economically viable levels for specific farm enterprises.

In intensive agricultural systems the objective is to attain maximum production which, in turn, requires significant P inputs to ensure that plant productivity is not limited by soil P availability. This is commonly achieved through the continued application of inorganic P in the form of soluble mineral fertilizers. Most agricultural soils have a large capacity to sequester this P via adsorption on to soil mineral surfaces and the formation of sparingly soluble P minerals, and as organic P (Fig. 5.1). This 'buffer capacity' ensures a continued supply of inorganic P to soil solution, which in effect confers a degree of resilience in the soil with regard to the provision of P to satisfy ongoing and future plant requirements. However, in many areas the continued application of fertilizer P, together with inadequate consideration of P inputs in imported feed and the fertilizer value of P in animal manure have resulted in the accumulation of significant amounts of soil P. Accordingly, in intensive agricultural systems levels of plant available P commonly exceed those required for maximum plant growth. Furthermore, it has become apparent that diffuse P transfer in drainage from agricultural land poses a significant risk to surface water quality, and that there is a direct link between high soil P status and P transfer.

The main operational objectives of reduced-input agricultural systems such as organic farming include protection of long-term soil fertility and quality with minimum adverse impact on the wider environment. Accordingly, organic farming systems seek to generate economically viable levels of production by using soil and applied nutrients as efficiently as possible. In terms of P, this

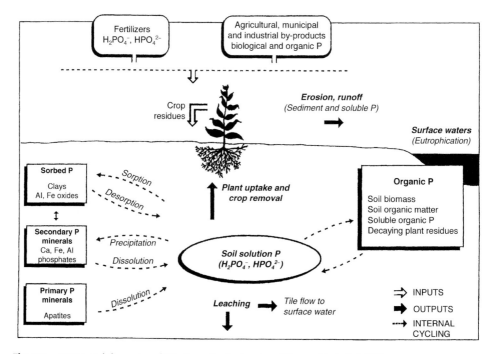

Fig. 5.1. Forms and dynamics of P in the soil–plant system (Pierzynski *et al.*, 2000).

involves the adoption of specific management practices designed to optimize soil P availability and utilization and thereby minimize levels of P inputs required to sustain production.

This chapter considers aspects of soil P dynamics that are most relevant to effective long-term management of soil quality in modern agriculture. It includes an overview of current understanding of soil P dynamics in agroecosystems, focusing on the key properties and processes that influence P availability in relation to P inputs and land-use intensity. The major challenges facing modern agriculture with respect to P and soil quality in high- and low-input agricultural systems are described and discussed. In high-P-input systems, diffuse P transfer from soil and its associated potential impacts on water quality are a major cause for concern, while the maintenance of adequate levels of available P to sustain production is a potentially important issue in reduced-P-input organic systems.

Soil Phosphorus Dynamics in Agroecosystems

Concentrations of P in topsoil vary widely (100–3000+ mg P/kg) and in many soils over 50% of the P is present in organic forms or associated with the microbial biomass (Frossard *et al.*, 2000). Phosphorus transformations and mobility in the soil–plant system are controlled by a combination of biological, chemical and physical processes (Fig. 5.1). The amounts, forms and associated dynamics of soil P are influenced by a number of factors including soil type and environmental conditions, as well as land-use and management practices. In natural ecosystems plant growth is often limited by P availability, while P is generally recycled and retained efficiently (Cole *et al.*, 1977; Attiwell and Adams, 1993). However, in managed ecosystems continued inputs of P in the form of fertilizers and imported fodder profoundly affect the quantity, availability and dynamics of soil P.

The chemical nature and dynamics of P in soil have been subject to extensive investigation over many years. Advances in our understanding of various aspects of soil P dynamics

have been highlighted in a number of review articles, particularly since the late 1970s (e.g. Dalal, 1977; Anderson, 1980; Stewart and McKercher, 1982; Tate, 1985; Harrison, 1987; Stewart and Tiessen, 1987; Wild, 1988; Sanyal and DeDatta, 1992; Cross and Schlesinger, 1995; Magid *et al.*, 1996; Frossard *et al.*, 2000; Pierzynski *et al.*, 2000). The following is a brief synopsis of the major properties and processes that influence the availability of soil P in relation to land use and management.

The soil solution is the primary source of P for plants (and microorganisms) and most P is taken up as inorganic orthophosphate (HPO_4^{2-}, $H_2PO_4^-$). The concentration of inorganic P present in soil solution at any time is generally very low ($< 5\,\mu M$). Phosphorus removed from the soil solution by biological uptake must be replenished by release from inorganic and organic forms of P associated with the solid phase. Inorganic P can be released to solution by desorption and/or dissolution of mineral P associated with aluminium, iron or calcium, together with mineralization of organic P and release of P from the microbial biomass. The amounts of inorganic, organic and microbial P in soil and the equilibrium concentration of inorganic P in soil solution are determined by a combination of factors. These include soil chemical properties (e.g. clay/hydrous oxide mineralogy, pH) and soil P status as determined by the type, duration and intensity of land use and the associated P inputs. The processes involved in adsorption–desorption and precipitation–dissolution of inorganic P in soil have been studied extensively and are well understood. For example, the physicochemical nature of adsorbed inorganic P on mineral surfaces changes with time such that the P becomes less exchangeable with soil solution. This reduction in exchangeability is often termed 'P fixation' and it is a major contributor to the need for continued application of P fertilizer to maintain a given level of plant production.

Organic and microbial phosphorus dynamics

Despite extensive investigation, our understanding of the properties and processes that determine the dynamics and availability of organic and microbial P in soil is limited (Stewart and Tiessen, 1987; Magid et al., 1996; Frossard et al., 2000). A significant proportion of the P added to soil in plant and animal detritus is present in organic forms and the transformations and fate of this P (mainly mediated by microorganisms) has a major influence on overall P availability. The mineralization of soil organic P is determined by its chemical nature and associated reactivity. For example, the predominance of orthophosphate monoester inositol phosphates in soil may be partly attributed to its association with the structural components of senescent plant material, which limits its susceptibility to mineralization (Stewart and Tiessen, 1987; Gressel et al., 1996). In contrast, orthophosphate diester forms of organic P (nucleic acid and phospholipid P) are relatively soluble in the soil and are therefore rapidly mineralized (Magid et al., 1996). In addition, the susceptibility of soil organic P to mineralization is influenced by physical factors, including particle size, aggregate stability and the associated effects of wetting and drying (Perrott et al., 1999; Rubæk et al., 1999; Chepkwony et al., 2001). Analytical constraints have limited progress in elucidating the nature and dynamics of organic and microbial P in soil. For example, the presence of a wide range of organic P compounds in soil and the nature of organo-mineral interactions mean that quantitative extraction and identification of soil organic P is very difficult. Our understanding of key processes such as organic P mineralization and the role of extracellular phosphatase enzymes is limited by the complex nature of interactions between organic P, carbon and nitrogen cycling, and the lack of appropriate methodologies (Olander and Vitousek, 2000; Chen et al., 2002).

Notwithstanding the constraints described above, significant progress has been made in the quantitative determination of soil organic P mineralization and microbial P dynamics. Isotope techniques have been used extensively for many years to study P transformations and availability in soil (Di et al., 1997). One of these techniques, isotopic exchange kinetics (IEK), can be used to characterize soil inorganic P availability by combining the three main determining factors, namely intensity (concentration of inorganic P in soil solution), quantity (amount of inorganic P exchangeable with soil solution over a given time) and capacity (rate of disappearance of added radioactive inorganic P from soil solution) (Frossard and Sinaj, 1997). This approach has been widely used to examine soil inorganic P availability in relation to land use and management in a variety of agroecosystems (Frossard et al., 2000). Other studies have also demonstrated that the principles of IEK can be used to measure soil organic P mineralization and P flux through the microbial biomass. Oehl et al. (2001a) measured basal P mineralization in a batch incubation experiment by separating the physicochemical and biological/biochemical processes that determine the concentration of inorganic P in soil solution. After 7 days they determined a basal daily mineralization rate of 1.7 mg P/kg for a soil that had been managed under a bioorganic (biodynamic) cropping system for 20 years. This was found to be equivalent to the equilibrium concentration of inorganic P in the soil solution and demonstrates the importance of P mineralization in maintaining P supply to crop plants. Using a similar approach, Lopez-Hernandez et al. (1998) measured basal daily mineralization rates of 0.22–0.90 mg P/kg for a series of Mollisols in the USA. Both Lopez-Hernandez et al. (1998) and Oehl et al. (2001a) acknowledged limitations associated with the use of IEK to determine gross mineralization when applied to P-deficient or high P-sorption capacity soils. Oehl et al. (2001b) used IEK to examine the kinetics of P uptake by the soil microbial biomass. This study involved comparing soils taken from a long-running field experiment in Switzerland that includes contrasting fertility management regimes (no fertilizer, mineral fertilizer, manure (organic), composted manure (bio-organic)) (Mäder et al., 1999). The specific radioactivity of soil solution and microbial P released by chloroform treatment over a 73-day incubation period were determined; it was found that P cycled through the microbial biomass faster in soils from the bioorganic (70 days) and organic (120 days) treatments than from the mineral fertilizer treatment (160 days). This study demonstrated that P turnover and availability are influenced by the form of P input, and also that organic P released following lysis of microbial cells in soil is rapidly hydrolysed and can make a significant contribution to plant P requirements. These findings confirm that organic- and microbial-P dynamics play an important role in the control and maintenance of inorganic P in soil solution.

Phosphorus Surplus – High-input Systems

Diffuse (non-point source) P transfer from soil to surface water (and groundwater) by overland flow (runoff) and subsurface flow has been the subject of extensive research since the early 1990s in particular. This has been prompted by the realization that declining water quality in many watersheds, rivers, lakes and estuaries may be partly attributed to increased P transfer from agricultural land (e.g. Watson and Foy, 2001). Many aspects of the processes, pathways and forms of diffuse P transfer have been studied in detail and the rapid progress of this important area of environmental research has been documented in several reviews (e.g. Sharpley et al., 1995, 2000; Tunney et al., 1997; Haygarth and Jarvis, 1999; Nash and Halliwell, 1999; Pierzynski et al., 2000; Sharpley and Tunney, 2000; Haygarth et al., 2000; McDowell et al., 2001a).

Phosphorus and water quality

There is mounting evidence that increased P transfer from agricultural land to water bodies can have adverse impacts on water and environmental quality (Daniel et al., 1998; Withers et al., 2000; de Clercq et al., 2001; Watson and Foy, 2001). The presence of elevated concentrations of P in streams, rivers and lakes in particular can lead to accelerated eutrophication. The process of eutrophication can be defined as 'an increase in the fertility status of natural waters that causes accelerated growth of algae or water plants' (Pierzynski et al., 2000). The eutrophication

threshold is commonly 20–100 µg total P/l, and accelerated eutrophication of fresh water is a widely recognized environmental issue that can result in significant limitations on water use for drinking and fishing, as well as for industrial and recreational use (Carpenter et al., 1998). For example, the US Environmental Protection Agency estimates that 45% of US waterways have impaired water quality due to nutrient enrichment (CEEP, 2001). In Europe, 55% of river stations reported annual average dissolved P concentrations in excess of 50 µg P/l over the period 1992–1996 (Crouzet et al., 1999). A report on the state of New Zealand's environment revealed that approximately 10% of shallow lakes were classified as eutrophic (20–50 µg total P/l) or hypereutrophic (> 50 µg total P/l) (Cameron et al., 2002). Accelerated europhication has also been linked with large-scale fish kills in some estuaries caused by increased populations of the dinoflagellate Pfiesteria piscicida, which in turn has the potential to adversely affect human health (Pierzynski et al., 2000). The protection of water quality is an important environmental issue in many countries. For example, the European Union Water Framework Directive (2000/60/EC) aims to restore all waters to 'good status' by 2015, which will require specific action to reduce and control diffuse pollution from agriculture, including P transfer.

Phosphorus accumulation and transfer

Most of the research on diffuse P transfer has been carried out in intensively managed agroecosystems where continued inputs of P in the form of mineral fertilizers and imported animal feed have resulted in significant accumulation of P in topsoil. For example, Haygarth et al. (1998a) compiled comprehensive P budgets for representative intensive dairy and extensive upland sheep farming systems in the UK and determined an annual accumulation rate of 26 kg P/ha under dairying compared to only 0.28 kg P/ha under sheep. Withers et al. (2001) determined an average P surplus on arable and grassland farms in the UK of 1000 kg P/ha over the 65 years from 1935 to 2000 (15 kg P/ha/year). High levels of P accumulation in soil have also been reported under intensive farming systems in other parts of Europe and North America (Sibbesen and Runge-Metzger, 1995; Sims *et al.*, 2000; de Clercq *et al.*, 2001), together with consequent increases in plant-available P (Tunney *et al.*, 1997; Fig. 5.2). The accumulation of P in soil from imported feed is particularly important in areas of intensive livestock production (e.g. pigs, poultry, dairy) where large quantities of manure are applied to land (Sharpley and Tunney, 2000).

Diffuse P transfer from soil is mainly controlled by a combination of hydrological factors that include the intensity and duration of rainfall (or irrigation) events, together with the spatial variables of scale and pathways (Haygarth *et al.*, 2000). Slope (topography) and drainage (substrate permeability) mainly influence P transfer pathways at the field scale. Phosphorus transfer to surface water occurs via overland flow and/or subsurface flow (interflow, drainflow) as base-flow and storm-flow, whereas transfer to groundwater occurs by a combination of throughflow

(percolating water) and preferential flow. Energy provided by water (as the carrier) results in mobilization of soil P and applied P in various physicochemical forms via dissolution, physical (erosion) and incidental processes (Haygarth and Jarvis, 1999; Haygarth and Sharpley, 2000). The process of dissolution or solubilization refers to the release of inorganic or organic P molecules from mineral surfaces and soil biota, and is clearly related to soil P status and biological activity. Physical P transfer involves detachment of soil mineral particles (including colloids < 0.45 μm) containing inorganic and organic P, and is associated with soil erosion processes. The 'incidental' mode refers to short-term transfer of P applied to soil in mineral fertilizer or animal manure (dung, slurry) and occurs mainly during contemporaneous P application and rainfall (Preedy *et al.*, 2001).

In general it has been found that the quantities of P lost annually from soil via overland and subsurface flow pathways are very small. Data from a wide range of field and catchment studies have shown that in most cases annual total P transfer from soil is less than 1 kg/ha, although higher rates of transfer

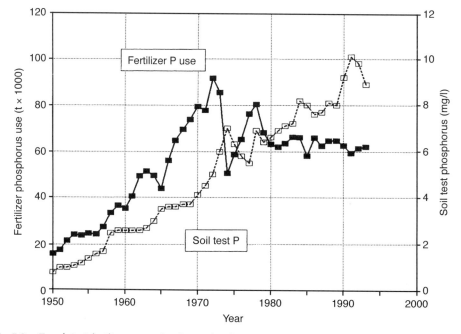

Fig. 5.2. Trends in P fertilizer use and soil test P levels in Ireland (Tunney *et al.*, 1997).

(2–6 kg P/ha/year, up to 17 kg P/ha/year) have been recorded from soil under intensive pastoral or arable farming, especially when animal manure is applied (Haygarth and Jarvis, 1999; Nash and Halliwell, 1999; Gillingham and Thorrold, 2000; Hooda et al., 2000; Turner and Haygarth, 2000; McDowell et al., 2001a). Phosphorus transfer is generally small compared with the amounts of P added to soil as mineral fertilizer and organic manure, which together exceed 25 kg P/ha/ year in many agroecosystems (Sibbesen and Runge-Metzger, 1995; Haygarth et al., 1998a; de Clercq et al., 2001; Cameron et al., 2002). Phosphorus transfer losses of less that 1 kg P/ha/year are generally considered to be of little significance from a direct agronomic or economic perspective (Nash et al., 2000). Apart from the environmental risk associated with accelerated P transfer, the continued application of P to soil in excess of crop requirements represents an inefficient use of a finite resource.

Relationships between P transfer in overland or subsurface flow and P inputs and soil P status have also been examined extensively. This has been prompted by the desire to establish appropriate management thresholds for the mitigation of P transfer in agricultural watersheds (Gburek et al., 2000; Sims et al., 2000). This is a very difficult task given the complex combination of properties and processes that control P transfer. Heckrath et al. (1995) observed that concentrations of P in drainage water from long-established arable field plots at Rothamsted (UK) were related to topsoil P status. They found that significant P transfer in drainage water (> 150 µg P/l) occurred during winter only when levels of plant available P (Olsen P) in topsoil exceeded 60 mg P/kg. This critical level of soil P was termed the 'change point', and is essentially a chemical phenomenon that describes the relationship between the quantity of P held on adsorption sites and its release to soil solution. Change point is closely related to the degree of P saturation which can also be used to assess the potential for P transfer from soil in subsurface flow (McDowell and Condron, 2000; Schoumans and Groenendijk, 2000). A number of studies have examined the change point and P saturation phenomena in different

soils and agroecosystems (Leinweber et al., 1997; Hesketh and Brookes, 2000; Hughes et al., 2000; Daly et al., 2001; McDowell and Sharpley, 2001a, b; McDowell et al., 2001b; Blake et al., 2002; Maguire and Sims, 2002). McDowell and Sharpley (2001a) examined relationships between extractable soil P (water, calcium chloride (0.01 M), soil tests (Olsen P, Mehlich-3 P)) and P loss by overland flow and subsurface drainage from a range of soils. Comparison of water or calcium chloride-extractable P with soil test P revealed change points that were closely related to P loss. Daly et al. (2001) examined P sorption and desorption dynamics for a range of soils in Ireland and confirmed that accelerated P loss was likely to be greater from peat and high organic matter soils. These findings indicate that soil test P levels and the degree of P saturation can be used to predict the risk of P transfer by overland or subsurface flow.

When considering the relationships between P transfer, P inputs and soil P status, it is important to recognize the influence of physical and biological properties and processes, in addition to the chemical factors described above. In particular, preferential flow through root channels and earthworm burrows may effectively bypass a significant proportion of P sorption surfaces in the soil and thereby facilitate rapid transfer of P from the soil surface and topsoil (Haygarth et al., 2000; Heathwaite and Dils, 2000; Simard et al., 2000). Several studies have also demonstrated that biological processes play an important role in determining the amounts and forms of P transfer from soil. For example, Turner and Haygarth (2001) found that the process of wetting and drying resulted in accelerated release of soluble organic P from soil biota and organic matter, which indicates that P loss may be influenced by environmental conditions that affect biological activity in the soil. This is confirmed by findings from several studies which revealed that a significant proportion of soluble P and P in subsurface and overland flow from grassland soils is present as organic P (Haygarth and Jarvis, 1997; Heathwaite and Dils, 2000; Turner and Haygarth, 2000; Turner et al., 2002). Further work is required to improve our understanding of the dynamics and associated mobility of organic and microbial P in the soil environment, as well as the reactions and bioavailability of soluble and particulate organic P in aquatic ecosystems.

Management of phosphorus transfer

Achieving the correct balance between maintaining plant productivity and minimizing P transfer from soil will be a vital component of future strategies for effective soil quality management in intensively managed agroecosystems. Closer alignment of P inputs with P requirements is an obvious way of reducing continued accumulation of P in soil and thereby reducing the extent and risk of P transfer. In situations where levels of plant available P in soil exceed those required for maximum potential plant production, P fertilizer inputs could be reduced or withdrawn for a time. Efficient storage and recycling of animal manure will also contribute to improved P utilization, which in turn may reduce P fertilizer requirements (Haygarth and Jarvis, 1999). This re-balancing of P requirement necessitates the development of appropriate farm P budgeting and management systems that can be easily understood and utilized by land managers. The combination of robust farm P budget systems with established knowledge of the relationships between soil P levels and agronomic performance (Hedley et al., 1995) will contribute to the development of P management systems that will enable production to be maintained with appropriate P inputs. Problems associated with balancing P inputs with P requirements are particularly difficult in regions of North America and Western Europe, where large quantities of feed (mainly grain) are imported to sustain intensive production of meat and dairy products (Sharpley and Tunney, 2000). Land application (effectively disposal) of animal manure from these operations is the principal cause of soil P accumulation. Options for better management of P in these systems include the development and adoption of improved strategies for timing and placement of manure application, as well as treatment and transport of manure to

other areas (Gburek et al., 2000). In the longer term, it may be necessary to relocate intensive animal production systems closer to their feed sources to facilitate efficient P recycling.

The effective management of P at the farm scale requires consideration of a wide range of factors including grazing and manure management, as well as P inputs (Haygarth and Jarvis, 1999; McDowell et al., 2001a; Watson and Foy, 2001). Haygarth and Jarvis (1999) developed a comprehensive agronomic model that integrates various farm management practices with the process of P transfer (Fig. 5.3). This model highlights the risks associated with continued accumulation of P in soil from fertilizer inputs, the timing of fertilizer and manure application, together with the importance of tillage management. For example, Nash et al. (2000) monitored storm-flow from grazed pasture in Victoria, Australia over 3 years (34 events) and found that P concentrations were most closely related to the timing of P fertilizer application relative to a storm event. They concluded that improved timing of P fertilizer application relative to rainfall using long-range meteorological forecasts could effectively reduce P transfer by overland flow. In a related study, Preedy et al. (2001) examined P transfer from grassland in the UK following application of P (29 kg/ha) as soluble mineral fertilizer (triple superphosphate (TSP)) or animal manure (dairy slurry). They found that significant quantities of P (1.8–2.3 kg/ha) were transferred by overland and subsurface flow as a result of 49 mm of rainfall over 169 h following application. The total P concentration in overland flow peaked at 11,000 and 7000 µg P/l from plots treated with TSP and dairy slurry, respectively, whereas concentrations in excess of 3000 µg P/l were determined in subsurface flow. The findings of this study illustrate the importance of short-term accelerated P transfer processes in determining overall P loss from soil, and also highlight the opportunities for significant reduction in P transfer by careful management of P application in relation to expected rainfall and consequent overland flow and drainage. Haygarth et al. (1998b) showed that the inclusion of mole drains reduced P transfer from grazed pasture in overland and subsurface flow by 30% due

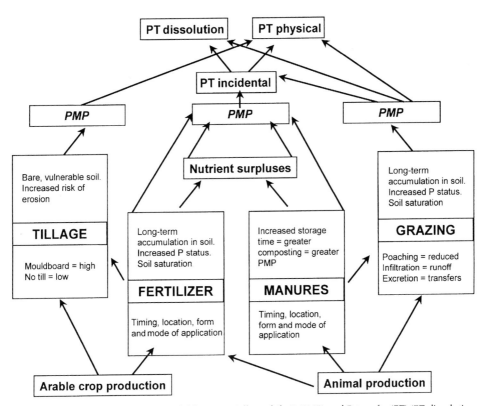

Fig. 5.3. Agronomic management model for potentially mobile P (PMP) and P transfer (PT) (PT dissolution, soluble mode of transfer; PT physical, particulate mode of transfer; PT incidental, transfer when applied P is removed from the soil surface coincident with hydrological factors) (Haygarth and Jarvis, 1999).

to increased sorption of P released from topsoil by the soil mineral matrix.

Other land management practices can influence the extent and pathway of P transfer from soil. Several studies have demonstrated that conservation tillage can reduce soil erosion and associated P loss (McDowell *et al.*, 2001a). Riparian buffer strips have also been shown to intercept sediment and thereby reduce P transfer to water bodies by overland flow, although their effectiveness may decrease with time (Gillingham and Thorrold, 2000). McDowell and Sharpley (2001b) showed that whereas a land-use change from arable cropping to grassland reduced P transfer by overland flow, long-term P loss in subsurface drainage may increase.

In addition to the above measures, application of specific minerals to manure and soil can also be used to reduce P transfer in agroecosystems. The addition of aluminium

sulphate to poultry and pig manure has been shown to decrease P solubility and thereby significantly reduce P transfer by overland flow following land application (Moore *et al.*, 2000; Smith *et al.*, 2001). Other studies have shown that application of selected industrial waste materials to soil can increase P sorption and thereby reduce P transfer. Stout *et al.* (1998, 2000) found that addition of coal combustion by-products to high P status soils markedly reduced P transfer by overland flow without adversely affecting plant P availability. The long-term effectiveness of continued addition of these waste materials to soil remains to be determined, and there are potential problems associated with increased transfers of toxic constituents such as aluminium.

Concern about the potential environmental risks associated with increased rates of diffuse P transfer has resulted in several US states recommending threshold levels of soil test P, especially for identified critical source areas (Sharpley and Tunney, 2000). These environmental threshold values are commonly two to four times greater than the corresponding values for optimum crop production (agronomic thresholds). For example, the recommended environmental threshold levels of Bray-1 P are 75 mg/kg in Michigan and Wisconsin, compared with 150 mg/kg in Ohio. The corresponding Mehlich-3 P values are 130 and 150 mg/kg in Oklahoma and Arkansas, respectively. Land application of pig manure in Ireland is prohibited when the soil test P (Morgan P) value is greater than 15 mg/kg (equivalent to an Olsen P of 60 mg/kg) (Sharpley and Tunney, 2000). Furthermore, a wide range of guidelines and regulations designed to improve on-farm P management practices and mitigate P transfer have been developed in many European countries (de Clercq et al., 2001). For example, currently in The Netherlands a maximum of 37 kg P/ha can be applied annually to arable land and grassland in the form of animal manure, while the corresponding maximum permissible annual P surplus is 15 kg P/ha.

The concept of P indexing is designed to establish how P transfer is determined by the combined influences of hydrology (as determined by soil type (texture) and topography) and soil P status and inputs within a watershed (Gburek et al., 2000; Coale et al., 2002). The objective of this approach is to identify areas with a high risk of accelerated P transfer within a watershed and focus monitoring and mitigation measures in these 'critical source areas'. Pionke et al. (2000) found that 80–90% of the total P transfer by overland flow from an agricultural catchment originated from stream channels that accounted for only 20% of total flow. It is important to note that the immediate and long-term effectiveness of using soil test P threshold values and/or P indexing to assess and manage P transfer in agroecosystems remains to be established.

Most of the work carried out on P transfer has focused on assessing the extent of immediate P loss from soil, and the development and evaluation of management strategies designed to minimize P transfer and thereby reduce the downstream environmental impacts. However, although it may be possible to reduce P loss from soil, the residual environmental impact of P enrichment of aquatic ecosystems also requires ongoing consideration and investigation. This will be determined to a large extent by the nature and dynamics of interactions between soluble and particulate P in water and sediments in streams, rivers and estuaries in particular. Baldwin et al. (2002) highlighted the importance of P–sediment interactions in determining the fate and environmental impact of P transfer from soil to water, which in turn are determined by a complex combination of chemical, physical and biological properties and processes. McDowell et al. (2001c) clearly demonstrated that base-flow concentrations of dissolved reactive P and total P in stream water were influenced by the sediment P sorption properties in an agricultural watershed. It is therefore important to examine the nature and associated chemical and biological processes that influence P retention and release in sediments.

Phosphorus Deficiency – Reduced-input Systems

Actual or potential P deficiency in modern developed agroecosystems is most likely to arise in reduced-input systems such as organic farming. Organic farming is essentially a prescribed and accredited system of farm management practices that is primarily designed to ensure product quality and enhance social and environmental sustainability (Condron et al., 2000; Stockdale et al., 2001). Organic systems currently account for 1–10% of farm production in Western Europe, and the area under organic farming is increasing steadily in these and other developed countries such as the USA and Japan (United Nations Food and Agriculture Organization, 1999). Much of the ongoing expansion in organic farming is being driven by consumer demand for food produced to an accredited standard that involves minimal use of synthetic inputs such as pesticides and soluble fertilizers, and consumers are prepared to pay substantial premiums for organic produce. The growth in organic farming has been prompted by concerns about the quality of food produced by conventional farming methods and perceived adverse environmental impacts associated with modern intensive land and nutrient management regimes. Organic production systems are seen by many as having a reduced impact on the wider environment compared with conventional farming, although the scientific evidence to support this perception is limited at present (Conacher and Conacher, 1998; Condron et al., 2000; Tinker, 2000; Stockdale et al., 2001).

The principal aims of organic farming with respect to soil and nutrient management are to protect and maintain soil fertility by enhancement of biological activity, nitrogen self-sufficiency and crop diversification, and to minimize impacts on the wider environment (Condron et al., 2000; Stockdale et al., 2001, 2002). Within the International Federation of Organic Agriculture Movements (IFOAM) there are many types of organic farming systems based on different philosophical and practical approaches to soil, plant and animal husbandry (Stockdale et al., 2001). The most commonly adopted organic farming accreditation schemes allow the use of sparingly soluble mineral fertilizers such as phosphate rock, in addition to appropriate forms of compost and animal manure. However, the amounts and forms of nutrient addition allowed under the 'biodynamic' system ('Demeter' trademark) are more restricted, especially since one of the aims of biodynamic farming is to achieve 'a self-sustaining system where inputs of almost all kinds gravitate to nil' (Condron et al., 2000).

As described previously, in many agro-ecosystems continued inputs of P in excess of crop requirements has resulted in the accumulation of significant quantities of plant available and total P in topsoil. Thus, the reduction or cessation of P fertilizer application as a consequence of conversion to organic farming may have a limited short-term impact on plant production due to the residual effect of previous P inputs. This is evident in results obtained from a long-term field experiment in Switzerland which showed that during the first crop rotation (1978–1984) production decreased by only 26% on plots which

received no nutrients compared with plots which received high rates of mineral fertilizer annually (Mäder et al., 1999). However, during the subsequent crop rotation (1985–1991) relative production on the unfertilized treatment declined to 50% compared with the fertilized plots. Mäder et al. (1999) also reported that crop production levels over 14 years on the organic farming treatments included in the same trial, which received nutrient inputs in the form of rotted or composted farmyard manure, were 76–88% compared with the complete mineral fertilizer treatment. Further detailed investigation of changes in soil P on this trial showed that during the first 21 years the average annual net loss of P from the organically managed treatments was 5–8 kg P/ha, compared with an annual accumulation of 5 kg P/ha for the mineral fertilizer treatment (Oehl et al., 2002). Nutrient budget data for 47 organic dairy farms in Western Europe showed that most (25) had an annual P surplus (1–21 kg P/ha), while the remainder were either in P balance or had net P losses of 1–7 kg P/ha/year (Watson et al., 2002).

It is clear that continued application of P in some form is necessary to maintain crop and animal production in organic farming systems, especially in the long term, if economically viable returns are to be sustained. Several studies have shown that P inputs in the form of fodder, compost and animal manure can be used to maintain high levels of production on organic farms (Clark et al., 1998; Mäder et al., 1999; Aarts et al., 2000; Watson et al., 2002), although there are potential problems associated with the continued long-term availability of sufficient quantities of compost and manure (Stockdale et al., 2001). Although the residual effect of previous P inputs will help to sustain production at reduced levels, data from several long-term comparative studies have confirmed that amounts of plant available P are generally lower in soil under organic management compared with conventional management (Aarts et al., 2000; Løes and Øgaard, 2001; Oehl et al., 2002). Although this may reflect reduced P inputs, it may also be attributed to increased P inputs in the form of compost, manure and phosphate rock (Stockdale et al., 2002).

Introduction

Potassium is the cation required in the largest amounts by plants. In the plant, K has an irreplaceable role in the activation of enzymes, e.g. in protein synthesis (Blevins, 1985), but for these biochemical roles only small amounts of K are required. Large amounts of K are required for its biophysical role in osmoregulation, cation–anion balance and water balance. Because of its particular mobility within the plant, K plays a major part in solute movement within the plant, especially of photosynthates to storage organs like grains and tubers. Additionally, K is associated with alleviating abiotic stress such as frost, drought, high light intensity and heat (Cakmak, 2003), biotic stress caused by fungal and pest attacks (Huber and Arny, 1985; Härdter, 2003). In its biophysical roles, K can be replaced by other cations to a limited extent. For general reviews see Marschner (1995) and Mengel *et al.* (2001). With the exception of possible local environmental issues associated with mining K-bearing ores (Scharf, 1990), there have been no reports of harmful effects of K to the environment, nor with respect to human health (Anonymous, 1984, 1998).

For plants in unamended ecosystems the main source of K is the weathering of soil minerals (Mengel *et al.*, 2001). Potassium is the seventh most abundant element in the earth's crust (Sparks and Huang, 1985). It is usually the most abundant nutrient element in soils (Reitemeier, 1951); concentrations vary between < 0.1% and > 3% K with 1–2% being most frequent (Schroeder, 1974). Despite the large amount of total K, only a small part is immediately available to plants. Although soil organic matter (OM) can supply appreciable amounts of nitrogen, phosphorus and sulphur, which are held in organic molecules, it plays only a very minor role in the K supply because K is readily leached from dead plant material. However, soil OM provides cation exchange sites, which can hold K.

Most commercially exploitable K deposits, which can be processed into K fertilizers, originate from the evaporation of seawater many millennia ago (Stewart, 1985). Globally, known estimates of high-grade K reserves that can be recovered at current market prices range from 8 to 17 billion t, and total resources that potentially could be available amount to about 120 billion t K (Sheldrick, 1985). Current use is about 20 million t K annually (Anonymous, 2001). The annual use should increase, especially in the developing countries where nutrient imbalances are severe (Krauss, 2001) and globally in response to feeding the expected increase in the world population by 3 billion people within the next 50 years (Anonymous, 2002). However, the assessable K reserves may allow for exploitation on a time-scale of hundreds of years.

When addressing 'sustainability' in relation to plant nutrition, the hierarchical level (e.g. field, farm, region) and the time-scale (e.g. growing season, generations, millennia) determine the degree of sustainability. If, for example, the nutrient balance of a field is zero (inputs equal outputs), nutrient availability and crop yield may not be affected. The system therefore seems sustainable. However, at a higher hierarchical level the system may not be sustainable if food production is insufficient to feed the population or if a substantial part of the output consists of leaching losses (Janssen, 1999). The Brundtland Commission (WCED, 1987) suggests 'generations' as the time-scale but different interpretations of this term may result in very different conclusions as to whether a particular farming practice and plant nutrient management is sustainable or not.

Cropping systems with continuous large negative K balances are not sustainable in the long term because a continued K deficit leads to impoverishment of the soil and to decreased crop production (Oenema and Heinen, 1999). Therefore, this chapter focuses on maximizing K use efficiency in the soil–plant system and on soil K thresholds for sustainable K management at the farm and field level in cropping systems with reduced K input.

Potassium in Soil

Soils differ greatly in their ability to supply K for plant growth. Some soils can release K so that the K status, measured by plant growth and chemical tests, does not decline appreciably over many years; other soils are rapidly exhausted (Arnold, 1962a). These differences in the ability of soils to supply K are related to the mineral type and amount, which is influenced by the parent material and the stage of weathering (Jackson and Sherman, 1953). Minerals at an intermediate weathering stage occur most abundantly in soils in temperate regions, and K is found mainly in the primary silicates, K feldspars and mica, and in the secondary 2:1 clay minerals, for example illite (Arnold, 1960). In the weathering of mica to illite and further to smectite or vermiculite, K ions (K^+) are exchanged by other cations and this process reduces the K content from about 10% in mica to less than 1% in smectite and vermiculite (Schroeder, 1978).

Various attempts have been made to clarify soil K but in a simple conceptual frame it can be considered to be in four fractions or pools (e.g. Sparks and Huang, 1985; Barber, 1994; Syers, 2003):

- K in soil solution (K_{sol}), which is the immediate source of K for plants.
- Exchangeable K (K_{ex}) held on the cation exchange sites on clay and OM. It exchanges readily with other cations and is readily available to plants.
- Slowly exchangeable K (K_{slex}) held in interlayer positions with high selectivity for K (2:1 clay minerals like illite or vermiculite). It is not readily accessible for exchange with other cations in the soil solution and, therefore, is slowly available to plants. It is sometimes referred to as fixed K.
- Lattice K (K_{latt}) (or native or mineral K, e.g. in K feldspar and unweathered mica). Lattice K is normally considered to be of little significance for the K supply to plants during a single growing season.

This classification gives an acceptable picture of the K pools in soil relative to the availability of the soil K to plants, with the possible exception that K_{latt} may be more available than assumed here (see section 'Exploiting Soil Potassium Reserves').

The release of K_{latt} is normally considered an irreversible process whereas the other pools constitute a dynamic system with continuous transfer between the fractions and with no distinct boundaries (Sparks, 1987; Sharpley, 1990; Syers, 2003). The interrelationship between these soil K pools and inputs and outputs is shown in Fig. 6.1.

When K fertilizer or organic manure is applied to a soil, the K enters the soil solution and then equilibrates between K_{sol}, K_{ex} and K_{slex} depending on the amount and type of clay, the total cation exchange capacity (CEC) (Sharpley, 1990), soil pH (Warren and Johnston, 1962), wetting and drying cycles (Zeng and Brown, 2000) and the soil K status. Using isotopic exchange, Fardeau et al. (1992) showed that more than 70% of fertilizer K applied in a given year may remain in the soil after the first growing season because of these exchange processes. The distribution of applied K between the different pools occurs quickly, and because the K in the K_{slex} pool is not determined as the exchangeable K, it has often been considered as fixed K, and the process has traditionally been termed fixation. However, we prefer the term 'retention' because the K held as K_{slex} can normally be released again over periods of tens of years as shown in long-term experiments (Johnston and Poulton, 1977).

Of the total soil K, most is K_{latt}, and K_{ex} and K_{slex} are only a few per cent of the total (Mengel and Kirkby, 1987). The K_{sol} concentration is usually 10^{-3} to 10^{-4} M (Wild, 1988) and the amount is insufficient to meet crop demand unless it is rapidly replaced by K from the other K pools. Whereas K_{latt} relates to the soil type and the mineralogy, K_{ex} and K_{slex} relate more to the K balance during the cropping and fertilization history (Johnston, 1986). Continued plant removal of K will in the short- or long-term reduce K_{ex} to a minimum value characteristic of the soil (Reitemeier, 1951; Ogaard et al., 2002). This minimum value will be in dynamic equilibrium with K_{slex} and both will be supported by release of K_{latt} (Fig. 6.1).

Roots take up K from the soil solution as the positively charged K ion (Nye and Tinker, 1977). Potassium depletion at the root surface

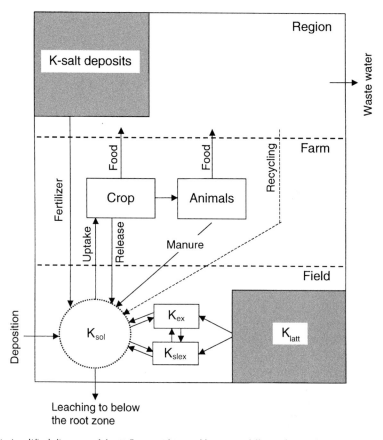

Fig. 6.1. A simplified diagram of the K flow, within and between different hierarchical levels.

creates a concentration gradient between the bulk soil and the root surface and this gradient is the driving force for the diffusive flow of K to the root surface (Jungk and Claassen, 1989). A high flow rate can be attained by the addition of sufficient K in fertilizer or manure, or by a sufficiently high buffering power of the soil. The buffering power, normally quantified by the slope of a quantity (Q)/intensity (I) curve, is the ability of a soil to maintain an adequate K_{sol} by the release of K from other K fractions. For optimum growth, K_{sol} must be above a plant-specific critical level and the lower the K buffering capacity, the higher the critical level (Mengel and Busch, 1982). Within soils of the same mineralogical composition, K_{sol} and K_{ex} are highly correlated (Warren and Johnston, 1962; Grimme, 1976). The buffering power is related to clay content (Sharpley,

1990) and clay mineralogy (Mengel and Busch, 1982). Using the concept of 'stability' outlined by Schjønning *et al.* (Chapter 1, this volume), the degree of stability of crop production to K availability is proportional to the soil buffering power, which may be synonymous with the 'resilience' of the soil – the ability of the soil to maintain K_{sol} under stress from K uptake by plants. In the short term (e.g. a growing season) and medium term (e.g. 1–10 years) the extent to which K_{sol} is maintained will depend on the extent to which K, added in past applications of fertilizers and manures, has been retained in the K_{ex} and K_{slex} soil pools. In the longer term (e.g. > 10 years), for cropping systems with low K input the stability of crop production with respect to K availability is mainly related to the release of K from soil minerals.

Potassium Flows and Balances

The flow of K at different hierarchical levels (field, farm and region, which may range from a country to the world) is simplified in Fig. 6.1. The K_{sol} fraction constitutes a transition phase and among the sources that contribute to K_{sol} are K inputs in manures and fertilizers. A minor input, between 1 and 8 kg/ha/year, comes from atmospheric deposition, soil dust and sea spray (Cooke, 1981; Grundahl and Hansen, 1990) with occasional much larger values in coastal areas. The main outputs from K_{sol} are crop uptake and leaching. Some of the K taken up by crops is returned to the soil in animal manure or in leaching losses from the crop before harvest or from crop residues after harvest, e.g. straw left in the field. Another part is exported in farm products and there is little recirculation of this K to farms. Potassium lost by leaching, in surface runoff and in the effluent discharge from sewage treatment must be considered as an irreversible loss from the current production system. Soil erosion is very variable but if large amounts of soil are lost appreciable quantities of K may be removed from a field.

Within and between each hierarchical level (Fig. 6.1) there are K flows. For each level, the K balance is the sum of the inputs minus the sum of the outputs. A positive balance indicates enrichment and a negative balance indicates depletion of the soil K reserves. Comparison of national and international nutrient balances shows nutrient enrichment in countries importing feed and/or fertilizer and depletion in countries exporting agricultural products without significant inputs of K. In several of the Western European countries approximately 80% of the primary plant production on the farms is converted into meat and milk, and the K output in these products is small (Beringer, 1992). A 2-year investigation on 60 farms in Denmark (27 dairy farms, 17 pig farms and 16 arable farms) showed positive K balances ranging from an average of 43 kg K/ha/year for dairy farms to 9 kg K/ha/year for arable farms (Holbeck and Hvid, 2002). On 17 organic farms, where import of nutrients was restricted, K balances

were smaller than on 19 conventional dairy farms, the average K surplus being 33 and 82 kg K/ha/year, respectively (Kristensen and Halberg, 1995). The K surplus increased with increasing livestock density because of increased feed imports (Kristensen and Halberg, 1995; Holbeck and Hvid, 2002). In a study comprising organic farms located in six Western European countries (34 mixed farms and 13 dairy farms), Watson et al. (2002) found a slightly negative K balance on 12 of the mixed farms and only on four farms did the loss exceed 5 kg/ha/year.

Crop uptake is related to the type of crop, its yield potential and the soil K status. At harvest the K content is usually less than at the maximum vegetative stage because of the loss of leaves and K being leached from mature plants as rainfall passes through the standing crop (Mengel, 1978). Table 6.1 gives some average data for K removal by crops. Depending on the yield, the K offtake ranges from 15–50 kg K/ha/year in cereal grains to 200–300 kg/ha/year or more in grass

Table 6.1. Guidelines for the calculation of K removal by crops (modified after PDA, 1997).

		kg K/t of fresh material
Cereals	Grain only	4.6
	Grain and straw	
	Winter wheat/barley	9.8[a]
	Spring wheat/barley	11.4[a]
	Winter/spring oats	14.4[a]
Oilseed rape	Seed only	9.1
	Seed and straw	14.5[a]
Peas	Seed	8.3
Field beans		10.0
Potatoes		4.8
Sugarbeet	Roots only	1.4
	Roots and tops	6.6
Grass	Fresh grass, 15–20% DM	4.0
Maize	Silage, 30% DM	3.7
Kale		4.2
Cabbage		3.0
Carrots		2.5
Onions	Bulb	1.5

[a]Offtake value is per tonne of grain or seed, but includes nutrients in straw.
DM, dry matter.

harvested for silage. There is a larger amount of K in cereal straw than in cereal grain and, therefore, the disposal of straw at harvest has a large effect on the K balance (Johnston, 1986).

The level of plant-available soil K influences K uptake and therefore the K balance. In comparison with Danish standard values for the K content in plant material, Askegaard and Eriksen (2000) found 20–40% less K in some crops grown organically on a soil with little available K compared with those grown on conventional farms.

Positive K balances at a farm level may hide large differences at a field level because large amounts of K may be relocated within farms through the movement of K, e.g. in silage, straw and manure. In calculating field and farm K balances it is important to consider leaching losses from fields (see below), from silage effluents and from manure heaps.

Leaching losses

The movement of K from the topsoil and through the soil profile varies with soil texture, clay mineralogy, current and past K inputs, and drainage (Munson and Nelson, 1963). Johnston and Goulding (1992) and Simmelsgaard (1996), reviewing data for mainly loams and clayey soils found a strong correlation between drainage and K leaching, and estimated an average loss of 1 kg K/100 mm drainage. Leaching can be much larger on sandy soils. In 1997 in an organic crop rotation experiment replicated at four locations in Denmark, K leaching below 1 m depth varied from 1 kg/ha on a soil with 24% clay at Holeby to 46 kg/ha on a soil with only 5% clay at Jyndevad (Table 6.2). These K leaching losses in 1997 corresponded to 0.5 kg K/100 mm drainage at Holeby and 7 kg K/100 mm drainage at Jyndevad.

On the coarse sandy soil at Jyndevad, K inputs were decreased in 1996 when the site was converted to organic farming. Before 1996, K inputs varied between 50 and 150 kg/ha/year, depending on the crop, and these declined to 18–34 kg/ha/year after 1996. In consequence, K leaching losses decreased by 50% over a 4-year period. Despite a considerable decline in K_{ex} at Foulum (9% clay) and Flakkebjerg (16% clay), K leaching losses remained fairly constant (Table 6.2), perhaps because K_{ex} was still large enough for K losses to be affected more by the amount of clay than by K_{ex}. Continued application of small amounts of K would probably reduce K leaching significantly (Askegaard and Eriksen, 2000). Such changes over time suggest that precise K balances require accurate K leaching losses, which

Table 6.2. Leaching of K (K_{leach}, kg/ha), drainage (mm), K input in slurry (K_{slu}, kg/ha) and K_{ex} (mg/kg) in the topsoil (1996: 0–25 cm; 2000: 0–20 cm) of a four-course crop rotation (barley, grass–clover, winter wheat, pea–barley mixture) as affected by location and year (Askegaard et al., 2003; M. Askegaard, 2002, unpublished results).

Year	Jyndevad (5% clay)			Foulum (9% clay)			Flakkebjerg (16% clay)			Holeby (24% clay)		
	K_{leach} (kg/ha)	drain. (mm)	K_{slu} (kg/ha)	K_{leach} (kg/ha)	drain. (mm)	K_{slu} (kg/ha)	K_{leach} (kg/ha)	drain. (mm)	K_{slu} (kg/ha)	K_{leach} (kg/ha)	drain. (mm)	K_{slu} (kg/ha)
1997	46	623	18	14	319	15	2	192	24	1	204	19
1998	41	890	34	16	392	20	2	316	12	1	337	11
1999	28	684	34	18	444	18						
2000	21	470	34	12	310	25						
Avg.	34	667	30	15	366	20	2	254	18	1	270	15

	K_{ex} (mg/kg)											
1996	43			135			105			109		
2000	35			87			87			109		

should include the time lag between changes in K input and changes in K leaching losses.

A study in the UK showed that there are smaller K losses on heavier textured soils because, depending on the amount and type of clay, K can be retained in the subsoil (Table 6.3). On the silty clay loam soil (25% clay), large amounts of K have been applied since 1843. Compared to K_{ex} in each horizon on the plot without K, the increase in K_{ex} declined with depth and the enrichment depended on the amount of K added. On the sandy loam soil (10% clay), the soil at each horizon has been uniformly enriched with K_{ex}, although less K was added than on the silty clay loam. The subsoil K_{ex} can be recovered by deep-rooted crops. However, even clay soils may suffer from an immediate loss of K by preferential flow through macropores as shown for urine patches by Williams et al. (1990).

Recycling of waste products

After the introduction of mineral fertilizers, the recycling of nutrients from urban areas to farms declined, and is now marginal (Eilersen et al., 1998). Potentially, urban waste might substitute about 10% of the mineral fertilizer consumption in Denmark (Eilersen et al., 1998). Compared to all other effluents from a household, human urine contains the largest amount of nutrients, 44–50% of the K, 70–80% of the N and 50–55% of the P (Holtze and Backlund, 2002; Jönsson, 2002).

However, most of the K entering wastewater treatment plants is lost to rivers. Efforts have been made to improve the utilization of nutrients in human urine as a fertilizer (Holtze and Backlund, 2002; Jönsson, 2002).

Potassium Availability to Crops

Soil analyses and crop response

Numerous laboratory methods for the assessment of plant-available K have been tested against crop response data from glasshouse and field experiments on different soils, without a universal solution emerging. Clearly, laboratory soil tests will never simulate the extent to which roots grow and absorb nutrients under field conditions (Arnold, 1962b; Nair, 1996).

Many methods for estimating readily available soil K determine K_{ex}, i.e. K that will exchange with another cation, and include the small amount of K_{sol}. Widely used reagents are 1 M ammonium acetate and 1 M ammonium nitrate but there are others. In general, the amounts of K extracted are strongly correlated (Johnston and Goulding, 1990).

Soil analysis frequently accounts for less than half of the variance in crop response to K (Cooke, 1982). For example, in Denmark in 1969–1977, there was no clear relationship between the yield of spring barley and topsoil K_{ex} in 528 field experiments even when the soils were grouped by clay content (Fig. 6.2).

Table 6.3. Amount of K added, kg/ha, and exchangeable K (K_{ex}), mg/kg, at four depths down the profile on a silty clay loam (25% clay) at Rothamsted (1843–1959), and a sandy loam soil (10% clay) at Woburn (1942–1961) (adapted from Johnston, 1986).

	Silty clay loam				Sandy loam				
K added	0	16,700	20,500	28,800	950	4,900	6,400	8,800	11,800
Treatment	No K	FYM	PK	FYM and PK	NPK	Veg. compost	FYM	Veg. compost	FYM
Depth (cm)	K_{ex}		Gain in K_{ex}		K_{ex}		Gain in K_{ex}		
0–23	119	174	380	641	111	144	163	205	266
23–30	129	104	179	386	101	137	169	213	265
30–46	139	23	102	225	93	134	174	226	268
46–61[a]	152	−2	79	152	83	91	143	186	226

[a]46–54 cm on the silty clay loam.

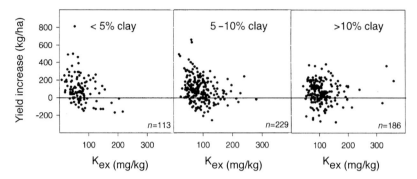

Fig. 6.2. The effect of K_{ex} in three soil groups (< 5%, 5–10% and > 10% clay) on yield increase of spring barley (kg grain/ha) following application of 50 kg K/ha. The experiment was carried out at farms throughout Denmark (n = 528 field trials) (after Anonymous, 1969–1977).

Conversely, there is often a good relationship in glasshouse experiments, especially for individual soils (Johnston and Mitchell, 1974), so the reasons for poor relationships in field experiments need to be sought.

Soil for routine analysis is typically sampled in the plough layer and air dried, which may change the balance between the K fractions (Schneider, 1997). Subsequently the soil sample is homogenized, and a specific fraction of soil K extracted. Plant roots, on the other hand, meet a complex soil system with a high degree of spatial variability in mineral particle size and composition, aggregation, pore size and tortuosity, K buffer capacity and water content (Gäth *et al.*, 1989). Also, replenishment of K_{sol} may come from K_{ex}, K_{slex} and K_{latt}, only the first normally being measured by soil analysis. Some of these factors are discussed here, although not all are specific to K.

Soil and plant factors influencing potassium uptake

Soil texture and mineralogy

The behaviour of K in the soil depends on the clay content and clay mineralogy (Mengel and Kirkby, 1987). Rao and Khera (1994) found that the minimum K_{ex} and the K replenishment rate were controlled by the nature of clay rather than the amount of clay. Soils with a high content of illite showed a high K-releasing power under intensive cropping compared to soils dominated by montmorillonite, vermiculite, chlorite and kaolinite. The diffusion of K in the soil is highly influenced by the soil texture; Gäth *et al.* (1989) showed that at all moisture tension levels, a sandy soil had lower K diffusion fluxes than a loess soil.

Temperature and soil moisture content

Low temperatures limit root extension and K uptake rates by roots (Nelson, 1968; Schimansky, 1981; Barber, 1994) and affect the availability of K in the soil (Sparks and Liebhardt, 1982; Barber, 1985). This may cause temporary symptoms of K deficiency in the spring. Soil moisture affects diffusive transport of K to the roots and, as a consequence, crop response to K fertilizers is often greater in dry than in wet seasons (Kuchenbuch *et al.*, 1986).

Soil compaction and aeration

Potassium deficiency has been observed in crops growing on compacted soils, even those with a high K status (Nelson, 1968). In soils with increasing compaction or mechanical impedance, root elongation is impaired and diffusion of the less-mobile nutrients, including K, is lessened (Marschner, 1995; Bennie, 1996). Waterlogging of compacted soils impedes oxygen transfer and, because K uptake is highly dependent on oxygen-controlled metabolic processes, the inhibitory effect of poor aeration is more pronounced with K than with other nutrients (Nelson, 1968).

K status

When roots take up K from the soil solution it can be replaced from K_{ex}, K_{slex} and K_{latt}. Frequently, over a period of a few years, small positive and negative K balances have little effect on measured K_{ex} values due to equilibration with K_{slex}. In the long term when there is K offtake but no input, any change in K_{ex} depends on the initial level of K_{ex} (Table 6.4). On soils exhausted to their minimum K_{ex} level, K offtake apparently comes from K_{latt}, assuming a constant ratio between K_{ex} and K_{slex}. In low-K-input systems, where a considerable amount of K is expected to come from the K_{slex} and the K_{latt} fractions, these fractions should be included in the assessment of the soil K status, but currently there are no generally accepted routine methods of analysis for them. However, it is not only the amount but also the rate of release of available K that is important (Grimme and Németh, 1979). A high growth rate can only be maintained if the nutrient supply to the roots keeps pace with the potential growth rate (Grimme, 1976), and this is usually the case when K_{ex} is above the critical level (see section 'Managing Low-potassium-input Systems'). The rate of mobilization is important on soils where K_{ex} is below the critical level.

Status of other plant nutrients

The interaction between K and other nutrients, N being the most important, has to be taken into account when the K demand is assessed (Loué, 1978). According to Liebig's 'Law of the Minimum', deficiency of other nutrients will decrease the effectiveness of applied K. Interactions between two nutrients are especially important when the contents of both are near the deficiency range (Marschner, 1995).

Crop species and varieties

In general, monocotyledons exploit soil K reserves better than dicotyledons. Cereals and grass frequently do not respond to K fertilizer (Mengel, 1982), and there is strong competition for K when ryegrass (*Lolium perenne*) is grown with red clover (*Trifolium pratense*) on low-K soils (Mengel and Steffens, 1985). Also, Rejado (1978) found that wheat was more sensitive to K deficiency than oats and barley, and cultivars of the same species may differ in their ability to exploit soil K resources (Mengel, 1978). Siddiqi and Glass (1983) found substantial differences between barley cultivars in their response to K_{sol} levels.

Crop uptake from the subsoil

Exchangeable K is usually determined only in the topsoil, but plants with adequate root growth can take up considerable amounts of K from the subsoil. Plant roots apparently develop independently of the spatial distribution of available K in the soil, and they take up K all along the root axis (Marschner, 1995). Kuhlmann (1990), using a split-root technique, showed that wheat recovered 9–70% of its K from the subsoil, whereas Haak (1981) found a contribution of subsoil K of 25–50% in cereals. The uptake of subsoil K depends

Table 6.4. Changes in exchangeable K (K_{ex}) in the 0–23 cm topsoil with negative potassium balances. Exhaustion Land, Rothamsted (adapted from Johnston and Poulton, 1977).

Plot	Initial K_{ex} mg/kg	Negative K balance, kg/ha		K_{ex}, kg/ha		Change in K_{ex} as a % of K balance
Period 1903–1951			1903	1951	Decrease	
A	95	594	288	223	65	11
B	265	848	801	321	480	57
Period 1951–1974			1951	1974		
A		317	223	208	15	5
B		640	321	264	57	9

on the amount of plant available K in the topsoil and subsoil, and on root distribution between the two layers (Haak, 1981; Kuhlmann, 1990).

Exploiting Soil Potassium Reserves

Many soils in early or intermediate stages of weathering contain very large K reserves. The ability of some soils to provide enough K for optimum yields has been demonstrated in long-term experiments (Johnston, 1986), but it is not possible to go on 'mining' soil K without eventually decreasing crop production (Johnston *et al.*, 2001a,b). However, focusing on the thresholds for sustainable crop production with reduced K inputs, the soil K reserves should be included as a potential K source. Currently most long-term field experiments compare 'with' and 'without' K input in combination with different levels of N and P. The ability of soils to release indigenous soil K under reduced K inputs has not been examined in detail.

Crop growth accelerates the chemical weathering of soil by lowering K_{sol} in the rhizosphere (Claassen and Jungk, 1982) or by acidification in the rhizosphere (Hinsinger *et al.*, 1993). Root-induced weathering probably involves the release of interlayer K (K_{slex} and K_{latt}) and may occur on a scale of days (Hinsinger and Jaillard, 1993), indicating the intensity of root–mineral interaction in the rhizosphere. Potassium depletion has been shown to decrease the content of illite and increase that of vermiculite and especially smectite (Nielsen and Møberg, 1984; Tributh *et al.*, 1987). These studies also showed increased soil cation exchange capacity (CEC) with K depletion.

In model calculations based on long-term experiments on Northern European soils, the weathering release rates for K varied between 3 and 82 kg/ha/year at 0–40 cm depth, and were least on sandy soils and most on clay soils (Holmqvist *et al.*, 2003). Specific surface area, mineralogy, soil temperature and moisture content significantly affected the K release rate. Chemical weathering proceeds more quickly at elevated temperatures,

as in the tropics, but nearly ceases during cold winter periods in temperate regions (Jackson and Sherman, 1953).

In long-term experiments without K input for a century or more it can be assumed that K_{ex} and K_{slex} have reached the minimum level associated with soil type, and that K_{latt} is the only plant-available K (Johnston, 1986). In the Askov long-term experiment (1894–1994), the annual K offtake in spring barley in the NP treatment was estimated at 18 kg/ha on a loamy soil (12% clay) and 12 kg/ha on a coarse sandy soil (4% clay) (Christensen *et al.*, 1994). For winter cereals the offtake was 16 kg/ha in wheat on the loamy soil and 22 kg/ha in rye on the coarse sandy soil. These results suggest that rye is a better 'K miner' than winter wheat. The maximum K release in the Askov experiment may have been overestimated due to soil movement between plots with and without K (Sibbesen, 1986). The average annual removal of K in long-term experiments at Rothamsted (25–30% clay), Woburn (10% clay) and Saxmundham (25% clay) were 18, 7 and 47 kg/ha/year, respectively (Johnston, 1986). The release of K_{latt} is not constant but varies with growing conditions, which include N supply and crop type. Increased N supply encourages growth and the uptake of K from K_{latt} (Johnston, 1986). In all these experiments, plots relying on weathering from K_{latt} had smaller crop yields than plots with adequate K.

Potassium in feldspar constitutes a very large K reserve in soils but it is generally assumed that the annual K release is small (Rasmussen, 1972). However, the abundance of K in feldspar and considerable variations in weathering rate suggest that, in some cases, they can supply significant amounts of K (Rich, 1972). Sparks (1987) considered that K release from feldspar could explain anomalous responses by maize to K on Atlantic Coastal Plain soils. Wulff *et al.* (1998) and Askegaard *et al.* (2003) suggested that feldspar K might explain the lack of crop response to K on some coarse sandy soils.

There seems to be a potential for including indigenous soil K in the management of K. However, in order not to jeopardize soil productivity, the degree of 'mining' should be carefully assessed with respect to soil type and mineralogy, cropping system and time-scale.

Managing Low-potassium-input Systems

So far there has been little research to develop a successful K management strategy for agricultural systems. Here we attempt, somewhat speculatively, to identify those management tools that could be used. They include soil analyses comprising a quantity and a kinetic parameter, K balance estimates and plant analyses to ascertain whether or not K has limited yield.

Management tools

Soil analysis

Exchangeable K correlates with the K balance on a field basis (Johnston, 1986), and K_{ex} will be a suitable tool in K management for a crop rotation where fluctuations caused by individual crops and management elements are reduced (Askegaard and Eriksen, 2002; Askegaard et al., 2003). Assessing the critical level of K_{ex}, i.e. the minimum level of K_{ex} at which the soil, without addition of fertilizer K, can fully meet crop K demand, is crucial for the management of low-K-input systems. Using of the critical level of K_{ex} as a criterion for K management allows for the release of K_{slex} and K_{latt}, and also prevents excessive additions of K in fertilizers and manures. Experiments show large differences in the critical K_{ex} levels: less than 30 mg K/kg for cereals grown on a coarse sandy soil (Askegaard et al., 2003), less than 45 mg K/kg for a barley–pea mixture on sandy loam (Askegaard and Eriksen, 2002), less than 80 mg K/kg for spring barley and about 200 mg K/kg for field beans, potatoes and sugarbeet all grown on a silty clay loam (Johnston and Goulding, 1990; Johnston, 2001). The critical K_{ex} level varies with soil type, farming system and input of nutrients other than K. Figure 6.3 shows that the potential of N to increase the yield of spring barley was limited at a K_{ex} of 60 mg/kg on a silty

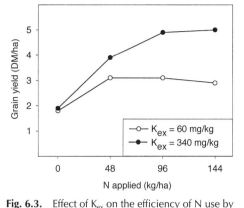

Fig. 6.3. Effect of K_{ex} on the efficiency of N use by spring barley on a silty clay loam at Rothamsted (Johnston, 2001).

clay loam; with excess N there was the risk of nitrate leaching losses.

A precondition for the use of K_{ex} in K management is that soil samples are taken within the same period each year at the same depth and are representative for the entire field. The K content of an aboveground plant biomass, e.g. in a catch crop, should be included when evaluating the K status of sandy soils especially, where the plant K content may be relatively large compared with the soil K_{ex} content. Refining the use of soil analysis may depend on the level of K_{ex}. For example on soils with little K_{ex} it may be necessary to quantify K_{slex}; on soils with much K_{ex} it may be important to quantify K_{sol}. Measurements of the buffering power could be valuable especially in the K management of crops with large K uptake, but it would probably be necessary to determine the buffering capacity for each soil because it is affected by past K fertilization (Wild, 1988).

Measuring long-term K release rates, which reflect soil mineralogy and fertilizer history, would help to group soils into different K release categories. This grouping would also help in the determination of critical K_{ex} values for soil types with different mineralogy. Goulding and Loveland (1986) used sequential extraction of soil with cation exchange resins to determine cumulative K release curves. These curves were very similar to cumulative K uptake curves by ryegrass grown in pots in the glasshouse (Johnston and Goulding, 1990).

Both methods are very time consuming, but confirm the concept of different pools of soil K that relate to the availability of K to plants. The shape of the curve also varies with soil type and past K balances. Other methods have been used to attempt to estimate K release rates. These include soil extraction with sodium tetraphenyl boron (NaTPB) (e.g. Dhillon and Dhillon, 1992) and the electro-ultrafiltration (EUF) technique (e.g. Ziadi *et al.*, 2001).

K balances

Although K_{ex} correlates with the K balance, the change in K_{ex} is usually smaller than the change in the K balance due to K transfer between the K fractions (Johnston, 1986). Positive K balances in field experiments gave less than 50% of the theoretical changes in K_{ex} (Johnston and Goulding, 1990; Askegaard and Eriksen, 2002) (see also Table 6.3). To evaluate soil fertility with respect to K, it is necessary to combine K_{ex} measurements with balance calculations. With K in fertilizer, organic manure and atmospheric deposition as inputs, and K lost by leaching or removed in plant material as outputs, balance estimates provide information on the size and direction of a change in the soil K reserve. However, nutrient budgets can be flawed by biases and errors in the estimates of inputs and especially outputs in leaching on sandy soils (Oenema and Heinen, 1999). In studies on organic farms, Watson *et al.* (2002) showed that nutrient budgets were significantly different when measured values for the K content of the different crops were substituted for estimates based on standard values. Consequently, K budgets to be used for management recommendations should rely on data recorded on individual farms, preferably over several years, in order to account for site and seasonal variability. Most often K leaching losses have to be based on calculated data but they should be adjusted according to the previous K management (Askegaard and Eriksen, 2002; Askegaard *et al.*, 2003). Spatial soil variability within and between fields due to past K applications can be a problem in K use efficiency. However, spatial variability may be assessed from yield maps that provide an estimate of crop K offtake, and from

replacing K offtake using GPS (global positioning system) to know where to vary the rate of application.

Crop analyses

The evaluation of whether K constrains yield cannot rely on visual plant deficiency symptoms because the relationship between K application and deficiency symptoms is poor. The K concentration in plant dry matter is also a poor indicator of plant available soil K because the concentration changes throughout growth (Leigh and Johnston, 1983a,b). However, these authors showed that for cereals, plant K concentrations on a tissue water basis were independent of growth stage and N supply, and clearly differentiated between crop growth with and without an adequate K supply. This has been confirmed for sugarbeet (G.M. Milford, Rothamsted, 2002, personal communication). Further work might provide the opportunity to develop a practical plant-based test for use in the field developed on the basis of these observations (Barraclough and Leigh, 1993).

Application of different potassium sources

Different K sources are available for the farmer and, whether it is mineral fertilizers (normally KCl or K_2SO_4 salts) or organic sources as manure or crop residues, they have comparable effects, because all contain K as the K^+ ion (Hinsinger, 2002). Occasional small differences in the K effect of these K sources have been measured and are typically ascribed to effects of the accompanying anions or organic material. Powdered rocks, e.g. feldspar and biotite, release K slowly but experiments have shown positive crop response to application (Hinsinger *et al.*, 1996; Bakken *et al.*, 1997). The use of rock powders will probably be limited to locations near their production.

Different approaches for different soil types

Differences in soil type call for different approaches in K management because the K dynamics in soils are highly influenced by soil type. For example, on coarse sandy soils, a key issue is the reduction of leaching losses. Such soils do retain some K but only at low levels of K_{ex} (Wulff *et al.*, 1998). Thus K_{ex} should be kept as low as possible to minimize leaching losses. In organic crop rotations on a coarse sandy soil, Askegaard *et al.* (2003) found a positive correlation between K_{ex} and leaching loss. They suggested that K_{ex} after harvest should not exceed 30 mg/kg on this soil type. At this K_{ex} level, the crop probably utilizes some of the K_{slex}, and during autumn and winter part of the K leached from the crop residues may be retained as K_{slex} instead of being lost by leaching. The application of fertilizer or manure K should always be carefully adjusted with respect to amount and time of application on sandy soils.

On clayey soils the reserves of K_{latt} may, to some extent, be included in the K supply of crops by lowering K_{ex}. However, utilization of these reserves requires reference to soil K status, texture and mineralogy, crop species and N level. There are major differences in the ability of crops to maintain yields at a low soil K status. The ranking of the different crops or varieties depends on the soil and growing conditions. Crop rotations should be designed so that crops highly sensitive to K deficiency do not follow crops that are able to rely on the release of indigenous soil K, because the soils may need time to re-equilibrate. Potassium can be applied to clayey soils once in a rotation without the risk of leaching losses. However, if large quantities of K are applied to crops where the residues are removed (e.g. cereal straw), there may be a larger than intended export of K.

including K, are used, and to maintain sustainable plant production with decreased external nutrient inputs. Irrespective of the farming system, a low K input may, however, compromise soil fertility and crop production, and this makes it important to explore the limits of K availability.

K management under low K inputs requires exploitation of soil K reserves, reduced K leaching losses on sandy soils and, if possible, increased return of K from urban areas. The potential for exploiting indigenous soil K reserves depends on soil type, mineralogy and crop type, but will always require the maintenance of a critical level of exchangeable K appropriate to the crop to be grown. To develop appropriate management strategies will require: an assessment of the soil's ability to release K and changes in exchangeable K over time; improved K balances based on field measurements of K uptake and leaching losses; an improved knowledge of the ability of different crops to exploit soil K; measurements of the effects of soil structure, soil N status and K supply from the subsoil on the crop's requirement for K; and development of a practical plant-based test to evaluate whether K constrains yield.

To test such strategies will require long-term experiments on selected soil types with appropriate cropping systems. The experiments must generate data required to produce models for K flows within the system. These well-controlled field experiments have to be supplemented by carefully monitored, on-farm studies to validate the models with data from a wider range of farming systems on soil types with different profile characteristics, mineralogy and fertilizer history.

Conclusions

Adverse environmental effects of K in agriculture have not been reported, and globally the known reserves of K are likely to last for hundreds of years. Although there is no fundamental scientific evidence against using K fertilizer, one persistent challenge facing research in modern agriculture is to increase the efficiency with which all plant nutrients,

References

Anonymous (1969–1977) Annual reports of plant production [in Danish]. Farmers and Smallholders Associations, Denmark.

Anonymous (1984) *Quality Criteria for Selected Substances in Drinking Water* [in Danish]. Danish Environmental Protection Agency, Danish Ministry of the Environment, Copenhagen, 43 pp.

Anonymous (1998) *Guide to Efficient Plant Nutrition Management*. Land and Water Development Division. Food and Agriculture Organization of the United Nations, Rome, 19 pp.

Anonymous (2001) *Current World Fertilizer Trend and Outlook to 2005/2006*. Food and Agriculture Organization of the United Nations, Rome, 14 pp.

Anonymous (2002) *World Development Report 2003*. The World Bank, Washington, DC, 272 pp.

Arnold, P.W. (1960) Nature and mode of weathering of soil potassium reserves. *Journal of the Science of Food and Agriculture* 11, 285–292.

Arnold, P.W. (1962a) *The Potassium Status of some English Soils Considered as a Problem of Energy Relationships*. Proceedings No. 72, The International Fertiliser Society, York, UK, pp. 25–43.

Arnold, P.W. (1962b) Soil potassium and its availability to plants. *Outlook on Agriculture* 263–267.

Askegaard, M. and Eriksen, J. (2000) Potassium retention and leaching in an organic crop rotation on loamy sand as affected by contrasting potassium budgets. *Soil Use and Management* 16, 200–205.

Askegaard, M. and Eriksen, J. (2002) Exchangeable potassium in soil as indicator of potassium status in an organic crop rotation on loamy sand. *Soil Use and Management* 18, 84–90.

Askegaard, M., Eriksen, J. and Olesen, J.E. (2003) Exchangeable potassium and potassium balances in organic crop rotations on a coarse sandy soil. *Soil Use and Management* 19, 96–103.

Bakken, A.K., Gautneb, H. and Myhr, K. (1997) The potential of crushed rocks and mine tailings as slow-releasing K fertilizers assessed by intensive cropping with Italian ryegrass in different soil types. *Nutrient Cycling in Agroecosystems* 47, 41–48.

Barber, S.A. (1985) Potassium availability at the soil–root interface and factors influencing potassium uptake. In: Munson, R.D. (ed.) *Potassium in Agriculture*. American Society of Agronomy, Madison, Wisconsin, pp. 309–324.

Barber, S.A. (1994) *Soil Nutrient Bioavailability*. John Wiley & Sons, New York, 414 pp.

Barraclough, P.B. and Leigh, R.A. (1993) Critical plant K concentrations for growth and problems in the diagnosis of nutrient deficiencies by plant analysis. *Plant and Soil* 155/156, 219–222.

Bennie, A.T.P. (1996) Growth and mechanical impedance. In: Waisel, Y., Eshel, A. and Kafkafi, U. (eds) *Plant Roots. The Hidden Half*, 2nd edn. Marcel Dekker, New York, pp. 453–470.

Beringer, H. (1992) *Environmental Aspects of Potash Needs, Use and Production*. Proceedings No. 328.

The International Fertiliser Society, York, UK, 24 pp.

Blevins, D.G. (1985) Role of potassium in protein metabolism in plants. In: Munson, R.D. (ed.) *Potassium in Agriculture*. American Society of Agronomy, Madison, Wisconsin, pp. 413–424.

Cakmak, I. (2003) Potassium nutrition and generation of reactive oxygen species in crop plants under abiotic stress. In: Johnston, A.E. (ed.) *Feed the Soil to Feed the People*. International Potash Institute, Basel, Switzerland.

Christensen, B.T., Petersen, J., Kjellerup, V. and Trentemøller, U. (1994) *The Askov Long-term Experiments on Animal Manure and Mineral Fertilizers: 1894–1994*. SP Report 43, Danish Institute of Plant and Soil Science, Tjele, Denmark, 85 pp.

Claassen, N. and Jungk, A. (1982) Kaliumdynamik im wurzelnahen boden in Beziehung zur Kaliumaufnahme von Maispflanzen. *Zeitschrift für Pflanzenernährung und Bodenkunde* 145, 513–525.

Cooke, G.W. (1981) The fate of fertilizers. In: Greenland, D.J. and Hayes, M.H.B. (eds) *The Chemistry of Soil Processes*. John Wiley & Sons, London, pp. 563–592.

Cooke, G.W. (1982) *Fertilizing for Maximum Yield*, 3rd edn. Granada Publishing, London, 465 pp.

Dhillon, S.K. and Dhillon, K.S. (1992) Kinetics of release of potassium by sodium tetraphenyl boron from some topsoil samples of red (Alfisols), black (Vertisols) and alluvial (Inceptisols and Entisols) soils of India. *Fertilizer News* 32, 135–138.

Eilersen, A.M., Tjell, J.C. and Henze, M. (1998) The potential of household waste as a resource in agriculture. In: Magid, J. (ed.) *Recirculation from City to Country?* Internal Report [in Danish], Royal Veterinary and Agricultural University, Denmark, pp. 11–40.

Fardeau, J.C., Poss, R. and Saragoni, H. (1992) Effect of potassium fertilization on K-cycling in different agrosystems. In: Anonymous (ed.) *Potassium in Ecosystems: Biochemical Fluxes of Cations in Agro-Forest-systems*. Proceedings of the 23th Colloquium of the International Potash Institute. International Potash Institute, Basel, Switzerland, pp. 59–78.

Gäth, S., Meuser, H., Abitz, C.A., Wessolek, G. and Renger, M. (1989) Determination of potassium delivery to the roots of cereal plants. *Zeitschrift für Pflanzenernährung und Bodenkunde* 152, 143–149.

Goulding, K.T.W. and Loveland, P.J. (1986) The classification and mapping of potassium reserves in soils of England and Wales. *Journal of Soil Science* 37, 555–565.

Grimme, H. (1976) Soil factors of potassium availability. *Bulletin of the Indian Society of Soil Science* 10, 144–163.

Grimme, H. and Németh, K. (1979) The evaluation of soil K status by means of soil testing. In: Anonymous (ed.) *Potassium Research – Review and Trends. Proceedings of the 11th Congress of the International Potash Institute.* International Potash Institute, Basel, Switzerland, pp. 99–108.

Grundahl, L. and Hansen, J.G. (1990) *Atmospheric Deposition of Nutrient Salts in Denmark.* [in Danish]. NPo-research A6, Department of the Environment, Copenhagen, 60 pp.

Haak, E. (1981) Nutrient uptake from subsoil by cereals. In: Anonymous (ed.) *The Agricultural Yield Potential in Continental Climates. Proceedings of the 16th Colloquium of the International Potash Institute.* International Potash Institute, Basel, Switzerland, pp. 87–94.

Härdter, R. (2003) Potassium and biotic stress in plants. In: Johnston, A.E. (ed.) *Feed the Soil to Feed the People.* International Potash Institute, Basel, Switzerland.

Hinsinger, P. (2002) Potassium. In: Lal, A. (ed.) *Encyclopedia of Soil Science.* Marcel Dekker, New York, pp. 1035–1039.

Hinsinger, P. and Jaillard, B. (1993) Root-induced release of interlayer potassium and vermiculitization of phlogopite as related to potassium depletion in the rhizosphere of ryegrass. *Journal of Soil Science* 44, 525–534.

Hinsinger, P., Elsass, F., Jaillard, B. and Robert, M. (1993) Root-induced irreversible transformation of a trioctahedral mica in the rhizosphere of rape. *Journal of Soil Science* 44, 535–545.

Hinsinger, P., Bolland, M.D.A. and Gilkes, R.J. (1996) Silicate rock powder: effect on selected chemical properties of a range of soils from Western Australia and on plant growth as assessed in a glasshouse experiment. *Fertilizer Research* 45, 69–79.

Holbeck, H.B. and Hvid, S.K. (2002) *Nutrient Surplus at Dairy, Pig and Arable Farms in a Demonstration Project with 'Green' Budgets 1999 and 2000* [in Danish]. Plant Production Information, Danish Agricultural Advisory Centre, 11–011, 10 pp.

Holmqvist, J., Øgaard, A.F., Öborn, I., Edwards, A.C., Mattson, L. and Sverdrup, H. (2003) Application of the PROFILE model to estimate potassium release from mineral weathering in Northern European agricultural soils. *European Journal of Agronomy* (in press).

Holtze, A. and Backlund, A. (2002) *Collection, Keeping and Application of Urine From the*

Møns Museumsgård. Danish Environmental Protection Agency, Danish Ministry of the Environment 23, 97 pp.

Huber, D.M. and Arny, D.C. (1985) Interaction of potassium with plant disease. In: Munson, R.D. (ed.) *Potassium in Agriculture.* American Society of Agronomy, Madison, Wisconsin, pp. 467–488.

Jackson, M.L. and Sherman, G.D. (1953) Chemical weathering of minerals in soils. *Advances in Agronomy* 5, 219–318.

Janssen, B.H. (1999) Basics of budgets, buffers and balances of nutrients in relation to sustainability of agroecosystems. In: Smaling, E.M.A., Oenema, O. and Fresco, L.O. (eds) *Nutrient Disequilibria in Agroecosystems.* CAB International, Wallingford, UK, pp. 27–56.

Johnston, A.E. (1986) Potassium fertilization to maintain a K-balance under various farming systems. In: Anonymous (ed.) *Nutrient Balances and the Need for Potassium. Proceedings of the 13th Congress of the International Potash Institute.* International Potash Institute, Basel, Switzerland, pp. 199–226.

Johnston, A.E. (2001) *Principles of Crop Production for Sustainable Food Production.* Proceedings No. 459, International Fertiliser Society, York, UK, 40 pp.

Johnston, A.E. and Goulding, K.W.T. (1990) The use of plant and soil analyses to predict the potassium supplying capacity of soil. In: Anonymous (ed.) *Development of K-fertilizer Recommendations. Proceedings of the 22nd Colloquium of the International Potash Institute.* International Potash Institute, Basel, Switzerland, pp. 77–204.

Johnston, A.E. and Goulding, K.W.T. (1992) Potassium concentrations in surface and ground waters and the loss of potassium in relation to land use. In: Anonymous (ed.) *Potassium in Ecosystems: Biochemical Fluxes of Cations in Agro–Forest-systems. Proceedings of the 23rd Colloquium of the International Potash Institute.* International Potash Institute, Basel, Switzerland, pp. 35–158.

Johnston, A.E. and Mitchell, J.D.D. (1974) The behaviour of K remaining in soils from the Agdell experiment at Rothamsted, the results of intensive cropping in pot experiments and their relations to soil analysis and the results of field experiments. *Rothamsted Experimental Station, Report for 1973,* Part 2, 74–97.

Johnston, A.E. and Poulton, P.R. (1977) Yields on the exhaustion land and changes in the NPK contents of soils due to cropping and manuring 1852–1975. *Rothamsted Experimental Station Report for 1976,* part 2, 53–85.

Johnston, A.E., Goulding, K.W.T., Poulton, P.R. and Chalmers, A.G. (2001a) *Reducing Fertilizer Inputs: Endangering Arable Soil Fertility.* Proceeding No. 488, The International Fertiliser Society, York, UK, 44 pp.

Johnston, A.E., Poulton, P.R., Dawson, C.J. and Crawley, M.J. (2001b) *Inputs of Nutrients and Lime for the Maintenance of Fertility of Grassland Soils.* Proceedings No. 486. The International Fertiliser Society, York, UK, 40 pp.

Jönsson, H. (2002) Urine separation – Swedish experiences. In: Magid, J., Lieblein, G., Granstedt, A., Kahiluoto, H. and Dýrmundsson, O. (eds) *Urban Areas – Rural Areas and Recycling – The Organic Way Forward?* NJF seminar 327, DARCOF Report no. 3, pp. 117–124.

Jungk, A. and Claassen, N. (1989) Availability in soil and acquisition by plants as the basis for phosphorus and potassium supply to plants. *Zeitschrift für Pflanzenernährung und Bodenkunde* 152, 151–157.

Krauss, A. (2001) *Regional Nutrient Balances in View of Food Security for Future Generations.* International Potash Institute, Basel, Switzerland, 15 pp.

Kristensen, I.S. and Halberg, N. (1995) Field net-yields, nutrient-supply and crop characteristics on organic and conventional mixed dairy farms. In: Kristensen, E.S. (ed.) *Organic Agriculture with Focus on the Dairy Farm* [in Danish]. Internal report, National Institute of Animal Science, 42, 33–51.

Kuchenbuch, R., Classen, N. and Jungk, A. (1986) Potassium availability in relation to soil moisture. *Plant and Soil* 95, 221–231.

Kuhlmann, H. (1990) Importance of the subsoil for the K nutrition of crops. *Plant and Soil* 127, 129–136.

Leigh, R.A. and Johnston, A.E. (1983a) Concentrations of potassium in the dry matter and tissue water of field-grown spring barley and their relationships to grain yield. *Journal of Agricultural Science* 101, 675–685.

Leigh, R.A. and Johnston, A.E. (1983b) The effects of fertilizers and drought on the concentrations of potassium in the dry matter and tissue water of field-grown spring barley. *Journal of Agricultural Science* 101, 741–748.

Loué, A. (1978) The interaction of potassium with other growth factors, particularly with other nutrients. In: Anonymous (ed.) *Potassium Research – Review and Trends. Proceedings of the 11th Congress of the International Potash Institute.* International Potash Institute, Basel, Switzerland, pp. 407–433.

Marschner, H. (1995) *Mineral Nutrition of Higher Plants,* 2nd edn. Academic Press, London, 889 pp.

Mengel, K. (1978) A consideration of factors which affect the potassium requirements of various crops. In: Anonymous (ed.) *Potassium Research – Review and Trends. Proceedings of the 11th Congress of the International Potash Institute.* International Potash Institute, Basel, Switzerland, pp. 225–237.

Mengel, K. (1982) Factors and processes affecting potassium requirements of crops. *Potash Review* No. 9, International Potash Institute, Basel, Switzerland, 12 pp.

Mengel, K. and Busch, R. (1982) The importance of the potassium buffer power on the critical potassium level in soils. *Soil Science* 133, 27–32.

Mengel, K. and Kirkby, E.A. (1987) *Principles of Plant Nutrition,* 4th edn. International Potash Institute, Basel, Switzerland, 687 pp.

Mengel, K. and Steffens, D. (1985) Potassium uptake of ryegrass (*Lolium perenne*) and red clover (*Trifolium pratense*) as related to root parameters. *Biology and Fertility of Soils* 1, 53–58.

Mengel, K., Kirkby, E.A., Kosegarten, H. and Appel, T. (2001) *Principles of Plant Nutrition,* 5th Edn. Kluwer Academic Publishers, Dordrecht, 849 pp.

Munson, R.D. and Nelson, W.L. (1963) Movement of applied potassium in soils. *Agricultural and Food Chemistry* 11, 193–201.

Nair, K.P.P. (1996) Buffering power of plant nutrients. *Advances in Agronomy* 57, 237–287.

Nelson, W.L. (1968) Plant factors affecting potassium availability and uptake. In: Kilmer, V.J., Younts, S.E. and Brady, N.C. (eds) *The Role of Potassium in Agriculture.* ASA-CSSA-SSSA, Madison, Wisconsin, pp. 355–383.

Nielsen, J.D. and Møberg, J.P. (1984) The influence of K-depletion on mineralogical changes in pedons from two field experiments and in soils from four pot experiments. *Acta Agriculturae Scandinavica* 34, 391–399.

Nye, P.H. and Tinker, P.B. (1977) *Solute Movement in the Soil–Root System.* Alden Press, Oxford, 342 pp.

Oenema, O. and Heinen, M. (1999) Uncertainties in nutrient budgets due to biases and errors. In: Smaling, E.M.A., Oenema, O. and Fresco, L.O. (eds) *Nutrient Disequilibria in Agroecosystems.* CAB International, Wallingford, UK, pp. 75–97.

Ogaard, A.F., Krogstad, T. and Lunnan, T. (2002) Ability of some Norwegian soils to supply grass with potassium: soil analyses as predictors of K supply from soil. *Soil Use and Management* 18, 412–420.

PDA (1997) *Phosphate and Potash Removal by Crops*. The Potash Development Association, Langharne, UK, 4 pp.

Rao, C.S. and Khera, M.S. (1994) Potassium replenishment capacity of illitic soils at their minimal exchangeable K in relation to clay mineralogy. *Zeitschrift für Pflanzenernährung und Bodenkunde* 157, 467–470.

Rasmussen, K. (1972) Potash in feldspars. In: Anonymous (ed.) *Potassium in Soil. Proceedings of the 9th Colloquium of the International Potash Institute*. International Potash Institute, Basel, Switzerland, pp. 57–60.

Reitemeier, R.F. (1951) Soil potassium. *Advances in Agronomy* 3, 113–164.

Rejado, P.Q. (1978) Potassium requirements of cereals. In: Anonymous (ed.) *Potassium Research – Review and Trends. Proceedings of the 11th Congress of the International Potash Institute*. International Potash Institute, Basel, Switzerland, pp. 239–257.

Rich, C.I. (1972) Potassium in soil minerals. In: Anonymous (ed.) *Potassium in Soil. Proceedings of the 9th Colloquium of the International Potash Institute*. International Potash Institute, Basel, Switzerland, pp. 15–31.

Scharf, H.J. (1990) Environmental aspects of K-fertilizers in production, handling and application. In: Anonymous (ed.) *Development of K-fertilizer Recommendations. Proceedings of the 22nd Colloquium of the International Potash Institute*. International Potash Institute, Basel, Switzerland, pp. 395–402.

Schimansky, C. (1981) Die Aufnahme von ^{28}Mg, ^{86}Rb und ^{45}Ca durch Gerstenpflanzen bei unterschiedlichen Wurzeltemperaturen. *Zeitschrift für Pflanzenernährung und Bodenkunde* 144, 356–365.

Schneider, A. (1997) Release and fixation of potassium by a loamy soil as affected by initial water content and potassium status of soil samples. *European Journal of Soil Science* 48, 263–271.

Schroeder, D. (1974) Relationships between soil potassium and the potassium nutrition of the plant. In: Anonymous (ed.) *Potassium Research and Agricultural Production. Proceedings of the 10th Congress of the International Potash Institute*. International Potash Institute, Basel, Switzerland, pp. 53–63.

Schroeder, D. (1978) Structure and weathering of potassium containing minerals. In: Anonymous (ed.) *Potassium Research – Review and Trends. Proceedings of the 11th Congress of the*

International Potash Institute. International Potash Institute, Basel, Switzerland, pp. 43–63.

Sharpley, A.N. (1990) Reaction of fertilizer potassium in soils of differing mineralogy. *Soil Science* 149, 44–51.

Sheldrick, W.F. (1985) World potassium reserves. In: Munson, R.D. (ed.) *Potassium in Agriculture*. American Society of Agronomy, Madison, Wisconsin, pp. 3–28.

Sibbesen, E. (1986) Soil movement in long-term field experiments. *Plant and Soil* 91, 73–85.

Siddiqi, M.Y. and Glass, A.D.M. (1983) Studies of the growth and mineral nutrition of barley varieties. I. Effect of potassium supply on the uptake of potassium and growth. *Canadian Journal of Botany* 61, 671–678.

Simmelsgaard, S.E. (1996) *Plant Nutrients in Subsurface Drainage Water and Soil Water* [in Danish with English abstract]. The Danish Institute of Agricultural Sciences, Report no. 7, 77 pp.

Sparks, D.L. (1987) Potassium dynamics in soils. *Advances in Soil Science* 6, 1–63.

Sparks, D.L. and Huang, P.M. (1985) Physical chemistry of soil potassium. In: Munson, R.D. (ed.) *Potassium in Agriculture*. American Society of Agronomy, Madison, Wisconsin, pp. 202–276.

Sparks, D.L. and Liebhardt, W.C. (1982) Temperature effects on potassium exchange and selectivity in Delaware soils. *Soil Science* 133, 10–17.

Stewart, J.A. (1985) Potassium sources, use and potential. In: Munson, R.D. (ed.) *Potassium in Agriculture*. American Society of Agronomy, Madison, Wisconsin, pp. 83–98.

Syers, J.K. (2003) Potassium in soil – current concepts. In: Johnston, A.E. (ed.) *Feed the Soil to Feed the People*. International Potash Institute, Basel, Switzerland.

Tributh, H., Boguslawski, E., Lieres, A., Steffens, D. and Mengel, K. (1987) Effect of potassium removal by crops on transformation of illitic clay minerals. *Soil Science* 143, 404–409.

Warren, R.G. and Johnston, A.E. (1962) *The Accumulation and Loss of Soil Potassium in Long-term Experiments at Rothamsted and Woburn*. Proceedings No. 72. The International Fertiliser Society, York, UK, pp. 5–24.

Watson, C.A., Bengtsson, H., Ebbesvik, M., Løes, A.-K., Myrbeck, A., Salomon, E., Schroeder, J. and Stockdale, E.A. (2002) A review of farm-scale nutrient budgets for organic farms as a tool for management of soil fertility. *Soil Use and Management* 18, 264–273.

WCED (1987) *Our Common Future: the Brundtland Report*. Report from the World Commission on Environment and Development (WCED). Oxford University Press, Oxford.

Wild, A. (1988) Potassium, sodium, calcium, magnesium, sulphur, silicon. In: Wild, A. (ed.) *Russell's Soil Conditions and Plant Growth*, 11th edn. Longman Scientific and Technical, Harlow, UK, pp. 743–779.

Williams, P.H., Gregg, P.E.H. and Hedley, M.J. (1990) Fate of potassium in dairy cow urine applied to intact soil cores. *New Zealand Journal of Agricultural Research* 33, 151–158.

Wulff, F., Schulz, V., Jungk, A. and Claassen, N. (1998) Potassium fertilization on sandy soils in relation to soil test, crop yield and K-leaching. *Zeitschrift für Pflanzenernährung und Bodenkunde* 161, 591–599.

Zeng, Q. and Brown, P.H. (2000) Soil potassium mobility and uptake by corn under differential soil moisture regimes. *Plant and Soil* 221, 121–134.

Ziadi, N., Simard, R.R. and Tran, T.S. (2001) Models for potassium release kinetics of four humic gleysols high in clay by electroultrafiltration. *Canadian Journal of Soil Science* 81, 603–611.

Chapter 7
Developing and Maintaining Soil Organic Matter Levels

W.A. Dick[1] and E.G. Gregorich[2]

[1]The Ohio State University, School of Natural Resources, 1680 Madison Avenue, Wooster, Ohio 44691, USA; [2]Agriculture and Agri-Food Canada, Central Experimental Farm, Ottawa K1A 0C6, Canada

Summary

Soil organic matter (OM) is a key attribute of soil quality. It affects, directly or indirectly, many physical, chemical and biological properties that control soil productivity and resistance to degradation. Changes in the quantity of soil OM and the equilibrium level of soil OM depend on the interaction of five factors: climate, landscape, texture, inputs and disturbance. Some of these factors, called soil OM capacity factors, can be managed, whereas others cannot. As OM enters and resides in soil, it is subjected to processes that alter its composition and quantity. Fundamental soil processes such as humification, aggregation, translocation, erosion, leaching and mineralization are driven by the capacity factors. These capacity factors and soil processes, in turn, largely dictate the management system imposed on a soil. Understanding how soil OM capacity factors and fundamental soil processes interact with management over time allows the comparison of management systems that affect soil OM levels. We propose a ranking of general benchmark management systems for the purpose of identifying how a particular system might influence OM levels relative to a new or different system.

©CAB International 2004. *Managing Soil Quality: Challenges in Modern Agriculture*
(eds P. Schjønning, S. Elmholt and B.T. Christensen)

Soil Organic Matter – a Key Attribute of Soil Quality

Organic matter (OM) is a key component in the creation and maintenance of a high quality soil. Many soil properties (e.g. microbial activity, cation exchange capacity and soil aggregation) are directly affected by the presence of OM. Indirectly, soil OM influences the quality of the air and water that interact with the soil. As water infiltrates the soil, the biological activity supported by soil OM transforms substances contained in the water. The production and emission from soils of greenhouse gases are largely controlled by how soil OM is managed (Rochette *et al.*, 2002; Lal, Chapter 17, this volume).

Soil quality has been defined as the capacity of the soil to function effectively, both in the present and in the future (Doran and Parkin, 1994). Effective functioning of soil varies depending on the required end use of the soil. For example, soils with increased levels of OM require higher application rates of pesticides for effective pest control, and pesticides may become more resistant to degradation (Loux *et al.*, 2002). However, it is generally true that it is better to have more, not less, OM in the soil.

The OM content in soil can range from less than 1% in coarse textured or highly oxidized soils to nearly 100% in wetland bogs. Farm soils have concentrations of soil OM that generally range between 1 and 10%. The chemical composition of soil OM can vary depending on when and where soil is sampled but typically contains about 50% carbon (C), 40% oxygen (O), 5% hydrogen (H), 4% nitrogen (N) and 1% sulphur (S). Most of the other mineral nutrients are also associated in some way with OM – some as exchangeable cations and anions, as cation bridges or simply adsorbed on the surface of organo-mineral particles. A comprehensive review of solid state ^{13}C-NMR (nuclear magnetic resonance) studies indicated that there is remarkable similarity in the distribution of carbon forms in soils under a wide range of climate, land use, cropping practice and fertilizer amendments (Mahieu *et al.*, 1999). O-alkyls were the most abundant functional groups (45%), followed by alkyls (25%), aromatics (20%) and carbonyls (10%).

Table 7.1 lists some of the benefits that OM contributes to the overall quality of a soil. Soil OM is a repository of nutrients (particularly N, phosphorus (P), S and micro-nutrients) and, during its turnover, contributes to fertility at times and in locations in the soil profile that are difficult to achieve with inorganic nutrient amendments. It increases the cation exchange capacity, particularly in coarse-textured soils, and increases the soil's available water-holding capacity (Hudson, 1994). Soil OM enhances aggregation, which in turn improves soil tilth, the infiltration of water and the exchange of gases between soil and atmosphere.

Soil OM can have a positive effect on the yield of crops. Bauer and Black (1994) estimated that each tonne of OM in the surface 30-cm layer of a hectare of soil increased wheat (*Triticum aestivum*) grain yields by 16 kg/ha. Similar relationships between crop yield and soil OM content probably exist for other crops, but it is difficult to quantify these effects (Johnston, 1986). The difficulty arises, in part, because in temperate climates it takes many years to establish soils with different levels of OM and then show that extra soil OM benefits yields. However, there are indications that maximum crop yields can only be realized on soils high in soil OM (Johnston, 1986, 1987) and that the effect of soil OM cannot be fully replaced by increased levels of N fertilizer (Christensen and Johnston, 1997).

Increased levels of soil OM can affect the economics of farming practices in ways other than directly influencing production. For example, farmers often report a noticeable decrease in power requirements for tillage machinery after several years of manure application or conservation tillage practices. This is usually attributed to the build-up of soil OM, and some research has corroborated this. McLaughlin *et al.* (2002) measured mouldboard draught and tractor fuel consumption during autumn tillage operations over several years on unamended plots or plots that received different levels of inorganic fertilizers (100 or 200 kg N/ha/year) and either fresh stockpiled or rotted manure (each at 50 and 100 Mg wet weight/ha/year).

Table 7.1. Soil properties affected by organic matter (OM) (adapted from Pierzynski *et al.*, 2000; Baldock and Nelson, 2000).

Property	Remarks	Function in soil
Biological properties		
Mineralization–immobilization	Decomposition of OM yields NH_4^+, NO_3^-, PO_4^{3-}, SO_4^{2-}, micronutrients and CO_2	Provides nutrients for plant growth and is essential for the global cycling of elements such as C and N
Stimulation and inhibition of enzyme activities and of plant and microbial growth	Soil enzyme activity and the growth of plants and microorganisms can be stimulated or inhibited by the presence of humic materials	Controls the size, growth and activity of plant and microbial communities in soil
Biological diversity	OM supports life processes for a wide range of species of microbes and fauna	Contributes to the functional integrity/resilience of ecosystems
Reservoir of metabolic energy for microorganisms and store of atmospheric C	Provides metabolic energy for soil microorganisms and fauna	Provides energy to drive biological processes (e.g. ammonification, nitrification, denitrification, mineralization, immobilization)
Chemical properties		
Cation exchange	Total exchange capacities of isolated OM fractions range from 300 to 1400 cmol/kg	Enhances retention of cations
Chelation of metals	Forms stable complexes with Cu^{2+}, Mn^{2+}, Zn^{2+} and other polyvalent cations	Enhances dissolution of soil minerals; reduces loss of micronutrients; enhances the availability of micro- and macronutrients to higher plants; reduces the potential toxicity of metals
Low solubility in water	Insolubility is due to association of OM with clay and the hydrophobic nature of its constituents; also, salts of divalent and trivalent cations with OM are insoluble	Reduces the loss of OM by leaching
Buffer action	Acts as a buffer in slightly acid and alkaline soils	Helps to maintain many chemical and biological properties in acceptable range
Combination with xenobiotics	Affects bioactivity, persistence and biodegradability of pesticides in soil	Modifies application rates of pesticides for effective control
Physical properties		
Combination with clay minerals	Cements soil particles into aggregates	Stabilizes structure, thereby reducing erosion and improving tilth; increases permeability of soil to gases and water
Water retention	OM can hold up to 20 times its weight in water and indirectly contributes to water retention through its effects on soil structure	Improves moisture-retaining properties of coarse-textured soils; helps prevent drying and shrinking
Colour	The typical dark colour of many soils is caused by OM	Facilitates warming

After 8 years of application, plots receiving manure amendments at the high rates exhibited 27–37% lower draught and 13–18% lower tractor fuel consumption than those receiving inorganic fertilizer (Fig. 7.1). These results imply potential savings in tillage costs due to lower capital costs (in terms of smaller tractor requirements) and lower fuel costs.

Organic Matter Capacity Factors

Jenny (1941) introduced the classical equation that related the formation of a soil to several factors. Similarly, one can develop a relationship between various factors and the capacity of a soil to hold (contain) soil OM.

$$OM_{soil\ capacity} = f\ (climate, landscape,\\ texture, inputs,\\ disturbance) \qquad (7.1)$$

This relationship introduces five factors that affect the OM carrying capacity of the soil. Given sufficient time and constant values for each of the five variables, an equilibrium OM capacity for a soil is established.

The five factors divide into two groups. The first three factors (i.e. climate, landscape and texture) establish boundary conditions that are associated with a particular location and parent material. These are defined primarily by nature and cannot be managed. The last two factors are associated with management and can be affected by choices made during crop production. For example, we may not be able to manage climate, but we can manage the level of soil disturbance by our choice of tillage. The contribution of each of the factors to the OM capacity of a soil is discussed below.

Climate

Climate affects the level of OM in soil through temperature, moisture and solar radiation, which, in turn, influence the type and growth of plants and the rate of decomposition of soil OM. Chronosequence studies have shown that rates of soil OM accumulation for different ecosystems are primarily controlled by climate (Schlesinger, 1990; Scharpenseel and Pfeiffer, 1998).

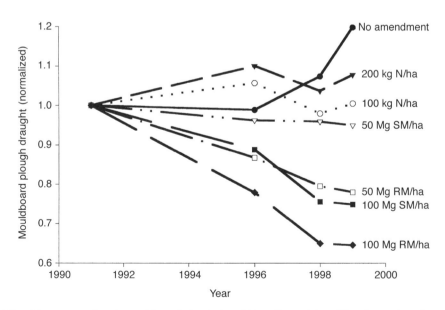

Fig. 7.1.   Changes over time in mouldboard plough draught in soils receiving different organic and inorganic fertilizers. Soil type = Orthic Humic Gleysol (Canadian System of Soil Classification); texture = 39% sand, 27% silt and 34% clay; N = inorganic NH_4NO_3 fertilizer; SM = stockpiled fresh manure; and RM = rotted manure. Both manures were applied on a wet-weight basis. Ploughing depth was approximately 15–17 cm (from McLaughlin *et al.*, 2002).

OM content in soil is strongly affected by the rate at which organic inputs (e.g. plant residues) decompose. For example, the average decomposition rate in soil located in areas with cold winters and dry summers (Canada) is 10 times slower than in soils in tropical areas (Nigeria) and is about 20 times slower than that in the laboratory (Table 7.2).

Interest has grown in determining the interaction among climate, management practices, change in net primary productivity resulting from higher temperatures, and the effects that higher temperatures might have on soil OM dynamics and storage. Global warming probably reduces soil OM by stimulating decomposition rates (Kirschbaum, 2000). However, the increasing atmospheric carbon dioxide level is likely to simultaneously have the effect of increasing soil OM through increases in net primary production. Thus, the net effect of atmospheric carbon dioxide loading on changes in soil organic carbon over the next decades is assumed to be slow, and the scale of change modest. Using an agroecosystem model, Paustian et al. (1998) determined that, for well-characterized agricultural systems in the central USA and southern Canada, soil OM may be more affected by changes in management than by projected climate changes. Rounsevell et al. (1999) also concluded that land-use change, arising from climate change, will probably be the main factor to affect soil OM content in the future.

Landscape position

Microclimatic differences and the movement of water drive the soil and biological processes that affect the OM content within a soil profile. Landscape position exaggerates these differences in hydrology and microclimate, resulting in soils with widely different OM contents within a few metres. For example, soils in depressional areas may remain waterlogged and accumulate OM. If we consider hillslope position as a key feature (Manies et al., 2001), we find that the amount of OM is generally relatively low near the summit or upper slope positions, and relatively high near the toeslope or lower slope positions. These differences in OM across the hillslope are related to several factors. Plant production (and the C inputs from residues) may be larger in toeslope areas because of higher soil water and nutrient contents. Also, soil OM may decompose more slowly in toeslope areas due to a reduction in aeration by accumulated clay and higher water levels. Erosion processes may also move OM downslope to lower slope positions.

Texture

Texture affects the capacity of soil to retain OM by influencing chemical adsorption, physical protection mechanisms (Six et al., 2002), net primary production, and by regulating the oxygen and water supplies that control the decay and formation of soil OM (Schjønning et al., 1999). Textural properties that optimize net primary production while minimizing OM degradation would provide the greatest amount of soil OM storage. It is probably impossible to optimize both production and minimize degradation at the same time, so a trade-off must be found that provides the most OM storage based on the balance between these two parameters.

The characteristics and quantity of OM associated with different particle size classes

Table 7.2. Half-lives ($t_{1/2}$) and decomposition rate constants (k) for plant residues in different environments (from Paul and Clark, 1989).

Residue, location	$t_{1/2}$ (day)	k (days^{-1})	Relative rate of decomposition
Wheat straw, laboratory	9	0.08	1
Rye straw, Nigeria	17	0.04	0.5
Rye straw, England	75	0.01	0.125
Wheat straw, Saskatchewan, Canada	160	0.003	0.05

are known to vary. The clay content of soil strongly influences retention (Jenkinson and Rayner, 1977), and this is widely attributed to the greater surface area of clay compared to sand and silt particles. A soil maintained in long-term bare fallow was used to develop an equation that relates texture and C in soil OM (C_{SOM}, %) content (Rühlmann, 1999).

$$C_{SOM} = a*\text{text} - b*\exp(c*\text{text}) \qquad (7.2)$$

where a, b and c are regression coefficients and 'text' is the clay + silt content (%) of the soil. The C_{SOM} pool was thought to be very similar to the stabilized OM pool and this pool was related to the capacity of soils to sorb OM. The extent of this capacity depends on textural and structural soil properties.

In contrast, Thomsen et al. (1999), using a well-defined set of 12 differently textured soils of similar mineralogical composition and cropping history, found that the content of native soil OM was not related to texture. In a related decomposition study of ^{15}N-labelled ryegrass (Lolium multiflorum) residues using this same suite of soils, textural composition again appeared to play only a minor role in residue turnover (Thomsen et al., 2001). Indeed, land use and soil management appeared to be more important that soil texture in affecting the accumulation of soil OM.

Other evidence suggests that stabilization of OM is not solely a function of clay content or surface area. The silt-size fraction of some soils has a greater concentration of OM than the clay fraction (Tiessen and Stewart, 1983; Christensen, 2001). A study of 167 pedons under different ecosystems in New Zealand concluded that clay content related poorly to long-term soil OM storage and that chemical stabilization of OM by pyrophosphate-extractable aluminium (Al) and allophanes was the key process controlling accumulation in the soils (Percival et al., 2000). Curtin (2002) observed that OM concentration was greatest in the fine-silt and coarse-clay fractions, and not in the fine clay as would be expected if specific surface area was the major factor controlling OM content. Instead, there was a strong relationship between OM in the size fractions and extractable Al, suggesting that Al played a role in determining the OM storage capacity of the size fractions.

Clay type is also a factor in the formation and turnover of OM in soil. In soils dominated by 2:1 clay mineralogy, the OM content of macroaggregates was 1.65 times greater compared to that in microaggregates (Six et al., 2000). In soil with mixed mineralogy, aggregate OM content did not increase with increasing aggregate size.

Texture indirectly affects the OM carrying capacity of a soil by affecting soil structure, which in turn controls oxygen and water supply, thereby regulating the rate and extent of decomposition of organic material. Decomposition has been concluded to be maximum when about 66% of the pore space is filled with water (Doran et al., 1990; Skopp et al., 1990). More recent studies, however, have indicated that the percentage of water-filled pore space that defines optimum microbial activity may differ between structurally intact field soil and the homogenized soil typically used in incubation studies (Schjønning et al., 1999). Medium-textured soils have the broadest range of available water content, whereas sandy soils or soils with high clay contents have the narrowest ranges.

Inputs

Organic inputs can influence soil OM levels as a result of their quantity, biodegradability and placement within the soil profile. Inputs to soil include crop residues (including roots), animal manures, composts and, in some cases, by-products such as biosolids and other organic wastes. In most models of soil OM decomposition, there is a distinction in the rate of decomposition between highly degradable and newly added plant residues or organic amendments, and the more stable chemically or physically protected organic matter fractions in soil. However, for a given soil and climate, there is generally a direct relationship between the amount of OM inputs and the level of OM measured in a soil (Paustian et al., 1997). A summary of experiments in which residue additions were carefully controlled supports the theoretical proportionality between inputs and soil OM levels (Fig. 7.2).

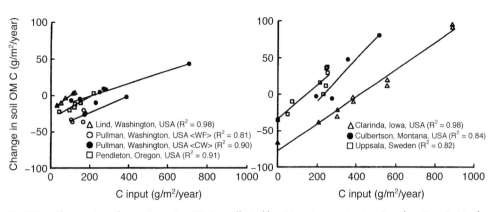

Fig. 7.2. Changes in soil organic matter (OM) as affected by C input amounts at various locations in North America. <WF> represents a wheat–fallow rotation system, while <CW> represents a continuous wheat cropping system (Paustian et al., 1997).

Management of crops and soils is an important parameter in controlling inputs. A study in the cerrado region of Brazil showed that improving soil quality by adding nutrients and applying no-tillage increased soil OM levels above those originally typical of this region (Sá et al., 2001). This was attributed to increased OM inputs to soil brought about by agriculture compared to the native ecosystem inputs. Improved forage management strategies, which result in increased biomass yields, were also found to increase soil OM in the southeastern USA over a 5-year period (Franzluebbers et al., 2001). Most of the net change in soil OM occurred in the 0- to 2-cm depth. The amount of soil OM after 12 years was correlated with the amount of residues returned to soil by the crop (spring wheat, *Triticum aestivum*; soybean, *Glycine max*; sunflower, *Helianthus annus*; and maize, *Zea mays*) (Studdert and Echeverría, 2000). Return of residues helped to maintain OM levels in soil as compared to when all of the residues were removed for forage silage (Dick et al., 1998).

Increases in the quantity of OM added to soil can be accomplished by increasing the net primary production and/or the ratio of that production that is returned to soil. One way to do this is by changing the harvest index of crops so more material is returned to soil. However, this would result in a reduction in harvestable yield and is not practical. Other ways include the use of fertilizers, cover crops, green manures, water or organic amendments (i.e. animal manures, biosolids and other organic waste materials). Water can increase production but, in some cases, also increase decomposition so that the net impact on the amount of OM stored in soil is negative. In general, manures and especially composts are more efficient in building OM than are fresh plant residues, because the OM in these amendments has already undergone some level of decomposition prior to their application to soil. For manures, this takes place during the digestive process, but the decomposition is generally not as complete as that which takes place during composting. It is important to note, however, that prestabilization of OM by composting (for example) causes a reduction in the amount of OM found in the original material. Thus the amount of OM that enters the soil after composting, compared to that of the uncomposted organic material, may be much less. A study to determine which is the more efficient approach to increasing soil OM – applying raw organic material and allowing stabilization to occur in soil or prestabilizing the organic material before applying to soil – would be of interest.

The influence of nitrogen fertilizer on OM storage in soil requires special attention. Nitrogen can be added as fertilizer or by other means (e.g. N_2 fixation by legumes or free-living microorganisms, atmospheric deposition). Because the carbon and nitrogen cycles are so closely linked and their proportionality (quantified as a C : N ratio) is relatively constant in many different soils, it is reasonable to assume that a limitation of nitrogen in a soil would also reduce the OM carrying capacity of that soil. In a modelling study, Schimel *et al.* (1997) found a strong correlation between carbon, nitrogen and water dynamics. Nitrogen inputs increased the flux of OM into soil by increasing net primary production, and the added nitrogen was then recycled to support further biomass production.

Studies have clearly shown that, in agricultural systems, fertilizer additions increase the amount of OM in soil (Paustian *et al.*, 1997). Additions of mineral fertilizer and manures to maize grain and silage plots in Michigan, USA, helped maintain soil OM concentrations, but could not negate the declines caused by tillage over a 20-year period (Dick *et al.*, 1998). The combination of mineral fertilizer and animal manure was most effective in maintaining OM levels compared to the mineral-only fertilized plots, with values being as much as 90% higher for the combination treatment. Nitrogen fertilization was also found to have a greater impact on OM concentrations than did crop sequence after 16 years in the western Corn Belt of the USA (Liebig *et al.*, 2002).

Besides input quantity, the quality (chemical composition) of the inputs is also recognized as having an effect on OM concentrations in soil. Table 7.3 shows the average concentrations of various organic chemical components in different plant species. These data indicate the differences in quality (chemical recalcitrance) among various plant residues. Cellulose and hemicellulose are more rapidly decomposed than lignin. Thus, the addition to soil of sawdust, which has the highest lignin content, would be expected to eventually lead to higher soil OM levels than would addition of lucerne stems. Paustian *et al.* (1997) observed that differences in soil OM in field plots in Sweden amended with different organic residues were predicted by the lignin contents of the residues (Fig. 7.3).

Legume-based rotations tend to conserve more of their OM inputs in the soil than continuous cropping systems with non-leguminous crops (Drinkwater *et al.*, 1998; Gregorich *et al.*, 2001). Qualitative differences in crop residues were found to be important in these studies because quantitative differences

Table 7.3. Average contents of major organic components in plant materials (Haider, 1992).

Material	Percentage of dry weight			
	Cellulose	Hemicellulose	Lignin	Protein[a]
Ryegrass (mature)	19–26	16–23	4–6	12–20
Lucerne (stem)	13–33	8–11	6–16	15–18
Wheat straw	27–33	21–26	18–21	3
Pinus sylvestris (sawdust)	42–49	24–30	25–30	0.5–1
Beech wood	42–51	27–40	18–31	0.6–1

[a]Nitrogen content times 6.25.

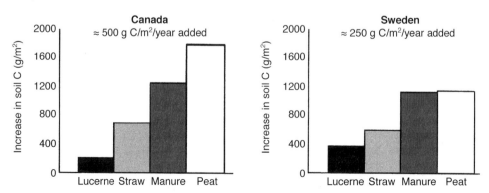

Fig. 7.3. Increases in soil carbon levels as affected by residue type (Paustian *et al.*, 1997).

in net primary productivity and OM inputs could not account for the differences observed.

Residue quality also affects soil OM content by influencing the process of aggregation. A study by Martens (2000) showed that mean weight diameter of soil aggregates was related to the phenolic acid content of residues added to soil, and to carbohydrate content that existed in the soil at the time of amendment, which, in turn, was related to the soil OM concentration. This supports the concept that phenolic compounds, like lignin, play a key role in OM storage in soil. Transient aggregate stability and the carbon stored in the aggregate as a result of microbial decomposition of the carbohydrate and amino acid content of residues is then strengthened by the interaction with phenolic acids released by microbial decomposition from the residue's structural components.

Genetically modified plants and their residues may have direct effects on decomposition processes, thereby affecting the quantity of OM stabilized in soil. Genetic modification of the amount, conformation and composition of lignin is being explored with a view to enhancing the digestibility of forage or improving paper-pulping properties. Increased decomposition of material from plants with genetic modifications to lignin biosynthesis was attributed to differences in the degree of protection from microbial attack afforded to the polysaccharides and other relatively labile plant components by the lignin (Hopkins *et al.*, 2001). Chemical analysis of maize that had been genetically modified to produce a larvicidal toxin (*Bt* toxin) to kill lepidopteran pests indicated that the lignin content of *Bt* maize hybrids was significantly (33–97%) higher than that of their respective non-*Bt* isolines (Saxena and Stotzky, 2001). In this experiment, maize had been genetically modified to express the *cry1Ab* gene from *Bacillus thuringiensis* (*Bt*). *In vitro* and *in situ* studies indicated that the toxin released in root exudates and from the biomass of *Bt* maize adsorbs and binds rapidly on surface-active particles, protecting it from biodegradation, and remains larvicidal for extended periods (Saxena and Stotzky, 2000). The implication of these findings is that the persistance of the *Bt* endotoxin could pose a hazard to non-target organisms (e.g. invertebrates) involved in the primary stages of decomposition of crop residues. The results of field and laboratory studies led Hopkins and Gregorich (2003) to conclude that much of the *Bt* endotoxin in crop residues is highly labile and quickly decomposes in soil, but a small fraction may be protected from decay in relatively recalcitrant residues.

Disturbance

Probably the two greatest disturbance events that affect soil OM concentrations are tillage and erosion. Many studies have shown that a rapid decrease in OM concentrations takes place when native soils are brought under the plough. In general the decline in soil OM occurs in two phases. An initial rapid decrease occurs in the first 10–20 years as the aggregates are broken down and the readily available OM is decomposed. Then a slower phase occurs as more recalcitrant OM is decomposed, along with any newly deposited materials. This phase can continue for many more years until a new equilibrium level is obtained.

Losses of OM when a native ecosystem is converted into arable agriculture can be very high (60–70%), but generally range between 20 and 50% in the zone of cultivation during the first 40–50 years of cultivation (Lal *et al.*, 1998). The loss of OM that occurs upon conversion of native ecosystems into arable agriculture is attributed primarily to four factors.

1. The amount of OM returned to the soil in plant litter is often lower in agricultural systems than in native systems. This is because under agricultural production a portion of organic material in the crop is harvested and removed from the system.

2. There is a change in plant species when a native system is converted into agricultural production. This change often results in a change in shoot and root OM deposition within the soil profile.

3. A change in the soil climatic conditions (temperature and moisture) and soil conditions (disruption of stable, protected OM by tillage) occurs under agricultural production,

particularly cultivated systems, leading to greater mineralization of OM than under native, uncultivated systems.

4.   There is a redistribution and subsequent loss of soil, with preferential losses of the finer particles rich in labile OM, due to water, wind and tillage erosion processes.

Most studies have shown that if all climatic and cropping system properties are held constant, except for tillage, there is little change in OM levels in soil. It seems that only the most extreme form of conservation tillage, i.e. no-tillage, has any great potential to reverse soil OM losses that have previously occurred. No-tillage affects the level of soil OM in a number of fundamental ways. These include: (i) affecting the water storage properties of a soil thus affecting crop production and OM decomposition rates; (ii) reducing the physical disturbance that ruptures soil aggregates, thus exposing new OM to decomposition; (iii) returning all residues to the soil, causing a redistribution of OM from below ground to the surface soil layers; and (iv) reducing the soil–residue contact that occurs when soil is tilled.

The mechanisms by which no-tillage increases OM storage in soil have been studied. Six *et al.* (1999) found that the proportions of crop-derived OM in macroaggregates were similar in no-tillage and conventional tillage, but were three times greater in microaggregates from no-tillage. This indicates a slower rate of microaggregate formation under conventional tillage. The beneficial effects of no-tillage on accrual of soil OM were thought to be primarily due to the increased retention of root-derived OM in the soil (Gale and Cambardella, 2000).

No-tillage can increase the potential for growing more crops with fewer fallow years in dryland areas such as that of the Great Plains of North America (Halvorson *et al.*, 1999). This more frequent cropping requires more nitrogen inputs and produces more biomass on a mean annual basis; here, the increased residue input increased soil OM levels in the 0–7.5 cm soil layer after 11 crops. Wheat producers on medium-textured soils in the semiarid Canadian prairies who switch from conventionally tilled fallow systems to

continuous no-till cropping could potentially sequester 5–6 Mg C/ha in soil OM in approximately 14 years (Curtin *et al.*, 2000). Similar results were reported by Schomberg and Jones (1999) in the southern High Plains of North America where, under dryland conditions, OM conservation was greater with no-tillage winter wheat because of less soil disturbance and shorter fallow periods.

The Conservation Reserve Program (CRP) in the USA has idled large amounts of land, resulting in increased soil OM levels. Post-CRP management of the soil using no-tillage for growing crops can conserve much of this accumulated OM (Wienhold and Tanaka, 2001; Dao *et al.*, 2002).

Erosion processes can have a large effect on the distribution of OM across the agricultural landscapes. Soil OM losses in agricultural systems due to erosion have been estimated to be greater than 75% of the native soil OM levels (Janzen *et al.*, 1998). Erosion can do this by selectively removing OM. Severe erosion can strip away surface soil and subsequent tillage mixes the thinner surface soil with subsoil (of relatively low OM), thereby diluting the OM content of the surface soil. The OM removed from one area is transported to another, causing a thicker, OM-rich horizon and profile in the area of deposition. Thus erosion changes both the lateral and vertical distribution of OM in the landscape (Pennock, 1997). Another consequence of erosion is how it affects primary productivity and total OM input into the soil (Gregorich *et al.*, 1998). If erosion reduces crop productivity and the amount of residues returned to the soil is also thus reduced, it will be difficult to stabilize or increase OM levels in soil.

The Effect of Time

The length of time it takes for soil OM levels to reach a new equilibrium following a change in management is controlled by the capacity factors of climate, texture, inputs, disturbance and landscape. Compared to the many short-term experiments that can affect soil OM concentrations in soil, there are far fewer long-term experiments (e.g. > 50 years) that can be used to determine what the equilibrium levels are for specific combinations of the OM capacity factors (Equation 7.1). A summary of work related to long-term soil and cropping system experiments in Canada, the USA and Europe is given by Powlson *et al.* (1996) and Paul *et al.* (1997). From this, it is clear that few management systems are ever maintained on an agricultural soil long enough to achieve equilibrium. Therefore, to predict soil OM levels and how they change with time, knowledge of the baseline conditions at the time a new management system is imposed is required. Baseline information is not needed, however, if we are only interested in the final capacity of a soil to store OM, because by extending the time variable out far enough, the baseline conditions become irrelevant. Please also see the three 'caveats' highlighted by Kay and Munkholm (Chapter 11, this volume).

Soil OM contents can be raised by maintaining, over time, land-use practices in which rates of OM inputs exceed losses. Inputs can be increased by increasing cropping intensity, especially with legumes, and by soil amendments. Results from a study of the long-term Breton plots in Canada showed that the decay rates in a wheat–fallow rotation required double the amount of OM input compared to a wheat–oat–barley–hay–hay rotation, just to maintain the original soil OM levels (Izaurralde *et al.*, 2001). Even after 51 years, the soil in some of the rotations in this study appear to be gaining OM.

In Ohio (USA), continuous application of no-tillage for 33 years to a silt loam or silty clay loam soil did not achieve equilibrium levels in the distribution and amount of soil OM (Dick *et al.*, 1991; Dick and Durkalski, 1998). It may be almost impossible to predict with absolute certainty what the new equilibrium level will be in these soils because our period of observation has not been long enough, even though these long-term plots have now been maintained for more than 40 years.

In the context of time, it is important to note that the length of observation time is critical to accurate assessment of rate of OM accumulation. As noted by Paustian *et al.* (1997), since the instantaneous rate of change in soil OM decreases over time following a change in

inputs, short-duration experiments would be expected to exhibit higher rates of change than long-duration experiments. Since data from published studies often do not cover the entire duration needed to reach a new threshold or equilibrium of soil OM level, the predicted rate of accumulation may be substantially higher than the average rate (Fig. 7.4)

In addition to the ambiguity caused by not evaluating a management practice for a long enough period of time, ambiguity can also result from different rates of OM accumulation under different management systems. The initial rate of increase in OM under one management practice may be different from that of another management practice (Fig. 7.5).

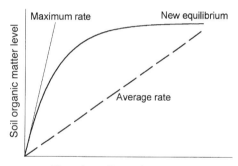

Fig. 7.4. Rates of change in soil C calculated at different times after a change in management. The rate of C accumulation calculated in studies of short duration may estimate higher rates than the average rate calculated over a relatively longer period of time (adapted from IPCC, 2000).

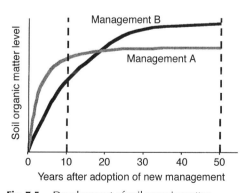

Fig. 7.5. Development of soil organic matter accumulation under different management systems (adapted from Ingram and Fernandes, 2001).

Some management practices (e.g. Management A in Fig. 7.5) result in larger gains in soil OM in the short term relative to other management practices (e.g. Management B), but lower gains over the long term.

Simulation models are a powerful tool to describe changes in soil OM over time. Provided that they are properly calibrated and validated against a time series of experimental data, models are the only practical means of making projections of management effects at places other than the sites of long-term experiments (Powlson *et al.*, 1996). Process-based models are most commonly used to characterize soil biology and biochemistry in agricultural systems (Paustian, 1994). These models have some common elements: (i) decomposition processes are based on first-order kinetics; (ii) soil OM contains discrete fractions that have characteristic rates of decomposition; and (iii) carbon and nitrogen dynamics are interconnected. Most soil OM models operate in the time-scale of months to centuries and many include texture (e.g. clay content) as a control to protect or slow decomposition of soil organic components,

and to regulate the partitioning of C among compartments (McGill, 1996).

Establishing a New Soil Capacity Level for Organic Matter

As OM enters and resides in soil, there are processes that occur in and on the soil that can alter it. These fundamental soil processes include humification, aggregation, translocation, erosion, leaching and mineralization (Table 7.4) and are driven by the five soil OM capacity factors identified in Equation 7.1. Humification, for example, is a fundamental process that occurs when OM is added to soil and is acted upon over time by climate, soil texture, disturbance and landscape.

For agricultural systems, the fundamental soil processes are affected when various types of management systems are imposed on the soil. Thus the integration of the fundamental soil processes with the five soil OM capacity factors is set into place by the establishment of a management system. Therefore, we can potentially change soil OM if we understand

Table 7.4. Fundamental soil processes that lead to increases or decreases in soil organic matter (OM).

Fundamental process	Brief description of process	Impact on soil OM
Humification	Conversion of biomass into humic substances that are relatively resistant to microbial decomposition and have a very long (hundreds to thousands of years) turnover time	Increases soil OM levels
Aggregation	Formation of stable organo-mineral complexes in which OM is bonded with clay colloids and metal elements to form aggregates. The OM enclosed within microaggregates or micropores in larger aggregates is protected against microbial decomposition	Increases soil OM levels
Translocation within the pedosphere	Eluviation and deposition of humus and stable microaggregates to lower horizons within the soil profile so that they are less disturbed by biotic and abiotic processes and are not exposed to climate	Increases soil OM levels
Erosion	Selective process that involves preferential removal/depostion of OM and clay fractions from soil	Decreases soil OM levels in eroded areas; increases soil OM in deposition areas
Leaching	Translocation of dissolved organic carbon to a lower horizon. The soluble soil OM may be either bound to metals (e.g. Al and Fe) or minerals in lower horizons and thereby stabilized (and sequestered), or it may be leached out of the profile	Increases/decreases soil OM levels
Mineralization	Decomposition of organic compounds into CO_2 by soil microbial activity	Decreases soil OM levels

how changes in one or more of the five soil OM capacity factors affect these fundamental soil processes. Together these factors affect choice of management and ultimately control what the new equilibrium soil OM will be. The integration of these various levels is illustrated in Fig. 7.6.

The effect of a management system on soil OM levels is not always immediately evident. For example, soil under tall fescue (*Festuca arundinacea*) with high endophyte (*Neotyphodium coenophialum*) levels had 13% greater concentration of soil OM to a depth of 15 cm than where low infection rates occurred (Franzluebbers *et al.*, 1999). This was due to reduced basal soil respiration in the soils where high infection rates were found. Thus any management system that reduces infection would also reduce soil OM levels.

Allowing the soil OM to reach a new and higher equilibrium level is not always an option, because there is a continuing need to use the soil resource for the production of food and fibre. In the preceding example of forage production (Franzluebbers *et al.*, 1999), it is not realistic to maintain a fescue field with high endophyte infection rates simply for the sake of increasing the soil OM levels. The establishment of a management practice is thus ultimately governed by economic, environmental and social constraints. With these constraints it is important to ask several questions. What are the various management practices that allow production to continue while maintaining or actually building soil OM levels? What is the current soil OM level

relative to the threshold or capacity of the soil to store OM? Can we predict how a change in management will affect the potential capacity of a soil to store OM? Models have been developed to predict such changes in the short term, but the effects in the long term are not as well defined and easy to predict. Indeed, in many agricultural systems a new equilibrium level of soil OM may not be reached before another management change occurs. Thus, in terms of the soil OM capacity factors, the time variable is critical in our assessment of the change in soil OM that a given management practice will bring about.

There is an almost infinite number of combinations of climate, inputs, soil texture, disturbances and landscapes that define and create the working of a specific management system. An understanding of the weight each of these factors has in changing the soil OM capacity of a soil will remain a topic of study for many years. We can, however, within certain constraints, begin to provide a relative listing of some major benchmark management systems that affect soil OM levels (West and Post, 2002). Listing these from the most aggrading to the most degrading system provides a framework within which other management systems can be assessed. Table 7.5 is an attempt to suggest ten benchmark management systems. They are ranked in the order of the management system most likely to result in an increase in soil OM levels (assigned a value of 100) to the management practice that is most likely to cause a decrease (assigned a value of 10). A management practice not listed can be placed within the framework of the table it most closely matches to identify how this practice might affect OM levels relative to a new practice that is to be imposed on a soil.

It is important to realize that such a listing is strictly comparative and cannot be used to predict actual levels of soil OM. This requires knowledge of the impacts of the soil OM capacity factors at a more fundamental level. However, identifying benchmark management practices should be useful for assessing the direction in OM content (positive or negative) caused by a change in management. Our conceptual model is analogous to the approach used by the Intergovernmental

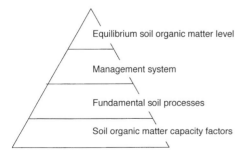

Fig. 7.6. The integration of soil organic matter capacity factors, fundamental soil processes and management systems in establishing a new equilibrium level of soil organic matter.

Panel on Climate Change (IPCC) to categorize and describe the activities for cropland and grassland management that affect potential rates and duration of carbon sequestration (e.g. Table 4–4 in IPCC, 2000).

Ingram and Fernandes (2001) presented a framework to characterize factors that determine soil C storage in agroecosystems. Applying this framework, Fig. 7.7 shows how the management systems listed in Table 7.5 can affect soil OM levels. Soil OM capacity factors (i.e. climate, landscape and texture) combine to define the potential OM level inherent for an individual soil. In managed agricultural systems, two sets of factors serve to limit or reduce the potential level of soil OM, one related to inputs of OM and the other to disturbances. Input-limiting factors reduce the OM levels that can be attained in soil. Net primary productivity (NPP) is the underlying control, modified by other plant characteristics such as shoot–root ratio and residue quality. Management systems that maximize NPP (e.g. by fertilization) and leave all the organic residues will achieve the highest levels of soil OM. In contrast, disturbance factors, such as erosion and tillage, generally lead to much lower OM levels.

Table 7.5. Management systems that affect soil organic matter levels from the most aggrading (assigned a value of 100) to the most degrading (assigned a value of 10).

100.	Improved permanent pastures with animal grazing (this basically leaves all material on site)
90.	No-tillage rotation with row crops alternating with legumes and soil treated with manure
80.	No-tillage with continuous row crops and manure additions
70.	Conservation tillage with long rotation sequences that include green manures and animal manures
60.	Plough tillage with rotation sequences that include green manures and animal manures
50.	No-tillage with grain and residue (for fuel and feed) harvested
40.	Conservation tillage with continuous row crops
30.	Intensive tillage with continuous row crops
20.	Intensive tillage with continuous row crops on sloping lands (grain and residues harvested)
10.	Intensive tillage with mechanical summer fallow in alternate years with little nutrient input

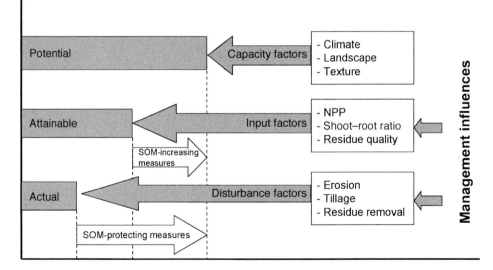

Equilibrium SOM level

Fig. 7.7. Management influences of input-limiting and disturbance-reducing factors that can affect the equilibrium soil organic matter (SOM) level (after Ingram and Fernandes, 2001). NPP, net primary production.

The management systems towards the top of Table 7.5 (e.g. pasture) serve to maintain or increase soil OM levels by both optimizing C inputs and protecting the soil from disturbances. For example, no-tillage protects the soil from erosion and increased soil temperatures which promote oxidation of soil OM. Management systems towards the bottom of Table 7.5 (e.g. summer fallow in cropping systems) both reduce C inputs into the soil and feature high levels of disturbance.

There is a large body of published information on the effects of specific management practices on soil OM levels. Several conclusions can be drawn from these studies.

1. Management impacts on soil OM levels affect the tillage layer or, for no-tillage, a soil layer that is often 10 cm or less (Dick *et al.*, 1991; Liebig *et al.*, 2002).
2. A relatively long period is required for the management to be maintained before changes in soil OM levels can be recorded with confidence, although there are attempts to identify early indicators of such change.
3. Practices that cause an increase in the level of soil OM are, in general, those that are also most sustainable.

It is not possible here to review all of the studies that have investigated the effect of various elements of management systems on soil OM levels. However, attempts have been made to provide information on how management practices and soil OM levels interact at a field, farm and regional level. Some trends that have been identified as being most effective in bringing about increased soil OM levels in broad regions of agricultural soils in the USA include: (i) the Conservation Reserve Program (i.e. the conversion of unproductive croplands to perennial grassland); (ii) expanded use of no-till agriculture; (iii) improved productivity caused by new plant varieties; and (iv) increased fertilizer use (Pacala *et al.*, 2001).

Conclusions

In order to maintain or even increase soil OM levels, we must have a much better understanding of the soil OM capacity factors identified in Equation 7.1 and the ways they interact. Use of the soil OM capacity factor equation then would be analogous to how the universal soil loss equation (Wischmeier and Smith, 1978) quantifies the prediction of erosion losses based on a set of values for each of the soil loss parameters in the equation. It is important to note that the OM levels measured at any one time and the capacity of a soil to store OM are not the same. Equation 7.1 is meant to predict the potential amount of OM that can be stored in soil, i.e. the soil's OM storing capacity. The OM capacity of a soil is a dependent variable and thus not an absolute fixed value. As our knowledge of how each of the storing-capacity variables affects OM grows, our predictive powers will also grow.

Fundamental soil processes can alter OM that enters and resides in soil. The mix of soil OM capacity factors at a given location affects the rate and extent of that alteration. The integration of the soil processes with the soil OM capacity factors is controlled by the management practices imposed on a soil. Together these processes and factors control what the new equilibrium soil OM will be. Therefore we have the potential to change soil OM levels if we understand how changes in one or more of the five soil OM capacity factors affect these fundamental soil processes.

References

Baldock, J.A. and Nelson, P.N. (2000) Soil organic matter. In: Sumner, M.E. (ed.) *Handbook of Soil Science*. CRC Press, Boca Raton, Florida, pp. B24–84.

Bauer, A. and Black, A.L. (1994) Quantification of the effect of soil organic mater content on soil productivity. *Soil Science Society of America Journal* 58, 185–193.

Christensen, B.T. (2001) Physical fractionation of soil and structural and functional complexity in organic matter turnover. *European Journal of Soil Science* 52, 345–353.

Christensen, B.T. and Johnston, A.E. (1997) Soil organic matter and soil quality: lessons learned from long-term experiments at Askov and Rothamsted. In: Gregorich, E.G. and Carter, M.R. (eds) *Soil Quality for Crop Production and*

Ecosystem Health. Elsevier Science Publishers, Amsterdam, pp. 399–430.

Curtin, D. (2002) Possible role of aluminum in stabilizing organic matter in particle size fractions of Chernozemic and Solonetizic soils. *Canadian Journal of Soil Science* 82, 265–268.

Curtin, D., Wang, H., Selles, F., McConkey, B.G. and Campbell, C.A. (2000) Tillage effects on carbon fluxes in continuous wheat and fallow–wheat rotations. *Soil Science Society of America Journal* 64, 2080–2086.

Dao, T.H., Stiegler, J.H., Banks, J.C., Boerngen, L.B. and Adams, B. (2002) Post-contract land use effects on soil carbon and nitrogen in conservation reserve grassland. *Agronomy Journal* 94, 146–152.

Dick, W.A. and Durkalski, J.T. (1998) No-tillage production agriculture and carbon sequestration in a Typic Fragiudalf soil of northeastern Ohio. In: Lal, R., Kimble, J.M., Follett, R.R. and Stewart, B.A. (eds) *Management of Carbon Sequestration in Soil*. CRC Press, Boca Raton, Florida, pp. 59–71.

Dick, W.A., McCoy, E.L., Edwards, W.M. and Lal, R. (1991) Continuous application of no-tillage to Ohio soils. *Agronomy Journal* 83, 65–73.

Dick, W.A., Blevins, R.L., Frye, W.W., Peters, S.E., Christenson, D.R., Pierce, F.J. and Vitosh, M.L. (1998) Impacts of agricultural management practices on C sequestration in forest-derived soils of the eastern Corn Belt. *Soil & Tillage Research* 47, 235–244.

Doran, J.W. and Parkin, T.B. (1994) Defining and assessing soil quality. In: Doran, J.W., Coleman, D.C., Bezdicek, D.F. and Stewart, B.A. (eds) *Defining Soil Quality for a Sustainable Environment*. SSSA Special Publication No 35. Soil Science Society of America, Madison, Wisconsin, pp. 3–21.

Doran, J.W., Mielke, L.N. and Power, J.F. (1990) Microbial activity as regulated by soil water filled pore space. *Transactions 14th International Congress of Soil Science*, 12–18 August, Kyoto, Japan, Vol. 3, pp. 94–99.

Drinkwater, L.E., Wagoner, P. and Sarrantonio, M. (1998) Legume-based cropping systems have reduced carbon and nitrogen losses. *Nature* 396, 262–264.

Franzluebbers, A.J., Nazih, N., Stuedemann, J.A., Fuhrmann, J.J., Schomberg, H.H. and Hartel, P.G. (1999) Soil carbon and nitrogen pools under low- and high-endophyte-infected tall fescue. *Soil Science Society of America Journal* 63, 1687–1694.

Franzluebbers, A.J., Stuedemann, J.A. and Wilkinson, S.R. (2001) Bermudagrass management in the Southern Piedmont USA. I. Soil and surface residue carbon and sulfur. *Soil Science Society of America Journal* 65, 834–841.

Gale, W.J. and Cambardella, C.A. (2000) Carbon dynamics of surface residue- and root-derived organic matter under simulated no-till. *Soil Science Society of America Journal* 64, 190–195.

Gregorich, E.G., Greer, K.J., Anderson, D.W. and Liang, B.C. (1998) Carbon distribution and losses: erosion and deposition effects. *Soil & Tillage Research* 47, 291–302.

Gregorich, E.G., Drury, C.F. and Baldock, J.A. (2001) Changes in soil carbon under long-term maize in monoculture and legume-based rotation. *Canadian Journal of Soil Science* 81, 21–31.

Haider, K. (1992) Problems related to the humification processes in soils of temperate climates. In: Stotzky, G. and Bollag, J.-M. (eds) *Soil Biochemistry*, Vol. 7. Marcel Dekker, New York, pp. 55–94.

Halvorson, A.D., Reule, C.A. and Follett, R.F. (1999) Nitrogen fertilization effects on soil carbon and nitrogen in a dryland cropping system. *Soil Science Society of America Journal* 63, 912–917.

Hopkins, D.W. and Gregorich, E.G. (2003) Detection and decay of the Bt endotoxin in soil from a field trial with GM maize. *European Journal of Soil Science* (in press).

Hopkins, D.W., Webster, E.A., Chudek, J.A. and Halpin, C. (2001) Decomposition in soil of tobacco plants with genetic modifications to lignin biosynthesis. *Soil Biology and Biochemistry* 33, 1455–1462.

Hudson, B.D. (1994) Soil organic matter and available water capacity. *Journal of Soil and Water Conservation* 49, 189–194.

Ingram, J.S.I. and Fernandes, E.C.M. (2001) Managing carbon sequestration in soils: concepts and terminology. *Agriculture, Ecosystems and Environment* 87, 111–117.

Intergovernmental Panel on Climate Change (IPCC) (2000) Chapter 4: Additional human-induced activities – Article 3.4. In: Watson, R.T., Noble, I.R., Bolin, B., Ravindranath, N.H., Verardo, D.J. and Dokken, D.J. (eds) *Land Use, Land-Use Change, and Forestry: Special Report of the Intergovernmental Panel on Climate Change (IPCC)*. Cambridge University Press, Cambridge, 375 pp.

Izaurralde, R.C., McGill, W.B., Robertson, J.A., Juma, N.G. and Thurston, J.J. (2001) Carbon balance of the Breton classical plots over half a century. *Soil Science Society of America Journal* 65, 431–441.

Janzen, H.H., Campbell, C.A., Gregorich, E.G. and Ellert, B.H. (1998) Soil carbon dynamics in Canadian agroecosystems In: Lal, R., Kimble, J., Follet, R.F. and Stewart, B.A. (eds)

Soil Processes and the Carbon Cycle. CRC Press, Boca Raton, Florida, pp. 57–80.

Jenkinsen, D.S. and Rayner, J.H. (1977) The turnover of soil organic matter in some of the Rothamsted classical experiments. *Journal of Soil Science* 123, 298–305.

Jenny, H. (1941) *Factors of Soil Formation – a System of Quantitative Pedology*. McGraw-Hill, New York, 241 pp.

Johnston, A.E. (1986) Soil organic matter, effects on soils and crops. *Soil Use and Management* 2, 97–105.

Johnston, A.E. (1987) Effects of soil organic matter on yields of crops in long-term experiments at Rothamsted and Woburn. *INTECOL Bulletin* 15, 9–16.

Kirschbaum, M. (2000) Will changes in soil organic carbon act as a positive or negative feedback on global warming? *Biogeochemistry* 48, 21–51.

Lal, R., Kimble, J.M., Follett, R.F. and Cole, C.V. (1998) *The Potential of U.S. Cropland to Sequester Carbon and Mitigate the Greenhouse Effect*. Ann Arbor Press, Chelsea, Michigan, 128 pp.

Liebig, M.A., Varvel, G.E., Doran, J.W. and Wienhold, B.J. (2002) Crop sequence and nitrogen fertilization effects on soil properties in the Western Corn Belt. *Soil Science Society of America Journal* 66, 596–601.

Loux, M., Stachler, J.M. and Harrison, S.K. (2002) *Weed Control Guide for Ohio Field Crops*. Bulletin 789 of the Ohio State University Extension Service, Columbus, Ohio, 157 pp.

Mahieu, N., Powlson, D.S. and Randall, E.W. (1999) Statistical analysis of published carbon-13 CPMAS NMR spectra of soil organic matter. *Soil Science Society of America Journal* 63, 307–319.

Manies, K.L., Harden, J.W., Kramer, L. and Parton, W.J. (2001) Carbon dynamics within agricultural and native sites in the loess region of western Iowa. *Global Change Biology* 7, 545–555.

Martens, D.A. (2000) Plant residue biochemistry regulates soil carbon cycling and carbon sequestration. *Soil Biology and Biochemistry* 32, 361–369.

McGill, W.B. (1996) Review and classification of ten soil organic matter (SOM) models. In: Powlson, D.S., Smith, P. and Smith, J.U. (eds) *Evaluation of Soil Organic Models*. NATO ASI Series, Vol. I, 38, Springer-Verlag, Berlin, pp. 111–132.

McLaughlin, N.B., Gregorich, E.G., Dwyer, L.M. and Ma, B.L. (2002) Effect of organic and inorganic soil nitrogen amendments on mouldboard plow draft. *Soil & Tillage Research* 64, 211–219.

Pacala, S.W., Hurtt, G.C., Baker, D., Peylin, P., Houghton, R.A., Birdsey, R.A., Heath, L., Sundquist, E.T., Stallard, R.F., Ciais, P., Moorcroft, P., Caspersen, J.P., Shevliakova, E., Moore, B., Kohlmaier, G., Holland, E., Gloor, M., Harmon, M.E., Fan, S.-M., Sarmiento, J.L., Goodale, C.L., Xhimel, D. and Field, C.B. (2001) Consistent land- and atmosphere-based U.S. carbon sink estimates. *Science* 292, 2316–2320.

Paul, E.A. and Clark, F.E. (1989) *Soil Microbiology and Biochemistry*. Academic Press, San Diego, California, 273 pp.

Paul, E.A., Paustian, K., Elliot, E.T. and Cole, C.V. (1997) *Soil Organic Matter in Temperate Agroecosystems*. CRC Press, Boca Raton, Florida, 414 pp.

Paustian, K. (1994) Modelling soil biology and biochemical processes for sustainable agriculture research. In: Pankhurst, C.E., Doube, B.M., Gupta, V.V.S.R. and Grace, P.R. (eds) *Soil Biota Management in Sustainable Farming Systems*. CSIRO Information Services, Melbourne, Australia, pp. 182–193.

Paustian, K., Collins, H.P. and Paul, E.A. (1997) Management controls on soil carbon. In: Paul, E.A., Paustian, K., Elliot, E.T. and Cole, C.V. (eds) *Soil Organic Matter in Temperate Agroecosystems*. CRC Press, Boca Raton, Florida, pp. 15–49.

Paustian, K., Elliot, E.T. and Killian, K. (1998) Modeling soil carbon in relation to management and climate change in some agroecosystems in central North America. In: Lal, R., Kimble, J.M., Follett, R.F. and Stewart, B.A. (eds) *Soil Processes and the Carbon Cycle*. CRC Press, Boca Raton, Florida, pp. 459–471.

Pennock, D.J. (1997) Effects of soil redistribution on soil quality: pedon, landscape, and regional scales. In: Gregorich, E.G. and Carter, M.R. (eds) *Soil Quality for Crop Production and Ecosystem Health*. Elsevier, Amsterdam, pp. 167–185.

Percival, H.J., Parfitt, R.L. and Scott, N.A. (2000) Factors controlling soil carbon levels in New Zealand grasslands. *Soil Science Society of America Journal* 64, 1623–1630.

Pierzynski, G.M., Sims, J.T. and Vance, G.F. (2000) *Soils and Environmental Quality*, 2nd edn. CRC Press, Boca Raton, Florida, 459 pp.

Powlson, D.S., Smith, P. and Smith, J.U. (1996) *Evaluation of Soil Organic Matter Models Using Existing Long-Term Datasets*. NATO ASI Series, Vol. I, 38, Springer-Verlag, Berlin, 429 pp.

Rochette, P., Lemke, R. and McGuinn, S. (2002) Soil quality and emissions of carbon and nitrogen gases. In: Lal, R. (ed.) *Encyclopedia of Soil Science*. Marcel Dekker, New York, pp. 1073–1077.

Rounsevell, M.D.A., Evans, S.P. and Bullock, P. (1999) Climate change and agricultural soils: impacts and adaptation. *Climatic Change* 43, 683–709.

Rühlmann, J. (1999) A new approach to estimating the pool of stable organic matter in soil using data from long-term field experiments. *Plant and Soil* 213, 149–160.

Sá, J.C. de M., Cerri, C.C., Dick, W.A., Lal, R., Venske Filho, S.P., Piccolo, M.C. and Feigl, B.E. (2001) Organic matter dynamics and carbon sequestration rates for a tillage choronosequence in a Brazilian oxisol. *Soil Science Society of America Journal* 65, 1486–1499.

Saxena, D. and Stotzky, G. (2000) Insecticidal toxin from *Bacillus thuringiensis* is released from roots of transgenic Bt corn *in vitro* and *in situ*. *FEMS Microbiology Ecology* 33, 35–39.

Saxena, D. and Stotzky, G. (2001) Bt corn has higher lignin content than non-Bt corn. *American Journal of Botany* 88, 1704–1706.

Scharpenseel, H.W. and Pfeiffer, E.M. (1998) Carbon turnover in different climates and environments. In: Lal, R., Kimble, J.M., Follett, R.F. and Stewart, B.A. (eds) *Soil Processes and the Carbon Cycle*. CRC Press, Boca Raton, Florida, pp. 577–590.

Schimel, E.S., Braswell, B.H. and Parton, W.J. (1997) Equilibration of the terrestrial water, nitrogen, and carbon cycles. *Proceedings of the National Academy of Sciences, USA* 94, 8280–8283.

Schjønning, P., Thomsen, I.K., Møberg, J.P., de Jonge, H., Kristensen, K. and Christensen, B.T. (1999) Turnover of organic matter in differently textured soils. I. Physical characteristics of structurally disturbed and intact soils. *Geoderma* 89, 177–198.

Schlesinger, W.H. (1990) Evidence from chronosequence studies for a low carbon storage potential of soils. *Nature* 348, 232–239.

Schomberg, H.H. and Jones, O.R. (1999) Carbon and nitrogen conservation in dryland tillage and cropping systems. *Soil Science Society of America Journal* 63, 1359–1366.

Six, J., Elliot, E.T. and Paustian, K. (1999) Aggregate and soil organic matter dynamics under conventional and no-tillage systems. *Soil Science Society of America Journal* 63, 1350–1358.

Six, J., Paustian, K., Elliot, E.T. and Combrink, C. (2000) Soil structure and organic matter: I. Distribution of aggregate-size classes and aggregate-associated carbon. *Soil Science Society of America Journal* 64, 681–689.

Six, J., Conant, R.T., Paul, E.A. and Paustian, K. (2002) Stabilization mechanisms of soil organic matter: implications for C-saturation of soils. *Plant and Soil* 241, 155–176.

Skopp, J., Jawson, M.D. and Doran, J.W. (1990) Steady-state aerobic microbial activity as a function of soil water content. *Soil Science Society of America Journal* 54, 1619–1625.

Studdert, G.A. and Echeverría, H.E. (2000) Crop rotations and nitrogen fertilization to manage soil organic carbon dynamics. *Soil Science Society of America Journal* 64, 1496–1503.

Thomsen, I.K., Olesen, J.E., Schjønning, P., Jensen, B., Kristensen, K. and Christensen, B.T. (1999) Turnover of organic matter in differently textured soils. II. Microbial activity as influenced by soil water regimes. *Geoderma* 89, 199–218.

Thomsen, I.K., Olesen, J.E., Schjønning, P., Jensen, B. and Christensen, B.T. (2001) Net mineralization of soil N and ^{15}N-ryegrass residues in differently textured soils of similar mineralogical composition. *Soil Biology & Biochemistry* 33, 277–285.

Tiessen, H. and Stewart, J.W.B. (1983) Particle-size fractions and their use in studies of soil organic matter: II. Cultivation effects on organic matter composition in size fractions. *Soil Science Society of America Journal* 47, 509–514.

West, T.O. and Post, W.M. (2002) Soil organic carbon sequestration rates by tillage and crop rotation: a global data analysis. *Soil Science Society of America Journal* 66, 1930–1946.

Wienhold, B.J. and Tanaka, D.L. (2001) Soil property changes during conversion from perennial vegetation to annual cropping. *Soil Science Society of America Journal* 65, 1795–1803.

Wischmeier, W.H. and Smith, D.D. (1978) *Predicting Rainfall Erosion Losses: a Guide to Conservation Planning*. USDA-ARS Agriculture Handbook No. 537, Washington, DC, 58 pp.

Chapter 8
Microbial Diversity in Soil – Effects on Crop Health

C. Alabouvette,[1] D. Backhouse,[2] C. Steinberg,[1] N.J. Donovan,[3] V. Edel-Hermann[1] and L.W. Burgess[4]

[1]CMSE-INRA, U.M.R. Biochimie, Biologie Cellulaire et Ecologie des Interactions Plantes Microorganismes, F 21065 Dijon, France; [2]University of New England, School of Environmental Sciences and Natural Resources Management, Armidale, New South Wales 2351, Australia; [3]New South Wales Agriculture, Elizabeth Macarthur Agricultural Institute, Private Mail Bag 8, Camden, New South Wales 2570, Australia; [4]University of Sydney, School of Land, Water and Crop Sciences, New South Wales 2006, Australia

©CAB International 2004. *Managing Soil Quality: Challenges in Modern Agriculture*
(eds P. Schjønning, S. Elmholt and B.T. Christensen)

Summary

Plant pathologists have traditionally viewed soil as a hostile environment, harbouring pathogens that have adverse effects on plant health. The emerging concept of soil quality sees pathogens as components of total soil biological diversity, with disease resulting from disturbances to the balance between functional groups in soil. Experimental support for this concept is beginning to be accumulated. Molecular methods based on polymerase chain reaction (PCR) of DNA extracted from soil allow rapid assessment of genetic diversity and will increasingly be used to measure functional diversity as well. This will enable the relationships between diversity and disease suppression to be characterized. Suppressiveness of soils to disease is biological in nature, although modified by abiotic factors, and is of two types. General suppression depends on overall diversity and activity of the soil biota and acts against a broad range of pathogens. Specific suppression is due to particular antagonists or functional groups, and acts against single pathogens. Studies on the effects of management practices on disease suppression are still limited in scope, and are often difficult to interpret because most practices have direct effects on pathogen populations as well as on suppressiveness. Continuous cropping of a plant species selects for microflora adapted to its rhizosphere, which may suppress the activities of some pathogens. Rotations and reduced tillage should increase microbial diversity and increase suppressiveness, but evidence for these causal links is hard to find. Treatments that increase soil organic matter, such as residue retention and application of manure, may increase general suppression, and certain types of manures may also increase specific suppression. Soil biodiversity, disease suppression and management practices are yet to be fully synthesized in models that have general applicability. This means that management thresholds remain speculative in most instances.

Introduction

From the plant pathologist's point of view, the soil not only provides nutrients and water for plant growth, but is also considered a hostile environment harbouring plant pathogenic fungi, bacteria and nematodes. Until recently, the most common attitude was to try to eliminate the plant pathogenic microorganisms by biocidal or other treatments. These treatments are frequently hazardous for humans and the environment, but are also inefficient, since they have to be applied every year to control soilborne pathogens and parasites. The most effective treatments kill off a large part of the microbial communities, which are essential for limiting the expression of the parasitic activities of the pathogens.

Contrary to this traditional viewpoint is the emerging concept of soil quality, including 'soil health' (Van Bruggen and Semenov, 2000). A healthy soil is not only a fertile soil providing the best physicochemical environment for plant nutrition and growth, but also the best biotic environment enabling the plant to resist pathogens. Healthy soils reflect an adequate balance among functional microbial communities resulting in the suppression of plant diseases. According to this concept of soil health, a disease due to a soilborne plant pathogen originates from an imbalance among microbial communities enabling the development of the deleterious populations.

Soil health implies ecosystem stability based on the largest biodiversity leading to functional connectedness and resilience in response to various disturbances, possibly resulting from anthropogenic activities.

Thus the first question to be addressed in this chapter is related to soil biodiversity: how can we assess this biodiversity, and at which level should we characterize it? The second question relates to the relationship between soil biodiversity and soil health. For this purpose we shall review some known examples of soil suppressive to soilborne diseases and discuss the mechanisms responsible for disease suppression. This chapter focuses on microbial diversity even if other soil organisms such as arthropods, earthworms and nematodes have important functions in relation to soil health (Doube and Schmidt, 1997). In the last section of this chapter we address the difficult question of how management practices affect the microbial balance and the level of soil receptivity to diseases and how to characterize management thresholds in relation to soil suppressiveness to diseases.

Soil Microbial Diversity

Plant pathologists tend to forget that the soil harbours a great diversity of living organisms besides the plant pathogens they are trying to control. The microbial biomass of a soil is mainly made up of bacteria and fungi but also of protozoa, algae, nematodes and micro-arthropods. Bacteria and fungi constitute major components of this microbial biomass, but estimates of this biomass vary. Gobat et al. (1998) gave estimates for bacteria of 1500 kg/ha, fungi 3500 kg/ha and algae 10–1000 kg/ha. Besides microorganisms, the soil is also the habitat for larger organisms such as earthworms, springtails and mites. These play an important role in the functioning of the soil ecosystem, especially through regulating the activities of microbial communities. Here we shall focus on microbial diversity, mostly bacteria and fungi.

Different methods have been proposed to assess microbial biomass, which is expressed as mg of C per kg soil (Jenkinson and Powlson, 1976; Anderson and Domsch, 1978). This global estimation of microbial biomass reveals differences from one soil to another, but gives no information on the microbial communities that make up this biomass.

The classical soil dilution plate technique, using more or less selective media, gives an idea of the relative importance of bacterial versus fungal communities and can identify, based on morphological characters, some of the species present in the soil. For a long time this family of techniques was the only one available to describe soil microbial diversity.

Molecular methods to characterize the diversity of the soil microbiota

Today, molecular-based techniques enable a better discrimination among soil microbial communities and can be used to compare different soils or to follow shifts in the microbial populations resulting from application of various treatments. The techniques are already in common use to describe bacterial communities, but improvements are needed to address the diversity affecting the communities of fungi and protozoa.

The methods used to assess the composition of microbial communities are mainly based on the direct extraction of nucleic acids from soil samples and the amplification by polymerase chain reaction (PCR) of ribosomal DNA (rDNA) sequences (Miller et al., 1999; Tiedje et al., 1999). These sequences contain both conserved and more variable regions, allowing the discrimination of microorganisms at different taxonomic levels. Such methods are also used to generate fingerprints of the whole microbial community of a soil. As the molecular patterns obtained can be affected by several parameters such as the DNA recovery method or the PCR amplification procedure used (Wintzingerode et al., 1997; Martin-Laurent et al., 2001), they should not be considered as an absolute description of the microbial diversity of a given soil. But these direct molecular methods are very useful in comparative situations, for example to monitor the diversity of microbial communities in soils in response to a perturbation. Initially, the methods were developed for bacteria, and most of the studies addressing soil microbial diversity investigate only bacterial communities (Marsh et al., 2000; Ranjard et al., 2001). Similar fingerprinting approaches are starting to be used for fungal communities (Smit et al., 1999; Borneman and Hartin, 2000), but some developments are still required. Finally, protozoa also represent an important component of the soil microbiota but are generally overlooked by soil microbiologists. Recently, Rasmussen et al. (2001) proposed an approach to investigate their diversity in environmental samples, based on the development of primers that specifically amplify protozoan rDNA sequences.

With such fingerprinting approaches based on PCR, the same soil DNA can be used to analyse bacterial, fungal and protozoan communities, depending on the primers used. In addition, given species or genera of interest can also be targeted from the total soil DNA by using specific primers. Finally, functional gene diversity can be assessed after PCR amplification of a given gene with conserved primers. For example, this strategy was developed to compare in different soils the pools of the genes involved in nitrogen cycling (Poly *et al.*, 2001; Wu *et al.*, 2001). A basic question is still unresolved: what is the appropriate level of resolution to assess the diversity affecting this microbial community? There is probably no general answer, with the resolution depending on the type of phenomenon being studied. Generally, techniques based on the study of the structure of the ribosomal DNA give information on the presence of genera and species, but not on the activity of these populations. In many cases it would be necessary not only to characterize the structure of the microbial populations but also to determine their functions, i.e. to assess the functional diversity of the microbial populations. Knowing that some essential microbial functions are redundant in soil, it is probably more important to make sure that a function exists in a given soil sample rather than to determine the existence of a given population without knowing if it is expressing this function.

Physiological methods to characterize the diversity of the soil microbiota

A representative estimate of microbial diversity and community structure is a prerequisite for understanding the functional activity of microorganisms in soil (Ovreas, 2000). It is therefore essential to combine studies on genotypic diversity with studies on phenotypic diversity to get a better insight into the microbial processes within the ecosystem. These include decomposition and nutrient cycling, but also impact of anthropic perturbations and management practices, and resilience (Ovreas, 2000). Garland and Mills (1991) proposed the use of the Biolog method (Biolog Inc., Hayward, California, USA) to characterize metabolic behaviours of whole soil bacterial communities. This Biolog method was initially developed for the characterization of microorganisms on the basis of their substrate utilization pattern. It has since been used to compare different ecological environments (Kaiser *et al.*, 1998)

and different agricultural practices (Fritze *et al.*, 1997), and to evaluate the impact of chemicals and of introduced microorganisms on ecosystems (Vahjen *et al.*, 1995; Kelly and Tate, 1998). It has not yet been applied widely in studies related to plant health.

Substrate utilization or production profiles evaluated at the soil microbial community level may provide useful information on functional biodiversity. They offer a suite of physiological indicators of the microflora to complement the suite of indicators revealed through targeting of genes coding for enzymes related to specific agronomic functions such as degradation of pesticides and denitrification, or the measurement of global functions such as dehydrogenase activity and soil microbial respiration.

Biodiversity among populations of soilborne *Fusarium* species

The case of the soilborne populations of *Fusarium* spp. will illustrate the different approaches that can be followed to describe the diversity of soilborne populations of fungi. The genus *Fusarium* is ubiquitous in soils all over the world. This genus is heterogeneous and includes many different species, some of them being important plant pathogens. Thus, in relation to crop health it is necessary to characterize the different species present in a given soil sample. Many different taxonomic systems have been proposed. They are based on morphological characters such as the size and the shape of macroconidia, the presence or absence of microconidia and chlamydospores, the coloration of the colonies and the structure of the conidiophores (Windels, 1992). Due to the very large variations affecting a single species, the different taxonomic systems distinguish from nine species (Snyder and Hansen, 1940) to 78 (Nelson *et al.*, 1983) or even more than 100 different species (Booth, 1971).

Other methods are based on physiological traits. They take into account the capacity of the strains to utilize specific nutrients and to produce different types of secondary metabolites such as enzymes and toxins. All these methods reveal a great diversity in *Fusarium* both at the inter- and intraspecific levels (Thrane and Hansen, 1995). Characterization of enzyme polymorphisms has been used to identify *Fusarium* species (Huss *et al.*, 1996), and attempts have been made to establish an identification key based on the Biolog system (Seifert and Bissett, unpublished workshop presentation). However, none of these methods is commonly used.

There was a need to base the taxonomy on other criteria and during the last 10 years different methods based on molecular tools have been developed for the identification of *Fusarium* species. For example, Edel *et al.* (1997a) have proposed a PCR–restriction fragment length polymorphism (RFLP) method to identify most of the *Fusarium* species.

In relation to soil health, identification of the species is not sufficient. It is also necessary to characterize the diversity within a species. Three molecular methods showing different levels of discrimination were proposed to describe the diversity within populations of *F. oxysporum* in soils (Edel *et al.*, 1995). The characterization of the vegetative compatibility (Puhalla, 1985) and of the trophic patterns (Steinberg *et al.*, 1997) of strains provided additional information on the diversity of the populations. These different methods, which are complementary, provided consistent results when applied to a population of *F. oxysporum* isolated from a soil. They showed the great diversity within this population, with trophic characterization and vegetative compatibility grouping being more discriminating than the genotypic characterization. The use of such methods revealed the influence of plant species and cultural practices on the structure of the population of *F. oxysporum* (Edel *et al.*, 1997b, 2001). However, none of these methods enables the characterization of the pathogenicity of the isolates. When interested in pathogenic *F. oxysporum*, it is necessary to identify the *formae speciales* and races, or the non-pathogenic ability of strains isolated from soil. At the present time, the only valid method to identify *formae speciales* or races is through a bioassay in which the putative host plants are confronted with the fungus.

This example shows the complexity of evaluating the diversity of a single population

of soilborne fungus and how difficult this approach is when considering the diversity among the whole soil microflora. The concept of soil health obliges us to consider not a single population but the main populations interacting to contribute to suppression of diseases and to crop health. The necessity to take into account the whole microflora is illustrated by the well-known example of soil naturally suppressive to some important diseases.

Soil Suppressiveness to Diseases

On the one hand, agronomists characterize soil quality mostly as physical and chemical factors potentially involved in plant productivity (fertility, texture, pH, water-holding capacity, soil bulk density, organic matter (OM) content, C : N, availability of micronutrients). On the other hand, plant pathologists consider pathogens and the soil inoculum potential as key factors determining disease incidence and crop losses, thus indicating poor soil quality. Soil inoculum potential is the soil-dependent capacity of pathogens to incite disease. It results from the interactions between inoculum density and soil suppressiveness. Soil suppressiveness corresponds to the global effects of the soil microbiota interacting with the pathogens. Two main mechanisms are responsible for soil suppressiveness: general suppression, which is correlated with the activity of the total microbial biomass at critical times for the pathogen, and specific suppression, which is due to the activity of specific microorganisms that are antagonistic to the pathogen (Cook and Baker, 1983). In fact, soil suppressiveness could be considered as one indicator of a microbiologically well-buffered soil.

The global approach of soil health relies on observations of natural field situations where the incidence of diseases induced by soilborne pathogens remains very limited, despite the presence of the pathogens. Soils suppressive to diseases induced by the most important soilborne pathogens have been described. They include both fungal and bacterial pathogens but also nematodes (Schneider, 1982; Schippers, 1992). Root rot and wilt diseases induced by *Aphanomyces euteiches*, *Cylindrocladium* spp., several *formae speciales* of *F. oxysporum*, *Gaeumannomyces graminis*, *Pythium* spp., *Phytophthora* spp., *Rhizoctonia solani*, *Ralstonia solanacearum*, *Streptomyces scabies*, *Verticillium dahliae* and *Thielaviopsis basicola* (*Chalara elegans*) are sensitive to suppression in various soils. This large diversity of pathogens controlled by suppressive soils shows that soil suppressiveness is not a rare phenomenon. On the contrary, every soil has some potential for disease suppression, leading to the concept of soil receptivity to diseases. The receptivity of a soil is the capacity of this soil to control more or less the activity of the microbial populations present in the soil. In the case of pathogens, it is the capacity to control the pathogenic activity. Soil receptivity is a continuum ranging from highly conducive soil in which disease incidence is very high to strongly suppressive soils (Alabouvette *et al.*, 1982).

Properties of suppressive soils

In nature, suppressive soils can be detected by the observation that disease incidence in a crop remains low despite the presence of a susceptible host plant, climatic conditions favourable to disease expression and ample opportunity for the pathogen to be present. It is quite easy to demonstrate experimentally that a soil is suppressive to a given disease. The pathogen has to be produced in the laboratory and introduced into the soil at increasing inoculum densities. A susceptible host plant is sown and cultivated under standardized conditions favourable to disease expression. Observations of symptom appearance enable disease progress curves to be drawn with respect to time and inoculum concentrations. Appropriate statistical methods (Baker *et al.*, 1967; Höper *et al.*, 1995) enable these curves to be compared with those obtained from another soil known to be conducive to the disease. All experimental conditions being similar, differences in disease incidence must be attributed to differences in the soil environment, i.e. differences in the level of soil receptivity.

Disease suppression does not necessarily imply suppression of the pathogen. In most cases the inoculum is still present but does not provoke the disease. Therefore, Cook and Baker (1983) distinguished pathogen-suppressive soils, in which the pathogen does not survive, from disease-suppressive soils, in which inoculum is present but does not induce the disease. Only studies of the mechanisms of suppression enable the distinction between the two types of suppressiveness to be made.

From a theoretical point of view, both the abiotic characteristics of a soil and its biological properties can be responsible for disease suppression. However in most cases, suppressiveness is fundamentally microbial in nature. Disease suppression results from more or less complex interactions between the pathogen and all or a part of the soil microbiota. Indeed, the suppressive effect disappears upon destruction of organisms by biocidal treatments such as steam or methyl bromide, and can be restored by mixing a small quantity of suppressive soil into the previously disinfested soil (Alabouvette, 1986). Suppressiveness can also be restored in the steamed disinfested soil by re-introduction of a mixture of microorganisms previously isolated from the suppressive soil (Alabouvette, 1986).

This demonstration of the essential role of the saprophytic microflora does not establish that soil physical and chemical properties do not play any role in the mechanisms of suppressiveness. On the contrary, early studies on *Fusarium* suppressive soils established correlation between soil type, presence of smectite clays and soil suppressiveness to *Fusarium* wilt in Central America (Stover, 1962; Stotzky and Martin, 1963). In the case of Swiss soils suppressive to black root rot of tobacco, Stutz *et al.* (1989) showed that only soils derived from moraine and containing vermiculitic clay minerals were suppressive to *Thielaviopsis basicola*. Many abiotic factors such as soil texture, water potential, aeration, OM content and cation availability (Al, Fe, Mn) are indirectly involved in the mechanisms of suppression, but it is impossible to generalize these findings from one soil to another (Höper and Alabouvette, 1996).

Microbial interactions responsible for soil suppressiveness

Cook and Baker (1983) proposed three criteria for disease suppressiveness in soils: 'the pathogen does not establish; it establishes but fails to produce disease; or it establishes and causes disease at first but then disease severity diminishes with continued growing of the same crop'.

The well-known and widespread phenomenon of take-all decline is the best example of soils becoming suppressive with continuous cropping of the susceptible host plant. The disease increases in severity during the first years of wheat cropping, then decreases to an economically acceptable threshold (Hornby, 1998).

Fusarium wilt suppressive soils provide a good example of soils where the pathogen is present in the soil but fails to produce the disease (Scher and Baker, 1980; Alabouvette, 1986). It was established that the dynamics of a marked inoculum of *F. oxysporum* f.sp. *melonis* were similar in a conducive soil and in a suppressive soil from Châteaurenard, France; thus the difference in disease incidence had to be attributed to a reduced activity of the pathogen in the suppressive soil. Indeed, the percentage of germinating chlamydospores is always extremely low in the suppressive soil. This limited germination of chlamydospores was attributed to the general phenomenon of soil fungistasis (Lockwood, 1977), which is related to competition for nutrients. Addition of increasing concentrations of available carbon, in the form of glucose, resulted in increasing percentages of germinating chlamydospores in both conducive and suppressive soils (Sneh *et al.*, 1984; Alabouvette, 1986). These results suggest that fungistasis and competition for nutrients are much more intense in suppressive than in conducive soils, and contribute to reducing the activity of the fungal pathogens. Indeed, glucose amendments that induce chlamydospore germination of the pathogen also induce disease in the suppressive soils. Competition for nutrients, especially competition for energy among heterotrophic microorganisms, is due to all the soil microorganisms active at any given time and therefore should be linked to the activity of the microbial biomass of the soil.

The microbial biomass, measured by Jenkinson's method (Jenkinson and Powlson, 1976) is always greater in the Châteaurenard suppressive soil than in a conducive control soil. Studies on the kinetics of soil microbial respiration after glucose amendment (Alabouvette, 1986; Amir and Alabouvette, 1993) showed further that the soil microflora of the suppressive soil is more responsive to carbon than that of the conducive soil. Consequently, carbon is utilized more quickly and the development of any given organism is stopped more rapidly after glucose amendment in the suppressive than in the conducive soil.

This type of phenomenon corresponds to the 'general suppression' described by Cook and Baker (1983) as the inhibition of the pathogen in soil in relation to the total amount of the microbial activity acting as a nutrient sink. A high microbial biomass combined with a very intense competition are responsible for a permanent state of starvation leading to fungistasis inhibiting the growth of the pathogen. This general suppression had been proposed by Gerlach (1968) as an explanation for take-all decline of wheat in polders.

Competition for nutrients other than carbon, especially nitrogen and iron, has been involved in the limitation of germination of fungal propagules in the soil (Cook and Snyder, 1965; Benson and Baker, 1970; Scher and Baker, 1982). Consequently, the population of pathogens faces general competition resulting from the activity of the microbial biomass but also competition exerted by a specific population. For instance, the siderophore-iron competition achieved by fluorescent pseudomonads is responsible for the reduced growth of *Fusarium* spp. *in vitro* and in suppressive soils (Sneh *et al.*, 1984; Elad and Baker, 1985). Addition of ethylenediaminedi-*o*-hydroxyphenyl-acetic acid (EDDHA), which limits the concentration of iron available for *Fusarium*, results in a lower percentage of diseased plants in a conducive soil. In contrast, addition of Fe-ethylenediaminetetraacetic acid (FeEDTA), which provides iron available for *Fusarium*, results in a higher percentage of diseased plants in suppressive soils (Lemanceau, 1989). General competition occurs simultaneously for both carbon and iron in the suppressive soil from Châteaurenard.

Competition for nutrients is not the only mechanism by which antagonistic populations interact with pathogens in soil. Today, antibiosis has been shown to be involved in the inhibition of the pathogen activity in suppressive soils. Indeed, Raaijmakers and Weller (1998) were able to correlate the suppressiveness of soils to take-all with the density of the population of *Pseudomonas fluorescens* producing 2'-4-diacetyl phloroglucinol. But it must be emphasized that this 'specific suppression' always operates against a background of general suppression as stated by Cook and Baker (1983). The high intensity of general competition enhances or increases the significance of specific interactions, either competition or antibiosis, between pathogens and antagonists sharing the same ecological niches in the soil and the rhizosphere. The choice of focusing on specific populations of antagonists is justified by the objective of developing biological control agents. But this approach has still not been as successful as expected, and it is time to think again of the general suppression in order to ensure soil health through integrated management.

Much more research is needed because generally a soil suppressive to a given type of disease, for example *Fusarium* wilts, is conducive to other types of diseases, for example *Pythium* damping-off or *Rhizoctonia* root rot, and the nature of suppressiveness can be different, as shown in Fig. 8.1. Thus, promoting soil health by increasing suppressiveness to the main diseases affecting a given crop is a real challenge. It will probably be necessary to stimulate both general suppression and several specific suppressions simultaneously.

Management Practices and Disease Suppression

Because soil suppressiveness is essentially biological, it should be possible to increase soil suppressiveness by cultural practices that influence aspects of soil biology including

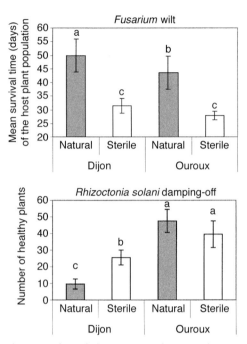

Fig. 8.1. The soils from Dijon and Ouroux, heat-treated (sterile) or not (natural) were inoculated with *Fusarium oxysporum* f.sp. *lini* and *Rhizoctonia solani* and then cultivated with a susceptible host plant, respectively flax and *Pinus nigra*. Disease incidence expressed as mean survival time and number of healthy plants, respectively, showed that the soil from Dijon is suppressive to *Fusarium* wilt but conducive to *Rhizoctonia* damping-off, and reciprocally for the soil from Ouroux. In both soils the suppressiveness to *Fusarium* wilt disappeared after heat treatment, demonstrating the biological nature of this suppressiveness. In contrast, destruction of the soil biota does not affect the suppressiveness to *Rhizoctonia* damping-off.

diversity (see Brussaard *et al.*, Chapter 9, this volume). Cultural management of plant diseases has been used for a long time and undoubtedly many of these practices act at least partly by their effects on suppressiveness.

There is not yet a good synthetic, or systems, framework for understanding the links between management, soil biodiversity and disease suppression (see also Carter *et al.*, Chapter 15, this volume). There are two major reasons for this. First, the techniques needed to study diversity at the required resolution have only recently been developed and have not yet been applied widely to studies that include both management effects and measurements of disease suppression. Secondly, most management practices have direct effects on pathogen populations, as well as effects mediated through the other soil biota, and there are still formidable technical difficulties in separating these effects.

In this section we consider the likely outcomes of common management practices on disease suppression.

Cropping sequence

The effects of cropping sequence on soil receptivity are confounded by their effects on inoculum density of the pathogens. As a rule, continuous cropping of the same plant species will lead to an increase in incidence of soilborne diseases, whereas rotation with non-hosts should lead to a decrease in incidence. The effects of crop sequence on biological suppression of pathogens can sometimes be recognized when unexpected phenomena occur, as is the case with disease declines.

Continuous cropping

For a small number of diseases, continuous cropping can lead to a decrease in disease incidence over a number of years, in what is known as disease decline. The best-characterized decline phenomenon is take-all decline (TAD) in cereals, where the causal pathogen is the fungus *Gaeumannomyces graminis* var. *tritici*. This is a clear example of continuous cropping altering disease suppressiveness of soil through its effects on specific components of the soil microflora.

Most hypotheses to explain TAD are based on increases in populations of antagonistic microorganisms. These may be selected for by the increased population of the pathogen, but are more likely to be selected for by the host. Continuous presence of wheat roots would encourage increased populations of rhizosphere bacteria and fungi adapted to wheat roots that could then compete with *Gaeumannomyces*. Evidence for this is that growing *Holcus lanatus*, a host of the take-all

fungus, disrupts the development of TAD in wheat, suggesting that the antagonistic microflora associated with TAD in wheat is specific to wheat (Hornby, 1998).

Increases in populations of antagonistic fluorescent pseudomonads on the rhizoplane have commonly been associated with the onset of TAD (Weller, 1983). Attempts to identify changes in fungal communities on wheat roots associated with TAD have been less successful (Bateman and Kwasna, 1999). *G. graminis* var. *tritici* is known to be sensitive to antagonism by a wide range of fungi and bacteria (Hornby, 1998) and suppression of take-all has been shown to be associated with different types of microorganisms in different soils (Andrade *et al.*, 1994). TAD may therefore be due to a suite of antagonistic microorganisms, the composition of which could vary from place to place. There is some evidence that TAD involves a reduction in saprophytic survival of the fungus as well as inhibition of pathogenic activity (Simon and Sivasithamparam, 1990), and it is likely that both specific and general suppression are involved in most cases of TAD.

Decline phenomena appear to result when the microbial populations that are selectively stimulated by healthy or diseased roots of the host are strongly antagonistic to a pathogen. The decline can be readily detected when the effect on pathogen activity is large, such as with TAD. However, it is likely that some form of increased suppression of pathogens is widespread, but cannot be detected by observation if the effect of suppression is less than the effect of increased inoculum of the pathogen caused by continuous cropping of the host. It would be worthwhile to develop techniques for assessing disease suppressiveness that are independent of resident inoculum levels to test the effect of continuous cropping on a wider range of diseases.

Crop rotation

The main effect of crop rotation is to allow time for decrease, through natural mortality, of inocula of pathogens that are poor saprophytic competitors. Clean fallows have the same mechanism. However, since mortality of pathogen propagules in soil is frequently due to the effects of other organisms, the stimulation of microbial activity by the growing rotational crop should make rotation more effective than fallowing. Although this principle is often stated, it is very difficult to demonstrate. For example, Gardner *et al.* (1998) found brassica break crops to be no more effective than fallows of equivalent length in controlling take-all in several experiments. However, rotation can sometimes be used to promote the activity of specific antagonists. Grass leys can increase the populations of *Phialophora* species that are associated with delayed build-up of take-all in subsequent wheat crops (Slope *et al.*, 1979), although this has not yet been deliberately exploited as a management tool for take-all (Hornby, 1998).

Crop rotation increases the diversity of plants within an agricultural system, which may have effects on the diversity of soil biota. For example, Lupwayi *et al.* (1998) found that bacterial diversity in the rhizosphere of wheat following legumes was higher than in wheat following wheat or wheat following fallow. The implications of this for plant health remain uncertain until we fully understand the relationship between microbial diversity and disease suppression. Experiments by Edel *et al.* (1997b; Fig. 8.2) showed that different crops stimulated different components of the community of non-pathogenic *F. oxysporum* in soil. Since non-pathogenic strains are involved in suppression of *Fusarium* wilt, this indicates the potential for manipulating soil suppressiveness by cropping sequence.

Tillage

Soil disturbance by tillage has been shown to have a variety of effects on diseases. Root rots of many crop plants caused by *R. solani* are generally less severe after tillage than with direct drilling, and this has been exploited as a means of disease management (Roget *et al.*, 1996). The mechanism is believed to be mechanical disruption of the hyphae. On the other hand, common root rot of wheat caused by *Cochliobolus sativus* may be more severe in tilled soils (Mathieson *et al.*, 1990),

a phenomenon associated with increases in spore populations or changes in the distribution of inoculum in soil. These well-characterized effects of tillage on disease seem to act directly on the pathogen, with no evidence yet of effects on other components of the soil biota.

Tillage would be expected to have some influence on soil suppressiveness because it does alter the activity and diversity of soil microflora. Typically, tillage reduces bacterial biomass and diversity in soil, possibly through its effects on soil aggregation (Lupwayi *et al.*, 1998, 2001). Reduced tillage systems should therefore have more diverse and active microflora, and greater general suppression of diseases. However, there appear to be no experimental data that clearly demonstrate this. Determining the effects of tillage on soil receptivity or suppressiveness is complicated, because different tillage treatments are usually confounded with other aspects of crop management. Comparisons of 'conventional' cultivation with organic, low-input or conservation tillage systems usually confound tillage and residue management effects. More experimentation is required that allows the determination of the effects of tillage by itself on biological suppression of soilborne diseases.

Residue management

Residue management, particularly retention of crop residues *in situ*, can have conflicting effects on disease. Retaining residues increases the inoculum potential of pathogens that survive in the residue. However, there is increasing evidence that residue retention can boost the levels of general suppression in soils. This is to be expected, since general suppression has been linked to high levels of microbial activity, which in turn depend on high levels of OM input into soils. Demonstrating suppressiveness has been difficult because of the confounding effect of increased inoculum. There have, however, been some examples of decline phenomena associated with residue retention that have been large enough to be observed directly in the field.

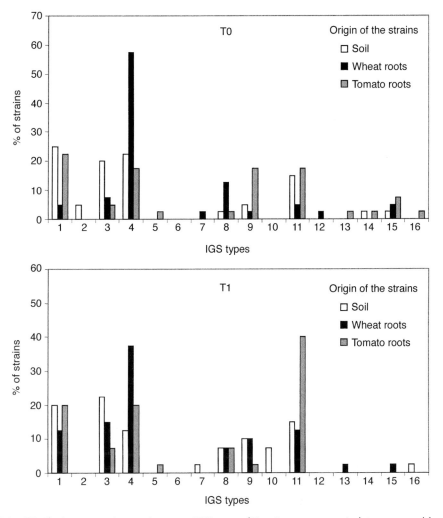

Fig. 8.2. Distribution among intergenic-spacer (IGS) types of *Fusarium oxysporum* isolates recovered from soil and roots of wheat and tomato, at two sampling times (T0 and T1). The distribution of soil isolates was similar at both sampling times. The structure of *F. oxysporum* populations associated with roots of wheat and tomato differed from the structure of soilborne populations, and also differed at each sampling time.

Roget (1995) reported decline in root rot of wheat caused by *R. solani* AG 8 in a long-term tillage and rotation experiment in South Australia. This was subsequently shown to be of biological origin (Wiseman *et al.*, 1996). Suppression of take-all and crown rot, caused by *Fusarium pseudograminearum*, could also be demonstrated in these soils (Mussared, 1996). Suppressiveness developed both under continuous wheat and under wheat in rotation with other crops, and it appeared that stubble retention was the most important factor

in development of suppression (Mussared, 1996).

Residue retention may favour specific antagonists. For example, populations of cellulolytic organisms tend to be higher in soils where crop residues are retained (Toresani *et al.*, 1998), and high cellulolytic activity has been correlated with suppression of disease such as *Fusarium* seedling blight in barley (Rasmussen *et al.*, 2001). Residue retention should also favour cellulolytic *Trichoderma* species, many of which are antagonistic to plant pathogens (Papavizas, 1985). Although organic amendments with high C:N ratios do increase populations of *Trichoderma* in soils (Kwasna *et al.*, 2000), evidence for a similar effect of crop residues is hard to find. This is certainly worth exploring further.

Organic amendments

Addition of organic amendments such as animal manures and industrial by-products is the best-documented strategy for increasing disease suppression in soils (Cook and Baker, 1983). Manures and other amendments tend to increase microbial biomass and biological activity in soil, and this is believed to enhance general suppression. However, high biological activity by itself does not necessarily lead to enhanced suppressiveness. Aryantha *et al.* (2000) found that a range of fresh and composted animal manures increased OM and total microbial activity in soil, but only chicken manure stimulated suppression of diseases of lupins caused by *Phytophthora cinnamomi*. This was associated with specific stimulation of endospore-forming bacteria by chicken manure. Similarly, Bulluck and Ristaino (2002) reported differences in effectiveness of various organic amendments, and also showed differences in the spectrum of microbial groups in soil that were stimulated by each amendment.

The foregoing suggests that the effectiveness of manuring against any disease will depend on which components of the microflora are stimulated. This will depend on the type of manure or other amendment, and also on soil type and other factors as illustrated in Fig. 8.3. The present state of knowledge does not allow the effectiveness of any organic amendment to be predicted, so each has to be tested empirically in each new situation.

Cover crops

Cover crops are widely used in agriculture and can take a range of forms. The most common are green manure crops that are grown to be cultivated back into the soil to build up

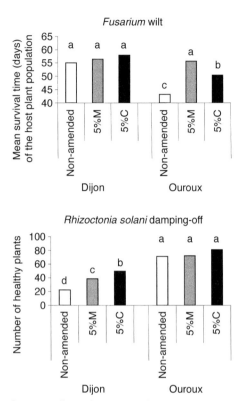

Fig. 8.3. Effects of two types of organic amendments: manure (M) and spent mushroom compost (C) on the receptivity level of two soils to *Rhizoctonia* damping-off and *Fusarium* wilt of flax. Organic amendments did not change the suppressiveness of the soil from Ouroux to *Rhizoctonia* damping-off but decreased its receptivity to *Fusarium* wilt. Manure was more effective than compost. Similarly, organic amendments did not change the suppressiveness of the soil from Dijon to *Fusarium* wilt but decreased its receptivity to *Rhizoctonia* damping-off, compost being more efficient than manure.

OM and nutrients. The effects of cover crops on the soil biota will be similar to those of rotation crops, residue retention or organic amendments. These include the provision of additional substrate and niche diversity for the soil biota, and other effects such as the release of ammonia. However, cover crops should be considered as a potential soil health management strategy in their own right.

Abawi and Widmer (2000) have reviewed work on the use of green manure crops to manage diseases of green beans. Root rot severity due to a complex of fungi, and populations of several nematodes, could be reduced by a number of different green manure crops. However, cover crop species varied widely in their effectiveness, even between closely related plants, so the effect was not a general one due solely to OM inputs. As well as stimulating microbial activity, green manure crops often have a biofumigant effect, as in brassicas and some legumes.

Soil chemistry

Alterations to soil chemistry, by addition of fertilizers or manipulation of pH, are major crop management tools. These can have significant effects on general suppressiveness of soil and on particular components of the antagonistic microflora.

Liming soil reduced the severity of cavity spot of carrots, caused by *Pythium* species (El-Tarabily *et al.*, 1996). This did not appear to be due to changes in either soil or tissue levels of calcium, but was associated with an increase in soil pH from 5.1 to 6.9. Limed soils had higher total microbial activity and populations than unlimed soils. Liming tended to increase the populations of bacteria and actinomycetes, but reduce the populations of fungi. El-Tarabily *et al.* (1996) suggested that the changes to the soil microbial community were responsible for the reduced disease severity, although their experimental design could not prove a causal link.

Reducing pH by annual applications of ammonium sulphate appeared to increase suppressiveness of a soil from Western Australia to take-all, and this increased suppressiveness could be reversed by addition of lime (Simon and Sivasithamparam, 1990). The suppressive soil had lower microbial activity than conducive soil, higher levels of fungi and lower levels of bacteria. A much higher proportion of fungal isolates from the suppressive soil were *Trichoderma* species, and Simon and Sivasithamparam (1990) suggested that these might have played a role in suppressiveness. *Trichoderma* species are known to be favoured by acid soils (Papavizas, 1985), possibly because

antagonism of *Trichoderma* by bacteria is reduced (Simon and Sivasithamparam, 1990).

Nitrogen is one of the largest inputs in modern agriculture, and has several possible interactions with plant diseases. Nitrogen nutrition may affect the susceptibility of the host, the growth and other activities of the pathogen, or other components of the microflora that are antagonistic to pathogens. Although one could speculate about the roles of nitrogen in biological disease suppression, there is insufficient evidence at present to separate the effects of nitrogen on suppressive organisms from other effects of nitrogen fertilizer. Almost all studies on the effects of nitrogen on disease suppression have been confounded by, or shown to be caused by, the effects of nitrate and ammonium ions on rhizosphere pH (Smiley and Cook, 1973). The other common confounding effect of nitrogen is ammonia toxicity (Lazarovits, 2001).

There is a clear need for detailed experimental work on the role of soil nutrients on disease suppression that distinguishes nutrient effects on the soil biota from indirect effects on plant physiology, soil pH and other factors.

Pesticides

With the exception of seed treatments to protect against damping-off and similar diseases, fungicides have relatively limited use against soilborne diseases. However, a wide range of biocidal compounds are applied to soil in many agricultural systems. These have demonstrable effects on susceptible organisms and the soil processes that these organisms mediate. Although it may be expected that use of some pesticides may affect suppressiveness of soil, this has yet to be investigated in any depth.

Management Thresholds

At this stage it is difficult to make generalizations about the effects of cultural techniques on soil quality as measured by suppressiveness to plant diseases. It is a common experience that observations made in one year or at one site can not be repeated else-where. There are a number of interacting, confounding factors at work in most situations (Fig. 8.4).

Management practices alter the physical and chemical environment of the soil and the resources, most simply thought of as OM inputs, available to soil microorganisms including pathogens. The populations of organisms will respond to these changes by increases or decreases in activity or by changes in mortality or reproduction rates. The ultimate effect on pathogen activity, expressed as plant disease, will depend on the balance of effects on, and interactions between, each member of the system.

Consider the role of antagonistic *Trichoderma* species in soil. Populations of *Trichoderma* can be increased by addition of suitable OM (Kwasna *et al.*, 2000), possibly increasing suppressiveness of soil to some pathogens. However, *Trichoderma* species are themselves sensitive to general suppression by high microbial activity (Papavizas, 1985), which could also result from addition of OM. Under some conditions addition of OM will favour *Trichoderma*, which may suppress some pathogens, whereas under other conditions addition of the same OM may favour general microbial activity at the expense of *Trichoderma* – and this high microbial activity may suppress other patho-gens. It is also possible that addition of the same OM may stimulate the activity of the pathogen. This implies that it will be difficult to predict the effect of a management practice on suppression of a particular disease in a specific situation from general principles.

In terms of management thresholds, we can state that practices that reduce total biological activity of soil will reduce general suppression of plant diseases. Indeed, this is how suppressiveness was shown to have a biological component. The converse proposi-tion, that treatments that increase biological activity will also enhance suppressiveness, is less certain, as was seen in the discussion of manuring. However, optimum levels of gen-eral suppression are likely to be maintained by practices that maximize biological activity and functional diversity of the soil biota.

The best-characterized examples of man-agement practices enhancing disease suppres-sion are usually cases of specific suppression. Generalizations about management thresh-olds for specific suppression are difficult to make, because each group of antagonists involved will have different ecological requirements. However, the potential for specific suppression will be maximized by practices that maintain diversity of functional groups. Thresholds for stimulation of specific suppression will be more locally variable, since they will depend on the suite of potential antagonists present in each soil.

Conclusion

Examples presented in this chapter show how complex the interactions between soil abiotic characteristics, soil microbiota and soil suppressiveness to diseases are. They also show how difficult it is to characterize the diversity of the soil biota, and thus how far we are from being able to manage this biodiversity to preserve or restore soil health. At the end of this chapter, we want to stress that much more research is needed to clearly understand the effects of management prac-tices on the diverse components of soil health, and therefore to determine when and what kind of management is necessary so that farmers can be helped to make the right decisions. We must admit that we are still

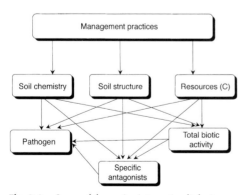

Fig. 8.4. Some of the major interactive links in the effect of management practices on pathogen populations or activity.

at the descriptive stage, and that we have difficulties addressing the question of soil health following a holistic approach.

The rapid development of molecular and physiological tools is enabling the characterization of the structure of the microbiota as a whole. Until recently, we were obliged to focus our attention on a very limited number of microbial populations, either pathogens or antagonists, but were unable to detect changes affecting the soil microbiota without having, a priori, a specific hypothesis. Thus the development of new techniques enabling the evaluation of biodiversity of the soil microbiota will totally change our views of the microbial balance. If, as expected, these new methods can be run automatically, they will make possible the characterization of many samples and thus enable comparison of the microbial communities in different soils under different cropping systems or in a single soil subjected to different management practices.

These new techniques will be useful for correlating changes affecting the level of soil suppressiveness with shifts affecting microbial communities. We will be able, for example, to detect and characterize shifts in the microbial communities following application of any treatment to the soil (fertilizer, pesticide, biocontrol agents, OM, etc.) and to correlate these changes with variations in the level of soil suppressiveness to a set of diseases. Moreover, the molecular techniques should enable one, by consulting a gene data bank, to determine which populations are affected by the treatments, and then to study their role or their function in the ecosystem. It will be possible to determine if these populations are really involved in mechanisms controlling soil health or if they are only indicators (markers) of soil health. However, it is obvious that several or probably many indicators will be needed to determine when management thresholds are exceeded. Therefore mathematical modelling will be necessary to organize all these data and to follow the dynamics of the measured parameters, whether biotic or abiotic. The resulting and evolving models will allow management techniques to be proposed that are useful for farmers and for the preservation of the environment.

References

Abawi, G.S. and Widmer, T.L. (2000) Impact of soil health management practices on soilborne pathogens, nematodes and root diseases of vegetable crops. *Applied Soil Ecology* 15, 37–47.

Alabouvette, C. (1986) Fusarium-wilt suppressive soils from the Châteaurenard region: review of a 10-year study. *Agronomie* 6, 273–284.

Alabouvette, C., Couteaudier, Y. and Louvet, J. (1982) Comparaison de la réceptivité de différents sols et substrats de culture aux fusarioses vasculaires. *Agronomie* 2, 1–6.

Amir, H. and Alabouvette, C. (1993) Involvement of soil abiotic factors in the mechanisms of soil suppressiveness to fusarium wilts. *Soil Biology and Biochemistry* 25, 157–164.

Anderson, J.P.E. and Domsch, K.H. (1978) A physiological method for the quantitative measurement of microbial biomass in soils. *Soil Biology and Biochemistry* 10, 215–221.

Andrade, O.A., Mathre, D.E. and Sands, D.C. (1994) Natural suppression of take-all disease of wheat in Montana soils. *Plant and Soil* 164, 9–18.

Aryantha, I.P., Cross, R. and Guest, D.I. (2000) Suppression of *Phytophthora cinnamomi* in potting mixes amended with uncomposted and composted animal manures. *Phytopathology* 90, 775–782.

Baker, R., Maurer, C.L. and Maurer, R.A. (1967) Ecology of plant pathogens in soil. VII. Mathematical models and inoculum density. *Phytopathology* 57, 662–666.

Bateman, G.L. and Kwasna, H. (1999) Effects of number of winter wheat crops grown successively on fungal communities on wheat roots. *Applied Soil Ecology* 13, 271–282.

Benson, D.M. and Baker, R. (1970) Rhizosphere competition in model soil systems. *Phytopathology* 60, 1058–1061.

Booth, C. (1971) *The Genus Fusarium*. Commonwealth Mycological Institute, Kew, UK, 237 pp.

Borneman, J. and Hartin, R.J. (2000) PCR primers that amplify fungal rRNA genes from environmental samples. *Applied and Environmental Microbiology* 66, 4356–4360.

Bulluck, L.R. III and Ristaino, J.B. (2002) Effect of synthetic and organic soil fertility amendments on southern blight, soil microbial communities, and yield of processing tomatoes. *Phytopathology* 92, 181–189.

Cook, R. and Baker, K.F. (1983) *The Nature and Practice of Biological Control of Plant Pathogens*. American Phytopathological Society, St Paul, Minnesota, 539 pp.

Cook, R.J. and Snyder, W.C. (1965) Influence of host exudate on growth and survival of germlings of *Fusarium solani* f. *phaseoli* in soil. *Phytopathology* 55, 1021–1025.

Doube, B.M. and Schmidt, O. (1997) Can the abundance or activity of soil macrofauna be used to indicate the biological health of soils. In: Pankhurst, C., Doube, B.M. and Gupta, V.V.S.R. (eds) *Biological Indicators of Soil Health*. CAB International, Wallingford, UK, pp. 265–295.

Edel, V., Steinberg, C., Avelange, I., Laguerre, G. and Alabouvette, C. (1995) Comparison of three molecular methods for the characterization of *Fusarium oxysporum* strains. *Phytopathology* 85, 579–585.

Edel, V., Steinberg, C., Gautheron, N. and Alabouvette, C. (1997a) Evaluation of restriction analysis of polymerase chain reaction (PCR)-amplified ribosomal DNA for the identification of *Fusarium* species. *Mycological Research* 101, 179–187.

Edel, V., Steinberg, C., Gautheron, N. and Alabouvette, C. (1997b) Populations of non-pathogenic *Fusarium oxysporum* associated with roots of four plant species compared to soilborne populations. *Phytopathology* 87, 693–697.

Edel, V., Steinberg, C., Gautheron, N., Recorbet, G. and Alabouvette, C. (2001) Genetic diversity of *Fusarium oxysporum* populations isolated from different soils in France. *FEMS Microbiology Ecology* 36, 61–71.

Elad, Y. and Baker, R. (1985) Influence of trace amounts of cations and siderophore-producing pseudomonads on chlamydospore germination of *Fusarium oxysporum*. *Phytopathology* 75, 1047–1052.

El-Tarabily, K.A., Hardy, G.E.S.J., Sivasithamparam, K. and Kurtböke, I.D. (1996) Microbiological differences between limed and unlimed soils and their relationship with cavity spot disease of carrots (*Daucus carota* L.) caused by *Pythium coloratum* in Western Australia. *Plant and Soil* 183, 279–290.

Fritze, H., Pennanen, T. and Vanhala, P. (1997) Impact of fertilizers on the humus layer microbial community of Scots pine stands growing along a gradient of heavy metal pollution. In: Insam, H. and Rangger, A. (eds) *Microbial Communities. Functional versus Structural Approaches*. Springer-Verlag, Berlin, pp. 68–83.

Gardner, P.A., Angus, J.F., Pitson, G.D. and Wong, P.T.W. (1998) A comparison of six methods to control take-all in wheat. *Australian Journal of Agricultural Research* 49, 1225–1241.

Garland, J.L. and Mills, A.L. (1991) Classification and characterization of heterotrophic microbial communities on the basis of patterns of community-level sole-carbon-source utilization. *Applied and Environmental Microbiology* 57, 2351–2359.

Gerlach, M. (1968) Introduction of *Ophiobolus graminis* into new polders and its decline. *Netherlands Journal of Plant Pathology* 74 (Suppl. 2), 1–97.

Gobat, J.M., Aragno, M. and Matthey, W. (1998) *Le Sol Vivant: Base de Pédologie, Biologie des Sols*. Presses Polytechniques et Universitaires Romandes, Lausanne, 519 pp.

Höper, H. and Alabouvette, C. (1996) Importance of physical and chemical soil properties in the suppressiveness of soils to plant diseases. *European Journal of Soil Biology* 32, 41–58.

Höper, H., Steinberg, C. and Alabouvette, C. (1995) Involvement of clay type and pH in the mechanisms of soil suppressiveness to fusarium wilt of flax. *Soil Biology and Biochemistry* 27, 955–967.

Hornby, D. (1998) *Takeall Disease of Cereals: a Regional Perspective*. CAB International, Wallingford, UK, 384 pp.

Huss, M.J., Campbell, C.L., Jennings, D.B. and Leslie, J.F. (1996) Isozyme variation among biological species in the *Gibberella fujikuroi* species complex (*Fusarium* section Liseola). *Applied and Environmental Microbiology* 62, 3750–3756.

Jenkinson, D.S. and Powlson, D.S. (1976) The effects of biocidal treatments on metabolism in soil. *Soil Biology and Biochemistry* 8, 209–213.

Kaiser, S.K., Guckert, J.B. and Gledhill, D.W. (1998) Comparison of activated sludge microbial communities using Biolog™ microplates. *Water Science and Technology* 37, 57–63.

Kelly, J.J. and Tate, R.L. III (1998) Use of Biolog for the analysis of microbial communities from zinc-contaminated soils. *Journal of Environmental Quality* 27, 600–608.

Kwasna, H., Sierota, Z. and Bateman, G.L. (2000) Fungal communities in fallow soil before and after amending with pine sawdust. *Applied Soil Ecology* 14, 177–182.

Lazarovits, G. (2001) Management of soil-borne plant pathogens with organic soil amendments: a disease control strategy salvaged from the past. *Canadian Journal of Plant Pathology* 23, 1–7.

Lemanceau, P. (1989) Role of competition for carbon and iron in mechanisms of soil suppressiveness to fusarium wilts. In: Tjamos, E.C. and Beckman, C.H. (eds) *Vascular Wilt Diseases of Plants – Basic Studies and Control*. Springer Verlag, NATO ASI Series, Berlin, pp. 386–396.

Lockwood, J.L. (1977) Fungistasis in soils. *Biological Review* 52, 1–43.

Lupwayi, N.Z., Rice, W.A. and Clayton, G.W. (1998) Soil microbial diversity and community structure under wheat as influenced by tillage and crop rotation. *Soil Biology and Biochemistry* 30, 1733–1741.

Lupwayi, N.Z., Arshad, M.A., Rice, W.A. and Clayton, G.W. (2001) Bacterial diversity in water-stable aggregates of soils under conventional and zero tillage management. *Applied Soil Ecology* 16, 251–261.

Marsh, T.L., Saxman, P., Cole, J. and Tiedje, J. (2000) Terminal restriction fragment length polymorphism analysis program, a web-based research tool for microbial community analysis. *Applied and Environmental Microbiology* 66, 3616–3620.

Martin-Laurent, F., Philippot, L., Hallet, S., Chaussod, R., Germon, J.C., Soulas, G. and Catroux, G. (2001) DNA extraction from soils: old bias for new microbial diversity analysis methods. *Applied and Environmental Microbiology* 67, 2354–2359.

Mathieson, J.T., Rush, C.M., Bordovsky, D., Clark, L.E. and Jones, O.R. (1990) Effects of tillage on common root rot of wheat in Texas. *Plant Disease* 74, 1006–1008.

Miller, D.N., Bryant, J.E., Madsen, E.L. and Ghiorse, W.C. (1999) Evaluation and optimization of DNA extraction and purification procedures for soil and sediment samples. *Applied and Environmental Microbiology* 65, 4715–4724.

Mussared, D. (1996) A natural mechanism for controlling soil disease. *Rural Research* 170, 4–6.

Nelson, P.E., Toussoun, T.A. and Marasas, W.F.O. (1983) Fusarium *species. An Illustrated Manual for Identification.* The Pennsylvania State University Press, University Park, Pennsylvania, 193 pp.

Ovreas, L. (2000) Population and community level approaches for analysing microbial diversity in natural environments. *Ecology Letters* 3, 236–251.

Papavizas, G.C. (1985) *Trichoderma* and *Gliocladium*: biology, ecology and potential for biocontrol. *Annual Review of Phytopathology* 23, 23–54.

Poly, F., Ranjard, L., Nazaret, S., Gourbiere, F. and Monrozier, L.J. (2001) Comparison of *nifH* gene pools in soils and soil microenvironments with contrasting properties. *Applied and Environmental Microbiology* 67, 2255–2262.

Puhalla, J.E. (1985) Classification of strains of *Fusarium oxysporum* on the basis of vegetative compatibility. *Canadian Journal of Botany* 63, 179–183.

Raaijmakers, J.M. and Weller, D.M. (1998) Natural plant protection by 2,4-diacetylphloroglucinol-producing *Pseudomonas* spp. in take-all decline soils. *Molecular Plant–Microbe Interactions* 11, 144–152.

Ranjard, L., Poly, F., Lata, J.C., Mougel, C., Thioulouse, J. and Nazaret, S. (2001) Characterization of bacterial and fungal soil communities by automated ribosomal intergenic spacer analysis fingerprints: biological and methodological variability. *Applied and Environmental Microbiology* 67, 4479–4487.

Rasmussen, L.D., Ekelund, F., Hansen, L.H., Sørensen, S.J. and Johnsen, K. (2001) Group-specific PCR primers to amplify 24S alpha-subunit rRNA genes from Kinetoplastida (Protozoa) used in denaturing gradient gel electrophoresis. *Microbial Ecology* 42, 109–115.

Roget, D.K. (1995) Decline in root rot (*Rhizoctonia solani* AG-8) in wheat in a tillage and crop rotation experiment at Avon, South Australia. *Australian Journal of Experimental Agriculture* 35, 1009–1013.

Roget, D.K., Neate, S.M. and Rovira, A.D. (1996) Effect of sowing point design and tillage practice on the incidence of Rhizoctonia root rot, take-all and cereal cyst nematode in wheat and barley. *Australian Journal of Experimental Agriculture* 36, 683–693.

Scher, F.M. and Baker, R. (1980) Mechanism of biological control in a *Fusarium*-suppressive soil. *Phytopathology* 70, 412–417.

Scher, F.M. and Baker, R. (1982) Effect of *Pseudomonas putida* and a synthetic iron chelator on induction of soil suppressiveness to fusarium wilt pathogens. *Phytopathology* 72, 1567–1573.

Schippers, B. (1992) Prospects for management of natural suppressiveness to control soilborne pathogens. In: Tjamos, E.C., Papavizas, G.C. and Cook, R.J. (eds) *Biological Control of Plant Diseases*. Plenum Press, New York, pp. 21–34.

Schneider, R.W. (1982) *Suppressive Soils and Plant Disease*. American Phytopathological Society, St Paul, Minnesota, 96 pp.

Simon, A. and Sivasithamparam, K. (1990) Effect of crop rotation, nitrogenous fertilizer and lime on biological suppression of the take-all fungus. In: Hornby, D. (ed.) *Biological Control of Soil-borne Plant Pathogens*. CAB International, Wallingford, UK, pp. 215–226.

Slope, D.B., Prew, R.D., Gutteridge, R.J. and Etheridge, J. (1979) Take-all, *Gaeumannomyces graminis* var. *tritici*, and yield of wheat grown after ley and arable rotations in relation to the occurrence of *Phialophora radicicola* var. *graminicola. Journal of Agricultural Science, Cambridge* 93, 377–389.

Smiley, R.W. and Cook, R.J. (1973) Relationship between take-all of wheat and rhizosphere pH in soils fertilized with ammonium vs. nitrate-nitrogen. *Phytopathology* 63, 882–890.

Smit, E., Leeflang, P., Glandorf, B., van Elsas, J.D. and Wernars, K. (1999) Analysis of fungal diversity in the wheat rhizosphere by sequencing of cloned PCR-amplified genes encoding 18S rRNA and temperature gradient gel electrophoresis. *Applied and Environmental Microbiology* 65, 2614–2621.

Sneh, B., Dupler, M., Elad, Y. and Baker, R. (1984) Chlamydospore germination of *Fusarium oxysporum* f.sp. *cucumerinum* as affected by fluorescent and lytic bacteria from a Fusarium-suppressive soil. *Phytopathology* 74, 1115–1124.

Snyder, W.C. and Hansen, H.N. (1940) The species concept in *Fusarium*. *American Journal of Botany* 27, 64–67.

Steinberg, C., Edel, V., Gautheron, N., Abadie, C., Valleys, T. and Alabouvette, C. (1997) Phenotypic characterization of natural populations of *Fusarium oxysporum* in relation to genotypic characterization. *FEMS Microbiology Ecology* 24, 73–85.

Stotzky, G. and Martin, R.T. (1963) Soil mineralogy in relation to the spread of *Fusarium* wilt of banana in Central America. *Plant and Soil* 18, 317–337.

Stover, R.H. (1962) Fusarial wilt (Panama disease) of bananas and other *Musa* species. CMI, *Phytopathological Papers* 4, 117.

Stutz, E., Kahr, G. and Défago, G. (1989) Clays involved in suppression of tobacco black root rot by a strain of *Pseudomonas fluorescens*. *Soil Biology and Biochemistry* 21, 361–366.

Thrane, U. and Hansen, U. (1995) Chemical and physiological characterization of taxa in the *Fusarium sambucinum* complex. *Mycopathologia* 129, 183–190.

Tiedje, J.M., Asuming-Brempong, S., Nüsslein, K., Marsh, T.L. and Flynn, S.J. (1999) Opening the black box of soil microbial diversity. *Applied Soil Ecology* 13, 109–122.

Toresani, S., Gomez, E., Bonel, B., Bisaro, V. and Montico, S. (1998) Cellulolytic population dynamics in a vertic soil under three tillage systems in the humid pampa of Argentina. *Soil and Tillage Research* 49, 79–83.

Vahjen, W., Munch, J.C. and Tebbe, C.C. (1995) Carbon source utilization of soil extracted microorganisms as a tool to detect the effects of soil supplemented with genetically engineered and non engineered *Corynebacterium glutamicum* and a recombinant peptide at the community level. *FEMS Microbiology Ecology* 18, 317–328.

Van Bruggen, A.H.C. and Semenov, A.M. (2000) In search of biological indicators for soil health and disease suppression. *Applied Soil Ecology* 15, 13–24.

Weller, D.M. (1983) Colonization of wheat roots by a fluorescent pseudomonad suppressive to take-all. *Phytopathology* 73, 1548–1553.

Windels, C.E. (1992) *Fusarium*. In: Singleton, L.L., Mihail, J.D. and Rush, C.M. (eds) *Methods for Research on Soilborne Phytopathogenic Fungi*. American Phytopathological Society, St Paul, Minnosota, pp. 115–128.

Wintzingerode, F.V., Göbel, U.B. and Stackenbrandt, E. (1997) Determination of microbial diversity in environmental samples: pitfalls of PCR-based rRNA analysis. *FEMS Microbiology Reviews* 21, 213–229.

Wiseman, B.M., Neate, S.M., Keller, K.O. and Smith, S.E. (1996) Suppression of *Rhizoctonia solani* anastomosis group 8 in Australia and its biological nature. *Soil Biology and Biochemistry* 28, 727–732.

Wu, L., Thompson, D.K., Li, G., Hurt, R.A., Tiedje, J.M. and Zhou, J. (2001) Development and evaluation of functional gene arrays for detection of selected genes in the environment. *Applied and Environmental Microbiology* 67, 5780–5790.

Chapter 9
Biological Soil Quality from Biomass to Biodiversity – Importance and Resilience to Management Stress and Disturbance

L. Brussaard,[1] T.W. Kuyper,[1] W.A.M. Didden,[1] R.G.M. de Goede[1] and J. Bloem[2]

[1]Wageningen University and Research Centre, Sub-department of Soil Quality, PO Box 8006, NL-6700 EC Wageningen, The Netherlands; [2]Wageningen University and Research Centre, Alterra Green World Research, PO Box 47, NL-6700 AA Wageningen, The Netherlands

Summary

We think of the soil as a living system, where many, if not most, physical and chemical properties and processes are mediated by the soil biota, affecting soil quality. Various aspects of the soil biota react sensitively to changes in the environment, including agricultural management. We conclude that changes in soil biodiversity, measured in terms of community structure of microbes and nematodes, give early warnings of long-term changes in organic matter, nutrient status and soil structure, which cannot be easily observed directly. These parameters are easy to measure and they are responsive to agricultural management.

©CAB International 2004. *Managing Soil Quality: Challenges in Modern Agriculture* (eds P. Schjønning, S. Elmholt and B.T. Christensen)

We also conclude that diversity confers stability/resilience on the ecosystem if (management) stress and disturbance reduce the number of species. However, at the level of the entire (soil) community, i.e. beyond the level of diversity *within* taxonomic groups (such as nematodes), a causal relationship between soil biodiversity and ecosystem functioning and stability does not seem to exist, and the existing knowledge at this level is not yet sufficiently complete and quantitative to be of practical value for management. Nevertheless, reductions of the soil biota result in loss of stability of the soil community with possible loss of ecosystem functioning.

Finally, we provide scientific knowledge that can contribute to the process of establishment of reference values of indicators of soil quality and to agricultural management recommendations, but the assessment of values for management is something to be subjectively agreed upon in practical situations, rather than objectively assessed.

Introduction

We think of agricultural management as the management of an ecosystem. Natural ecosystems may be anywhere on a spectrum from very open to almost closed, i.e. with relatively large or small inputs and outputs of nutrients, respectively.

Agricultural systems are all found at the 'open' end of the spectrum of ecosystems, yet they may differ in the extent to which nutrients are going in and out. Reduction of inputs seems inevitable, but will this inevitably reduce output? In any case a shift to a less open ecosystem is required, for which novel, and perhaps also ancient, knowledge about the use of internal resources is needed. Is soil biodiversity one of those resources? Here, we face the challenge of evaluating this in the framework of (biological) soil quality in general. Therefore, we will first briefly deal with other measures of (biological) soil quality to address what soil biodiversity might add as an indicator and a management target.

Biological Indicators of Soil Quality Other than Biodiversity

Introduction

Although Doran and Parkin (1994) proposed a theoretical overall Soil Quality Index, the data to fill the equation are still largely lacking (Harris *et al.*, 1996). For the time being, soil quality therefore has to be based on various indicators relating to various soil properties/attributes/functions. Doran and

Parkin (1996) proposed a minimum data set (MDS), consisting of an array of chemical, physical and biological characteristics of soil as basic indicators of soil quality. The biological characteristics are:

- Microbial biomass C and N
- Potentially mineralizable N
- Soil respiration (SR)
- $C_{microbial} : C_{total}$ ratio
- Respiration : microbial biomass ratio

What is striking about this set of basic indicators is: (i) that it emphasizes organic matter (OM) decomposition and mineralization of nutrients over direct processes in the rhizosphere; (ii) that it emphasizes nitrogen and largely neglects phosphorus; (iii) that the identity of the indicator organisms remains hidden; and (iv) that the indicators are to some extent autocorrelated.

Yet, soil biological parameters are totally lacking from the list of 'indicators for producers' of 'crop performance and soil and environmental health' and the 'template of proposed indicators for measuring the sustainability of agricultural systems at the farm level' (Doran and Zeiss, 2000).

Rationale for biological indicators of soil quality and some examples

The first three of the indicators listed above were included in a comprehensive study of nine 'great groups' of New Zealand soils by Schipper and Sparling (2000). They made the interesting observation that, although many soil-quality indicators will be different between soil types differing in clay and OM contents, land use had an overriding effect on soil quality: agricultural systems could be clearly differentiated from managed and natural forests, and grassland and arable land were also clearly separated (Fig. 9.1).

In most unmanaged terrestrial ecosystems, soil quality and functional integrity are

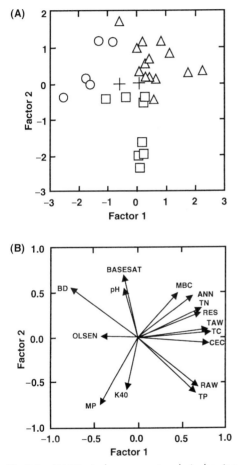

(A)

(B)

maintained via control of nutrient supply and losses and erosion. Therefore, it would seem that indicators related to control of *nutrients* and *soil structure* (both closely related to the control of water), would also have a relationship with the sustainability of managed agricultural systems (cf. Rapport *et al.*, 1997). These aspects have traditionally been dealt with by hydrologists (water), soil chemists (nutrients) and soil physicists (structure), so one might ask whether biological soil-quality indicators, such as the ones used by Schipper and Sparling (2000), are necessary at all. The soil biota, however, in translocating and transforming litter and soil OM and in changing the supply of nutrients to plant roots, fulfil crucial roles in the control of water, nutrients and soil structure (Fig. 9.2). This point is often overlooked, because in conventional agriculture, the soil biota are effectively by passed by soil tillage, artificial fertilizers and pesticides. However, if modern agriculture has to rely more on internal resources, restoring and sustaining the major functions of the soil biota is pertinent. In an illuminating comparison of organic and conventional arable agriculture, Mäder *et al.* (2002) observed a close association between higher aggregate stability on the one hand and microbial biomass and earthworm biomass and abundance on the other (Fig. 9.3). Soluble P was lower in the organic systems, but root length colonized by mycorrhizal fungi and phosphatase activity were higher, and

Fig. 9.1. (A) Principal component analysis showing that, irrespective of soil type of nine 'great groups', agricultural land and managed/original forest could be distinguished using a set of soil-quality indicators. (B) Loading diagram showing that biological soil-quality indicators (microbial biomass carbon (MBC), annual net nitrogen mineralization (ANN) and microbial respiration (RES)) contributed to further separating grassland from arable land. +, indigenous forest; ■, pine forest; △, pastures; ○, cropping and market gardening. Eigenvalues: 38% (axis 1) and 23% (axis 2) (from Schipper and Sparling, 2000).

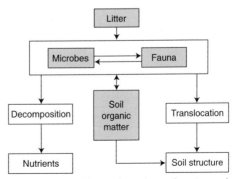

Fig. 9.2. The soil biota, through translocating and transforming litter and soil organic matter, acts as an invaluable internal resource, controlling nutrient turnover and soil structure (simplified after Swift and Woomer, 1993).

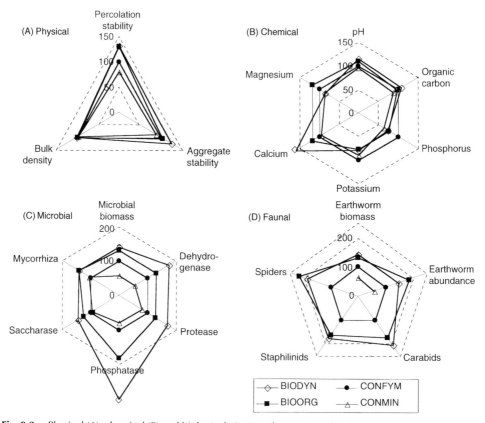

Fig. 9.3. Physical (A), chemical (B) and biological (C, D) soil properties after four 6-year cropping cycles in soils of four arable farming systems: CONMIN, conventional with mineral fertilizers; CONFYM, conventional with farmyard manure; BIOORG, bio-organic; BIODYN, biodynamic. Results are presented relative to CONFYM (= 100%) (from Mäder *et al.*, 2002).

microbial biomass P and P flux through the microbial biomass were also higher in the organic systems (Fig. 9.3). Microbial activity is often a more sensitive indicator of soil quality than biomass. Under increasing stress (in 18-year-old trial fields with installed pH values from 6.1 to 4.0) the bacterial biomass was not reduced, whereas the bacterial growth rate, measured as the incorporation of [^{3}H]thymidine and [^{14}C]leucine into bacterial DNA and proteins during incubation for 1 h (Michel and Bloem, 1993), had significantly decreased (Bloem and Breure, 2003). Mäder *et al.* (2002) found that under organic farming (considered to be a less-stressed system), the microbial specific activity, expressed as μg CO_2-C respired per mg C_{mic} (qCO_2), was reduced, mainly caused by an increase in microbial biomass. Both studies suggest that

microbes use less energy for maintenance, i.e. use their substrate more efficiently, under less-stressed conditions. Biomass specific activity often gives more significant results than biomass or activity alone.

A number of soil enzymes are also good measures of microbial activity, such as acid or alkaline phosphatase, amidase and catalase, and also correlate well with plant biomass production. Enzyme activity would also be a good biological indicator of soil quality, because it is sensitive to changes in various land management practices including residue management, soil compaction, tillage and crop rotation. Changes in certain enzymes often occur within months up to 1 year from the time a soil treatment is applied, long before there are measurable changes in soil OM (Dick, 1994).

Conclusions

1. Microbial biomass and earthworm biomass and abundance are indicators of soil aggregate stability.
2. Various measurements of microbial biomass specific activity are indicators of the physiological status of the microbial biomass.
3. These parameters are useful indicators of soil quality because they react sensitively to changes in agricultural management.

In the context of the present book it is noteworthy that there appears to be much potential to set threshold values and management targets for such indicators in agriculture, but we know of no published literature in which this has actually been done.

Soil Biodiversity

Introduction

Biodiversity in an agricultural context is often abbreviated as agrobiodiversity. Plant breeders argue that *ex situ* collections of as many crop and livestock varieties as possible are necessary for yet unforeseen future use, and discussions of agrobiodiversity among agricultural scientists often centre around the merits of genetic properties and maintenance of genetic variation of crop and livestock varieties in production systems ('planned diversity': Giller *et al.*, 1997). Agrobiodiversity is a matter of concern because of a very large and still ongoing genetic erosion in the world's major crops due to the spread of modern, commercial agriculture (FAO, 1996). Modern cultivars of various crops have partially lost the ability to associate with arbuscular mycorrhizal (AM) fungi or rhizosphere bacteria necessary in low-input agriculture (Hetrick *et al.*, 1992; Germida and Siciliano, 2001). Agrobiodiversity, however, is also about the diversity of non-crop and non-livestock species ('associated diversity': Giller *et al.*, 1997). Here we will leave 'rhizosphere biotechnology' aside and approach soil quality as related to agrobiodiversity from the perspective of crop diversity and associated biodiversity only. Does it serve

to make agricultural systems, in particular biological productivity, (more) sustainable and to provide resilience against stress and disturbance?

Community diversity of taxonomic groups

Perhaps due to the specialization of both taxonomists and ecologists in certain groups of soil biota, such as microorganisms or micro-, meso- or macrofauna, proposed biological indicators of soil quality are in many cases restricted to (subsets of) these groups. This is less unfortunate than it may seem, because (subsets of) these groups constitute meaningful aggregations of the soil biota in view of their functions in nutrient cycling, the dynamics of soil structure and rhizosphere processes (Table 9.1), and play important roles in these functions.

Recent developments in signature lipid biomarker (SLB) technology (White and Macnaughton, 1997) and the combination of ribosomal RNA (rRNA) methodology (to develop highly specific probes) with polymerase chain reaction (PCR) are beginning to provide the tools to quantify microorganisms at increasing levels of taxonomic detail. Such molecular methods are particularly promising in those cases where the level of taxonomic resolution adequately covers a well-characterized group that fulfils important biological processes, such as nitrogen-fixing and nitrifying bacteria and AM fungi. Molecular methods with Glomales-specific primers in root systems have shown a tremendous loss of AM fungal species diversity with agricultural intensification, resulting especially from the high degree of soil disturbance (Helgason *et al.*, 1998). In conditions where there is no close match between level of taxonomic resolution and specific functional roles (e.g. the higher basidiomycetes include both ectomycorrhizal fungi and saprotrophic ligninolytic fungi), such molecular methods are not without their problems. By selective sampling such problems can partly be overcome. Although such DNA-based methods have been tremendously helpful in discovering (new) taxa in ecosystems that have not been observed

Table 9.1. Influences of soil biota on soil processes in ecosystems (from Hendrix *et al.*, 1990).

	Nutrient cycling	Soil structure
Microflora	Catabolize organic matter	Produce organic compounds that bind aggregates
	Mineralize and immobilize nutrients	Hyphae entangle particles on to aggregates
Microfauna	Regulate bacterial and fungal populations	May affect aggregate structure through
	Alter nutrient turnover	interactions with microflora
Mesofauna	Regulate fungal and microfaunal populations	Produce faecal pellets
	Alter nutrient turnover	Create biopores
	Fragment plant residues	Promote humification
Macrofauna	Fragment plant residues	Mix organic and mineral particles
	Stimulate microbial activity	Redistribute organic matter and microorganisms
		Create biopores
		Promote humification
		Produce faecal pellets

before, quantification of occurrence of these taxa is still problematic, especially in the case of fungi. In cases where the soil microbiota form visible or microscopic structures that allow identification, such methods can still find their place next to these novel techniques.

Alternatively, a functional approach to microbial community diversity has been proposed. For example, community-level physiological profiling (CLPP), which at present is better developed for bacterial communities than for fungal communities, does not reflect the microbial community *per se*, but instead the variation of compounds that can be degraded by the community (Garland, 1997; Rutgers and Breure, 1999). The technique readily separates different combinations of soil type and land use, but its responsiveness to changes in land management is not known yet. Whereas CLPP measures the degradation of specific compounds in soil suspensions, microbial catabolic diversity analysis (Degens, 2001) measures the short-term respiration responses upon addition of 25 specific compounds into soil itself. Catabolic diversity is composed of catabolic richness, i.e. the range of substrates metabolized, and catabolic evenness, i.e. the variation between the use of different substrates within a soil. It is claimed to be a versatile indicator of land management on change in soil OM. Finally, stable isotope profiling (SIP) is a novel technique with a very high potential that allows assessment of which

species or taxonomic group of the microbial community is capable of metabolizing a particular substrate. After labelling that substrate with a stable isotope, followed by soil DNA extraction and density gradient separation, the labelled DNA, which can be sequenced, must have been derived only from the organisms that metabolized the substrate. SIP therefore combines the strength of species-directed and functional approaches (Radajewski *et al.*, 2000).

As mentioned above, Mäder *et al.* (2002) found a decreasing specific respiration (qCO_2) in arable agriculture in the following range: conventional with mineral fertilizers – conventional with farmyard manure – bio-organic – biodynamic. This indicates increasing substrate use efficiency. Interestingly, this coincided with a higher microbial diversity, measured as the Shannon index, H′. Nevertheless, most authors agree that many diversity indices (see Gupta and Yeates, 1997, for an overview) are not unambiguous indicators of change, because if a number of species is replaced by an equal number of other species, the index will remain the same. Instead, indicators have been proposed based on the relative abundance of taxonomic groups (e.g. ciliate, flagellate and amoebal protozoa), feeding groups (e.g. bacterivorous, fungivorous, predatory, etc. nematodes) or life-strategy groups (e.g. soil mites with different synchronization, reproduction and dispersal tactics). Mainly due to difficulties of quantification and/or identification (such as in protozoa, mites and springtails), few indicators of practical value have been developed as yet. An outstanding exception is the nematode community diversity. Without much difficulty, nematodes can be ascribed to groups with different feeding modes (bacterivorous, fungivorous, herbivorous = plant-parasitic, carnivorous and omnivorous) and life-strategies (ranging from fast to slow responding to favourable conditions, short to long generation time and short to long life-span; see Ferris *et al.*, 2001, for complete list of criteria). Of the various nematode community indicators, the Maturity Index (MI: Bongers, 1990; de Goede *et al.*, 1993) has been the most widely advocated as an indicator of stress (sometimes as a ratio with the Plant Parasite Index (PPI), Linden *et al.*, 1994; Bongers *et al.*, 1997; Gupta and Yeates, 1997). The MI is the weighted mean of the values, assigned to nematode taxa in a sample, according to life-strategy:

$$MI = \frac{\sum\limits_{i=1}^{n} (v(i). \, a(i))}{\sum\limits_{i=1}^{n} a(i)}$$

where $v(i)$ = life-strategy assigned to taxon i on scale 1–5 and $a(i)$ = abundance of taxon i in a sample.

A low MI may indicate nutrient enrichment (mainly due to dominance of life-strategy 1), or environmental stress (mainly due to absence of life-strategies 3–5). Therefore, when used as an indicator of environmental stress, the MI2–5 is used as an indicator of soil quality. An illuminating example is given by Korthals *et al.* (1996). Under increasing stress (in 10-year-old trial fields with installed pH values from 6.1 to 4.0 and added total Cu concentrations of 0–750 kg/ha under a rotation of silage maize, starch potatoes and oat), the MI2–5 significantly decreased (Fig. 9.4). The background total Cu concentration at pH 6.1 was 29 mg/kg, well below the target value (36 mg/kg; considered to be safe) in The Netherlands. At pH 4.0 and 750 kg/ha added Cu, the total Cu concentration was 168 mg/kg, still below the intervention value (190 mg/kg; considered to be critical) in The Netherlands. The clear change in MI2–5 indicates that the nematode community can be regarded

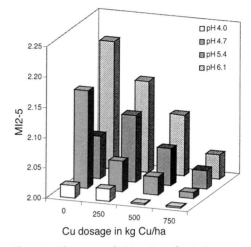

Fig. 9.4. The nematode Maturity Index 2–5 as an indicator of soil quality in trial fields 10 years after installing pH values of 4.0–6.1 and adding 0–750 kg Cu/ha. Fimic Anthrosol near Wageningen, The Netherlands, with a texture of 3% clay, 10% silt, 87% sand and an organic carbon content of 2.1% (w/w) (from Korthals *et al.*, 1996).

as an indicator of increasing environmental stress and underlines that the intervention value, based on total Cu concentration, is not related to ecological effects. In this case there were also significant negative effects of decreasing pH and increasing Cu and their interactions in every year on potato and maize yields (effects on oat not recorded). Cu is an important additive to pig feed. It has furthermore been extensively used in vineyards as an ingredient of fungicides. As a result, Cu concentrations in the topsoil of vineyards in southern France can be as high as 1500 mg/kg (P. Hinsinger, Perth, Australia, 2002, personal communication). In a study of 12 topsoils (0–15 cm) in 10 vineyards with total Cu ranging from 38 to 251 mg/kg, Brun *et al.* (2001) found high Cu concentrations in roots (90–600 mg/kg), but not in the aboveground parts (< 18 mg/kg) of their test plant maize. So, paradoxically, effects on the nematode community are visible well below the intervention value, whereas well above this value, aboveground plant parts of maize did not show an effect.

In addition to the MI, the fungivorous : bacterivorous nematode ratio (characterizing the bacterial or fungal domination of the decomposition pathway) or the trophic diversity (describing the relative abundance of feeding groups), are used for assessing the impact of (changes in) agricultural management (Gupta and Yeates, 1997; Ferris *et al.*, 2001).

Soil biodiversity and ecosystem functioning

Introduction

If one process is of primary importance in terrestrial ecosystems, namely photosynthesis, i.e. the fixation of inorganic to organic C, then its mirror image, namely decomposition, i.e. the degradation of organic to inorganic C, is to be ranked *ex aequo*. These processes occur largely above and below ground, respectively. Nutrients are recycled in the soil with repeated mineralization and immobilization during OM decomposition. Soil organisms decompose, but also re-synthesize organic compounds, thereby contributing to humification of OM. Such humified OM, along with root exudates, contributes to soil structure formation and is quite resistant to decomposition. Finally, parasites and predators in soil may be especially important as control

agents of soilborne diseases and pests (see Alabouvette *et al.*, Chapter 8, this volume).

The 'goods and services' (Table 9.2) are provided by a plethora of species. The species concept is difficult to apply to bacteria and fungi, but using the concept of independent genomes, their numbers are approximately 3000 and up to 35,000 (but see Hawksworth, 2001), respectively, but most are uncultivable and the real number may be orders of magnitude higher. Of the microfauna, i.e. protozoa and nematodes, 1500 and 5000 species have been described, respectively, but the majority await the attention of an ever-decreasing number of systematists. Approximately 30,000 mites, 6500 collembola and over 600 species of enchytraeid worms have been described. Finally, there are around 2000 species of termites, 8800 species of ants, 3600 species of earthworms and 40,000 root-herbivorous insect species (all figures from Brussaard *et al.*, 1997). In view of the bewildering diversity of life in soil ('the soil is the poor man's tropical rainforest': Giller, 1996), soil ecologists have long been intrigued by the causes and significance of soil biodiversity.

Functional groups

In the absence of sound taxonomic knowledge of much of the soil biota, it is difficult to

Table 9.2. Key biological functions, the groups of soil biota principally responsible for these functions and management practices most likely to affect them (from Giller *et al.*, 1997).

Biological function	Biological/functional group	Management practices
Residue comminution/ decomposition	Residue-borne microorganisms, meso/macrofauna	Burning, soil tillage, pesticide applications
Carbon sequestration	Microbial biomass (especially fungi), macrofauna building compact structures	Burning, shortening of fallow in slash-and-burn, soil tillage
Nitrogen fixation	Free and symbiotic nitrogen-fixers	Reduction in crop diversity, fertilization
Organic matter/nutrient redistribution	Roots, mycorrhizas, soil macrofauna	Reduction in crop diversity, soil tillage, fertilization
Nutrient cycling, mineralization/immobilization	Soil microorganisms, soil microfauna	Soil tillage, irrigation, fertilization, pesticide applications, burning
Bioturbation	Roots, soil macrofauna	Soil tillage, irrigation, pesticide applications
Soil aggregation	Roots, fungal hyphae, soil macrofauna, soil mesofauna	Soil tillage, burning, reduction in crop diversity, irrigation
Population control	Predators/grazers, parasites, pathogens	Fertilization, pesticide application, reduction in crop diversity, soil tillage

argue how much species diversity is needed to sustain the processes mentioned. In an effort to make this issue more tractable, the soil biota has, to the best of current knowledge, been assembled in so-called functional groups, mainly based on food preference (Moore *et al.*, 1988). Figure 9.5 conceptually shows the functional groups in a simplified soil food web. Each of the groups in the detrital-based part of the food web may contribute to ecosystem processes, as demonstrated in numerous experimental studies (references in Brussaard, 1998). In a food-web study on the role of functional groups of soil organisms in net nitrogen mineralization, de Ruiter *et al.* (1994) moreover found that model perturbations affecting specific functional groups often had quantitatively important effects on the simulated nitrogen mineralization, which exceeded the direct effect of that particular group under undisturbed conditions, indicating indirect effects of functional group interactions.

Hence, the question 'how much diversity is needed to sustain ecosystem processes?' may be narrowed down to *functional diversity*:

how many functional groups are needed? More recently, the unique nature of species has entered the discussion in the use of the term *functional composition*, which suggests that the presence of (a) particular species rather than other species of a functional group may be decisive for (the intensity of) an ecosystem process (Brussaard *et al.*, 1997). For example, dead OM accumulated on the surface of the mineral soil after land reclamation followed by grassland management in Dutch polder areas due to the absence of one earthworm species, *Lumbricus terrestris*, even though a plethora of other saprotrophic organisms was present (Hoogerkamp *et al.*, 1983; Marinissen and Bok, 1988). This example also underlines that the dispersal ability of 'key species' may be limiting for ecosystem processes if source areas are not within a distance that can be naturally overcome within reasonable time.

Above- and belowground biodiversity

The notion of functional groups also exists above ground, where the term 'functional

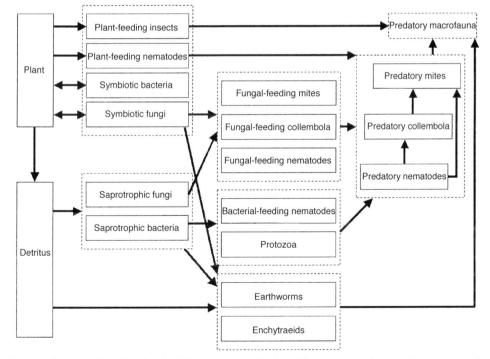

Fig. 9.5. Resource-based food web. All boxes enclosed in dotted lines represent functional groups, based on food preference.

study, restricted to microbial diversity (Griffiths *et al.*, 2001), various levels of diversity were created by inoculating sterile soil with serially diluted soil suspensions prepared from the parent soil. Here again, there was no consistent effect of biodiversity on a range of soil processes, such as bacterial activity, potential nitrification, CLPP and decomposition. These examples add to the evidence that most ecosystem processes are non-additive functions of the traits of two or more species (Chapin *et al.*, 2000). It is the interactions among species at various spatial and temporal scales, rather than simple species presence or absence, that determine ecosystem processes (Chapin *et al.*, 2000). Next to trophic interactions (consumption, predation and parasitism), such interactions include competition and mutualism.

Temporal and spatial variability

To understand the possible relationship between biodiversity and ecosystem functioning, it is furthermore important to consider the various temporal and spatial scales at which species or functional groups interact to isolate the 'signal' of individual species populations or functional groups and biological interactions from the 'noise' of all the other factors affecting ecosystem functioning. Recent analyses have shown that spatial patterning of the soil biota may occur over ranges of more than several hundreds of metres, even in the microflora (references in Ettema and Wardle, 2002). Such patterning may be determined by similar patterning of environmental variables and plants, but in other cases intrinsic population processes, such as dispersal, reproduction and competition, are more important. These factors together result in complex spatial patterns in soil communities, which are made still more complex by local stochastic disturbances. In fact this stochasticity may be one of the main prerequisites for species to co-exist, even in the absence of niche specialization, because not all species will always be present when and where a transient microhabitat is suitable. This spatial patterning of the soil biota (ecosystem engineers, decomposers, plant parasites and root mutualists) and that of

decomposing OM hotspots may in turn influence the spatial patterning of plant species, plant community structure and individual plant performance (Ettema and Wardle, 2002; Verschoor *et al.*, 2002). However, this relationship will not be straightforward either, because, as in soil organisms, also in plants intrinsic population processes will affect spatial patterning.

Although it follows that under natural conditions there are unexploited spatial and temporal gaps, Ekschmitt and Griffiths (1998) state that existing theories predict that a species-poor functional group or trophic level is likely to leave more unexploited niche space than a species-rich functional group or trophic level. This means that a species-poor assembly may not show the full range of microhabitat utilization, tolerance ranges (e.g. to abiotic soil factors) and response dynamics (related to the presence of certain life-strategies). Such species-poor assemblages occur to the best of our knowledge (Brussaard *et al.*, 1997) in:

- macrofauna among the litter transformers, such as isopods and millipedes (with effects on microbial activity and decomposition);
- nitrifying and denitrifying bacteria and bacteria involved in CH_4, hydrogen, iron and sulphur transformations (with effects on element cycling and greenhouse gases);
- mycorrhizal fungi (with effects on plant health and competitive relationships);
- ecosystem engineers such as earthworms among the macrofauna (with effects on soil structure and transport properties).

Soil biodiversity and ecosystem stability

Ecosystem stability has two aspects: resistance, i.e. the inherent capacity of a system to withstand disturbance, and resilience, i.e. the capacity to recover after disturbance (see also Schjønning *et al.*, Chapter 1, this volume). In a review on the diversity–stability relationship, McCann (2000) concludes that diversity tends to be correlated positively with ecosystem stability. This, however, does not imply a causal relationship; higher diversity increases the odds that at least some species or functional groups will respond differently to perturbations, which means that an ecosystem has functional redundancy by containing species that are capable of functionally replacing important other species, that may have gone lost. This notion has been called the insurance hypothesis (Yachi and Loreau, 1999).

At the *functional group* level it has become clear that increasing diversity can increase stability if the distribution of consumer–resource interaction strengths is:

- skewed towards weak interactions (representing the size of the effects of species on each other's dynamics near equilibrium);
- patterned in the sense of simultaneous top-down effects at lower trophic levels and bottom-up effects at higher trophic levels (de Ruiter *et al.*, 1995); and
- organized in trophic loops in such a way that long loops contain relatively many weak links, thereby reducing the amount of intraspecific interaction needed for stability (Neutel *et al.*, 2002).

The available evidence from field studies suggests that this is the case (de Ruiter *et al.*, 1995; McCann, 2000; Neutel *et al.*, 2002). In the field study of Wardle *et al.* (1999) the removal of functional groups of plants did not, as long as plants were present, affect the temporal variability or spatial heterogeneity of the properties considered, namely various functional groups of soil biota and decomposition rates of added substrates, with the exception of CO_2 release. Given time, even the latter effect may disappear, if another stabilizing mechanism, suggested by Ekschmitt *et al.* (2001) operates, whereby changes in the decomposition efficiency, following the loss of decomposer species, will be largely compensated by an adjustment of detritus mass, i.e. when soil OM has increased to the extent that CO_2 release resumes its previous level.

At the *within-functional-group* level, Griffiths *et al.* (2000) assessed the stability of the decomposition of grass residues in microcosms by microbial communities that had been made increasingly less diverse by fumigation for increasing periods of time, followed by imposing a transient (brief heating to 40°C) or a persistent (addition of $CuSO_4$) stress. Decomposition of grass residues was determined on three occasions afterwards. As shown in Fig. 9.7, the less diverse soils fumigated for 2 and 24 h had considerably less resistance (measured as the immediate effect 1 day after fumigation) to copper than the soils fumigated for 0 and 0.5 h, and showed no sign of resilience after 2 months. In contrast, the soils fumigated for 2 and 24 h tended to be more resistant against the heat stress than those fumigated for 0 and 0.5 h, but even so the unfumigated soil was the most resilient. The patterns for equal treatments were basically the same in the 2 years, except for the soil that had been fumigated for 24 h and underwent the heat stress. The results are consistent with the hypothesis that, following a transient perturbation, systems can regain their original level of functioning, but that recovery (resilience) is impaired by a loss of diversity. The presence of a persistent perturbation prevents resilience.

Soil biodiversity, ecosystem functioning and stability in agricultural systems

Agricultural systems differ to various degrees from the natural systems from which they were originally derived, and this supposedly has led to an accompanying loss of associated biodiversity. A distinction should be made between perennial and annual crop-based systems. Annual crop-based systems often have not been derived from species-rich tree-based systems, which are relatively buffered against environmental perturbations. Rather they were derived from species-poor grasslands, which underwent frequent natural disturbances such as fire and/or grazing, and where the precursors of annual crops, such as rice, wheat, sorghum, pearl millet, wild barley and wild oat, were dominant species (Wood and Lenné, 1999).

Given the impact of a reduction of plant and litter diversity and input, and the added impact of soil tillage and pesticide use on habitat structure, microclimate and food resources, the relationships between agricultural intensification and agroecosystem

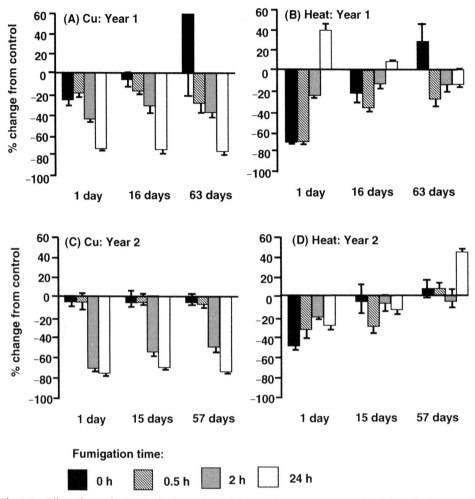

Fig. 9.7. Effect of perturbation, in the form of Cu addition or heat treatment, on the ability of soils fumigated for different times to decompose grass residues at increasing time intervals following the application of the perturbation. Values are the mean (*n* = 3; bars show SE) percentage change in decomposition relative to unperturbed soil that had originally been fumigated for the same time-span (from Griffiths *et al.*, 2000).

biodiversity have been hypothesized to be as in Fig. 9.8. Curves I and II represent extreme cases, where biodiversity already strongly declines at low levels (curve I) or only at high levels of disturbance (curve II). Curves III and IV are intermediate. Swift *et al.* (1996) suggest that curve IV may best represent the response of soil biodiversity, but given the observations of Wood and Lenné (1999), curve II may be more applicable for annual cropping systems derived from natural grasslands. The point of inflection in curve IV was further suggested

to differ for species and functional groups differing in body size: the macrofauna, such as earthworms, millipedes and centipedes, were considered to be the most susceptible, the microflora and -fauna, such as protozoa and nematodes, the least susceptible, and the mesofauna, such as mites and springtails, in between (Swift *et al.*, 1996). There is partial evidence for this supposition. Macrofauna species may differ markedly in preference for certain *litter* types (Tian *et al.*, 1997). Hence, differential loss of certain types will lead to

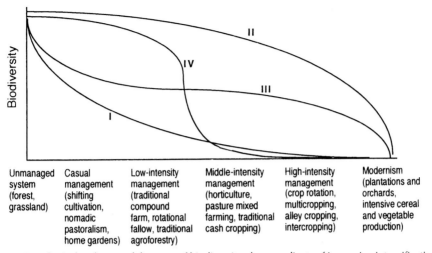

| Unmanaged system (forest, grassland) | Casual management (shifting cultivation, nomadic pastoralism, home gardens) | Low-intensity management (traditional compound farm, rotational fallow, traditional agroforestry) | Middle-intensity management (horticulture, pasture mixed farming, traditional cash cropping) | High-intensity management (crop rotation, multicropping, alley cropping, intercropping) | Modernism (plantations and orchards, intensive cereal and vegetable production) |

Fig. 9.8. Hypothetical pathways of decrease of biodiversity along gradients of increasing intensification of agricultural land use. For explanation see text (from Swift *et al.*, 1996).

differential loss of macrofauna species. Yet, the effects of plant and litter diversity were the most clearly observed in the community structure of the soil microflora and -fauna, not in the meso- and macrofauna (Wardle *et al.*, 1999). The quantitatively most-used *pesticides* are herbicides, in particular in reduced and no-tillage systems. The effects of herbicides are largely indirect, through their impact on plant cover and litter input, but earthworms are relatively strongly inhibited and nematodes stimulated, as compared to the effects on other groups (Wardle, 1995). The relationship between organism size and response to *tillage* was also found to differ. Wardle (1995) used the following index, V, to illustrate this:

$$V = 2M_{CT}/(M_{CT} + M_{NT}) - 1,$$

where M_{CT} and M_{NT} = abundance or biomass of organisms, or percentage C or N, under conventional (CT) and no-tillage (NT). V ranges between -1 and $+1$; values ±0.33 and ±0.67 express the boundaries between mild to moderate and moderate to extreme stimulating/inhibitory effects of tillage, respectively.

There was a significant negative relationship between mean organism width and the mean value of V; the variance of V was positively correlated with body width (Fig. 9.9). If the imminent relationship between agricultural intensification and diversity of soil biota according to size class is confirmed by additional research, this may have important consequences for community stability. First, species higher in the food chain generally have both a larger body size and a stabilizing effect on complex food webs, whereas omnivory, a trait which also confers community stability, is also more widespread among organisms higher in the food chain (e.g. Neutel *et al.*, 2002). Secondly, the four major components of biotic activity, i.e. habitat utilization, tolerance range to perturbation, functional and numerical response dynamics and contribution to ecosystem functioning, are conceptually independent, but in reality they are coupled according to body size class (Ekschmitt and Griffiths, 1998). From these observations it follows that, if disturbance differentially affects soil biota of a certain size class, community stability is negatively affected. Once again, it is implied that the composition of the assemblages concerned, not biodiversity *per se*, is decisive in its effect on community stability. At the same time, the consequences for ecosystem functioning are not straightforward, as they depend on the redundancy of the remaining species. However, at decreasing diversity, especially if the decrease differentially concerns specific assemblages of species, an increase in the *variability* of functioning may be expected.

In addition to having differential effects on organisms of different size classes, it is

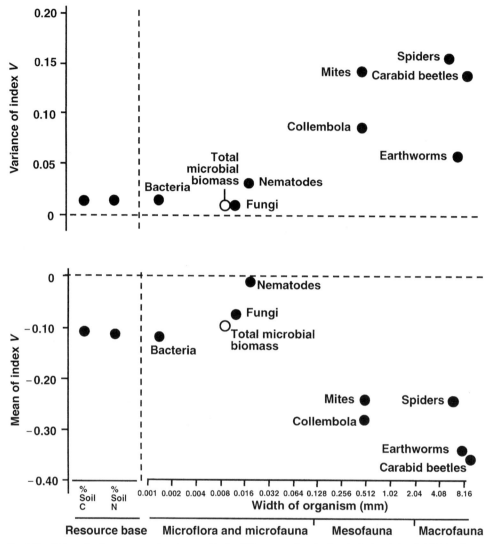

Fig. 9.9. Values for the mean and variance of index *V* (representing the level of stimulation/inhibition by conventional tillage vs. no-tillage systems) for various groups of organisms, plotted against mean organism width. Mean and variance of *V* for resource base (soil) carbon and nitrogen are included for comparative purposes (from Wardle, 1995).

well-known that no-tillage systems show a shift in the decomposition pathway from bacterial- to fungal-dominated as compared with conventional tillage (Beare, 1997). Whether this shift adds to community stability is unknown. There seems to be little experimental evidence from the field where disturbance, biodiversity and stability in agricultural systems were *concomitantly* studied regarding the two main roles of the soil biota in ecosystem functioning:

nutrient cycling and soil structure dynamics (Table 9.1). Numerous studies have shown the negative effects of tillage on burrowing earthworms, which henceforth negatively affected porosity and hydraulic conductivity. However, it seems likely that this, too, is related to the properties of one or a few species and not to earthworm biodiversity *per se*.

With increasing simplification of the ecosystem in agriculture, the integrity of decomposition, nutrient cycling and plant production is uncoupled. Various studies on the relationship between plant diversity and ecosystem functioning have also shown this phenomenon. It would seem, however, that it is not biodiversity *per se* that counts, but the presence of the right combination of plant species to exploit the soil for nutrients without major gaps in time and space, and to produce a range of qualities of litter to maintain soil OM levels. It is unknown to what extent soil biodiversity plays a role here, but it can be hypothesized that fluctuating resource availability and a low biodiversity (as in most agroecosystems) will bring about opportunistic responses of consumers, determining lack of synchronization between nutrient release and nutrient uptake (Ekschmitt and Griffiths, 1998).

Sustainable agriculture strongly depends on the synchronization of nutrient fluxes between trophic levels, including crops. A sustainable agricultural system may be defined as one in which output trend is non-declining and resistant, in terms of yield stability, to normal fluctuations of stress and disturbance, and has minimal effects on the environment (Brussaard, 1994). In the Western European context this leads us to intercropping and crop rotation systems. Intercropping has been reported to promote enhanced nutrient utilization, disease control, weed control and other agricultural functions (references in Swift and Anderson, 1993). Monocrops of maize and soybean have a lower diversity of AM fungi and a lower productivity than crop rotations (Johnson et al., 1992). But the causality between diversity and functioning has yet to be investigated.

Conclusions

Once more keeping in mind that biological indicators should be related to (changes in) OM, nutrient cycling, soil structure and/or biological productivity, we conclude that:

1. Microbial community characterization serves to assess the physiological status of parts of the community.

2. Community structure of the nematode fauna indicates nutrient status and management and environmental stress.

3. Determination of the community structures of other faunal groups is less practical due to laborious methods and difficulties of identification.

In contrast, the relationships between various diversity parameters at the level of the entire (soil) community and various aspects of ecosystem functioning and stability are not straightforward. The presence/absence of specific (assemblages of) species may be more important than the number of species or functional groups *per se*. We are only beginning to understand these relationships and we are not yet capable of quantitatively predicting the effects of disturbances or management. A possible relationship between biodiversity and ecosystem functioning may be elucidated by applying a hierarchical approach to discern the signal of species populations or functional groups and biological interactions in soil from the noise at coarser levels of observation. Yet, the spatial and temporal patterning of the distribution of species and functional groups, the multitude of trophic, competitive and mutualistic interactions among soil organisms, and the averaging out of their contributions to ecosystem processes across spatial and temporal scales, make it unlikely that a straightforward relationship between species diversity *per se* and ecosystem functioning will be found. If some kind of relationship between biodiversity and ecosystem functioning will be found anywhere, it is probably in species-poor assemblages. It should be noted, however, that within those assemblages once again it will be the (in)ability to fill niche space of each of the constituent species, i.e. the functional composition of the assemblage, that determines the outcome at the process level, rather than diversity *per se*.

Synthesis and Outlook

Biodiversity at the level of the entire soil community confers stability on the community and on ecosystem functioning: resistance against stress is higher in more diverse communities, whereas resilience only occurs when the stress is not permanent. This relationship is not straightforward, however, because it is the range of traits of individual species, not species richness or functional diversity *per se*, which determines functional redundancy, i.e. integrity of function in the face of stress and disturbance. In the absence of detailed knowledge of those traits for most species in soil and in the absence of knowledge of the susceptibility of most species to environmental change and management measures, it can at best be only qualitatively predicted what the effects of agricultural management on soil biodiversity, ecosystem stability and ecosystem functioning will be. There is no evidence that current agricultural practice has irreversibly stressed the soil community, but this is mainly due to lack of research. The study by Korthals *et al.* (1996) and the high level of Cu contamination in vineyard soils mentioned earlier, suggest that agricultural practices may have had irreversible effects on soil communities. Also, nutrient losses, decline of OM content and erosion are testimony to the fact that the functioning of the soil community has been affected to a degree that many would value as unacceptable.

So biodiversity changes at the level of the entire soil community are not operational in agricultural management. Fortunately, this is different for community diversity and activity of some taxonomic groups. For any parameter to be suitable as an indicator of soil quality some conditions must be met:

- spatial heterogeneity must be accounted for (Ettema and Wardle, 2002);
- it must be sufficiently stable over time under non-changing conditions, and annual fluctuations must be sufficiently predictable to discriminate the signal of human-induced change from the natural background (van Straalen, 1997); and
- it must be both specific for environmental factors and sensitive to agricultural management measures to indicate, at an early stage, changes in rhizosphere functioning, soil OM, nutrient cycling and soil structure affecting biological productivity.

Biological indicators of soil quality differ with respect to these conditions, so a combination would make most sense. Based on our analysis from a scientific and a practical perspective, and focusing on soil biodiversity, we propose that this combination consists of:

- microbial community diversity, activity and biomass;
- community diversity of nematodes; and
- 'incidence' of 'larger' soil fauna; the available evidence suggests that 'larger' means meso- and macrofauna (Wardle, 1995) and that with decreasing diversity the identity and abundance/biomass of species become more important.

The standard against which to judge the signal of biological indicators of soil quality is not straightforward. It has been argued that the standard should be soil-type specific. Because the overriding aspects in which soils differ are clay content and OM content, standardization would best be achieved by expressing an indicator relative to clay content or C_{total}, such as in the microbial quotient $C_{microbial} : C_{total}$ (Sparling, 1997). Parkin *et al.* (1996) suggest that a measurement of microbial respiration in a particular soil be compared to a standard, set as the respiration at 60% water-filled pore space at 25°C. Under such conditions the respiration is considered at its maximum in any soil. However, the possibility exists that the enzymatic breakdown of different carbon sources shows compound-specific temperature dependency, so the underlying assumption of the sufficiency of one standard temperature needs further underpinning. For similar reasons Parkin *et al.* (1996) suggest expressing soil respiration (as an indicator of C storage) relative to the C inputs. In evaluating enzyme activity (EA), differences between soil types and OM content can best be accounted for by expressing it as the ratio EA : clay content or EA : C_{org} (Dick, 1994). Yet, Schipper and Sparling (2000) found that the effects of land use on soil quality parameters overrode those of soil type in nine 'great groups' of soil. This may or may not hold true for microbial and nematode community diversity, but this has not, to the best of our knowledge, been investigated. It will not hold true for the soil meso- and macrofauna, which are well known to differ with humus/soil types.

In any case, analytical measurements do not inherently include a value-based dimension that can be used for assessing soil quality (Harris *et al.*, 1996). Soil quality is assessed with respect to specific soil functions, such as biological production, erosivity, the quality of air, ground water and surface water, food quality, human health, etc. Doran and Parkin (1994) rightly state that in their theoretical soil quality equation, weighting factors can be assigned to these functions, determined by geography, societal concerns, economic constraints, etc. This is, however, a matter of consensus or a result of negotiations among policy makers and (local) stakeholders, rather than a product of science (alone). Nevertheless, negotiations and consensus building can be greatly assisted if a baseline for comparison of indicator values of soil quality for a particular site with a particular soil type is available, e.g. the value of the indicator:

- at the same site, but on a previous occasion, or
- at the same time, but on a reference site,

in order to decide if the value of the indicator is acceptable or if management is required to improve it (Harris *et al.*, 1996).

Another possibility is to derive reference values from a range of observed values. For example, in the Dutch Soil Quality Monitoring Network, approximately 20 replicates of certain combinations of soil type and land use are sampled. From such a range one can agree on values of biological indicators, which would be the reference values for these types of systems, if temporal and spatial variability are accounted for.

In a provocative essay, Lancaster (2000) challenges the concept of ecosystem health and related concepts on the ground of subjectivity. She advocates the long-term, spatially and temporally explicit monitoring of ecological processes to obtain information on differences between and changes within similar ecosystems, but is not willing however, to consider the subjective problems of value judgements. At the same time, she recommends that such judgements be made by thoroughly trained and well-informed experts. We support

the plea for long-term monitoring and in this chapter we take up the challenge of contributing to the making of value judgements. We do not subscribe, however, to the technocratic view that value judgements are the prerogative of 'experts'. Reference values must be subjectively agreed upon between scientists and other stakeholders in practical situations; they cannot be objectively assessed. Once reference values have been agreed upon, they may serve as a basis for identifying management thresholds to be observed or aimed at whenever changes in management are considered.

In the absence of agreed standards against which to judge soil(-biodiversity-based)-quality indicators, and with few hard data on the reversibility of damage to the biological diversity of agricultural soils, but in the face of severe problems with nutrient losses and erosion in modern agriculture, the *directions* of desirable changes in agricultural management are meanwhile becoming apparent, namely (see also Table 9.2):

- to avoid/reduce persistent (use of) sources of ecosystem stress, such as pesticides and artificial fertilizers and animal feed (in particular if containing heavy metals), which affect the community diversity of soil organisms; and
- to avoid/reduce practices that selectively affect the soil macrofauna, such as tillage, affecting soil community stability.

This means a shift towards conservation agriculture. Pankhurst *et al.* (1997) raised the question: 'can the world afford the production penalty which can accompany conservation management?' We believe that the problem will not lie with the farmers. Farmers have always remained managers of the whole, however forceful the pressure has been on increasing grain, milk and meat output. As a result the sustainability issue has natural appeal to most of them. We hence observe a rapidly growing willingness among farmers to accept reduced production levels as long as their base of existence is not threatened. Many of them do not see this as a penalty, but as an investment for the future, if only because they wish to transfer viable farms to their sons and daughters. In many less-endowed countries the resource use efficiency can still be raised, starting from a deficit rather than a surplus situation, with concomitant *increase* of production levels. Even if, on a world scale, total production would have to decline, the question would not be 'can the world afford the production penalty?', but 'can the world afford *not to accept* the production penalty?' Clearly, this is not an issue that can be solved by agricultural scientists and farmers alone. Their role is to jointly explore the options to reconcile production with sustainability and to put those to society at large to make well-informed choices. Compared to chemical and physical aspects of soil quality, biological soil quality has an important role to play, because the soil biota are important causal agents of soil physical and chemical conditions and plant health.

Acknowledgements

This chapter was written when the senior author (L. Brussaard) was on sabbatical leave at the Laboratoire de Biologie et Organization des Sols Tropicaux, Martinique, French West Indies, a branch of the Institut de Recherches pour le Développement. Dr Eric Blanchart of this institute is kindly thanked for his hospitality and for his comments on the first draft of the chapter. We are also indebted to Esther van den Brug and Gerda Bijl for secretarial assistance, to Bertrand Urien for help with the figures and to Wouter Gerritsma for designing the template to transfer the literature references to the prescribed format.

References

Andrén, O., Brussaard, L. and Clarholm, M. (1999) Soil organism influence on ecosystem-level processes – bypassing the ecological hierarchy? *Applied Soil Ecology* 11, 177–188.

Beare, M.H. (1997) Fungal and bacterial pathways of organic matter decomposition and nitrogen mineralization in arable soils. In: Brussaard, L. and Ferrera-Cerrato, R. (eds) *Soil Ecology in Sustainable Agricultural Systems.* CRC/Lewis, Boca Raton, Florida, pp. 37–70.

Bengtsson, J. (1998) Which species? What kind of diversity? Which ecosystem function? Some problems in studies of relations between biodiversity and ecosystem function. *Applied Soil Ecology* 10, 191–199.

Bloem, J. and Breure, A.M. (2003) Microbial indicators. In: Markert, B.A., Breure, A.M. and Zechmeister, H.G. (eds) *Bioindicators/ Biomonitors – Principles, Assessment, Concepts.* Elsevier, Amsterdam, pp. 259–282.

Bongers, T. (1990) The maturity index, an ecological measure of environmental disturbance based on nematode species composition. *Oecologia* 83, 14–19.

Bongers, T., van der Meulen, H. and Korthals, G. (1997) Inverse relationship between the nematode maturity index and plant parasite index under enriched nutrient conditions. *Applied Soil Ecology* 6, 195–199.

Brun, L.A., Maillet, J., Hinsinger, P. and Pépin, M. (2001) Evaluation of copper availability to plants in copper-contaminated vineyard soils. *Environmental Pollution* 111, 293–302.

Brussaard, L. (1994) Interrelationships between biological activities, soil properties and soil management. In: Greenland, D.J. and Szabolcs, I. (eds) *Soil Resilience and Sustainable Land Use.* CAB International, Wallingford, UK, pp. 309–329.

Brussaard, L. (1998) Soil fauna, guilds, functional groups and ecosystem processes. *Applied Soil Ecology* 9, 123–135.

Brussaard, L., Behan-Pelletier, V.M., Bignell, D.E., Brown, V.K., Didden, W., Folgarait, P., Fragoso, C., Freckman, D.W., Gupta, V.V.S.R., Hattori, T., Hawksworth, D.L., Klopatek, C., Lavelle, P., Malloch, D.W., Rusek, J., Söderström, B., Tiedje, J.M. and Virginia, R.A. (1997) Biodiversity and ecosystem functioning in soil. *Ambio* 26, 563–570.

Chapin, F.S. III, Zavaleta, E.S., Eviner, V.T., Naylor, R.L., Vitousek, P.M., Reynolds, H.L., Hooper, D.U., Lavorel, S., Sala, O.E., Hobbie, S.E., Mack, M.C. and Diaz, S. (2000) Consequences of changing biodiversity. *Nature* 405, 234–242.

Degens, B.P. (2001) Microbial catabolic evenness: a potential integrative indicator of organic matter management? In: Rees, R.M., Ball, B.C., Campbell, C.D. and Watson, C.A. (eds) *Sustainable Management of Soil Organic Matter.* CAB International, Wallingford, UK, pp. 357–362.

de Goede, R.G.M., Bongers, T. and Ettema, C.H. (1993) Graphical presentation and interpretation of nematode community structure: C-P triangles. *Mededelingen Faculteit Landbouwkundige en Toegepaste Biologische Wetenschappen Universiteit Gent* 58, 743–750.

de Ruiter, P.C., Neutel, A.-M. and Moore, J.C. (1994) Modelling food webs and nutrient cycling in agro-ecosystems. *Trends in Ecology and Evolution* 9, 378–383.

de Ruiter, P.C., Neutel, A.-M. and Moore, J.C. (1995) Energetics, patterns of interaction strengths, and stability in real ecosystems. *Science* 269, 1257–1260.

Dick, R.P. (1994) Soil enzyme activities as indicators of soil quality. In: Doran, J.W., Coleman, D.C., Bezdicek, D.F. and Stewart, B.A. (eds) *Defining Soil Quality for a Sustainable Environment*. SSSA, Madison, Wisconsin, pp. 107–124.

Doran, J.W. and Parkin, T.B. (1994) Defining and assessing soil quality. In: Doran, J.W., Coleman, D.C., Bezdicek, D.F. and Stewart, B.A. (eds) *Defining Soil Quality for a Sustainable Environment*. Soil Science Society of America, Madison, Wisconsin, pp. 3–21.

Doran, J.W. and Parkin, B.P. (1996) Quantitative indicators of soil quality: a minimum data set. In: Doran, J.W. and Jones, A.J. (eds) *Methods for Assessing Soil Quality*. Soil Science Society of America Special Publication No. 49, pp. 25–37.

Doran, J.W. and Zeiss, M.R. (2000) Soil health and sustainability: managing the biotic component of soil quality. *Applied Soil Ecology* 15, 3–11.

Ekschmitt, K. and Griffiths, B.S. (1998) Soil biodiversity and its implications for ecosystem functioning in a heterogeneous and variable environment. *Applied Soil Ecology* 10, 201–215.

Ekschmitt, K., Klein, A., Pieper, B. and Wolters, V. (2001) Biodiversity and functioning of ecological communities: why is diversity important in some cases and unimportant in others? *Journal of Plant Nutrition and Soil Science* 164, 239–246.

Ettema, C.H. and Wardle, D.A. (2002) Spatial soil ecology. *Trends in Ecology and Evolution* 17, 177–183.

FAO (1996) *Report on the State of the World's Plant Genetic Resources for Food and Agriculture*. FAO, Rome, 82 pp.

Ferris, H., Bongers, T. and de Goede, R.G.M. (2001) A framework for soil food web diagnostics: extension of the nematode faunal analysis concept. *Applied Soil Ecology* 18, 13–29.

Garland, J.L. (1997) Analysis and interpretation of community-level physiological profiles in microbial ecology. *FEMS Microbiology Ecology* 24, 289–300.

Germida, J.J. and Siciliano, S.D. (2001) Taxonomic diversity of bacteria associated with the roots of modern, recent and ancient wheat cultivars. *Biology and Fertility of Soils* 33, 410–415.

Giller, P.S. (1996) The diversity of soil communities, the 'poor man's tropical rainforest'. *Biodiversity and Conservation* 5, 135–168.

Giller, K.E., Beare, M.H., Lavelle, P., Izac, A.M.N. and Swift, M.J. (1997) Agricultural intensification, soil biodiversity and agro-ecosystem function. *Applied Soil Ecology* 6, 3–16.

Griffiths, B.S., Ritz, K., Bardgett, R.D., Cook, R., Christensen, S., Ekelund, F., Sørensen, S.J., Bååth, E., Bloem, J., de Ruiter, P.C., Dolfing, J. and Nicolardot, B. (2000) Ecosystem response of pasture soil communities to fumigation-induced microbial diversity reductions: an examination of the biodiversity–ecosystem function relationship. *Oikos* 90, 279–294.

Griffiths, B.S., Ritz, K., Wheatley, R., Kuan, H.L., Boag, B., Christensen, S., Ekelund, F., Sørensen, S.J., Muller, S. and Bloem, J. (2001) An examination of the biodiversity–ecosystem function relationship in arable soil microbial communities. *Soil Biology and Biochemistry* 33, 1713–1722.

Gupta, V.V.S.R. and Yeates, G.W. (1997) Soil microfauna as bioindicators of soil health. In: Pankhurst, C.E., Doube, B.M. and Gupta, V.V.S.R. (eds) *Biological Indicators of Soil Health*. CAB International, Wallingford, UK, pp. 201–233.

Harris, R.F., Karlen, D.L. and Mulla, D.J. (1996) A conceptual framework for assessment and management of soil quality and health. In: Doran, J.W. and Jones, A.J. (eds) *Methods for Assessing Soil Quality*. Soil Science Society of America, Madison, Wisconsin, pp. 61–82.

Hawksworth, D.L. (2001) The magnitude of fungal diversity: the 1.5 million species estimate revisited. *Mycological Research* 105 Part 12, 1422–1432.

Hector, A., Schmid, B., Beierkuhnlein, C., Caldeira, M.C., Diemer, M., Dimitrakopoulos, P.G., Finn, J.A., Freitas, H., Giller, P.S., Good, J., Harris, R., Högberg, P., Huss Danell, K., Joshi, J., Jumpponen, A., Korner, C., Leadley, P.W., Loreau, M., Minns, A., Mulder, C.P.H., O'Donovan, G., Otway, S.J., Pereira, J.S., Prinz, A., Read, D.J., Scherer Lorenzen, M., Schulze, E.D., Siamantziouras, A.S.D., Spehn, E.M., Terry, A.C., Troumbis, A.Y., Woodward, F.I., Yachi, S. and Lawton, J.H. (1999) Plant diversity and productivity experiments in European grasslands. *Science* 286, 1123–1127.

Helgason, T., Daniell, T.J., Husband, R., Fitter, A.H. and Young, J.P.W. (1998) Ploughing up the wood-wide web? *Nature* 394, 431.

Hendrix, P.F., Crossley, D.A. Jr, Blair, J.M. and Coleman, D.C. (1990) Soil biota as components of sustainable agroecosystems. In: Edwards, C.A., Lal, R., Madden, P., Miller, R.H. and House, G. (eds) *Sustainable Agricultural*

Systems. Soil and Water Conservation Society, Ankeny, Illinois, pp. 637–654.

Hetrick, B.A.D., Wilson, G.W.T. and Cox, T.S. (1992) Mycorrhizal dependence of modern wheat varieties, landraces and ancestors. *Canadian Journal of Botany* 70, 2032–2040.

Hoogerkamp, M., Rogaar, H. and Eijsackers, H.J.P. (1983) Effect of earthworms on grassland on recently reclaimed polder soils in the Netherlands. In: Satchell, J.E. (ed.) *Earthworm Ecology, from Darwin to Vermiculture.* Chapman and Hall, London, pp. 85–105.

Huston, M.A., Aarssen, L.W., Austin, M.P., Cade, B.S., Fridley, J.D., Garnier, E., Grime, J.P., Hodgson, J., Lauenroth, W.K., Thompson, K., Vandermeer, J.H. and Wardle, D.A. (2000) No consistent effect of plant diversity on productivity. *Science* 289, 1255 (in Technical Comments).

Johnson, N.C., Copeland, P.J., Crookston, R.K. and Pfleger, F.L. (1992) Mycorrhizae: possible explanation for yield decline with continuous corn and soybean. *Agronomy Journal* 84, 387–390.

Korthals, G.W., de Goede, R.G.M., Kammenga, J.E. and Bongers, T. (1996) The maturity index as an instrument for risk assessment of soil pollution. In: van Straalen, N.M. and Krivolutsky, D.A. (eds) *Bioindicator Systems for Soil Pollution.* Kluwer Academic Publishers, Dordrecht, pp. 85–93.

Laakso, J. and Setälä, H. (1999) Sensitivity of primary production to changes in the architecture of belowground food webs. *Oikos* 87, 57–64.

Laakso, J., Setälä, H. and Palojärvi, A. (2000) Influence of decomposer food web structure and nitrogen availability on plant growth. *Plant and Soil* 225, 153–165.

Lancaster, J. (2000) The ridiculous notion of assessing ecological health and identifying the useful concepts underneath. *Human and Ecological Risk Assessment* 6, 213–222.

Lawton, J.H. (1994) What do species do in ecosystems? *Oikos* 71, 367–374.

Lawton, J.H. and Brown, V.K. (1994) Redundancy in ecosystems. In: Schultze, E.D. and Mooney, H.A. (eds) *Biodiversity and Ecosystem Function.* Springer Verlag, Heidelberg, pp. 255–270.

Linden, D.R., Hendrix, P.F., Coleman, D.C. and van Vliet, P.C.J. (1994) Faunal indicators of soil quality. In: Doran, J.W., Coleman, D.C., Bezdicek, D.F. and Stewart, B.A. (eds) *Defining Soil Quality for a Sustainable Environment.* Soil Science Society of America, Madison, Wisconsin, pp. 91–106.

Loreau, M., Naeem, S., Inchausti, P., Bengtsson, J., Grime, J.P., Hector, A., Hooper, D.U.,

Huston, M.A., Raffaeli, D., Schmid, B., Tilman, D. and Wardle, D.A. (2002) Biodiversity and ecosystem functioning: current knowledge and future challenges. *Science* 294, 804–808.

Mäder, P., Fliessbach, A., Dubois, D., Gunst, L., Fried, P. and Niggli, U. (2002) Soil fertility and biodiversity in organic farming. *Science* 296, 1694–1697.

Marinissen, J.C.Y. and Bok, J. (1988) Earthworm-amended soil structure: its influence on Collembola populations in grassland. *Pedobiologia* 32, 243–252.

McCann, K.S. (2000) The diversity–stability debate. *Nature* 405, 228–233.

Michel, P.H. and Bloem, J. (1993) Conversion factors for estimation of cell production rates of soil bacteria from tritiated thymidine and tritiated leucine incorporation. *Soil Biology and Biochemistry* 25, 943–950.

Mikola, J. and Setälä, H. (1998) Relating species diversity to ecosystem functioning: mechanistic backgrounds and experimental approach with a decomposer food web. *Oikos* 83, 180–194.

Moore, J.C., Walter, D.E. and Hunt, H.W. (1988) Arthropod regulation of micro- and mesobiota in below-ground detrital food webs. *Annual Review of Entomology* 33, 419–439.

Neutel, A.-M., Heesterbeek, J.A.P. and de Ruiter, P.C. (2002) Stability in real food webs: weak links in long loops. *Science* 296, 1120–1123.

Pankhurst, C.E., Doube, B.M. and Gupta, V.V.S.R. (eds) (1997) *Biological Indicators of Soil Health.* CAB International, Wallingford, UK, 451 pp.

Parkin, T.B., Doran, J.W. and Franco-Vizcaíno, E. (1996) Field and laboratory tests of soil respiration. In: Doran, J.W. and Jones, A.J. (eds) *Methods for Assessing Soil Quality.* Soil Science Society of America, Madison, Wisconsin, pp. 231–245.

Radajewski, S., Ineson, P., Parekh, N.R. and Murrell, J.C. (2000) Stable-isotope profiling as a tool in microbial ecology. *Nature* 403, 646–649.

Rapport, D.J., McCullum, J. and Miller, M.H. (1997) Soil health: its relationship to ecosystem health. In: Pankhurst, C.E., Doube, B.M. and Gupta, V.V.S.R. (eds) *Biological Indicators of Soil Health.* CAB International, Wallingford, UK, pp. 29–47.

Ritz, K. and Griffiths, B.S. (2001) Implications of soil biodiversity for sustainable organic matter management. In: Rees, R.M., Ball, B.C., Campbell, C.D. and Watson, C.A. (eds) *Sustainable Management of Soil Organic Matter.* CAB International, Wallingford, UK, pp. 343–356.

Rutgers, M. and Breure, A.M. (1999) Risk assessment, microbial communities, and pollution-induced community tolerance. *Human and Ecological Risk Assessment* 5, 661–670.

Schipper, L.A. and Sparling, G.P. (2000) Performance of soil condition indicators across taxonomic groups and land uses. *Soil Science Society of America Journal* 64, 300–311.

Setälä, H. (2000) Reciprocal interactions between Scots pine and soil food web structure in the presence and absence of ectomycorrhiza. *Oecologia* 125, 109–118.

Sparling, G.P. (1997) Soil microbial biomass, activity and nutrient cycling as indicators of soil health. In: Pankhurst, C.E., Doube, B.M. and Gupta, V.V.S.R. (eds) *Biological Indicators of Soil Health*. CAB International, Wallingford, UK, pp. 97–119.

Swift, M.J. and Anderson, J.M. (1993) Biodiversity and ecosystem function in agricultural systems. In: Schulze, E.D. and Mooney, H.A. (eds) *Biodiversity and Ecosystem Function*. Springer-Verlag, Berlin, pp. 15–41.

Swift, M.J. and Woomer, P.L. (1993) Organic matter and the sustainability of agricultural systems: definition and measurement. In: Mulongoy, K. and Merckx, R. (eds) *Soil Organic Matter Dynamics and Sustainability of Tropical Agriculture*. John Wiley & Sons, Chichester, UK, pp. 3–18.

Swift, M.J., van der Meer, J., Ramakrishnan, P.S., Anderson, J.M., Ong, C.K. and Hawkins, B.A. (1996) Biodiversity and agroecosystem function. In: Mooney, H.A., Hall Cushman, J., Medina, E., Sala, O.E. and Schulze, E.D. (eds) *Functional Roles of Biodiversity: a Global Perspective*. John Wiley & Sons, Chichester, UK, pp. 261–298.

Tian, G., Brussaard, L., Kang, B.T. and Swift, M.J. (1997) Soil-fauna mediated decomposition of plant residues under constrained environmental and residue quality conditions. In: Cadish, G. and Giller, K.E. (eds) *Driven by Nature – Plant Litter Quality and Decomposition*. CAB International, Wallingford, UK, pp. 125–134.

van Straalen, N.M. (1997) Community structure of soil arthropods as a bioindicator of soil health. In: Pankhurst, C.E., Doube, B.M. and Gupta, V.V.S.R. (eds) *Biological Indicators of Soil Health*. CAB International, Wallingford, UK, pp. 235–264.

Verschoor, B.C., Pronk, T.E., de Goede, R.G.M. and Brussaard, L. (2002) Could plant-feeding nematodes affect the competition between grass species during succession in grasslands under restoration management? *Journal of Ecology* 90, 753–761.

Wardle, D.A. (1995) Impacts of disturbance on detritus food webs in agro-ecosystems of contrasting tillage and weed management practices. *Advances in Ecological Research* 26, 105–185.

Wardle, D.A. and Giller, K.E. (1996) The quest for a contemporary ecological dimension to soil biology. *Soil Biology and Biochemistry* 28, 1549–1554.

Wardle, D.A., Verhoef, H.A. and Clarholm, M. (1998) Trophic relationships in the soil microfood-web: predicting the responses to a changing global environment. *Global Change Biology* 4, 713–727.

Wardle, D.A., Giller, K.E. and Barker, G.M. (1999) The regulation and functional significance of soil biodiversity in agroecosystems. In: Wood, D. and Lenné, J.M. (eds) *Agrobiodiversity: Characterization, Utilization and Management*. CAB International, Wallingford, UK, pp. 87–121.

Wardle, D.A., Huston, M.A., Grime, J.P., Berendse, F., Garnier, E., Lauenroth, W.K., Setälä, H. and Wilson, S.D. (2000) Biodiversity and ecosystem function: an issue in ecology. *Bulletin of the Ecological Society of America* 81, 235–239.

White, D.C. and Macnaughton, S.J. (1997) Chemical and molecular approaches for rapid assessment of the biological status of soils. In: Pankhurst, C.E., Doube, B.M. and Gupta, V.V.S.R. (eds) *Biological Indicators of Soil Health*. CAB International, Wallingford, UK, pp. 371–396.

Wood, D. and Lenné, J.M. (eds) (1999) *Agrobiodiversity: Characterization, Utilization and Management*. CAB International, Wallingford, UK, 464 pp.

Yachi, S. and Loreau, M. (1999) Biodiversity and ecosystem functioning in a fluctuating environment: the insurance hypothesis. *Proceedings of the National Academy of Sciences* 96, 1463–1468.

Chapter 10
Subsoil Compaction and Ways to Prevent It

J.J.H. Van den Akker[1] and P. Schjønning[2]

[1]Wageningen University and Research Centre, Alterra Green World Research, PO Box 47, NL-6700 AA Wageningen, The Netherlands; [2]Danish Institute of Agricultural Sciences, Department of Agroecology, PO Box 50, DK-8830 Tjele, Denmark

Summary

Subsoil compaction affects all aspects of soil quality and, contrary to topsoil compaction, it is persistent. Natural alleviation processes such as wetting/drying, freezing/thawing and biological activity, including root growth, decrease rapidly with depth. In compacted soil, these alleviation processes are moreover diminished because root growth and biological activity are reduced and soil water contents remain higher in compacted than in well-structured soil. Wheel loads are still increasing and, in consequence, so are the extent and severity of subsoil compaction. Sustainable soil management requires the uncompromising criterion that no subsoil compaction can be accepted. Consequently, only field traffic with wheel loads lower than the carrying capacity of the subsoil is allowed. This implies that subsoil stress caused by wheel load should not exceed the strength of the subsoil. Therefore, this chapter emphasizes the importance of soil strength and the calculation of soil stresses in the subsoil. One of the main constraints in using the carrying capacity concept proves to be the lack of data on soil strength. Existing recommended limits for wheel loads and inflation pressures are not adequate and can result in over- or underestimation of allowable wheel loads and subsequent uneconomical solutions or subsoil compaction. Adequate drainage of soils is a prerequisite for reduced subsoil compaction. Mouldboard ploughing with all tractor wheels on the non-ploughed 'land' and umbilical systems for applying manure slurry are realistic options to reduce compaction. Controlled traffic systems that limit the wheeled area may be implemented by the use of wide-span vehicles and by steering tractors along traffic lanes. Although we support such provisions, our chapter will advocate and focus primarily on adjusting wheel loads to the carrying capacity of the subsoils.

Introduction

Soil compaction is defined as an increase in the density of unsaturated soil. The top layer of agricultural soil experiences cycles of loosening and compaction during the year. This is due to tillage and traffic as well as weather conditions and inherent soil processes triggered by soil biota. In contrast, most subsoils display a rather constant density as determined by the long-term geological history of each specific soil. In this chapter, subsoil is defined as soil below ploughing depth (normally ~25 cm). Subsoil compaction will be harmful to the functions of importance to agricultural use of most soils because the subsoil is a habitat for plant roots and soil fauna and flora. Thus a high quality of the subsoil is an environmental aim in itself and a precondition for organic and integrated farming systems.

The compaction problem has been extensively reviewed in a number of monographs (e.g. Voorhees, 1992; Håkansson and Voorhees, 1998), conference proceedings (e.g. Van den Akker *et al.*, 1999; Arvidsson *et al.*, 2000b; Birkas *et al.*, 2000; Canarache *et al.*, 2002), special issues of scientific journals (Håkansson, 1994; Van Ouwerkerk and Soane, 1995; Van den Akker *et al.*, 2003) and books (e.g. Soane and Van Ouwerkerk, 1994; Horn *et al.*, 2000). How compaction influences soil is well described, at least in terms of the physical conditions. An important conclusion from these investigations is that subsoil compaction appears to be rather persistent if not permanent, i.e. the resilience (consult Chapter 1) regarding subsoil compaction is extremely low. Factors generally considered to alleviate compaction are wetting/drying, freezing/thawing, biological activity and tillage. The efficiency of freezing in alleviating subsoil compaction has previously been overestimated (Håkansson and Petelkau, 1994).

Further, soil compaction becomes more persistent the deeper it penetrates because the frequency and intensity of the factors mentioned above decrease rapidly with depth. This effect will be further accelerated because reduced root growth in compacted soil will reduce oscillations in water content (Whalley *et al.*, 1995) and biological activity. Thus, subsoil compaction may be regarded as one of the most severe threats to soil quality. It is often considered in terms of short-term economics due to reductions in yield. However, persistent subsoil compaction affects soil functions of importance for generations to come (Håkansson and Petelkau, 1994). In this chapter, we exclusively address subsoil compaction. The major emphasis will be on the possibilities of estimating the susceptibility of subsoils to compaction and the potentials for accommodating the loads from agricultural implements with the carrying capacity of subsoils.

Extent and Severity of Subsoil Compaction

According to Fraters (1996), about 32% of the subsoils in Europe are highly vulnerable to subsoil compaction and another 18% are moderately vulnerable. Oldeman *et al.* (1991) estimated that compaction is by far the most important type of physical deterioration of agricultural soils, being responsible for soil degradation of a worldwide area of 68,000 km^2. Approximately half of this acreage is in Europe and 25% in Africa. In countries of the former USSR, heavy equipment is used even on wet soils and yield losses up to 50% by soil compaction were reported in former Soviet agriculture (Libert, 1995). It is not possible to calculate what part of these losses can be attributed to subsoil compaction, but very persistent subsoil compaction going deeper than 80 cm has been registered in large areas of the former USSR (Libert, 1995). These figures are rough estimates because a thorough inventory was never performed. Farmers in many countries are unaware that subsoil compaction is a serious problem. It is a hidden form of soil degradation that affects all the agricultural area and results in gradually decreasing yields and gradually increasing problems with waterlogging. The impact of subsoil compaction is primarily prominent in years with extreme dry or wet periods – another reason that it is hardly recognized. In fact, all agricultural soils in developed countries display compacted subsoils. Some soils are naturally compacted, strongly cemented or have a thin topsoil layer on rock subsoil. Håkansson et al. (1996) measured 40% higher penetration resistance in fields trafficked with heavy slurry manure tankers in intensive potato and sugarbeet production than in fields never trafficked by farm machinery. In an area with less intensive traffic, the increase of penetration resistance was still 10%. Persistent crop yield reductions caused by subsoil compaction in the fields with intensive and moderate machinery traffic were estimated to be 6% and 1.5%, respectively. The detrimental effects of subsoil compaction on crop production are often compensated by improved drainage and by increased supply of nutrients and water (irrigation). These pseudo-solutions to the compaction problem lead to excessive use of water and nutrients and environmental pollution, and are no longer socially and politically accepted.

Due to the ever-increasing wheel loads in agriculture, compaction is increasingly expanding into the subsoil. In the field experiments initiated in the early 1980s (Håkansson et al., 1987), wheel loads of 50 kN were used for the initial subsoil compaction action. This was considered a very high wheel load at that time. Van de Zande (1991) considered 35–40 kN as the highest wheel-load class in an investigation in 1984 on traffic intensity on fields in The Netherlands. Nowadays self-propelled slurry tankers with injection equipment have wheel loads of 90–120 kN and are used in early spring on wet soils. Arvidsson (2001) investigated the impact on subsoil compaction

of wheel loads of 90 kN exerted by self-propelled, six-row sugarbeet harvesters, which were introduced in Sweden in 1993. Van der Linden and Vandergeten (1999, cited by Poodt et al., 2003) reported maximum wheel loads of 130 kN for sugarbeet harvesters. The high wheel loads were not only caused by the high total weight of the loaded harvester but also by the uneven distribution of this total weight over the wheels. The largest tyres available have to be used for such conditions. These tyres are claimed to be low pressure tyres, yet inflation pressures are in practice at least 140 kPa and on average 190 kPa. Only inflation pressures below 100 kPa should be considered 'low pressure'.

Impact of Subsoil Compaction on Soil Functions and Processes

Subsoil compaction exerts a pronounced effect on soil functions. Basically, compaction means reduction of soil volume, i.e. in effect a reduction in void volume. Pore continuity may also be affected. This reduces the ability of the soil to conduct water and air. A poor saturated hydraulic conductivity may trigger surface runoff and eventually water erosion. It may also induce preferential flow in macropores, which has been shown to facilitate transport of colloid-adsorbed nutrients and pesticides to deeper horizons. A poor aeration of soil may yield non-optimal plant growth and induce loss of soil nitrogen and production of greenhouse gases through denitrification in anaerobic sites. Plastic deformation of soil aggregates and higher bulk density resulting from compaction increase the mechanical strength of the soil matrix. This may limit root growth and crop exploitation of soil water and nutrients. Accordingly, this may increase leaching of nutrients. These and other compaction effects on soil functions are described in a number of scientific reports and reviews (e.g. Soane and van Ouwerkerk, 1994; van Ouwerkerk and Soane, 1995).

The functions above may be quantitatively related to soil compaction effects by the Least Limiting Water Range (LLWR) concept suggested by Da Silva *et al.* (1994) as based on an idea by Letey (1985). The LLWR concept is based on the fact that plant response to varying soil water contents is least limited inside a range and most limited outside the same range. The water-content-related limitations are expressed through available water, soil aeration and mechanical resistance. Four water content limiting soil functions are identified: θ_{fc}, θ_{wp}, θ_{afp} and θ_{pr}, representing field capacity (fc), wilting percentage (wp), 10% air-filled pore space (afp) and 2 MPa penetration resistance (pr), respectively. Figure 10.1a shows how compaction reduced the LLWR of differently tilled clay loam topsoils (Betz *et al.*, 1998). The LLWR is read as the water range between the two narrowest limitations. This may be done at the average bulk density (ρ_b, indicated by the vertical line in Fig. 10.1a and b) or in the range of densities observed for each particular situation (the shaded areas). For the non-tracked part of the chiselled and mouldboard ploughed soils, θ_{fc} formed the upper limit of the topsoil rooting environment, whereas for the tracked part of the same tillage systems, soil functions were limited at θ_{afp} (Fig. 10.1a). Note that the latter limit is active for the no-till soil irrespective of traffic. The lower water limit for optimal soil function in the no-till soil was a high mechanical strength, θ_{pr}, whereas it was the wilting percentage water content, θ_{wp}, that limited growth for the two other tillage systems (Fig. 10.1a).

Figure 10.1b shows the LLWR for the upper part of the subsoil for the same clay loam soil as in Fig. 10.1a. This horizon was interpreted by Betz *et al.* (1998) as being highly compacted and was labelled a plough pan. It should be noted that the LLWR was close to zero at the average bulk density. The upper and lower water limits for optimal plant growth were determined by the 10% air-filled pore space and the root-restricting penetration resistance, respectively. The compaction effect is also evident from the huge difference between the limit given by θ_{fc} and that by θ_{afp} (Fig. 10.1b).

The data presented in Fig. 10.1 offer a very clear illustration of compaction effects on soil functions. Betz *et al.* (1998) primarily considered root growth when evaluating the LLWR. However, the 10% air-filled pore space

suggested as a limit for adequate soil aeration for root growth (Grable and Siemer, 1968) has been shown also to indicate a threshold for aerobic turnover of organic matter in soil (Schjønning *et al.*, 2003). Hence, the compaction effects on the LLWR given in Fig. 10.1 may be interpreted also in relation to the fate of N in soil organic matter. An air-filled pore space below the 10% limit may trigger the production of greenhouse gases (N_xO) rather than nitrate (NO_3) for crop uptake. It is clear from Fig. 10.1 that the range of water content optimal for the key LLWR soil functions discussed may be very small for untilled topsoil and for the subsoil. Unsatisfactory conditions in the topsoil may be alleviated through a (temporary) change in tillage system, whereas poor conditions obtained in the subsoil are much more difficult to manage.

The crop integrates the effects of the soil functions. Alakukku (2000) measured significant crop and nitrogen yield losses 17 years after a single, high axle load compaction action (Fig. 10.2). Only moderate axle loads were allowed on the fields after the initial wheel-by-wheel compaction treatment. In practical agriculture, the subsoil will be partly compacted every year and the subsoil quality probably worse than in the long-term experiments. It should be noted that the compaction effect for all years was more pronounced for harvested nitrogen than for grain dry matter. As the crop was equally fertilized irrespective of compaction treatment, the decreased recovery of crop nitrogen probably indicates nitrogen loss to the environment (denitrification and/or leaching). The lower content of raw protein is a further expression of a reduced crop quality. Therefore, the results in Fig. 10.2 show that all three aspects of the soil quality concept (consult Chapter 1) are negatively influenced by subsoil compaction.

The potential impact of subsoil compaction on crop yield and environmental aspects may be much more severe than deduced from average results of even long-term field trials. Therefore, in addition to the field trials, models run with different scenarios of weather conditions may better reveal the damage exerted on soil functions (e.g. Feddes *et al.*, 1988; Simota *et al.*, 2000; Lipiec *et al.*, 2003; Stenitzer and Murer, 2003).

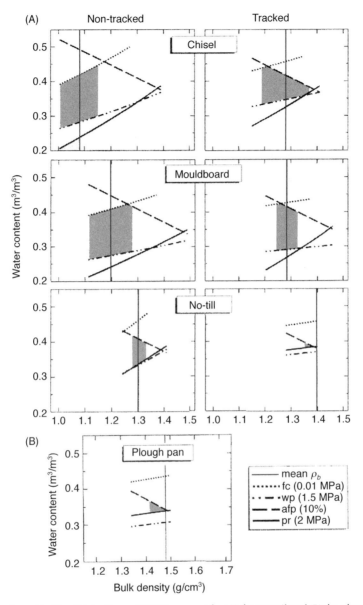

Fig. 10.1. The Least Limiting Water Range (LLWR) concept for (A) the topsoil and (B) the plough pan of a clay loam. Crop growth is constricted at water contents below and above the limits indicated by the shaded areas. Four limiting water contents for soil functions are identified: θ_{fc}, θ_{wp}, θ_{afp} and θ_{pr}, representing field capacity (fc), wilting percentage (wp), 10% air-filled pore space (afp) and 2 MPa penetration resistance (pr), respectively. ρ_b is soil bulk density (Betz *et al.*, 1998).

Sustainability Criterion

As discussed in Chapter 1 of this volume, sustainability criteria are expressions of societal priorities. In the context of subsoil compaction, we select as a criterion for sustainable soil management that *no* subsoil compaction should be accepted. This

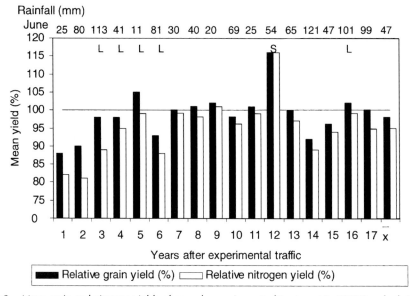

Fig. 10.2. Mean grain and nitrogen yields of annual crops in control treatment (= 100%) and relative to the control in loading treatment of four passes in 1981 on clay soil, for 17 successive years after the loading. L, lodging; S, sprouting (Alakukku, 2000).

uncompromising criterion was also advocated by Medvedev and Cybulko (1995) and is chosen due to the severity and persistence of subsoil compaction that is reported in the comprehensive literature on this subject and briefly reviewed above.

Although it has been shown in recent decades that compacted subsoil is not at all or only extremely slowly ameliorated by natural processes (e.g. Voorhees, 2000), it might be claimed that the soil resource could be reconstituted by other means. However, as reviewed by Larson *et al.* (1994) and Håkansson and Reeder (1994), several studies have shown that mechanical subsoiling is not able to fully remedy the malfunctions created by compaction. Further to that, it appears that the mechanical loosening effect seems to persist for only a few years (e.g. Bishop and Grimes, 1978; Kooistra *et al.*, 1984). To a large extent this is due to the loosened soil being extremely vulnerable to recompaction from traffic (Soane *et al.*, 1986). Another potential means of ameliorating compacted soil is through the action from actively growing plant root systems. However, Cresswell and Kirkegaard (1995) reviewed the literature and

concluded that it is still not clear whether 'biological drilling' is an effective process for ameliorating dense subsoils. In conclusion, we judge subsoil compaction so severe that *no subsoil compaction* is our criterion for sustainability.

Susceptibility of Subsoils to Compaction

Subsoil compaction can be prevented and hence the criterion for sustainability can be met if subsoil stresses caused by wheel loads do not exceed the strength of the subsoil. Prediction of stresses and access to data on soil strengths are thus essential.

All shear and normal stresses at a point in the soil can be expressed in just three normal stresses, which are called the principal stresses. S_1, S_2 and S_3 are the major, intermediate and minor principal stresses, which completely determine the stress situation at the considered point (Koolen and Kuipers, 1983). In the subsoil, the vertical soil stress is almost equal to the major principal stress and the two horizontal soil stresses are almost equal to the intermediate and minor principal stresses.

Strength parameters

Two soil strength properties can be distinguished: the 'preconsolidation load' or 'precompression stress' and the shear strength. We prefer precompression stress because consolidation does not occur at all in unsaturated soil.

Precompression stress

The precompression stress is the maximum major principal stress a soil can resist without major plastic deformation and compaction. It is determined with a uni-axial test by which a soil sample in a cylinder is loaded by a piston with increasing vertical stress σ_v. The side walls of the cylinder prevent horizontal deformation of the sample. Vertical soil deformation will be limited as long as the aggregates can withstand the load. When the aggregates fail and deform, the cracks are filled with soil and the vertical deformation of the sample increases, yielding an inflection point in the log (σ_v)–strain curve. The precompression stress is defined as the stress at the inflection point (Lebert and Horn, 1991). The drier the soil, the stronger the aggregates, and hence the higher the measured precompression stress. In agricultural soils, the precompression stress depends on soil moisture conditions, soil structure and development of physical and chemical bonds in time. Drying is the basis for ripening, shrinkage, structure formation and increase of precompression stress and shear strength of soils. If soil is loaded beyond the precompression stress, the soil is plastically deformed. Some drying and wetting periods may re-establish the soil structure with bigger and stronger aggregates and a higher precompression stress. Hence, there is an indirect relation between the historic load on a subsoil and its precompression stress. Restrictions in subsoil rooting and drainage caused by compaction will create wet subsoil conditions and a higher vulnerability to further compaction.

In sandy soils, structure and cracks are absent, and the described failure mechanism in the uni-axial test is not possible. However, many sandy soils have some structure due to cementation, humus content and biological activity. These soils can also show an inflection point in the stress–strain curve derived from a uni-axial test when the bonds are broken. However, in many sandy soils the precompression stress is not well defined.

Shear strength

Shear strength is the resistance of a soil against tangential movement between parts of the soil. Shear strength can be determined in a shear box, tri-axial test apparatus or shear annulus (see, e.g. Koolen and Kuipers, 1983). The shear box test subjects a sample to a vertical normal load resulting in a vertical stress σ_n, and a horizontal load resulting in a shear stress τ in the soil sample. The relation between the normal stress σ_n and the shear stress τ is given by the Mohr–Coulomb failure line:

$$\tau = C + \sigma_n \tan \varphi \qquad (10.1)$$

where C is cohesion and φ is the angle of internal friction.

In structured, unsaturated subsoil, the soil consists of structural elements (aggregates) fitting neatly together. When subjected to shearing, this soil will behave like a very dense sandy soil with the aggregates as 'sand grains'. Lebert and Horn (1991) distinguished an interaggregate and texture-dependent Mohr–Coulomb failure line (Fig. 10.3). In the interaggregate part, the aggregates are strong enough to resist the stresses and the soil acts like very dense sand with a high angle of internal friction and a low cohesion. In this part, the soil is very stable and relatively strong and often displays a low bulk density and good physical properties. Failure will not necessarily remove all of these good soil qualities. In the second, texture-dependent part, when the aggregates fail and crush and deform plastically, the structure will be lost, the macropores filled up, and continuous macropores disconnected. In the texture-dependent part, the strength properties resemble a non-structured clay soil, in which the angle of internal friction is smaller and cohesion higher than in the interaggregate part (Fig. 10.3). In weakly structured soils the interaggregate part is almost negligible and strength properties are mainly texture-dependent.

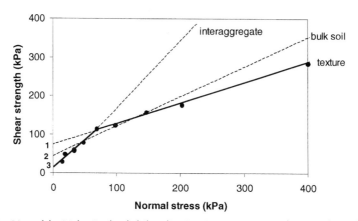

Fig. 10.3. Partition of the Mohr–Coulomb failure line into interaggregate and texture-dependent failure lines and the bulk soil failure line (load range 0–400 kPa) of a clay loam. Water potential is −30 kPa. Black dots show measured values: 1, intraaggregate cohesion; 2, cohesion of the bulk soil; 3, interaggregate cohesion (Lebert and Horn, 1991).

Water effects

The strength of the aggregates determines to a great extent the precompression stress and the shear strength of a structured soil. The soil moisture conditions influence the strength of the aggregates. During drying, the aggregates are subjected to large soil water suctions, which compact the aggregate under a hydrostatic stress that is many times the soil stresses exerted by wheel loads. The soil water suction does not act in the air filled pores, so the soil water potential must be multiplied with the saturation degree *of the aggregates* to derive the effective hydrostatic stress on the aggregates. In a heavy clay soil with a water potential of, e.g. $\psi = -400$ kPa, the saturation degree of the aggregates can still be almost 1, so the effective stress is about 400 kPa. Soil stresses induced by a wheel load of, for example, 100–200 kPa will not be able to crush the aggregates. Even when the soil is not that dry, the water potential will strengthen the aggregates considerably. Moreover the chemical and physical bonds between soil particles strengthen by drying. Another mechanism that strengthens the aggregates is dilatancy (volume increase) of the soil in the aggregate when it is sheared. The aggregate, being relatively compact, will increase its volume during shearing. This results in a decrease of the water potential and accordingly an

increase in aggregate strength. In a wet soil, the smaller macropores are filled with water. Fast compression of the soil results in compression of water and enclosed air, and the aggregates will weaken due to increase in the water potential in the aggregates (Baumgartl and Horn, 1991).

Fleige *et al.* (2002) determined the average precompression stress, cohesion and angle of internal friction of German subsoils of the texture groups sand, silt, loam and clay at soil water potentials of −6 kPa and −30 kPa. At a soil water potential of −6 kPa, the precompression stress was about 120 kPa for the sandy soil and about 95 kPa for the remaining three texture classes (Fig. 10.4). By drying to a water potential of −30 kPa, the precompression stress of the sandy soil increased 13% and that of the silty, loamy and clayey soils increased 34, 23 and 18%, respectively. The angle of internal friction decreased with increasing clay content (Fig. 10.5). The water potential had hardly any effect on the angle of internal friction. The cohesion of loamy and clayey soils was at least two times higher than that of sandy and silty subsoils (Fig. 10.5). By drying the soil from a water potential of −6 kPa to −30 kPa, the cohesion increased markedly, especially for the silty subsoil (85% increase). Sandy and silty soils with low clay contents develop no or only weak structure. A well-aggregated clayey subsoil is about

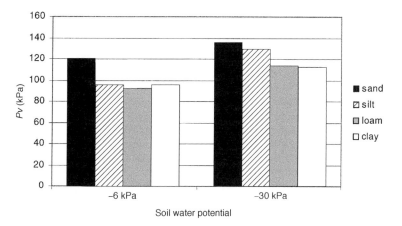

Fig. 10.4. Precompression stress Pv in kPa of German subsoils at a soil water potential of −6 kPa and −30 kPa for sand, silt, loam and clay (after Fleige *et al.*, 2002).

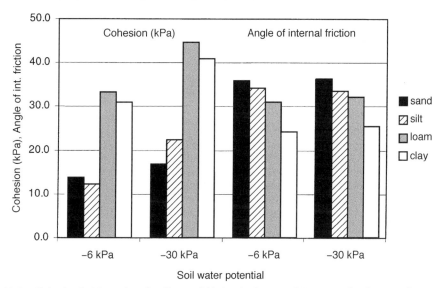

Fig. 10.5. Cohesion in kPa and angle of internal friction in degrees of German subsoils at a soil water potential of −6 kPa and −30 kPa for sand, silt, loam and clay (Fleige *et al.*, 2002).

25–30% stronger than a weakly aggregated clayey subsoil.

Indirect strength measures

The internationally approved uni-axial compression test and the direct shear method were adopted from civil engineering and have a low rate of applying the stress. However, stresses by wheel traffic are exerted within fractions of a second. It has proven difficult to include the dynamic aspects of loading and soil deformation into the classical methods. Moreover, these methods are very time consuming. Data on soil strength for different soil types, different depths (horizons) and at different water contents are therefore very scarce, and this is a major problem in combating the problem of (sub)soil compaction. We are therefore met with an urgent need to increase the knowledge on soil susceptibility to compaction by indirect measurements that include the dynamic aspects of soil deformation in agriculture.

Quick, semi-empirical methods

Koolen (1974) suggested a strain-controlled uni-axial compression test for estimating the stress–strain behaviour of agricultural soils. The loading time is similar to that under a moving wheel. The time consumption is about 15 min as compared to several days for the classical uni-axial test. Schjønning (1991, 1999) applied the simple test to a range of Danish soils and succeeded in identifing the precompression stress. Further studies are needed, however, to test the correlation between precompression stress, estimated by the classical and the simple methods, and to elucidate which of the two methods is the more relevant regarding the field situation during traffic.

The drop-cone (or Swedish fall-cone) penetrometer is a simple instrument consisting of a metal cone of well-defined mass and cone angle (Hansbo, 1957). In this test the penetration depth h is defined as the depth to which the cone sinks within 5 s. The drop-cone strength parameter, h^{-2}, is linearly correlated with the cohesional strength of puddled clay (Hansbo, 1957; Houlsby, 1982; Wood, 1985). h^{-2} has also frequently been used as an index of strength for more heterogeneous substances such as structured soil (Towner, 1973; Bradford and Grossman, 1982; Campbell and Hunter, 1986). Mullins and Fraser (1980) stressed that the drop-cone penetrometer gives a better simulation of the interaction of a wheel load with the soil than an instrument such as the shear vane. Schjønning (2000) showed that this simple instrument may have the potential to estimate the shear strength of soil and even to differentiate the strength into the cohesional and frictional strength components. However, as for the simple method of measuring precompression stress, a calibration of the results with those of the classical methods is needed. Included in such work should be an evaluation of the strength estimates by both methods in field situations with well-defined stress distributions.

Pedotransfer functions

Pedotransfer functions are quantitative relations between difficult-to-measure soil properties and quickly assessed properties. Lebert and Horn (1991) developed pedotransfer functions for precompression stress. Within each soil type investigated, parameters like bulk density, hydraulic conductivity and different water fractions explained 37–75% of the variation in data. For clayey soils, the inclusion of the shear parameters cohesion and angle of internal friction significantly increased model performance. For sandy soils, the precompression stress can be predicted from readily available parameters, whereas for clayey soils, it is necessary to measure shear strength or to use tables with average shear strength parameters classified according to soil type, soil water suction and structure development (DVWK Merkblätter zur Wasserwirtschaft, 1995).

Horn and Lebert (1994) claimed that it is impossible to estimate the strength of a soil in its structured condition from the plasticity limits obtained from remoulded soil samples. However, Veenhof and McBride (1996) showed that the precompression stress for a number of structurally intact, unsaturated Canadian subsoils correlated with the consolidation behaviour of saturated, remoulded soil. This created the basis for a pedotransfer function (McBride and Joosse, 1996) by which the precompression stress was estimated from characteristics obtained by simpler means than the series of uni-axial compression tests normally needed.

Soil Stresses versus Soil Strength

Stress propagation in soil

No subsoil compaction implies that soil deformation in the subsoil remains elastic. That is, the soil matrix bounces back to its original form after removal of stresses. In most cases, deformation and compaction of the topsoil cannot be prevented. Assuming that the deformed and compacted topsoil exhibits similar stiffness and elasticity to the uncompacted subsoil, it is possible to use relatively simple analytical models based on the theory of Boussinesq (1885, cited by Fröhlich, 1934). Vertical and horizontal normal and shear stresses can be calculated for any volume element in the soil. Fröhlich (1934) introduced a concentration factor (v) in the Boussinesq formulae to account for the anticipated tendency of the soil to concentrate the stresses around the load axis. The value of v is considered greater when the soil is softer. The simple form of the Boussinesq/Fröhlich equation predicts the vertical stress, σ_z, at depth z:

$$\sigma_z = (vP/2\pi r^2)\, cos^v(\beta) \qquad (10.2)$$

where P is the point load on the soil surface, r is the linear distance between the surface point and the point considered in the soil horizon, and β is the declination to vertical for this vector.

Because the equations of Boussinesq and Fröhlich are based on a linear elastic material, it is possible to superpose the stresses at any point in the soil caused by several point loads on the rut bottom. Most analytical models consider only vertical loads and vertical soil stresses. However, horizontal stresses are needed to check whether wheel-load stresses exceed the shear strength in the subsoil.

The socomo model

A model that includes horizontal loads and horizontal stresses is SOCOMO (Van den Akker and Van Wijk, 1987; Van den Akker, 1994, 1997, 1999). SOCOMO (SOil COmpaction MOdel) can be downloaded from the Internet site (http://www.subsoil-compaction.alterra.nl). In SOCOMO, the stress distribution in the soil–tyre interface is projected on a horizontal rectangular grid and the stresses are concentrated at the grid points. The input of the wheel load in SOCOMO consists of two corresponding matrices with vertical and horizontal point loads, respectively. All stresses exerted by these point loads on a given point in the soil can be superposed and converted into the principal stresses S_1, S_2 and S_3 (Koolen and Kuipers, 1983). The width of the tyre determines the width of the loading grid. The sum of all vertical point loads must be equal to the vertical wheel load. The distribution of the vertical and horizontal stresses in the tyre–soil contact area can be complicated, e.g. when stresses are considered to be

concentrated under the lugs (Hammel, 1994). However, in most cases: (i) an even, (ii) a trapezoic, or (iii) a parabolic stress distribution is assumed. In addition to the pressure distribution, the mean vertical pressure p_m over the tyre footprint is important. According to a rule of thumb given by Koolen and Kuipers (1983), the mean normal stress in the contact surface is 1.2 times the tyre inflation pressure. Van den Akker (1992) suggested that for low-pressure tyres, mean ground pressure is equal to inflation pressure plus 50 kPa. This accounts for the increased influence of the carcass stiffness on the ground pressure due to the large deformation of low-pressure tyres and will be used in the calculations below.

A comparison of Terra Tyres and tandem and dual wheel configurations

Wide low-pressure tyres (Terra Tyres) can be used to prevent soil compaction by heavy wheel loads. As an alternative, the load can be divided over two smaller tyres in dual wheel or tandem configuration. The major principal stresses were computed for these three alternatives and a tyre with normal inflation pressure (Fig. 10.6). The figure shows that the major principal stresses are much higher under a normal tyre than under low-pressure tyres. Further, the vertical pressure distribution over the tyre footprint has a strong effect on the major principal stress. Replacing a Terra Tyre with two smaller tyres will result in slightly higher stresses in the topsoil and much lower stresses in the subsoil. This positive effect strongly depends on the distance between the tyres.

The wheel-load carrying capacity

The SOCOMO model compares the predicted soil stresses at each point in the soil with the strength of the soil at that point. No failure (compaction) of the subsoil will occur when: (i) the major principal stress is smaller than the precompression stress, and (ii) the soil shear strength represented by the Mohr–Coulomb failure line is not exceeded.

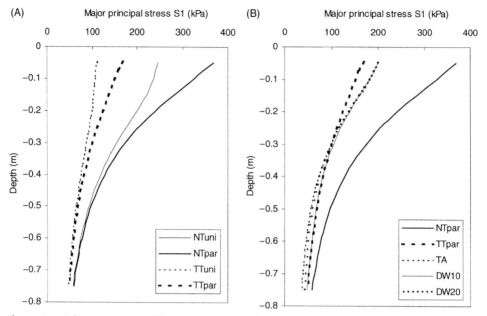

Fig. 10.6. Highest major principal stresses at a certain depth under (A) a Terra Tyre (code: TT; inflation pressure: 60 kPa; width: 1.1 m) and a normal tyre (code: NT; inflation pressure: 200 kPa; width: 0.6 m) with a wheel load of 50 kN with an assumed uniform (uni) or parabolic (par) vertical pressure distribution over the tyre footprint. These tyres were (B) compared with tyres (inflation pressure: 80 kPa; width: 0.5 m) in tandem (TA) and dual wheel configurations with spacings of 0.10 m (DW10) and 0.20 m (DW20) with wheel loads of 2 × 25 kN. A parabolic vertical pressure distribution was assumed.

The wheel-load carrying capacity of a subsoil can be defined as the maximum wheel load exerted by a specific tyre type and tyre inflation pressure, which does not exert stresses in the subsoil exceeding the strength of that subsoil. The lower of the two strength criteria determines the wheel-load carrying capacity. The wheel load and the tyre dimensions, properties and inflation pressure determine the stress distribution in the soil. Therefore, the wheel-load carrying capacity is always associated with the specific tyre dimensions and inflation pressure.

The carrying capacity of subsoils loaded with a Terra Tyre

Figure 10.7 presents the carrying capacity of sand, silt, loam and clay subsoils with the strength properties presented in Figs 10.4 and 10.5 for −6 kPa and −30 kPa water potentials when loaded with a Terra Tyre 73 × 44.00 − 32. In these calculations, the maximum allowable tyre deflection at 16 km/h

with variable load is assumed to be effective. The soil–tyre contact area on the hard surface of this tyre is 0.513 m² at this allowable tyre deflection. This means that the inflation pressure is taken as low as possible for a given wheel load. The lower the wheel load, the lower the allowable inflation pressure. Inflation pressures ranged between 40 and 80 kPa. This minimizes the vertical soil stresses and results in the highest carrying capacity with that tyre. Figure 10.7 shows a strong relation between the carrying capacity and the allowable deformation depth. The deformation depth is the depth below rut bottom, where compaction is predicted. To prevent subsoil compaction the deformation depth is not allowed to enter the subsoil. A thick topsoil and a minimal rut depth result in a higher carrying capacity.

At −6 kPa water potential, the carrying capacity on sand is governed by the precompression stress for allowable deformation depths less than 0.19 m and by the shear failure criterion if larger deformation depths

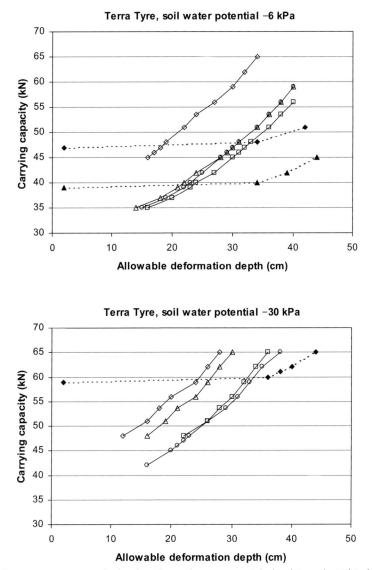

Fig. 10.7. The carrying capacity of subsoils with a soil water potential of −6 kPa and −30 kPa loaded with a Terra Tyre. The allowable deformation depth is in general the ploughing depth minus rut depth. P, the precompression stress is exceeded. MC, Mohr–Coulomb shear strength is exceeded. ◇, Sand P; ◆, Sand MC; △, Silt P; ▲, Silt MC; ■, Loam P; ○, Clay P.

are allowed (Fig. 10.7). However, large allowable deformation depths are needed to increase the carrying capacity with just a few kN. For silt at −6 kPa water potential, the carrying capacity is determined by the pre-compression stress for depths below ~21 cm, whereas it is the shear failure criterion that yields the carrying capacity in upper soil layers. At −30 kPa water potential, the carrying capacity on sand is determined by the shear failure criterion if allowable deformation depths >~0.24 m are considered (Fig. 10.7).

The most frequent ploughing depth is 25 cm. Hence, if taking 25 cm as an allowable deformation depth, the criterion of no subsoil compaction when using this specific Terra Tyre can be read from Fig. 10.7. For the wet soil at −6 kPa, the carrying capacity of the sand

and silt soils is limited by the shear strength. These soils may be exposed to a wheel load of ~47 and ~40 kN, respectively. The loam and clay soils may be loaded by ~42–43 kN. When drained to a matric potential of –30 kPa, the allowable loads with this tyre are approximately 55, 57, 50 and 50 kN for the sand, silt, loam and clay soils, respectively. The results indicate that the silt soil especially becomes stronger by drying out. The carrying capacity of the silt soil increases about 17 kN when dried from –6 to –30 kPa, whereas the carrying capacity of the other soils increases about 8–12 kN.

It appears that *when using the best tyres available* with the optimal inflation pressure, it is possible to avoid compaction of the subsoil (> 25 cm depth) at traffic events involving the wheel loads mentioned. This is an important finding as several tillage operations will be possible with wheel loads lower than these estimated management thresholds. However, the results also imply that traffic with heavy equipment like sugarbeet harvesters, etc. can only take place when such vehicles are equipped with more wheels than are often used at present. It is furthermore very important to stress that the depth of compaction especially for soil layers to ~50 cm is dependent on the contact pressure under the wheel, i.e. the inflation pressure and tyre characteristics.

Construction of maps with the carrying capacity of subsoils loaded with a specific tyre

SOCOMO was used to construct a map of The Netherlands with the carrying capacity of a tyre with a width of 0.50 m and an inflation pressure of 80 kPa on arable soil at –30 kPa matric potential (Van den Akker, 1997). The maximum allowable wheel load of this tyre with an inflation pressure of 80 kPa is 32 kN. Table 10.1 presents the strength properties of the major subsoils in The Netherlands used in the construction of this map. For each soil the set of strength parameters with the lowest values, representing the weakest structure, was chosen in order to derive the lowest, thus safest, carrying capacity of that particular soil. Pedotransfer functions derived by Lebert and Horn (1991) were used to calculate the precompression stress (given as Pv in Table 10.1). The effective thickness, D, was defined as the thickness of the topsoil minus an estimated rut depth of 0.03 m. Hence, D is equal to 'allowable deformation depth' (Fig. 10.7). Typical Dutch ploughing depths were used. The ploughing depth in sandy soils is usually deeper than in clay soils. The carrying capacities of sandy soils, sandy loams and clay loams were limited by the shear failure criterion, and those of the other soils by the precompression stress failure criterion. The carrying

Table 10.1. Carrying capacity of Dutch subsoils. Figures in bold denote whether the carrying capacity is limited by the precompression stress or by the Mohr–Coulomb shear failure criterion. Angle of internal friction φ, cohesion C and precompression stress Pv of the major subsoils in The Netherlands at a soil water potential of –30 kPa. The table is based on DVWK Merkblätter zur Wasserwirtschaft (1995) and Lebert and Horn (1991). D is the effective topsoil thickness = ploughing depth minus estimated 0.03 m rut depth.

| | Clay (%) | C (kPa) | φ (°) | Pv (kPa) | D (m) | Carrying capacity (kN) | |
						Precompression stress	Shear
Sandy soils	< 8	12	28	198	0.32	> 32	**17**
Coarse sand	< 8	10	32	240	0.32	> 32	**16**
Sandy loam 1	< 8	10	32	122	0.32	> 32	**16**
Sandy loam 2	8–18	10	32	140	0.27	> 32	**11**
Clay loam	18–25	14	31	79	0.27	20	**17**
Light clay	25–35	26	36	118	0.22	**22**	> 32
Medium clay	35–50	26	36	96	0.22	**22**	> 32
Heavy clay	> 50	34	38	114	0.22	**24**	> 32
Sandy silt	< 18	15	39	82	0.22	**18**	> 32
Silt loam (loess)	< 18	26	37	110	0.22	**28**	> 32

capacities are lower than the wheel loads commonly used during harvest operations. The soil strength properties used in the model simulations were those of a weakly structured or unstructured soil. Well-drained soils will often be structured and therefore much stronger in practice than in the simulations.

Recommendations to Prevent Subsoil Compaction

Generally, it can be stated that matching machine operations to the conditions of the soil is a major management tool for preventing soil compaction (Larson *et al.*, 1994). However, the real challenge is to implement this in practice. The soil water content is decisive for (sub)soil strength and knowledge of the soil water content is thus a prerequisite for decisions on allowable loads. If the carrying capacity of the (sub)soil as a function of water content is known from one of the approaches discussed in the section 'Soil Stresses versus Soil Strength', this may be combined with the soil water content to decide if a certain machine operation is allowable. A water balance model may interactively predict the water content for the subsoil. Arvidsson *et al.* (2000a, 2003) demonstrated how this approach may point out the risk of compaction at a given time of the year for a specific combination of climatic region and soil type. The same procedure may be used for assessing soil vulnerability at a given date for a given soil grown with a given crop. At the moment, due to the lack of quantitative data on (sub)soil strength for most soil types, the above scenario is unrealistic for most farmers. Instead, a number of precautionary measures should be regarded in order to decrease the risk of compaction. Below, we summarize some of the most important management tools to consider in this context.

Reducing soil stresses and increasing soil strength

Soil stresses in the subsoil can be decreased by decreasing tyre inflation pressures, wheel loads and rut depths and by using wider, larger and flexible tyres (cf. Fig. 10.6). Wheel loads can be decreased by lowering the payload or weight of the equipment or by increasing the number of wheels (e.g. dual wheels or tandem). A very effective way of decreasing soil stresses is to reduce the average ground pressure on the rut bottom by using wide tyres with low inflation pressures. These low-pressure tyres also have the advantage of decreased topsoil compaction and reduced rut depth, allowing for an increased distance to the subsoil. Tijink (1998) collected guidelines to prevent soil compaction, expressed as limits for inflation pressure, *average* ground pressure (given as p_c in Table 10.2) and vertical soil stresses at 50 cm depth (p_{50}) in spring or in summer/autumn (Table 10.2). However, these guidelines neglect the effect of high wheel loads on subsoil compaction. Håkansson and Petelkau (1994) mentioned recommended axle load limits of 60 kN (30 kN wheel load) in Sweden and 30–40 kN (15–20 kN wheel load) for machines with standard tyres in the former German Democratic Republic. These axle load limits are very simple, however, neglecting the considerable benefits of low tyre inflation pressures. The use of limits for axle load or inflation pressure as generalized management thresholds will often imply under- or overestimation of allowable wheel loads, leading to subsoil compaction or to too-low, uneconomical wheel loads. This aspect makes it controversial to introduce strict legislation on maximum allowable wheel loads or inflation pressures. However, the carrying capacity concept makes it possible to determine whether a certain combination of wheel equipment, machinery and payload causes subsoil compaction and should not be allowed. This would enable certification of allowable agricultural machinery when data on subsoil strength become available.

One of the most effective measures to increase soil strength is by improving drainage. In soils with a high water table, drainage is essential to permit field operations with little or no subsoil compaction risk. Here, a vicious circle may be triggered because a wet soil is likely to yield subsoil compaction, resulting in a further reduction in the soil

Table 10.2. Guidelines to prevent soil compaction, expressed in limits for inflation pressure (p_i), average ground pressure (p_c) and vertical soil stresses at 50 cm depth (p_{50}) in spring or in summer/autumn (Tijink, 1998).

Reference	p_i (kPa)	p_c (kPa) Spring	Summer/ autumn	p_{50} (kPa) Spring	Summer/ autumn	Remarks
Söhne (1953)	80					Normal moisture conditions
Perdok and Terpstra (1983)	100					
Petelkau (1984)		50	80[a]			Sand
		80	150[a]			Loam
		80	200[a]			Clay
USSR (1986)[b]		80	100	25	30	w.c. (0–30) > 90% f.c.[c]
Rusanov (1994)		100	120	25	30	w.c. (0–30) > 70–90% f.c.[c]
		120	140	30	35	w.c. (0–30) > 60–70% f.c.[c]
		150	180	35	45	w.c. (0–30) > 50–60% f.c.[c]
		180	210	35	50	w.c. (0–30) < 50% f.c.[c]
Vermeulen et al. (1988)	40	50				Early spring, arable land
	80		100			Arable land

[a]Moisture content < 70% of field capacity.
[b]Official standard for fine-grained soils for the whole former Soviet Union. For undriven wheels values are 10% higher. For two passes in the same rut the values are 10% lower; for 3 and more passes values are 20% lower.
[c]w.c. (0–30), water content (0–30 cm depth); f.c., field capacity.

hydraulic conductivity and hence the drainage to field capacity. But well-drained soils also have to develop strong structures in order to increase soil strength. Conservation tillage systems with a minimum of soil disturbance have been shown to increase soil strength, and stresses exerted on the subsoil are decreased because the topsoil is stronger and rut depth diminished (e.g. Horn, 2002).

Conventional mouldboard ploughing is one of the most vigorous means of compacting the subsoil. A substantial part of the tractor weight is loaded on the rear wheel running on top of the subsoil in the furrow. Recent research indicates that subsoil stresses may be significantly reduced by using ploughs allowing the tractor to drive with all wheels on the untilled land (on-land ploughing) (Anken et al., 2000; Weisskopf et al., 2000; Keller et al., 2002).

Many studies indicate that tracked vehicles are less vigorous regarding compaction than wheeled vehicles (e.g. Soane, 1973). In theory, the contact pressure below tracks is much lower than under wheels. However, recent studies have revealed non-uniform distribution of the weight of the machine below the tracks creating peaks of stress even higher than under wheels (e.g. Keller et al., 2002). Moreover, tracked vehicles are generally much heavier than comparable wheeled equipment. Soil stresses can therefore only be reduced with well-designed tracked vehicles. An important advantage of tracked vehicles versus wheeled tractors is their superior tractive power and steering stability, which makes them very suitable for on-land ploughing.

Umbilical systems for slurry application are effective in avoiding unnecessary traffic on wet soils. A significant part of (sub)soil compaction in humid areas is due to traffic by heavy slurry wagons in early spring. Paradoxically, this problem has increased dramatically in several countries due to environmental legislation banning autumn application of slurry, since it results in nitrate leaching (see Christensen, Chapter 4, this volume). In consequence, liquid manure has to be spread in early spring when the soil is very susceptible to compaction. Moreover legislation to reduce ammonia emission by inserting the slurry into the soil requires intensive and heavy traffic on the soil. Slurry separation in order to allow spreading the high amount of water in the manure by irrigation equipment is one means of reducing the need to traffic the soil with heavy equipment.

An effective way of reducing traffic on wet soil is to use cable traction of tillage implements. The system takes use of a traction device located outside the field and thus avoids traffic with heavy tractors in the field. Håkansson et al. (1985) have demonstrated reduced compaction in top- and subsoil and increased crop yields using this technique on heavy clay soils. The technique would seem realistic for large flat areas of compaction-sensitive soil types.

Another promising alternative is the use of light, unmanned tractors with low ground-contact stresses (Alakukku et al., 1997). The tractor can work 24 h a day and a single operator can control several of these self-guiding tractors, the only task being to respond to emergency situations. This system would fit very well into precision farming and further implementation of high technology in modern agriculture.

Reduction of the compacted area

Other ways to prevent subsoil compaction are limiting the wheeled area. In the zero- or controlled-traffic concept, mostly wide-span vehicles (gantries) are used for permanent separation of wheeled and cropped areas (Chamen et al., 1992; Taylor, 1994). Thereby soil conditions for both crops and tyres can be optimized. A typical width of a gantry is 12 m, hence per 12 m width of the field the width of one wheelway is sacrificed for traffic. In most cases the increase in crop production compensates the loss of cropped area, but to be cost effective, farm sizes of 400 ha minimum are needed (Chamen et al., 1994). The system is not suitable for crops like sugarbeet and potatoes due to the heavy transport also needed between the wheelways during harvest.

The compacted area can also be reduced by taking care that every year the same traffic lanes are used and strictly avoiding traffic outside these lanes. To find and follow these lanes, high technology systems like Global Positioning System (GPS), computer steering

or sensor technology are available or can be developed.

Conclusions

1. Subsoil compaction is a hidden and persistent form of soil degradation that affects all of the agricultural area. It results in gradually decreasing soil quality regarding crop yield, environmental aspects and habitat of roots and soil organisms. In modern agriculture, wheel loads are still increasing and this will inevitably result in an ongoing increase of subsoil compaction.

2. Subsoil compaction exerts pronounced effects on soil functions. The ability of soil pores to conduct water and air is reduced, and the mechanical strength of the soil matrix is increased. This may trigger surface runoff and erosion. Crop growth is inhibited due to poor aeration and/or restricted rooting. Increased loss of nitrogen may take place by leaching and denitrification. These compaction effects are well described through the least limiting water range concept.

3. Sustainable soil management requires that *no* subsoil compaction should be accepted. Prevention of subsoil compaction by field traffic requires that soil stresses in the subsoil caused by a wheel load do not exceed the strength of the subsoil. Data on soil strength and prediction tools for soil stresses are essential to find this balance.

4. There is a striking shortage of data on soil strength. The amount of soil strength data should be increased and soil strengths should be measured in a systematic and harmonized way. Pedotransfer functions should be developed and methods to measure soil strength in a quick way should be correlated with classical, more time-consuming methods and improved accordingly.

5. With soil strengths known, wheel-load carrying capacities of subsoils can be calculated rather easily with analytical models such as SOCOMO. However, this requires a good estimation of rut depth and ground pressure distribution on the rut bottom.

6. Maps with the carrying capacity of subsoils loaded with a specific tyre can be

an effective way to bring the problem to the attention of the general public, policy makers and farmers.

7. The carrying capacity concept provides a means of identifying management thresholds in relation to subsoil compaction. Existing limits for tyre inflation pressures, ground pressures and wheel loads are in many cases not adequate or universal, and neglect either the effect of wheel load or the effect of inflation pressure. New limits must be developed that include wheel load as well as inflation pressures and are adapted to the kind of subsoil and climate considered. Even better would be the development of a straightforward decision support system that requires limited input and has the wheel-load carrying capacity as output.

8. Designers of agricultural equipment must reduce wheel loads to the carrying capacity of subsoils by using lighter equipment, more wheels and wide, low-pressure tyres.

References

Alakukku, L. (2000) Response of annual crops to subsoil compaction in a field experiment on clay soil lasting 17 years. In: Horn, R., Van den Akker, J.J.H. and Arvidsson, J. (eds) *Subsoil Compaction: Distribution, Processes and Consequences*. Advances in GeoEcology 32, Catena Verlag, Reiskirchen, Germany, pp. 205–208.

Alakukku, L., Pöyhönen, A. and Sampo, M. (1997) Soil compaction control with a light, unmanned tractor in two tillage systems. In: Fotyma, M., Jozefaciuk, A., Malicki, L. and Borowiecki, J. (eds) *Fragmenta Agronomica TOM 2A/97. Proceedings 14th ISTRO Conference, 27 July–1 August 1997, Pulawy, Poland*. Polish Society of Agrotechnical Sciences, Poland, pp. 19–22.

Anken, T., Diserens, E., Nadlinger, M., Weisskopf, P., Zihlmann, U., Wiermann, C. and Horn, R. (2000) Small onland ploughs protect the subsoil. In: *Proceedings of the 15th International Conference of the International Soil Tillage Research Organization*. Published by P. Dyke, Texas Agricultural Experiment Station, Temple, Texas (CD-ROM).

Arvidsson, J. (2001) Subsoil compaction caused by heavy sugarbeet harvesters in southern Sweden. 1. Soil physical properties and crop yield in six field experiments. *Soil and Tillage Research* 60, 67–78.

Arvidsson, J., Trautner, A. and Van den Akker, J.J.H. (2000a) Subsoil compaction – risk assessment and economic consequences. In: Horn, R., Van den Akker, J.J.H. and Arvidsson, J. (eds) *Subsoil Compaction: Distribution, Processes and Consequences*. Advances in GeoEcology 32, Catena Verlag, Reiskirchen, Germany, pp. 3–12.

Arvidsson, J., Van den Akker, J.J.H. and Horn, R. (eds) (2000b) Experiences with the impact and prevention of subsoil compaction in the European Community. In: *Proceedings of the 3rd Workshop of the Concerted Action 'Experiences with the impact of subsoil compaction on soil, crop growth and environment and ways to prevent subsoil compaction', 14–16 June 2000, Uppsala, Sweden*. Report 100, Division of Soil Management, Department of Soil Sciences, Swedish University of Agricultural Sciences, Uppsala, Sweden, 219 pp.

Arvidsson, J., Sjöberg, E. and Van den Akker, J.J.H. (2003) Subsoil compaction by heavy sugarbeet harvesters in southern Sweden. III. Risk assessment using a soil water model. *Soil and Tillage Research* (in press).

Baumgartl, T. and Horn, R. (1991) Effect of aggregate stability on soil compaction. *Soil and Tillage Research* 19, 203–213.

Betz, C.L., Allmaras, R.R., Copeland, S.M. and Randall, G.W. (1998) Least limiting water range: traffic and long-term tillage influences in a Webster soil. *Soil Science Society of America Journal* 62, 1384–1393.

Birkas, M., Gyuricza, C., Farkas, C. and Gecse, M. (eds) (2000) *Proceedings of 2nd Workshop and International Conference on Subsoil Compaction, 29–31 May 2000*. Szent István University, Gödöllö, Hungary, 223 pp.

Bishop, J.C. and Grimes, D.W. (1978) Precision tillage effects on potato root and tuber production. *American Potato Journal* 55, 65–72.

Boussinesq, J. (1885) *Application des Potentiels à l' Étude de l' Equilibre et du Mouvement des Solides Élastique*. Gauthier-Villais, Paris.

Bradford, J.M. and Grossman, R.B. (1982) *In-situ* measurement of near-surface soil strength by the fall-cone device. *Soil Science Society of America Journal* 46, 685–688.

Campbell, D.J. and Hunter, R. (1986) Drop cone penetration *in situ* and on minimally disturbed soil cores. *Journal of Soil Science* 37, 153–163.

Canarache, A., Dumitru, E. and Enache, R. (eds) (2002) *Proceedings 3rd Workshop INCO COPERNICUS Concerted Action – Experiences with the Impact of Subsoil Compaction on Soil Nutrients, Crop Growth and Environment, and*

Ways to Prevent Subsoil Compaction, Busteni, Romania, 14–18 June 2001. Research Institute for Soil Science and Agrochemistry, Bucharest, Romania, 456 pp.

Chamen, W.C.T., Watts, C.W., Leede, P.R. and Longstaff, D.J. (1992) Assessment of a wide span vehicle (gantry), and soil and cereal crop responses to its use in a zero traffic regime. *Soil and Tillage Research* 24, 359–380.

Chamen, W.C.T., Audsley, E. and Holt, J.B. (1994) Economics of gantry- and tractor-based zero-traffic systems. In: Soane, B.D. and van Ouwerkerk, C. (eds) *Soil Compaction in Crop Production*. Developments in Agricultural Engineering 11, Elsevier, Amsterdam, 569–596.

Cresswell, H.P. and Kirkegaard, J.A. (1995) Subsoil amelioration by plant-roots – the process and the evidence. *Australian Journal of Soil Research* 33, 221–239.

Da Silva, A.P., Kay, B.D. and Perfect, E. (1994) Characterization of the least limiting water range of soils. *Soil Science Society of America Journal* 58, 1775–1781.

DVWK Merkblätter zur Wasserwirtschaft (1995) *Gefügestabilität Ackerbaulich Genutzter Mineralböden, Teil I, Mechanische Belastbarkeit*. Kommissionsvertrieb Wirtschafts- und Verlagsdesellschaft Gas und Wasser mbH, Bonn, Germany, Heft 234, 12 pp.

Feddes, R.A., De Graaf, M., Bouma, J. and van Loon, C.D. (1988) Simulation of water use and production of potatoes as affected by soil compaction. *Potato Research* 31, 225–239.

Fleige, R., Horn, R. and Stange, F. (2002) Soil mechanical parameters derived from the CA-database 'subsoil compaction'. In: Pagliai, M. and Jones, R. (eds) *Sustainable Land Management – Environmental Protection: a Soil Physical Approach*. Advances in GeoEcology 35, Catena Verlag, Reiskirchen, Germany, pp. 359–366.

Fraters, B. (1996) *Generalized Soil Map of Europe*. Aggregation of the FAO-Unesco soil units based on the characteristics determining the vulnerability to degradation processes. RIVM Report no. 481505006, National Institute of Public Health and the Environment (RIVM), Bilthoven, The Netherlands, 60 pp.

Fröhlich, O.K. (1934) *Druckverteilung im Baugrunde*. Verlag von Julius Springer, Vienna.

Grable, A.R. and Siemer, E.G. (1968) Effects of bulk density, aggregate size, and soil water suction on oxygen diffusion, redox potentials and elongation of corn roots. *Soil Science Society of America Journal* 34, 20–25.

Håkansson, I. (ed.) (1994) Subsoil compaction by high axle load traffic. *Special Issue, Soil and Tillage Research* 29, 105–306.

Håkansson, I. and Petelkau, H. (1994) Benefits of limited axle load. In: Soane, B.D. and van Ouwerkerk, C. (eds) *Soil Compaction in Crop Production*. Developments in Agricultural Engineering 11, Elsevier, Amsterdam, pp. 479–499.

Håkansson, I. and Reeder, R.C. (1994) Subsoil compaction by vehicles with high axle load – extent, persistence and crop response. *Soil and Tillage Research* 29, 277–304.

Håkansson, I. and Voorhees, W.B. (1998) Soil compaction. In: Lal, R., Blum, W.H., Valentine, C. and Stewart, B.A. (eds) *Methods for Assessment of Soil Degradation*. CRC Press, Boca Raton, Florida, pp. 167–179.

Håkansson, I., Henriksson, L. and Gustafsson, L. (1985) Experiments on reduced compaction of heavy clay soils and sandy soils in Sweden. *Proceedings International Conference on Soil Dynamics, 17–19 June, Auburn University*. Auburn University, Auburn, Alabama, pp. 995–1009.

Håkansson, I., Voorhees, W.B., Elonen, P., Raghavan, G.S.V., Lowery, B., Van Wijk, A.L.M., Rasmussen, K. and Riley, H. (1987) Effect of high axle load traffic on subsoil compaction and crop yield in humid regions with annual freezing. *Soil and Tillage Research* 10, 259–268.

Håkansson, I., Grath, I. and Olsen, H.J. (1996) Influence of machinery traffic in Swedish farm fields on penetration resistance in the subsoil. *Swedish Journal of Agricultural Research* 26, 181–187.

Hammel, K. (1994) Soil stress distribution under lugged tires. *Soil and Tillage Research* 32, 163–181.

Hansbo, S. (1957) *A New Approach to the Determination of the Shear Strength of Clay by the Fall Cone Test*. Proceedings No. 14. Royal Swedish Geotechnical Institute.

Horn, R. (2002) Soil mechanical properties and processes in structured unsaturated soils under various landuse and management systems. In: Pagliai, M. and Jones, R. (eds) *Sustainable Land Management – Environmental Protection: a Soil Physical Approach*. Advances in GeoEcology 35, Catena Verlag, Reiskirchen, Germany, pp. 305–318.

Horn, R. and Lebert, M. (1994) Soil compactability and compressibility. In: Soane, B.D. and van Ouwerkerk, C. (eds) *Soil Compaction in Crop Production*. Developments in Agricultural Engineering 11, Elsevier, Amsterdam, pp. 45–69.

Horn, R., Van den Akker, J.J.H. and Arvidsson, J. (eds) (2000) *Subsoil Compaction: Distribution,*

Processes and Consequences. Advances in GeoEcology 32, Catena Verlag, Reiskirchen, Germany, 462 pp.

Houlsby, G.T. (1982) Theoretical analysis of the fall cone test. *Geotechnique* 32, 111–118.

Keller, T., Trautner, A. and Arvidsson, J. (2002) Stress distribution and soil displacement under a rubber-tracked and a wheeled tractor during ploughing, both on-land and within furrows. *Soil and Tillage Research* 68, 39–47.

Kooistra, M.J., Bouma, J., Boersma, O.H. and Jager, A. (1984) Physical and morphological characterization of undisturbed and disturbed ploughpans in a sandy loam soil. *Soil and Tillage Research* 4, 405–417.

Koolen, A.J. (1974) A method for soil compactibility determination. *Journal of Agricultural Engineering Research* 19, 271–278.

Koolen, A.J. and Kuipers, H. (1983) *Agricultural Soil Mechanics.* Advanced Series in Agricultural Sciences 13. Springer-Verlag, Berlin, 241 pp.

Larson, W.E., Eynard, A., Hadas, A. and Lipiec, J. (1994) Control and avoidance of soil compaction in practice. In: Soane, B.D. and van Ouwerkerk, C. (eds) *Soil Compaction in Crop Production.* Developments in Agricultural Engineering 11, Elsevier, Amsterdam, pp. 597–625.

Lebert, M. and Horn, R. (1991) A method to predict the mechanical strength of agricultural soils. *Soil and Tillage Research* 19, 275–286.

Letey, J. (1985) Relationship between soil physical properties and crop production. In: Stewart, B.A. (ed.) *Advances in Soil Science,* Volume 1. Springer-Verlag, Berlin, pp. 277–294.

Libert, B. (1995) *The Environmental Heritage of Soviet Agriculture.* CAB International, Wallingford, UK, 228 pp.

Lipiec, J., Arvidsson, J. and Murer, E. (2003) Review of modelling crop growth, movement of water and chemicals in relation to topsoil and subsoil compaction. *Soil and Tillage Research* (in press).

McBride, R.A. and Joosse, P.J. (1996) Overconsolidation in agricultural soils: II. Pedotransfer functions for estimating preconsolidation stress. *Soil Science Society of America Journal* 60, 373–380.

Medvedev, V.V. and Cybulko, W.G. (1995) Soil criteria for assessing the maximum permissible ground pressure of agricultural vehicles on Chernozem soils. *Soil and Tillage Research* 36, 153–164.

Mullins, C.E. and Fraser, A. (1980) Use of the drop-cone penetrometer on undisturbed and remoulded soils at a range of soil-water tensions. *Journal of Soil Science* 31, 25–32.

Oldeman, L.R., Hakkeling, R.T.A. and Sombroeck, W.G. (1991) *World Map of the Status of Human-induced Soil Degradation: an Explanatory Note.* International Soil Reference and Information Centre (ISRIC), Wageningen, The Netherlands, 34 pp. + 3 maps.

Perdok, U.D. and Terpstra, J. (1983) Berijdbaarheid van landbouwgrond: Bandspanning en bodemverdichting [Trafficability of agricultural land: tyre inflation pressure and soil compaction]. *Landbouwmechanisatie* 34, 363–366.

Petelkau, H. (1984) Auswirkungen von Schadverdichtungen auf Bodeneigenschaften und Pflanzenertrag sowie Massnahmen zu ihrer Minderung [Effects of harmful compaction on soil properties and crop yield and measures for its reduction]. *Tagungsbericht Akademie der Landwirtschaftswissenschaften der Deutschen Demokratischen Republik* 227, 25–34.

Poodt, M.P., Koolen, A.J. and Van der Linden, J.P. (2003) FEM analysis of subsoil reaction on heavy wheel loads with emphasis on soil preconsolidation stress and cohesion. *Soil and Tillage Research* (in press).

Rusanov, V.A. (1994) USSR standards for agricultural mobile machinery: permissible influences on soils and methods to estimate contact pressure and stress at a depth of 0.5 m. *Soil and Tillage Research* 29, 249–252.

Schjønning, P. (1991) *Soil Mechanical Properties of Seven Danish Soils.* Danish Institute of Agricultural Sciences, Internal Report No. S2176, Research Centre Foulum, Tjele, Denmark, 33 pp.

Schjønning, P. (1999) Mechanical properties of Danish soils – a review of existing knowledge with special emphasis on soil spatial variability. In: Van den Akker, J.J.H., Arvidson, J. and Horn, R. (eds) *Proceedings of the First Workshop of the Concerted Action 'Experiences with the Impact of Subsoil Compaction on Soil, Crop Growth and Environment and Ways to Prevent Subsoil Compaction', 28–30 May 1998, Wageningen, The Netherlands.* Report 168, ISSN 0927–4499, DLO Winand Staring Centre, Wageningen, The Netherlands, pp. 290–303.

Schjønning, P. (2000) Proposing drop-cone penetration as a mean of estimating soil internal friction and cohesion. In: Horn, R., Van den Akker, J.J.H. and Arvidsson, J. (eds) *Subsoil Compaction: Distribution, Processes and Consequences.* Advances in GeoEcology 32, Catena Verlag, Reiskirchen, Germany, pp. 453–462.

Schjønning, P., Thomsen, I.K., Moldrup, P. and Christensen, B.T. (2003) Linking soil microbial activity to water- and air-phase contents and diffusivities. *Soil Science Society of America Journal* 67, 156–165.

Simota, C., Lipiec, J., Dumitru, E. and Tarkiewicz, S. (2000) SIBIL – a simulation model for soil water dynamics and crop yield formation considering soil compaction effects: model description. In: Horn, R., Van den Akker, J.J.H. and Arvidsson, J. (eds) *Subsoil Compaction: Distribution, Processes and Consequences*. Advances in GeoEcology 32, Catena Verlag, Reiskirchen, Germany, pp. 154–168.

Soane, B.D. (1973) Techniques for measuring changes in the packing state and cone resistance of soil after the passage of wheels and tracks. *Journal of Soil Science* 24, 311–321.

Soane, B.D. and van Ouwerkerk, C. (eds) (1994) *Soil Compaction in Crop Production*. Developments in Agricultural Engineering 11, Elsevier, Amsterdam, 662 pp.

Soane, G.C., Godwin, R.J. and Spoor, G. (1986) Influence of deep loosening techniques and subsequent wheel traffic on soil structure. *Soil and Tillage Research* 8, 231–237.

Söhne, W. (1953) Druckverteilung im Boden und Bodenverformung unter Schlepperreifen. *Grundlagen der Landtechnik* 5, 49–63.

Stenitzer, E. and Murer, E. (2003) Impact of soil compaction upon soil water balance and maize yield estimated by the SIMWASER model. *Soil and Tillage Research* (in press).

Taylor, J.H. (1994) Development and benefits of vehicle gantries and controled-traffic systems. In: Soane, B.D. and van Ouwerkerk, C. (eds) *Soil Compaction in Crop Production*. Developments in Agricultural Engineering 11, Elsevier, Amsterdam, pp. 521–537.

Tijink, F.G.J. (1998) Mechanisation strategies to reduce traffic-induced soil compaction. *Advances in Sugar Beet Research* IIRB, Vol. 1, 57–66.

Towner, G.D. (1973) An examination of the fall-cone method for the determination of some strength properties of remoulded agricultural soils. *Journal of Soil Science* 24, 470–479.

USSR State Committee for Standards (1986) *Standard 26955: Agricultural Mobile Machinery; Rates of Force by Propelling Agents on Soil*. Moscow, USSR, 7 pp.

Van de Zande, J.C. (1991) Computed reconstruction of field traffic patterns. *Soil and Tillage Research* 19, 1–15.

Van den Akker, J.J.H. (1992) Stresses and required soil strength under Terra tires and tandem and dual wheel configurations. *Proceedings International Conference on Soil Compaction and Soil Management, 8–12 June 1992, Tallinn, Estonia*, pp. 23–26.

Van den Akker, J.J.H. (1994) Prevention of subsoil compaction by tuning the wheel load to the bearing capacity of the subsoil. In: Jensen, H.E., Schjønning, P., Mikkelsen, S.A. and Madsen, K.B. (eds) *Proceedings of the 13th ISTRO Conference, Aalborg, Denmark, 1994*. The Royal Veterinary and Agricultural University and The Danish Institute of Plant and Soil Science, Tjele, Denmark, Vol. 1, 537–542.

Van den Akker, J.J.H. (1997) Construction of a wheel-load bearing capacity map of the Netherlands. In: Fotyma, M., Jozefaciuk, A., Malicki, L. and Borowiecki, J. (eds) *Fragmenta Agronomica TOM 2A/97. Proceedings 14th ISTRO Conference, 27 July–1 August 1997, Pulawy, Poland*. Polish Society of Agrotechnical Sciences, Poland, pp. 15–18.

Van den Akker, J.J.H. (1999) Development, verification and use of the subsoil compaction model SOCOMO. In: Van den Akker, J.J.H., Arvidsson, J. and Horn, R. (eds) *Proceedings of the Concerted Action 'Experiences with the Impact of Subsoil Compaction on Soil, Crop Growth and Environment and Ways to Prevent Subsoil Compaction', 28–30 May 1998, Wageningen, The Netherlands*. Report 168, ISSN 0927–4499, DLO Winand Staring Centre, Wageningen, The Netherlands, pp. 321–336.

Van den Akker, J.J.H. and Van Wijk, A.L.M. (1987) A model to predict subsoil compaction due to field traffic. In: Monnier, G. and Goss, M.J. (eds) *Soil Compaction and Regeneration. Proceedings of the Workshop on Soil Compaction: Consequences and Structural Regeneration Processes/Avignon 17–18 September 1985*. A.A. Balkema, Rotterdam, pp. 69–84.

Van den Akker, J.J.H., Arvidsson, J. and Horn, R. (eds) (1999) Experiences with the impact and prevention of subsoil compaction in the European Community. *Proceedings of the First Workshop of the Concerted Action 'Experiences with the Impact of Subsoil Compaction on Soil, Crop Growth and Environment and Ways to Prevent Subsoil Compaction', 28–30 May 1998, Wageningen, The Netherlands*. Report 168, ISSN 0927–4499, DLO Winand Staring Centre, Wageningen, The Netherlands, 344 pp.

Van den Akker, J.J.H., Arvidsson, J. and Horn, R. (eds) (2003) Experiences with the impact and prevention of subsoil compaction in the European Union. *Special Issue, Soil and Tillage Research* (in press).

Van der Linden, J.P. and Vandergeten, J.-P. (1999) Aandacht voor rooiwerk bespaart tot 200 gulden per hectare. Een terugblik op de rooidemonstratie in het Belgische Watervliet. *CSM-informatie* 521, 10–12.

van Ouwerkerk, C. and Soane, B.D. (eds) (1995) Soil compaction and the environment. *Special Issue, Soil and Tillage Research* 35, 1–113.

Veenhof, D.W. and McBride, R.A. (1996) Overconsolidation in agricultural soils: I. Compression and consolidation behaviour of remolded and structured soils. *Soil Science Society of America Journal* 60, 362–373.

Vermeulen, G.D., Arts, W.B.M. and Klooster, J.J. (1988) Perspective of reducing soil compaction by using a low ground pressue farming system: selection of wheel equipment. *Proceedings 11th Conference International Soil Tillage Research Organisation (ISTRO), Edinburgh, UK.* Scottish Centre Agricultural Engineering, Penicuik, UK, Vol. 1, pp. 329–334.

Voorhees, W.B. (1992) Wheel-induced soil physical limitations to root growth. In: Hatfield, J.L. and Stewart, B.A. (eds) *Limitations to Plant Root Growth. Advances in Soil Science*, Volume 19, Springer-Verlag, Berlin, pp. 73–95.

Voorhees, W.B. (2000) Long-term effect of subsoil compaction on yield of maize. In: Horn, R., Van den Akker, J.J.H. and Arvidsson, J. (eds) *Subsoil Compaction: Distribution, Processes and Consequences*. Advances in GeoEcology 32, Catena Verlag, Reiskirchen, Germany, pp. 331–338.

Weisskopf, P., Zihlmann, U., Wiermann, C., Horn, R., Anken, T. and Diserens, E. (2000) Influences of conventional and onland-ploughing on soil structure. In: Horn, R., Van den Akker, J.J.H. and Arvidsson, J. (eds) *Subsoil Compaction: Distribution, Processes and Consequences*. Advances in GeoEcology 32, Catena Verlag, Reiskirchen, Germany, pp. 73–81.

Whalley, W.R., Dumitru, E. and Dexter, A.R. (1995) Biological effects of soil compaction. *Soil and Tillage Research* 35, 53–68.

Wood, D.M. (1985) Some fall-cone tests. *Geotechnique* 35, 64–68.

Chapter 11
Management-induced Soil Structure Degradation – Organic Matter Depletion and Tillage

B.D. Kay[1] and L.J. Munkholm[2]
[1]University of Guelph, Department Land Resource Science, Guelph, Ontario N1G 2W1, Canada; [2]Danish Institute of Agricultural Sciences, Department of Agroecology, PO Box 50, DK-8830 Tjele, Denmark

---

Summary

Soil structure is an important element of soil quality since changes in structural characteristics can cause changes in the ability of soil to fulfil different functions and services. Emphasis in this chapter is placed on the role of soil structure in biological productivity of agroecosystems. Combinations of management practices in which the extent of the degradation of soil structure caused by one practice is balanced or exceeded by the extent of regeneration by other practices will help sustain the productivity of agroecosystems. Tillage and practices that change the organic matter (OM) content of soil are foremost among the many practices that influence soil structure. The links between management, OM content and soil structure and between tillage and soil structure are explored. The feasibility of defining management thresholds or soil-quality indicator thresholds with respect to soil structure is assessed.

Introduction

The ability of soils to fulfil the different functions and services described in Chapter 1 is strongly influenced by soil structure. The most important functions of soils in agro-ecosystems include supporting root growth and crop development, receiving, storing and transmitting water, cycling carbon and nutrients, and diminishing the dispersal of agricultural chemicals. Management practices can cause substantial changes in soil structure and can, therefore, alter the ability of soils to fulfil these functions.

Field experience and decades of soil inventory and research have given rise to a multitude of methods to characterize soil structure. However, irrespective of the method and the temporal and spatial scale at which measurements are made, each of the resulting characteristics reflects one of four general aspects of soil structure: form, stability, resiliency or vulnerability. The term structural form is used to describe the architecture of soil, i.e. the heterogeneous arrangement of solid and void space and their spatial patterns that exist in soil at a given time. Examples of characteristics of structural form include total porosity, pore size distribution, continuity and tortuosity of the pore system, and the orientation and shape of pores. Other examples include the morphology of peds or aggregates and the hierarchical structural states of aggregates arising from failure zones of different strengths. The concepts of stability, resiliency and vulnerability have been described in Chapter 1. From the perspective of soil structure, we use the term stability to describe the ability of the soil to retain its arrangement of solid and void space when exposed to different stresses. Structural resiliency describes the ability of a soil to recover its structural form through natural processes when the applied stresses are reduced or removed. Structural vulnerability describes the inability of the structural form of soil to cope with stress and reflects the combined characteristics of stability and resiliency.

The ability of soil to fulfil a function, at any given time, is determined by characteristics of structural form. For instance, structural form strongly influences the ability of soil to receive, store and transmit water and also determines the availability of the stored water to growing plants. Qualitative terms are often used to imply value to the direction of change in different characteristics of soil structure over time. Implicit in the use of such terms must be recognition of the role of soil structure with respect to a function of soil, e.g. root growth and crop development. The influence of soil structure on the crop is strongly dependent on climate, and therefore use of the term soil degradation with respect to the crop environment implies a change in soil structure, which, under prevailing climatic conditions, will lead to diminished capability to support root growth and crop development.

While some management practices used by farmers today can lead to a degradation of soil structure, others lead to a regeneration of soil structure. Farmers may, for different reasons, not have the option to employ practices that never lead to the degradation of soil structure. Management systems that sustain the ability of soils to fulfil different functions therefore require balancing the extent of structural degradation over time caused by one practice with the extent of regeneration by other practices, such that soil structure is maintained or enhanced. The extent of degradation or regeneration under a given practice is determined by the rate of change in soil structure and the duration over which the practice occurs.

Changes in soil structure arising from changes in management practices can be considered over time-scales ranging from hours to decades and spatial-scales ranging from micrometres to metres. It has been hypothesized (Kay, 1990) that the rate of change of a characteristic of structural form can be related to three functions that are additive: (i) a function related to stresses arising from the management practice; (ii) a biological-related function; and (iii) a weather-related function. Each of the functions include a variable characterizing the magnitude of the agent causing the change in structural form and a parameter that links the causal agent to the rate of change in structural form (i.e. a response characteristic). The individual functions may have opposing effects on the rate of change in soil structure. For instance, stress associated with tillage or traffic can readily destroy any change in structural form introduced by biological factors or weather. Under such conditions the rate of change in structural form is largely determined by the magnitude of the stress applied (the causal agent) and the stability (the response characteristic). If, however, stress is minimized, the rate of change in structural form will be much more strongly influenced by the biological and weather-related functions (both of which are linked to resiliency). In the longer term, management can change the organic matter (OM) content, which, in turn, can change stability and resiliency and therefore alter the response of soil structure to stress and weather. These few examples provide a glimpse of the multiplicity of mechanisms and time-scales over which management can influence changes in soil structure.

Impacts of Management on Soil Structure

Foremost among the many management practices that influence soil structure are: (i) practices that deplete the OM content of soil, and (ii) tillage.

Depletion of organic matter

Management effects

CAVEATS   Three caveats must be identified in formulating generalizations about management practices and the depletion of OM. First, a generalization about a given management practice is most applicable when other management practices or processes occurring within the landscape remain constant. For instance, a generalization about the effect of tillage on OM content may not be applicable to situations where a change in tillage has been accompanied by a change in crop or fertilization practices. Furthermore, a generalization about the effect of tillage on OM content that is based on studies in level landscapes may

not be applicable to variable landscapes (VandenBygaart *et al.*, 2002). Secondly, the generalizations are commonly based on research that has been directed to increasing OM content above levels found under 'current' management conditions and given soil, drainage and climatic conditions. For instance, if OM contents increase with increasing applications of manure, then, relative to the highest applications, a smaller application is interpreted as resulting in decreased OM content. This interpretation is based on the assumption that the increase in OM content beyond a relatively low level of OM with variations in a management practice will follow the same relation to management as if one started with a high level of OM and changed management in the reverse direction. That is, there is no hysteresis in the relation between OM and the direction of change in the management variable. This assumption has not been evaluated. Thirdly, the rates of change in OM content following introduction of a given management practice are often sufficiently small that long-term studies are required to accurately document rates of change. Long-term studies are expensive and the number of treatments is constrained by operating costs. Consequently the number of treatments of a given management practice are often too few to make it possible to identify critical management thresholds.

GENERAL EFFECTS   Keeping the above caveats in mind, the OM content of soil generally decreases with decreasing amount of OM returned to the soil, i.e. decreasing crop residue returned to soil and decreasing rates of application of manure and other organic amendments. The amount of crop residue returned to the soil decreases as the nutrient supply increasingly limits biomass production, and/or a decreasing proportion of the biomass that is produced is returned to the soil. The impact of limiting nutrient supply on OM contents is well documented in long-term studies (e.g. Johnston, 1986), but can be measurable when the treatment has continued for as few as 11 years (Solberg *et al.*, 1998). Crop residue can represent a major source of organic material returned to the soil and therefore, it is not surprising that removal of crop residue can result in a depletion of OM (e.g. Solberg et al., 1998). The positive impacts of addition of manure and other organic residues on OM contents have been documented in a multitude of studies. A decrease in the rate of application of manure with a subsequent depletion of OM is a logical consequence of the shift away from animal-based agriculture. The loss of this source of organic material may be partially offset by utilizing waste organic materials of urban or industrial origin. However, contamination of these materials with metals, pathogens and organic materials that have an unknown effect on food quality is expected to limit their use in agriculture (see Naidu et al., Chapter 13, this volume).

Increasing the intensity of tillage may also contribute to depletion of soil OM although the effects of tillage may vary with soils, climate and cropping system. Numerous studies since the early 1980s have assessed the effects of reducing tillage on OM content. A recent review (Kay and VandenBygaart, 2002) indicated that the impact of a reduction in tillage appears to be variable with changes in the amounts of organic carbon in the profile ranging up to more than 1 Mg/ha after 30 years.

The OM content of soils that have been in continuous grass have higher OM contents than arable soils (Johnston, 1986). Conversion of grassland to the production of arable crops can result in a progressive decline in OM contents with a subsequent deterioration in soil structure (Low, 1972). Forages may also be grown as part of a crop rotation and decreasing the frequency of forages in a crop rotation is associated with a shift away from animal agriculture (although forages may still be retained as a cover crop under seeded in cereals). Although there are few studies on rotations involving forages of different frequency in rotations or forages under seeded in cereals of different rotations, the evidence is not compelling that replacement of forages in a rotation necessarily results in depletion of OM within at least 20 years (Yang and Kay, 2001).

Organic matter and soil characteristics

Depletion of the OM content of soil can influence characteristics of structural form

in a multitude of ways but the influence is largely indirect. OM indirectly affects characteristics of structural form by changing their sensitivity to stress, i.e. by changing structural stability. Pores are created by the formation of ice lenses, shrinkage, root growth, the activity of soil animals or tillage. Once pores are created, their persistence is determined largely by the structural stability of the pore walls and surrounding soil, and the magnitude of the stresses experienced. Where the process of creating the pore does not substantially increase the OM content of the pore wall, the stability of the pore wall would be expected to be equivalent to the stability of the bulk soil and this stability is often inadequate to allow the pore to withstand the effective stress created by drying, rewetting or swelling pressures or stresses from overburden or traffic. For instance, Kay et al. (1985) found that macropores created through the formation of ice lenses over winter in a non-swelling soil disappeared as the soil dried and reconsolidated after thawing. Many of the macropores created by tillage are equally unstable. Gusli et al. (1994) showed that the structural collapse of beds of aggregates on wetting and draining was caused by the development of failure zones on wetting followed by consolidation on draining as a result of the development of effective stresses. The extent of collapse was greater for soils having lower OM contents.

The compactability of soils is a measure of structural stability. Soane (1990), in a comprehensive review of the role of OM on soil compactability, has noted that OM is likely to have a greater influence on the compaction of soil at low stress (e.g. 100 kPa) than at high stress. As compaction occurs, the larger pores are lost first and aeration capacity declines. Ball and Robertson (1994) found that even relatively small changes in air-filled porosity due to compaction were accompanied by large changes in relative diffusivity and air permeability. These authors showed that after application of a given stress, these properties decreased less in direct drilled than in ploughed soil, and attributed this trend to the structure associated with the higher OM content in the direct drilled soil.

The ability of soil aggregates to withstand disruption during wetting and sieving in water is another measure of stability. In a comprehensive study of (air dried) soils with organic carbon contents ranging from less than 0.6% to about 10%, Kemper and Koch (1966) found that the stability of vacuum-saturated aggregates from the western USA and Canada increased curvilinearly with organic carbon content. The stability changed most rapidly with organic carbon at carbon contents lower than 1.2–1.5%. A slightly higher threshold value was obtained by Greenland et al. (1975) for soils from England and Wales. Using the Emerson test that is based on the swelling, slaking and dispersion behaviour of soil aggregates, Greenland et al. (1975) found that the critical level of organic carbon was about 2%. However, the water stability of aggregates from a soil can exhibit large changes due to cropping treatments before changes in the total OM content are observed (e.g. Baldock et al., 1987). These changes have been attributed to changes in the amounts of fine roots and fungal hyphae, which strengthen failure zones through physical entanglement, and which also act as sources of carbon for bacteria, thereby contributing to increased production of microbial cementing materials (Tisdall and Oades, 1982; Miller and Jastrow, 1990). These stabilizing materials are very labile and represent only a small part of the total carbon content. Their amounts in the soil at any given time are determined by the rates of input of plant carbon and the mineralization of this carbon and the microbial by-products. Rasiah and Kay (1994) found that the aggregate stability increased exponentially over time after forages have been introduced on soils that had previously been used for the production of row crops and that the stability increased at a faster rate on finer textured soils.

OM may also influence structural form indirectly by changes in the sensitivity of structural form to weather. Structural form can be altered by freezing as well as by wetting and drying. Analyses of tensile strengths of air-dry aggregates from a depth of 0–5 cm measured on five different years on the long-term rotation plots at the Waite Institute, Australia, showed that the tensile strength decreased with increasing number of antecedent wetting events (Kay et al., 1994). The decline was greater, i.e. the sensitivity of the strength of failure zones to preceding weather conditions was greater, on the rotation with the lowest OM content.

Direct and indirect effects of OM on characteristics of structural form are integrated when studies relate OM contents to structural characteristics. When other factors are constant, the total porosity generally decreases with decreasing OM content whether the studies are site specific or based on pedotransfer functions. Pedotransfer functions commonly show that the effect of OM on total porosity diminishes with increasing clay content. An analysis by Manrique and Jones (1991) of 12,000 soil pedons in the USA that were grouped according to their Soil Taxonomy classification also showed that the influence of OM on bulk density varied with the soil order.

Depletion in OM content does not have a uniform effect on all sizes of pores. Pores larger than 30 µm are the most dynamic of any size class of pores and the volume fraction of this size class of pores is influenced by tillage, traffic, shrinkage and swelling as well as biological activity. Consequently, the volume fraction of this size class of pores is often not strongly correlated with OM contents. Pores ranging in effective diameter from 0.2 to 30 µm are less dynamic than larger pores. Water retained by these pores has been referred to as plant-available water (Veihmeyer and Hendrickson, 1927). Emerson (1995) considered several site-specific studies from various parts of the world and showed that a decrease in the OM content resulted in decreases in the available water content ranging from 1 to 10 g of water per g organic carbon. Management has less influence on pores with an effective diameter < 0.2 µm than on other pore size classes. Emerson (1995) found that the decrease in the water content at −1.5 MPa (where water is held in pores with an effective diameter < 0.2 µm) ranged from 1 to 3 g water per g organic carbon.

Other important characteristics of pores include pore tortuosity and continuity. Very little information is available on the influence of OM on these characteristics. However,

Schjønning *et al.* (2002) showed that the pore system of a soil with 3.4% OM was much more tortuous and complex than for a neighbouring soil with an OM content of 2.5%. The increased complexity of the former was considered beneficial for aeration of microsites within the soil.

An increase in OM can cause an increase in porosity with a concomitant loss of strength. Also, an increase in OM in soil with the same porosity and water content can cause a decrease in water potential (because of the effect on the water release curve), thereby increasing the effective stress and consequently the strength. Further, an increase in OM can cause a change in the cementation between mineral particles. Tensile strength has been found to increase with OM under wet conditions (Munkholm and Kay, 2002; Munkholm *et al.*, 2002). A similar trend has been inferred from measurements of penetration resistance by Emerson (1995) who noted that high levels of polysaccharide gels could lead to compacted parts of the mineral matrix becoming so strongly cemented that, at a potential of −100 hPa, the strength was sufficiently high to limit root penetration. However, it has also been found (Munkholm and Kay, 2002; Munkholm *et al.*, 2002) that the increase in strength with decreasing water potential was less in the soil with higher OM contents.

An important aspect of OM is its influence on the optimum and range in water content for tillage. Recent results for silt loams exposed to contrasting land use indicate that the optimum water content for tillage, and the range in water contents that are suitable for tillage, increase with the OM content (Dexter and Bird, 2001). Munkholm *et al.* (2002) showed similar effects in a study on a sandy loam soil exposed to different long-term fertilization treatments.

Tillage

Considerable research effort has been directed to reducing tillage intensity (no-, minimum-, reduced- and conservation-tillage) in order to decrease the risk of erosion, improve soil structure and reduce the costs in plant production. However, the manner of reducing tillage has varied around the world. Reduced tillage has been extensively adopted in subhumid and semiarid areas with high erosion risks. Conventional tillage with annual mouldboard ploughing is still the most common practice in humid areas such as northwestern Europe (Christian and Ball, 1994; Ehlers and Claupein, 1994; Rasmussen, 1999). Constraints relating to problems with soil compaction, workability, residue management and low soil temperatures at the time of seeding have hampered the adoption of reduced tillage in cool humid climates (Carter, 1994). Evidence even suggests that tillage intensity has increased in these areas during the last few decades (e.g. Ehlers and Claupein, 1994). Ehlers and Claupein (1994) found that the depth of tillage had increased by 10 cm in Germany over the last 30 years and that the degree of loosening of the ploughed layer also had increased. The increasing need for soil loosening was partly attributed to increased soil compaction related to the increasingly heavier machinery used in modern agriculture. The widespread application of power takeoff (PTO)-driven rotary cultivators has significantly increased the energy input in seedbed preparation. Spiess *et al.* (2000) found that the energy input in 'normal' and 'intensive' rotary harrowing was 28 and 45 Wh/m³, respectively, for a Swiss loam. 'Normal' rotary harrowing was equal to the intensity normally needed to prepare a proper seedbed, whereas the rotor speed was 425% higher for 'intensive' than for 'normal'. The energy input in rotary harrowing was between three and five times the energy input from a single tine cultivation (approximately 10 Wh/m³). Many have shown that the turnover rate of carbon increases with tillage intensity (e.g. Reicosky, 1997). Therefore, the long-term effects of tillage on soil structure cannot be separated from other management effects on soil OM contents. Soil degradation caused by depletion of soil OM was discussed above. Here, the focus is mainly on the short-term effects of tillage intensity on soil structure.

Tillage requirements

Tillage may be required to produce a desirable seedbed, incorporate OM and control weeds and pests. The latter is not necessarily required in modern conventional agriculture where effective pesticides are available. In organic farming, however, tillage is needed for weed and pest control (Lampkin, 1990). For instance, mouldboard ploughing may be needed in organic farming to control perennial weeds like couch grass, and intensive seedbed preparation is needed to provide optimal conditions for seedling growth and to reduce competition with weeds. Further, intensive seedbed preparation may be needed to provide suitable conditions for mechanical weeding.

Crops require ample aeration for the plant roots and for decomposition of OM, and at the same time an adequate soil–root contact to secure uptake of water and nutrients. Braunack and Dexter (1989) concluded in a review that the optimal seedbed in general is produced by 0.5–8 mm aggregates. A large fraction of small aggregates (less than 0.5–1 mm) is not desirable because of increased risk of wind and water erosion. Furthermore, a large fraction of aggregates larger than 8 mm is not desirable because of a reduction in the soil–root contact area and a higher impedance to root penetration. It has to be emphasized that crops require different seedbed quality. For instance, crops with small seeds such as sugarbeets, cruciferous crops and clover may require a finer and more uniform seedbed than cereals (Larney et al., 1988; Håkansson et al., 2002).

Many have reported effects of direct drilling on crop yield and physical properties in comparison with conventional tillage (including mainly mouldboard ploughing). Studies on direct drilling have resulted in rough classification systems of the suitability of soils for direct drilling (Christian and Ball, 1994). They classified well-drained soils with a stable structure in low rainfall areas as most suitable for direct drilling in Great Britain. Medium-textured soils with high compactability and low resiliency have been considered unsuitable for direct drilling (Ehlers and Claupein, 1994).

The data needed to define the minimum tillage-intensity threshold are lacking for most combinations of soil type, crop and climate. A good example is, however, the extensive studies carried out in eastern Canada. In Ontario, studies on fine-textured soils have shown that some tillage is recommended in the autumn (zone-tillage or disc harrowing) for maize and soybeans (Opoku et al., 1997; Vyn et al., 1998). In contrast, winter wheat performs equally well under no-tillage and conventional tillage (Vyn et al., 1991). For acidic sandy loams at Prince Edwards Island, Carter (1991) concluded that deep tillage was needed at least periodically to loosen the compact layer produced in the lower part of the Ap horizon under shallow cultivation. In Europe, the need for intensive seedbed preparation has been addressed by Larney et al. (1988) in a study on loamy Irish soils. They showed that a more intensive rotary cultivation was needed to produce a proper sugarbeet seedbed on poorly structured soils than on soils of inherently better structure. In another European study, Spiess et al. (2000) showed that intensive rotary cultivation was needed after a dry summer to produce a desirable seedbed for a winter cereal on a Swiss loam. They also showed that intensive rotary cultivation was not needed in a moist spring to produce a desirable seedbed for a spring cereal.

Observing tillage effects

REDUCED TILLAGE   Reduced tillage has been found to result in increased stability of surface soil and thereby reduce the risk of surface crusting and erosion (e.g. Douglas and Goss, 1982; Schjønning and Rasmussen, 1989; Carter, 1992). Changes may occur within a short time frame as shown by Schjønnir and Rasmussen (1989) and Carter (19). However, seasonal changes may be g iter than differences detected between illage systems (Bullock et al., 1988; Perf et al., 1990). Seasonal changes may be elated to changes in water content (P ect et al., 1990) and microbial activity erfect et al., 1990; Kandeler and Murer 993; Suwardji and Eberbach, 1999) and t reezing/thawing (Bullock et al., 1988; An s et al., 1993).

Reduced tillage has been shown to affect soil workability both directly and indirectly (as a consequence of carbon accumulation in surface soils under reduced tillage). On hard-setting Australian soils, reduced tillage has been found to cause increased soil friability (Chan, 1989; Macks et al., 1996). In contrast, Perfect and Kay (1994) found no significant effect of reduced tillage on soil friability for a Canadian silt loam. However, both Perfect and Kay (1994) and Macks et al. (1996) found lower tensile strength of air-dry aggregates for the reduced than for the conventionally tilled soil.

SEEDBED PREPARATION   Greater tractor power and improved tyres have made it possible to till the soil under wetter conditions and with higher intensity. This has led to concern, particularly in northwestern Europe (Ehlers and Claupein, 1994; Schjønning et al., 1997; Watts and Dexter, 1997). Few investigations have addressed the effect of different second-ary tillage tools on the strength and stability of soil. However, Bullock et al. (1988) found an immediate and profound decrease in wet stability of a loam soil following rotary cultivation. In addition, Schjønning et al. (1997) showed a larger destabilizing effect from rotary cultivation than from ordinary tine seedbed preparation in ploughed soil in a study on 80 predominantly sandy loams.

A number of studies have indicated that excessive seedbed preparation may yield a soil with decreased workability. Watts and Dexter (1997) showed that aggregate tensile strength increased immediately after rotary cultivation. Two to four passes with the rotary cultivator yielded aggregates (16–19 mm) with more than twice the strength that was found for uncultivated aggregates. Munkholm and Schjønning (2003) showed that intensive rotary cultivation of a wet sandy loam gave significantly stronger tensile strength of dry aggregates than did a soil that was tine cultivated at lower water content.

Understanding tillage effects

As shown above, tillage has a direct and immediate destabilizing effect on soil structure. Extensive laboratory and field studies by Watts et al. (1996a,b) showed that the level of readily dispersible clay decreased with increased energy input and water content. This indicates that intensive tillage decreases soil stability, which may manifest itself in surface crusting, surface erosion and clay migration in the soil profile. It may also manifest itself in decreased soil workability. Studies have indicated that the energy input of the tillage operation itself may explain the increase in aggregate tensile strength with tillage intensity when tilling wet soil (Watts et al., 1996b; Watts and Dexter, 1997; Munkholm and Schjønning, 2003). Watts et al. (1996b) showed that an increase in aggregate tensile strength was associated with a marked increase in dispersible clay. It was supposed that clay dispersed by tillage in wet conditions was concentrated behind menisci of cracks and would precipitate or flocculate under drying. This would result in increased tensile strength of dry aggregates.

Tillage may also have an indirect effect on soil stability. Improved stability of the surface soil under reduced tillage has been attributed to increased OM content and microbial activity (Carter, 1992). Tillage may also have a profound effect on the factors affecting the seasonal changes. For instance, direct drilling has been shown to maintain a higher resistance to slaking and freeze/thaw events (Angers et al., 1993).

The implications of long-term intensive tillage on soil stability may be a soil with pro-gressively lower workability and greater risk of surface crusting and erosion. It was shown that intensive tillage in wet conditions has an immediate negative effect on soil workability. If the soil has not fully recovered before the time of tillage for next season's crop, a vicious circle may develop where intensive tillage leads to stronger and less friable soil, which again increases the demand for more intensive tillage. Limited data are available on the resil-ience of soil structure after intensive tillage. In a recent study, Munkholm and Schjønning (2003) found little resilience of topsoil struc-ture within a 6-month summer period after intensive tillage in wet conditions on a sandy loam. That is, the increase in aggregate tensile strength was of the same magnitude (around 35%) for sampling in May and in November.

Soil degradation from loss of clay from the topsoil and accumulation in the subsoil also needs to be considered. Numerous reports deal with the effect of tillage intensity on surface erosion, whereas few deal with effects on clay migration. The question of whether intensive tillage significantly increases the natural process of clay migration in the profile merits greater attention, especially in humid regions.

Management Thresholds and Soil-quality Indicator Thresholds

Progressive changes in soil structural form that lead to decreased availability of water and oxygen, accessibility of nutrients or ease of penetration of the soil by plant organs, or increased susceptibility of the plant to disease can be avoided by a combination of management practices that balance structural degradation with regeneration. This balance could be obtained through knowledge of management thresholds and/or soil-quality indicator thresholds.

Management thresholds

There is little evidence that can be used to define threshold values below which the form or intensity of tillage, the amount of crop residue returned to soil, the frequency of forages in rotations or the rates of application of manure and other organic amendments result in OM depletion. There is also a paucity of data available to determine threshold values for tillage intensity that could be used to identify the onset of a deterioration of soil structure through mechanisms unrelated to OM. Threshold values for management will depend on soil characteristics and climate.

The adoption of different management practices at the farm level will be strongly influenced by economic considerations. For instance, the benefits from increasing tillage intensity may be a finer and more suitable seedbed, and saving of time due to fewer passes needed to produce a desirable seedbed. However, under all soil conditions the positive effects of high tillage intensity need to be balanced against short- and long-term negative effects of tillage intensity on biological productivity and environmental quality (Fig. 11.1). Farmers readily recognize the potential short-term benefits of intensive tillage on soil productivity. However, the long-term costs of intensive tillage on soil productivity are less obvious. Farm as well as environmental economics must be used to determine threshold values of management practices for specific soil and climatic conditions.

Soil-quality indicator thresholds

OM content is an important indicator of soil quality. Limited data are available to estimate the threshold OM contents below which structural form begins to seriously impact on biological productivity. Five factors contribute to this paucity of data. First, OM dynamics are strongly controlled by climate and soil characteristics (see Dick and Gregorich, Chapter 7, this volume). Secondly, the relationship between OM and structural form varies with soil properties such as texture and parent material, with climate and with the stresses associated with different management practices. Thirdly, the influence of soil structure on biological productivity is

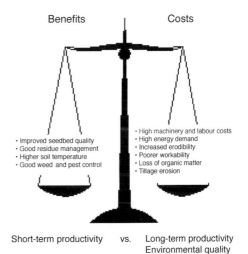

Benefits   Costs

• Improved seedbed quality
• Good residue management
• Higher soil temperature
• Good weed and pest control

• High machinery and labour costs
• High energy demand
• Increased erodibility
• Poorer workability
• Loss of organic matter
• Tillage erosion

Short-term productivity   vs.   Long-term productivity
                                  Environmental quality

Fig. 11.1.   Balancing benefits and costs of intensive tillage.

strongly dependent on climate. Adverse effects of soil structure on biological productivity are most obvious under extremes in weather conditions. Fourthly, the magnitude of adverse effects of soil structure on biological productivity that are linked to weather may depend on the stage of crop development in which the adverse effects are encountered. Finally, identification of threshold OM contents in relation to the effects of soil structure on biological productivity must be based on studies in which no factor other than soil structure limits productivity. This is often difficult to accomplish (Johnston, 1986).

An assessment of yields in the classical and long-term experiments at Rothamsted and Woburn, UK, some of which have remained unchanged for more than 100 years, has indicated that until the 1970s, increasing amounts of OM had little effect on yields of arable crops on a sandy loam and silty clay loam soil when N fertilization was adequate (Johnston, 1986). However, differences in yields of some crops began to be manifested in the 1970s with larger yields on soils with organic carbon contents of 1.0–2.0%, than on soils with organic carbon contents ranging from 0.7 to 0.8%. Some of the increase in yields was attributed to the availability of N in ways that could not be mimicked by addition of fertilizer N, some to extra water-holding capacity and some to improvements in other structural characteristics (Johnston, 1986). However, the range in organic carbon contents on these sites was not large enough to enable threshold OM contents to be conclusively identified. Greenland (1997) has suggested that the influence of OM contents will be most obvious when other management practices result in high yields and that in these cases, the return from the use of high levels of inorganic fertilizers and other inputs will only be fully realized if higher levels of OM can be established and maintained.

In the absence of appropriate site-specific data, models that predict crop yield from soil and climatic conditions may be used in conjunction with models linking soil structure and OM contents to identify threshold OM

contents. For instance, pedotransfer functions have been developed to describe the influence of soil properties on bulk density, the water-release curve and soil resistance to penetration (e.g. Kay et al., 1997). However, we are unaware of studies that assess the sensitivity of crop yields to variation in soil structural characteristics under a range of climatic conditions.

Conclusions

Sustainable productivity is jeopardized when management practices result in a progressive change in soil structural form leading to decreases in the availability of water and oxygen, accessibility of nutrients or ease of penetration of the soil by plant organs, or increased susceptibility of the plant to disease. A management practice can lead to a progressive degradation of structural form as a consequence of increasing stress, decreasing structural stability, decreasing population and activity of organisms, or decreasing resiliency with respect to weather. Practices that lead to a progressive loss of OM lead to a degradation of soil structure because of the important role of OM in structural stability, and to a lesser extent on the population of organisms and resiliency with respect to weather. Tillage can cause an increase in erodibility, susceptibility to crust formation and, in the long term, may yield a soil with diminished friability and lower workability.

There is very little information on threshold values for management practices or for OM contents. Management thresholds are expected to be specific to soil and climate conditions. Threshold values of OM content, below which soil structure has significant adverse effects on productivity, would also be expected to be soil and climate specific. In the absence of site-specific information, it is proposed that threshold values of OM contents be identified using crop models and pedotransfer functions relating soil structure to OM for specific soils. A modelling approach may also be helpful in defining threshold intensity for tillage.

References

Angers, D.A., Samson, N. and Légère, A. (1993) Early changes in water-stable aggregation induced by rotation and tillage in a soil under barley production. *Canadian Journal of Soil Science* 73, 51–59.

Baldock, J.A., Kay, B.D. and Schnitzer, M. (1987) Influence of cropping treatments on the monosaccharide content of the hydrolysates of a soil and it's aggregate fractions. *Canadian Journal of Soil Science* 67, 489–499.

Ball, B.C. and Robertson, E.A.G. (1994) Effects of uni-axial compaction on aeration and structure of ploughed or direct drilled soils. *Soil and Tillage Research* 31, 135–149.

Braunack, M.V. and Dexter, A.R. (1989) Soil aggregation in the seedbed: a review. II. Effect of aggregate sizes on plant growth. *Soil and Tillage Research* 14, 281–298.

Bullock, M.S., Kemper, W.D. and Nelson, S.D. (1988) Soil cohesion as affected by freezing, water content, time and tillage. *Soil Science Society of America Journal* 52, 770–776.

Carter, M.R. (1991) Evaluation of shallow tillage for spring cereals on a fine sandy loam. 2. Soil physical, chemical and biological properties. *Soil and Tillage Research* 21, 37–52.

Carter, M.R. (1992) Influence of reduced tillage systems on organic matter, microbial biomass, macro-aggregate distribution and structural stability of the surface soil in a humid climate. *Soil and Tillage Research* 23, 361–372.

Carter, M.R. (1994) A review of conservation tillage strategies for humid temperate regions. *Soil and Tillage Research* 31, 289–301.

Chan, K.Y. (1989) Friability of a hardsetting soil under different tillage and land use practices. *Soil and Tillage Research* 13, 287–298.

Christian, D.G. and Ball, B.C. (1994) Reduced cultivation and direct drilling for cereals in Great Britain. In: Carter, M.R. (ed.) *Conservation Tillage in Temperate Agroecosystems.* Lewis Publishers, Boca Raton, Florida, pp. 117–140.

Dexter, A.R. and Bird, N.R.A. (2001) Methods for predicting the optimum and the range of soil water contents for tillage based on the water retention curve. *Soil and Tillage Research* 57, 203–212.

Douglas, J.T. and Goss, M.J. (1982) Stability and organic matter content of surface soil aggregates under different methods of cultivation and in grassland. *Soil and Tillage Research* 2, 155–175.

Ehlers, W. and Claupein, W. (1994) Approaches toward conservation tillage in Germany. In: Carter, M.R. (ed.) *Conservation Tillage in Temperate Agroecosystems.* Lewis Publishers, Boca Raton, Florida, pp. 141–165.

Emerson, W.W. (1995) Water retention, organic C and soil texture. *Australian Journal of Soil Research* 33, 241–251.

Greenland, D.J. (1997) Inaugural Russell Memorial Lecture. Soil conditions and plant growth. *Soil Use and Management* 13, 169–177.

Greenland, D.J., Rimmer, D. and Payne, D. (1975) Determination of the structural stability class of English and Welsh soils, using a water coherence test. *Journal of Soil Science* 26, 294–303.

Gusli, S., Cass, A., MacLeod, D.A. and Blackwell, P.S. (1994) Structural collapse and strength of some Australian soils in relation to hardsetting: 1. Structural collapse on wetting and draining. *European Journal of Soil Science* 45, 15–21.

Håkansson, I., Myrbeck, Å. and Etana, A. (2002) A review of research on seedbed preparation for small grains in Sweden. *Soil and Tillage Research* 64, 23–40.

Johnston, A.E. (1986) Soil organic matter, effects on soils and crops. *Soil Use and Management* 2, 97–105.

Kandeler, E. and Murer, E. (1993) Aggregate stability and soil microbial processes in a soil with different cultivation. *Geoderma* 56, 503–513.

Kay, B.D. (1990) Rates of change of soil structure under different cropping systems. *Advances in Soil Science* 12, 1–52.

Kay, B.D. and VandenBygaart, A.J. (2002) Conservation tillage and depth stratification of porosity and soil organic matter. *Soil and Tillage Research* 66, 107–118.

Kay, B.D., Grant, C.D. and Groenevelt, P.H. (1985) Significance of ground freezing on soil bulk density under zero tillage. *Soil Science Society of America Journal* 49, 973–978.

Kay, B.D., Dexter, A.R., Rasiah, V. and Grant, C.D. (1994) Weather, cropping practices and sampling depth effects on tensile strength and aggregate stability. *Soil and Tillage Research* 32, 135–148.

Kay, B.D., da Silva, A.P. and Baldock, J.A. (1997) Sensitivity of soil structure to changes in organic carbon content: predictions using pedotransfer functions. *Canadian Journal of Soil Science* 77, 655–667.

Kemper, W.D. and Koch, E.J. (1966) *Aggregate Stability of Soils from Western United States and Canada.* Technical Bulletin. No. 1355, Agricultural Research Service, USDA, Washington, DC, 52 pp.

Lampkin, N. (1990) *Organic Farming*. Farming Press, Ipswich, UK, 701 pp.

Larney, F.J., Fortune, R.A. and Collins, J.F. (1988) Intrinsic soil physical parameters influencing intensity of cultivation procedures for sugar beet seedbed preparation. *Soil and Tillage Research* 12, 253–267.

Low, A.J. (1972) The effect of cultivation on the structure and other physical characteristics of grassland and arable soils (1945–1970). *Journal of Soil Science* 23, 363–380.

Macks, S.P., Murphy, B.W., Cresswell, H.P. and Koen, T.B. (1996) Soil friability in relation to management history and suitability for direct drilling. *Australian Journal of Soil Research* 34, 343–360.

Manrique, L.A. and Jones, C.A. (1991) Bulk density of soils in relation to soil physical and chemical properties. *Soil Science Society of America Journal* 55, 476–481.

Miller, R.M. and Jastrow, J.D. (1990) Hierarchy of root and mycorrhizal fungal interactions with soil aggregation. *Soil Biology and Biochemistry* 22, 579–584.

Munkholm, L.J. and Kay, B.D. (2002) Effect of water regime on aggregate-tensile strength, rupture energy, and friability. *Soil Science Society of America Journal* 66, 702–709.

Munkholm, L.J. and Schjønning, P. (2003) Structural vulnerability of a sandy loam exposed to intensive tillage and traffic in wet conditions. *Soil and Tillage Research* (in press).

Munkholm, L.J., Schjønning, P., Debosz, K., Jensen, H.E. and Christensen, B.T. (2002) Aggregate strength and soil mechanical behaviour of a sandy loam under long-term fertilization treatments. *European Journal of Soil Science* 53, 129–137.

Opoku, G., Vyn, T.J. and Swanton, C.J. (1997) Modified no-till systems for corn following wheat on clay soils. *Agronomy Journal* 89, 549–556.

Perfect, E. and Kay, B.D. (1994) Influence of corn management on dry aggregate tensile strength: Weibull analysis. *Soil and Tillage Research* 32, 149–161.

Perfect, E., Kay, B.D., van Loon, W.K.P., Sheard, R.W. and Pojasok, T. (1990) Factors influencing soil structural stability within a growing season. *Soil Science Society of America Journal* 54, 173–179.

Rasiah, V. and Kay, B.D. (1994). Characterizing changes in aggregate stability subsequent to introduction of forages. *Soil Science Society of America Journal* 58, 935–942.

Rasmussen, K.J. (1999) Impact of ploughless soil tillage on yield and soil quality: a Scandinavian review. *Soil and Tillage Research* 53, 3–14.

Reicosky, D.W. (1997) Tillage-induced CO_2 emission from soil. *Nutrient Cycling in Agroecosystems* 49, 273–285.

Schjønning, P. and Rasmussen, K.J. (1989) Long-term reduced cultivation. I. Soil strength and stability. *Soil and Tillage Research* 15, 79–90.

Schjønning, P., Thomsen, A. and Olesen, J.E. (1997) Effects of secondary tillage strategy on soil and crop characteristics. In: Fotyma, M., Jozefaciuk, A., Malicki, L. and Borowiecki, J. (eds) *Fragmenta Agronomica TOM 2B. Proceedings 14th ISTRO Conference, 27 July–1 August 1997, Pulawy, Poland*. Polish Society of Agrotechnical Sciences, Poland, pp. 579–582.

Schjønning, P., Munkholm, L.J., Moldrup, P. and Jacobsen, O.H. (2002) Modelling soil pore characteristics from measurements of air exchange: the long-term effects of fertilization and crop rotation. *European Journal of Soil Science* 53, 331–339.

Soane, B.D. (1990) The role of organic matter in soil compactibility: a review of some practical aspects. *Soil and Tillage Research* 16, 179–201.

Solberg, E.D., Nyborg, M., Izaurralde, R.C., Mahli, S.S., Janzen, H.H. and Molina-Ayala, M. (1998) Carbon storage in soils under continuous cereal grain cropping: N fertilizer and straw. In: Lal, R., Kimble, J.M., Follett, R.F. and Stewart, B.A. (eds) *Management of carbon sequestration in soils*. CRC Press, Boca Raton, Florida, pp. 235–254.

Spiess, E., Anken, T., Heusser, J., Weisskopf, P., Hogger, C. and Oberholzer, H.R. (2000) Soil tillage: energy input and seedbed structure. *Agrarforschung* 7, 348–353.

Suwardji, P. and Eberbach, P.L. (1999) Seasonal changes of physical properties of an Oxic Paleustalf (Red Kandosol) after 16 years of direct drilling or conventional cultivation. *Soil and Tillage Research* 49, 65–77.

Tisdall, J.M. and Oades, J.M. (1982) Organic matter and water-stable aggregates in soils. *Journal of Soil Science* 33, 141–163.

VandenBygaart, A.J., Yang, X.M., Kay, B.D. and Aspinall, J.D. (2002) Variability in carbon sequestration potential in no-till soil landscapes of southern Ontario. *Soil and Tillage Research* 65, 231–241.

Veihmeyer, F.J. and Hendrickson, A.H. (1927) Soil moisture conditions in relation to plant growth. *Plant Physiology* 2, 71–82.

Vyn, T.J., Sutton, J.C. and Raimbault, B.A. (1991) Crop sequence and tillage effects on winter wheat development and yield. *Canadian Journal of Plant Science* 71, 669–676.

Vyn, T.J., Opoku, G. and Swanton, C.J. (1998) Residue management and minimum tillage systems for soybean following wheat. *Agronomy Journal* 90, 131–138.

Watts, C.W. and Dexter, A.R. (1997) Intensity of tillage of wet soil and the effects on soil structural condition. In: Fotyma, M., Jozefaciuk, A., Malicki, L. and Borowiecki, J. (eds) *Fragmenta Agronomica TOM 2B. Proceedings 14th ISTRO conference, 27 July–1 August 1997, Pulawy, Poland*. Polish Society of Agrotechnical Sciences, Poland, pp. 669–672.

Watts, C.W., Dexter, A.R., Dumitru, E. and Arvidsson, J. (1996a) An assessment of the vulnerability of soil structure to destabilisation during tillage. Part I. A laboratory test. *Soil and Tillage Research* 37, 161–174.

Watts, C.W., Dexter, A.R. and Longstaff, D.J. (1996b) An assessment of the vulnerability of soil structure to destabilisation during tillage. Part II. Field trials. *Soil and Tillage Research* 37, 175–190.

Yang, X.M. and Kay, B.D. (2001) Rotation and tillage effects on soil organic carbon sequestration in a typic Hapludalf in Southern Ontario. *Soil and Tillage Research* 59, 107–114.

Chapter 12
Soil Erosion – Processes, Damages and Countermeasures

G. Govers, J. Poesen and D. Goossens

Catholic University of Leuven, Laboratory for Experimental Geomorphology,
Redingenstraat 16, B-3000 Leuven, Belgium

Summary

Soil erosion has been recognized for a long time as a major threat to soil quality and soil functions. Erosion occurs through various processes, which have different operating modes and result in different soil redistribution patterns. Some of these processes, such as rill and interrill erosion and wind erosion, have been studied for a long time and are relatively well understood. The importance of other processes, such as (ephemeral) gully erosion, net soil translocation due to tillage and dust emission due to tillage has only recently been recognized and data concerning rates and effects of these processes are often still lacking.

Studies on the effect of erosion on soil properties show that erosion negatively affects many soil-quality indicators such as solum depth, organic matter (OM) content, nutrient status and aggregate stability. Studies relating soil erosion to soil functions have hitherto mainly focused on the relationship between soil erosion and crop productivity. The alarming results from one-time desurfacing experiments need to be interpreted with care: gradual erosion has a much weaker effect on crop productivity than the sudden removal of a significant proportion of the topsoil. Experiments that have been carefully designed to isolate the effect of soil erosion from other confounding factors suggest that, in the case of intense, mechanized agriculture, soil erosion only leads to a significant reduction in crop productivity if the plant available water capacity and/or the plant rooting depth are negatively affected. Erosion also affects water quality, air quality and human health through various mechanisms.

Most of the eroded soil material is deposited in another place in the landscape: this may have additional negative effects but it may also lead to a local improvement in soil quality. When assessing the effects of soil erosion on soil quality it is therefore necessary to include a landscape-scale perspective: the detrimental effects experienced at one location may be (partly) compensated by beneficial effects at other locations.

For some soil-quality indicators, like soil OM and soil cover, it is possible to establish rather precise threshold values that should be reached in order to achieve sufficient protection against erosion. Specific techniques could be applied in order to achieve these threshold values. However, we believe that management approaches that are targeted at improving the various soil-quality indicators in an integrated and consistent way are most efficient at reducing excessive soil erosion risk. Management strategies aimed at maintaining a sufficient amount of residue cover at the soil surface and reducing tillage intensity appear to be most suitable for soils exposed to excessive erosion risks.

It is ultimately the farmer who has to take decisions on land use and management, but policy makers at various levels can help create an environment that stimulates farmers to adopt conservation strategies. A major barrier for doing this effectively is that we do not always have a clear understanding of the reasons why conservation strategies are (not) accepted by farmers. Future research will therefore have to combine further technical refinement of conservation technology with the development of efficient strategies for soil conservation adoption.

Introduction

Soil erosion has been considered as a threat to the soil resource long before the concept of soil quality was introduced. The threat imposed by droughts and erosion to agricultural production in the USA in the 1930s was the direct reason for the start of a major research effort on soil erosion processes and soil conservation strategies. It is, therefore, no surprise that as soon as the concept of soil quality was accepted by the research community, studies started to emerge to link soil quality to soil erosion. This chapter provides a brief overview of the various processes causing soil erosion, and the relationship between soil properties, soil functions and erosion is discussed. Most of the studies relating erosion to (an aspect of) soil quality do not explicitly use a definition of soil quality. The majority of these studies have an empirical, comparative design, whereby the response of soil properties or soil functions to erosion is assessed by comparing them on eroded and non-eroded sites. In our overview, we have adopted a functional approach which is in agreement with present-day concepts of soil quality. The information provided on the relation between erosion and soil functions other than food and fibre productivity is necessarily limited by the lack of existing research. As this chapter deals with the effect of surface processes on soil quality, it is important that the relationship between erosion and soil quality is approached not only from a pedon perspective but also from a landscape perspective. In a final section we discuss the relation between soil erosion, soil management and soil quality.

Soil Erosion Processes

Significant erosion on arable land occurs mainly because of water, wind and tillage. The effects of these processes on the soil are quite different due to their different modes of operation. The discussion below focuses on those aspects of erosion processes that are important for understanding their effects on soil quality.

Water erosion

When water erosion takes place, soil is detached by raindrop impact and/or flowing water. Although some net transport can occur due to splash alone, important soil loss only happens when soil material is transported over considerable distances by flowing water. Various types of water erosion can be distinguished. When splash detachment is dominant, *interrill erosion*, i.e. the sheetwise detachment and removal of soil material, takes place. Various studies have shown that interrill erosion is slope dependent (e.g. Govers and Poesen, 1988). However, there is less agreement on the effect of slope length on interrill erosion: often it is assumed to be independent of slope length (e.g. Govers *et al.*, 1996), but in some cases interrill erosion may well decrease with slope length, due to the progressive filling up of the transporting capacity of the flow and/or the loss of water due to increasing infiltration in the downslope direction (e.g. Dunne *et al.*, 1991). *Rill erosion* occurs when the detachment capacity of the flowing water becomes important: incision leads to the formation of small channels from which considerable quantities of soil can be removed. Rill erosion is, in normal conditions, both dependent on slope length or contributing area and on slope gradient (McCool and George, 1983; Govers, 1991). It is, therefore, most important on long, steep hillslopes. When rill erosion occurs, it is usually far more important than interrill erosion. This is due to the much higher efficiency of concentrated flow to transport sediment off the hillslope (Govers and Poesen, 1988; Whiting *et al.*, 2001). When water concentrates in hollows or in linear landscape elements such as parcel borderlines or wheel tracks, *ephemeral gullying* may occur, i.e. the formation of relatively large erosion channels that are often responsible for a large part of the sediment export out of a catchment. The distinction between rills and ephemeral gullies is primarily volumetric: an incision with a cross-section exceeding $0.09 \, \text{m}^2$ is considered to be an ephemeral gully (Poesen, 1993). As for rills, ephemeral gully formation is controlled by contributing area (slope

length is often irrelevant as ephemeral gullies are located in the converging zones of the landscape) and slope gradient (Nachtergaele *et al.*, 2001). It is possible to define topographical threshold conditions that are necessary (but often not sufficient) for the generation of ephemeral gullies: however, these threshold conditions vary between various agricultural regions (e.g. Vandaele *et al.*, 1996).

Wherever topographical steps are present in the landscape (e.g. lynchets or sunken lane banks), runoff crossing these steps might erode *bank gullies* (Poesen *et al.*, 1996). Although spectacular in nature, mean soil loss rates due to bank gully erosion are less important than ephemeral gully erosion rates.

Sediment deposition by water is also important. In many situations a large amount of the sediment eroded by water is re-deposited within the field or small catchment: conditions for deposition are related to flow hydraulic characteristics, which are mainly controlled by topography and vegetation. Therefore, deposition occurs mainly on low slope angles and/or in or just before highly vegetated zones, like buffer strips (Steegen *et al.*, 2000).

Wind erosion

Two distinct types of wind erosion exist: deflation and corrasion. *Deflation* is defined as the detachment (and subsequent transportation) of loose particles from the surface by wind forces. After having accomplished their aerial trajectory, particles will fall back to the surface, thereby transferring a large part of their energy to other particles, which may then be launched from the surface. *Corrasion* is the polishing of rock, soil aggregates or crusts by impacting airborne particles. Corrasion may also destroy vegetation, and in that case is usually known as *sandblasting*.

The mode in which wind-eroded particles are transported largely determines the nature as well as the extent of the effect wind erosion will exert on the environment. The largest particles (> 500 µm) roll or slide over the surface without losing contact with the latter, a process known as surface creep or surface traction (Pye and Tsoar, 1990). Smaller particles, between 500 and approximately 50–100 µm, are transported in saltation: particles jump and bounce over the surface, reaching a maximum height of approximately 1 m. When they fall back to the surface they not only eject other saltation-size grains but also induce surface creep, reptation (some small-scale saltation with only a limited displacement of grains near the points of impact) and surface deformation. They also cause the raising of dust particles, which are transported in suspension. A distinction is commonly made between short-term and long-term suspension depending on whether the particles will stay airborne for only a short time (normally a few hours) or longer (days or weeks). A transition mode, known as modified saltation, exists between saltation and suspension. Saltating particles are normally > 100 µm whereas suspended particles are usually < 50 µm. Certain particles may be moved by different transport modes, depending on particle density, wind speed and the level of turbulence in the airflow.

Particles may be moved over a wide range of distances. During a single storm, creep can move particles over distances from a few centimetres to several metres; saltating particles travel from a few metres to a few hundred metres, and suspension transport ranges from several tens of metres to thousands of kilometres (Sterk, 1998).

The physical threshold for wind erosion is proportionally higher than that of water erosion. A minimum wind speed of 6–7 m/s (or more, depending on the surface characteristics) is required to initiate wind erosion. These speeds represent only a small fraction (0.1–1%) of the potential erosion time, even in areas favourable to wind erosion (De Ploey, 1980; Goossens and Gross, 2002). Also, water erosion always results in a net downward displacement of sediment. This means that in the case of water erosion any erosion event will normally proceed with the work done by the previous event. The effects exerted by a wind erosion event are not necessarily a continuation of the effects created by the previous event, for example because of a change in wind direction. It follows that the long-term

effects caused by wind erosion are usually less easy to predict than for water, except if the wind regime stays unchanged for a long time.

Tillage erosion

Two forms of tillage erosion need to be distinguished. Both net soil translocation due to tillage and dust emission due to tillage are often referred to as tillage erosion (Govers *et al.*, 1999; Goossens *et al.*, 2001). Net soil translocation due to tillage is caused by variation in the magnitude of soil movement during tillage. The transport and resultant displacement of soil by tillage is referred to as tillage translocation (Govers *et al.*, 1999).

The basic nature of the process is illustrated in Fig. 12.1. During a downslope tillage operation, a downslope displacement of the plough layer occurs. Naturally, soil is moved upslope during an upslope operation, but generally over a smaller distance as a result of gravity. Consequently, there is net downslope displacement of soil on the hillslope. The net amount of soil translocation by a given tillage operation is primarily slope dependent (Govers *et al.*, 1994):

$$Q_s = kS \qquad (12.1)$$

where: Q_s is the net unit soil transport rate (kg/m/tillage pass); k is the tillage transport coefficient (kg/m/tillage pass); and S is the slope gradient (tan).

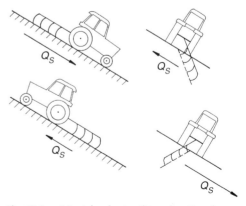

Fig. 12.1. Principle of net soil translocation due to tillage (Q_s: unit soil transport rate (kg/m/tillage pass)).

On a rectilinear slope, soil loss due to tillage will only occur near the upslope field boundary. However, if a convexity (shoulder) is present, soil loss due to tillage will occur all over the convexity as the unit soil transport increases in the downslope direction. Similarly, soil deposition will occur in concavities.

Thus, net soil translocation due to tillage is quite different from water erosion. It leads to soil erosion near upslope field boundaries and on convexities, whereas the most important soil losses due to water erosion occur at the base of long, steep slopes and in steep concavities where water concentrates. Whereas a single tillage operation will translocate the whole plough layer over a distance of 0.5–2 m, water erosion will translocate a relatively small amount of soil over distances that may exceed 1 km. Net soil translocation due to tillage leads to soil redistribution only within a field, whereas a significant part of the soil material eroded by water is usually exported from the field into the surface water system.

Under common agricultural practices in Western Europe, soil losses on convexities and near field borders are often 10–30 t/ha/year and deposition rates in concavities can exceed 100 t/ha/year. At the field scale, net soil translocation due to tillage is therefore often by far the most important soil redistribution process (Govers et al., 1994).

Tillage does not only result in a displacement of soil particles on the ground. During tillage considerable amounts of soil particles will become airborne for a distinct time. When the topsoil is sufficiently dry, part of these particles will be carried by the wind for a while and the fine particles (dust fraction) will usually settle at remote distances from the emission area. This process is referred to as dust emission due to tillage.

In the flat northern European lowland plains, where water erosion is small and wind is the major natural agent causing erosion, the evacuation of soil via the atmosphere is one order of magnitude higher for dust emission due to tillage than for natural wind erosion (Goossens et al., 2001). On soils relatively rich in silt and clay, soil loss via the atmosphere due to tillage operations can exceed 50 t/ha/year (Goossens et al., 2001; Goossens and Gross, 2002), i.e. the same order of magnitude as for tillage-induced soil redistribution on steep slopes.

Soil loss by root and tuber harvesting

When harvesting roots (e.g. sugarbeet) or tubers (e.g. potato), significant amounts of soil, sticking to the crop, can leave the field, and hence this constitutes an erosion process (Poesen et al., 2001). Large variations in soil loss due to root and tuber harvesting occur from plot to plot and over time because of variations in soil type, soil moisture at harvest time, harvest technique, etc., and losses exceeding 10 t/ha per harvest are not uncommon.

Effects of Erosion on Soil Quality

Erosion and deposition by wind and water are often selective and therefore have a direct impact on topsoil properties both in erosional and depositional areas. Net soil translocation due to tillage and soil loss due to root crop harvesting are, in general, non-selective. Nevertheless, these processes may also lead to important changes in soil properties as they cause a vertical as well as a lateral mixing of soil constituents.

Changes in soil properties due to selective erosion and deposition

Erosion

CHANGES IN SOIL TEXTURE   Due to the preferential erosion of fine-grained sand and coarse silt, wind erosion generally results in a gradual coarsening of the top layer leading to serious degradation for several reasons. First, the soil's nutrients are largely situated in the fine particle fractions. Secondly, due to the evacuation of these fractions, the water economy in the topsoil degrades. The coarse sandy top layer dries quickly, and this not only affects the crops, but also increases the soil's vulnerability to subsequent wind erosion. Thirdly, a lower silt and clay content enhances the eluviation of humus and other soil constituents and promotes acidification. Fourthly, to remain fertile, degraded sandy topsoil requires a sufficient amount of fertilizer, which is much less the case for topsoil rich in silt and clay.

The size selectivity of water erosion is limited. Various studies have demonstrated that, due to its limited transporting capacity, interrill erosion often leads to the selective removal of the fine soil fraction (e.g. Savat, 1982). Rill and gully erosion only occur when the runoff is capable of eroding all soil fractions, and are therefore non-selective. In cases where water erosion is a serious problem, rill and ephemeral gully erosion are generally much more important than interrill erosion.

LOSS OF ORGANIC MATTER   Since most of the humic matter in the soil is directly or indirectly bound to the finer soil fractions (< 150 μm; Zenchelsky *et al.*, 1976), soil particles lost in wind erosion are normally characterized by a much higher organic matter (OM) content than their parent soil. Recent measurements on German agricultural fields show that the OM percentage in the suspended fraction during wind erosion can exceed 30% (Goossens *et al.*, 2001). Medium and fine-textured soils are specially vulnerable to this type of degradation.

A small amount of soil OM exists in the form of plant residue. Because of their very low density, these residues are very easily eroded by wind and water. Their rapid evacuation precludes their transformation into humus, which could have been incorporated into the upper soil layer.

As OM is strongly bound to the clay fraction, the enrichment of OM in water-eroded sediment is generally directly proportional to its enrichment in clay. Enrichment is most important during low-intensity events: during such events the relative difference in erodibility and transportability of the different soil fractions is most important, and the eroded sediment may then be enriched in OM by a factor of three or more. Size selectivity decreases with increasing erosion intensity (e.g. Steegen *et al.*, 2001).

LOSS OF OTHER SOIL CONSTITUENTS   M a n y other soil constituents, such as N, P, K and other nutrients, are removed from the parent soil in wind and water erosion. Analyses of wind-blown sediments have clearly demonstrated the enrichment of these elements in the eroded particles (Duncan and Moldenhauer, 1968; Sterk *et al.*, 1996). Despite the fact that the highest nutrient fluxes are measured in the saltation layer, the suspended fraction is more important with respect to soil degradation. Saltation results only in a local redistribution (usually within the field itself), whereas most nutrients in suspension leave the parent field (Sterk *et al.*, 1996; Goossens and Gross, 2002).

Analysis of runoff samples shows a strong preferential export of P by water erosion, as this nutrient is strongly bound to the clay and OM fractions (Steegen *et al.*, 2001). Similarly to OM, the enrichment in P of the eroded sediment is most pronounced for small erosion events with low sediment concentrations. Although absolute values of P concentrations in eroded sediments vary greatly, the decrease of the P concentration with increasing sediment concentration appears to be relatively constant (Steegen *et al.*, 2001; see also Condron, Chapter 5, this volume).

Deposition

DEPOSITION BY WIND   A distinction should be made between the deposition of coarse aeolian sediment (sand) and fine aeolian sediment (dust, i.e. the silt + clay fraction).

Sand.   Many wind-eroded agricultural fields show a sand blanket several centimetres thick after a storm on the leeward field side. The quality of the accumulated sediment is usually very poor. Sand blankets consist of almost pure sand, contain very little OM and nutrients, and are very acid. When sand blankets are incorporated into the underlying soil, they strongly contribute to the general degradation of the topsoil in the settling areas, as discussed earlier.

Dust.   Contrary to sand, the deposition of silt and clay may positively affect the quality of the topsoil. A minimum of silt and clay is beneficial to the soil structure, especially in very sandy soil (Sterk *et al.*, 2001); it increases the water-holding capacity, serves as an important source of nutrients (Offer *et al.*, 1992; Littmann, 1997) and stimulates the formation of a thin soil crust, which protects the soil from further wind erosion (Offer and Goossens, 2001). There are also negative effects, however. Probably the most important is the risk of pollution with contaminants bound to the fine silt and clay particles.

DEPOSITION BY WATER   Sediment deposition by water is potentially very selective: laboratory experiments show that the deposition process can in most cases be described by the simple settling theory, which implies that the fine colloidal soil fraction would almost never be deposited (Beuselinck *et al.*, 1999). Nevertheless, sediment deposits in the field are often only marginally depleted in clay and OM. This is due to the fact that, in the case of soils containing a significant clay fraction, most sediment is eroded in the form of small aggregates having an internal composition very similar to the parent material (Meyer *et al.*, 1992). Although the deposition of these aggregates can be selective, the primary grain size composition of the sediment deposits is often very similar to that of the parent material (Beuselinck *et al.*, 2000). If water erosion occurs on a weakly structured, sandy soil, size selectivity during deposition is much more important and the colluvial deposits will be strongly depleted in clay and OM.

Changes in soil properties due to vertical and lateral mixing

At a point, soil erosion will lead to a gradual thinning of the A horizon. Due to tillage, more and more material from lower horizons will be incorporated into the plough layer. This will cause changes in topsoil properties, the nature of which depends on the properties of both the topsoil and the subsoil. As in many cases illuvial B horizons are present below the A horizon, strongly eroded sites are often characterized by a higher clay content in the topsoil (e.g. Olson *et al.*, 1999). In cases where a thin regolith is present over

rocky material, soil erosion will lead to an increased stoniness of the topsoil. On the other hand, topsoil properties in depositional areas will become progressively more similar to the properties of the deposited material.

Soil tillage not only leads to erosion and deposition, but also to lateral mixing of soil constituents within the plough layer, because translocation is non-uniform (e.g. Sibbesen *et al.*, 1985). On sloping land the dispersion is accompanied by a net downslope movement of the plough layer. Topsoil material from upslope will move gradually downslope and become more strongly mixed.

Soil erosion and soil productivity

Most studies documenting the relationship between soil erosion and soil quality focus on the productivity of the soil. The first studies on the relationship between crop productivity and erosion were published early in the 20th century. Many of these studies were mainly based on desurfacing experiments, whereby a pre-set amount of topsoil was artificially removed. Identical management was applied to both the desurfaced and non-desurfaced plots and the crop response was measured. The results of these desurfacing experiments point to a very strong relationship between soil erosion and crop productivity, with yield declines often over 50% when half of the topsoil is removed (e.g. Gantzer and McCarthy, 1987; Thompson *et al.*, 1991).

However, these results cannot be extrapolated to normal agricultural conditions. This is due to two mechanisms: dilution and compensation. When erosion occurs gradually, a limited amount of subsoil (equivalent to the erosion rate) is mixed with the topsoil each year. This causes a dilution of the topsoil constituents beneficial to crop production, such as nutrients and OM. Consequently when erosion occurs again during the next year at the same intensity, a smaller absolute amount (same relative amount) of these constituents is removed by erosion. Compensation occurs because soil erosion causes a disequilibrium: normally, the OM content of agricultural topsoils is in equilibrium with the soil type, current agricultural practices, crop rotation and climate (see Dick and Gregorich, Chapter 7, this volume). If some of the OM in the topsoil is removed by erosion, a disequilibrium between OM production and mineralization is created, leading to a net production of soil OM. Deliberate compensations by the farmer are also possible: the loss of nutrients due to erosion can be compensated for by extra inputs.

Even when different erosion phases of the same soil type are compared, it remains difficult to attribute variations in crop yield to erosion. Stone *et al.* (1985) suggested that, due to the strong covariation that exists between erosion phase and landscape position, many studies describing the effect of soil erosion on crop productivity are confounded by the effect of landscape position. They noticed that, in their study area, relationships between landscape position and productivity were far more consistent than those between soil erosion phase and productivity, and attributed this observation to the importance of soil water redistribution. Areas receiving water from upslope areas by surface and/or subsurface flow were more productive in most years.

Some recent studies on the relationship between soil erosion and soil productivity have been set up differently: crop yields have been monitored on different erosion phases of the same soil profile, preferably located within the same landscape position. The results are, in general, quite different from those of desurfacing experiments. The NC-174 project was designed to assess the impact of erosion on the productivity of 12 different soils in north central USA (Schumacher *et al.*, 1994; Olson *et al.*, 1999). It was found that maize yields on strongly eroded phases of topsoil with root-restricting subsoils were 18% lower on average than on slightly eroded phases. The average reduction on soils without root-restricting subsoils was 0%. Both on soils with and without root-restricting subsoils, erosion led to a decrease of the OM content and an increase of the clay content of the topsoil. However, erosion of soils with a root-restricting subsoil also caused a very significant reduction of the rooting depth and the soil-available water storage capacity, two properties which were hardly affected by

erosion on the soils without a root-restricting subsoil. Using a similar experimental design, Weesies *et al.* (1994) reported similar reductions in crop yield for maize and soybean on three soils with root-restricting horizons in Indiana. Langdale *et al.* (1979) reported even larger reductions of maize yield with increasing erosion on a sandy loam in the Southern Piedmont (USA).

The strong linkage between soil-available water capacity/soil depth and crop productivity is also found in Mediterranean and temperate environments in Europe. Kosmas *et al.* (2001) found a near-perfect curvilinear increase in grain yield with soil depth on a field near Thiva, Greece. Strauss and Klaghofer (2001) reported that yield reductions due to erosion in Lower Austria could mainly be attributed to a reduction in water-holding capacity of the rooting zone. In the much wetter environment of Mississippi, Rhoton and Lindbo (1997) found a very strong correlation between the topsoil depth above a root-restricting fragipan and soybean crop yields. They proposed use of the effective soil depth (ESD), i.e. the depth of the soil in which most processes take place, as a soil-quality indicator. Crop yield decreased by about 25% when the ESD was reduced from 0.6 to 0.2 m.

Thus, even under mechanized agriculture, soil erosion may lead to a significant reduction of crop productivity when relevant soil physical properties such as rooting depth and plant-available water capacity are affected by erosion. Soil erosion may also affect other soil properties, but in intensive, mechanized agricultural systems the negative effects of these changes are generally overcome by changes in management and appropriate (or excessive) nutrient inputs. This explains why no negative effects of soil erosion on crop productivity are noticed on sufficiently deep soils with an appropriate nutrient supply (Olson *et al.*, 1999; Lal *et al.*, 2000).

Soil erosion and other soil functions

Soil erosion and water quality

Soil erosion reduces the thickness of the solum and selectively removes soil OM, one of the most reactive soil components of the soil profile. This has a direct effect on the buffering and filtering capacity of the soil. Experimental studies on the effect of erosion on the water-buffering and filtering function of the soil are lacking (Lowery et al., 1999). Model simulations by Lowery et al. (1999) on the leaching potential of the different erosion phases of three soils showed no direct linkage between erosion phase and simulated leaching of the herbicide atrazine. However, leaching did relate to the OM content of the soil, a property clearly affected by soil erosion (e.g. Lal et al., 1999).

Water erosion has several effects on surface water quality. It leads to an increase of the suspended sediment content, and hence to a decrease of the light transmissivity. Erosion also causes surface water eutrophication due to a high input of particulate phosphorus (Sharpley and Smith, 1990; Steegen et al., 2001) and a decrease in chemical quality because the sediment exported into fresh water often contains (residues of) agrochemicals. In areas with mechanized, intense agriculture the direct cost of these off-site effects of erosion is very significant, and probably much higher than the cost of on-site effects (Uri and Lewis, 1999; Pretty et al., 2000). A reduction of soil-erosion rates through appropriate management will therefore not only have a beneficial effect on soil, but also on water quality.

Soil erosion and air quality

The relationship between wind erosion and air quality has been underestimated for a long time, in particular because emissions of aerosols originating from the earth's surface have long been neglected. In various aerosol emission inventories, both at the national and the international scales, wind erosion simply does not appear in the list of emission sources (see, for example, MAFF, 1998; Den Hartog, 1999). However, it was already reported more than 25 years ago that soil dust contributes almost 50% to the global atmospheric dust load (Fenelly, 1976). Research during the last few years has provided considerable evidence for the importance of soil as a source of fine aerosol. However, there is still a serious lack of experimental data, especially for Europe.

In a 2-year monitoring study in an agricultural area in northern Germany, Goossens et al. (2001) showed that, at least in the region investigated, approximately 40% of the total suspended particulate (TSP) aerosol was of agricultural (soil) origin. This is important with respect to air quality, for eroding soil dust does not only consist of quartz but also contains products that can be harmful to the environment, such as fertilizers, herbicides, pesticides and other pollutants. The spread of pollutants carried by fine particles eroded from contaminated arable land has not been fully quantified, but must be considered important. Schulz (1992) proposed a value of 100 ng TEQ (toxicity equivalents) per kg for dioxin-contaminated arable land due to wind-induced emission of fine soil particles. This rate of emission is comparable to that from modern incineration plants. The quantities of deposited matter are very large, particularly because of the very large scale of the deposition area (millions of km^2). The affected areas may be very far from the erosion source(s), thus making control problematic.

There is much controversy as to what extent emitted soil particles affect human health. Evidence for the effect of soil dust on respiratory diseases has been provided during the last few years (e.g. Pope et al., 1999). Also, effects of soil dust on skin diseases have been reported (see for example Leathers, 1981). Air quality has become a major issue in the USA since the 1990s. The 1990 Federal Clean Air Act made states responsible for monitoring and controlling the amount of particles smaller than 10 μm (PM$_{10}$). In the EU, guidelines have been published for various airborne components, among which are PM$_{10}$ dust and total dust. All this shows that the effects that fine aerosol exerts on human health are now being taken seriously, although there is still a considerable need for more documented information.

Finally, since fine aerosol can travel over very large distances, and since chemical substances (and also biotic material) primarily bind to fine particles, fine aerosol is one of the most efficient agents ensuring the dispersion of contaminants over the globe. The role of wind erosion, and that of wind-eroded particles as a transportation vector, is thus of utmost importance in environmental issues.

Soil Erosion and Soil Quality: the Importance of a Landscape-scale Approach

Nearly all studies dealing with the relationship between soil quality and erosion have been carried out at the plot scale: a relationship was sought between local soil functions or properties and the local erosion rate. Although this approach has provided important information and insights with respect to the relationship between soil quality and erosion, it is also limited to some extent. Erosion processes are spatially non-uniform and cause important lateral fluxes of soil particles within the landscape. Material that is eroded at one place is deposited somewhere else after a more or less important transport phase. Different soil erosion processes operate differently within a landscape and lead to different erosion and deposition patterns. Consequently, erosion effects on soil quality are highly non-uniform. In a landscape-scale investigation of soil quality, Pennock *et al.* (1999) noticed that erosion caused a significant decrease of soil quality on hillslope convexities (shoulders). At the same time, the soil quality on the footslopes and in the depressions was considerably improved due to the deposition of carbon-rich soil material. Other researchers already noted a (relative) increase in crop yield in deposition areas (e.g. Lal *et al.*, 2000). The effects of increased land pressure on soil quality are scale-dependent. In a study in Burkina Faso, Gray (1999) noted that although the increased land pressure certainly led to deforestation and an extension of the arable land areas, it led at the same time to improved land management within the arable areas. Consequently no soil degradation occurred within these zones. Thus the effect of erosion on soil functions needs to be assessed at different places in the landscape and a suitable strategy has to be employed to extrapolate the results over the whole landscape. This can be quite complicated as the

role of landscape positions acting as soil sources or sinks may change over time (Pennock *et al.*, 1999). Here, the use of spatially distributed models predicting the effect of different erosion processes on soil properties may be useful (e.g. Van Oost *et al.*, 2000, 2003; Rosenbloom *et al.*, 2001).

Management of Soils in Order to Reduce the Erosion Risk

In order to protect the soil from being eroded, various management practices can be recommended. Given that soil erosion processes and controlling factors (e.g. soil type, topography and land use) vary in space, management may be quite different from one site to another. Although soil conservation measures as described below are in the first place designed to control erosion, they often have additional beneficial effects for soil quality and soil functions. When applied for a time-span of at least several years, conservation tillage and/or proper residue management often leads to an increase in soil OM, aggregate stability, water infiltration and/or water retention capacity of the topsoil (e.g. Karlen *et al.*, 1994; Arshad *et al.*, 1999). This will, in turn, reduce further the erosion risk, although the reduction may be rather limited compared to the immediate effects of increased vegetation and/or residue cover (McGregor *et al.*, 1999).

Water erosion

Most management techniques aiming at the reduction of the water erosion risk apply some basic principles. These are: (i) the increase of the infiltration rate of rain and runoff where possible, so as to reduce runoff volumes and hence runoff erosivity; (ii) the increase of the topsoil resistance to detachment and transport of soil particles by the erosive forces of rain and runoff; and (iii) the protection of the soil surface with plant and residue cover. The list of soil or soil-related parameters that can be manipulated through management and that affect the infiltration rate or the resistance of the topsoil to erosion includes soil cover (with crop residues or rock fragments, i.e. mulch cover), soil OM content, aggregate stability, macroporosity, bulk density, soil surface roughness and cohesion through crop root reinforcement.

Interrill and rill erosion

Techniques to reduce interrill and rill erosion risk are applied at the field-plot scale. Here various soil conservation strategies can be followed and these can be subdivided into agronomic measures (crop management and mulching) and soil management (conservation tillage). For an overview, see Morgan (1995). These strategies aim at protecting the topsoil against the erosive forces by rain and runoff and to increase the water acceptance rate and the resistance against erosion of the topsoil.

SOIL COVER   A very efficient management practice to reduce sheet and rill erosion rates consists of increasing soil cover in space and time (both canopy cover and contact cover (mulch)). The increased soil cover leads to a reduction of the rate of physical degradation (surface sealing, crusting, compaction) of the topsoil resulting in a relatively higher infiltration rate compared to that of a bare soil. Furthermore, the presence of a soil cover retards runoff and reduces soil detachment by rainfall and overland flow. Many experimental studies indicate that rill and interrill erosion (Eir) decrease exponentially with vegetation or mulch cover (C, %). In other words, $Eir = e^{-bC}$, where b is a coefficient indicating the efficiency of the cover in reducing the erosion rate. Values for b range between 0.02 and 0.08 both for vegetation cover and for mulch cover (Bochet, 1996). Mean b values for contact cover are, on average, lower than those for vegetation (canopy) cover. Rock fragments at the soil surface behave like natural soil surface stabilizers and are therefore also considered to be important mulch elements. Poesen et al. (1994) reported b values for rock fragment cover ranging between 0.025 and 0.05 with a mean value of 0.04. Soil cover can be increased by applying various types of crop management (i.e. high-density planting, multiple cropping

through crop rotations, strip cropping, cover cropping) and by conservation tillage (i.e. any tillage or tillage and planting combination that leaves a 30% or greater cover of crop residue on the surface; Soil Science Society of America, 2002). If the soil is stony, rock fragment cover at the soil surface can be increased through tillage of the dry topsoil by tines or tine-like tillage implements (Oostwoud Wijdenes et al., 1997). These tillage operations lead to kinetic sieving or particle-size segregation whereby the largest rock fragments and clods concentrate at the soil surface.

SOIL CONSTITUENTS AND AGGREGATE STABILITY Various soil constituents control aggregate stability, e.g. clay content, soil OM content, free Fe_2O_3, free Al_2O_3, $CaCO_3$ and exchangeable Na. Kemper and Koch (1966) investigated the water-stable aggregates (WSA) from 500 soils sampled in the western part of the USA and Canada, and found that an increase in soil OM from 0 to 2% resulted in a strong increase of water stability of aggregates. Beyond a soil OM value of 2% the increase in WSA evolved in a degressive way. Later studies showed that this conclusion was too general. The role of soil OM in stabilizing soil aggregates is rather complex and ranges from entanglement by fungal hyphae and roots to binding by the decomposition products and secretions of roots, microorganisms and soil animals. The roles of different organic components in aggregate stabilization vary with time and with aggregate size considered (Tisdall and Oades, 1982). As different types of soil OM are produced by different land uses and/or crops, their effect on WSA may be quite different, even for a similar level of soil OM (Dinel and Gregorich, 1995). Furthermore, the effect that soil OM has on WSA depends to some extent on the process that is responsible for aggregate breakdown (Le Bissonnais and Arrouays, 1997).

Many experimental studies have conclusively shown that an increase in WSA leads to an increase in infiltration rates and increased resistance to physical degradation, interrill and rill erosion (e.g. Le Bissonnais et al., 2002). The contents of several WSA-increasing soil constituents can be raised through crop and soil management (application of OM (e.g. green manure or farmyard manure), lime (CaCO$_3$) or gypsum (CaSO$_4$.2H$_2$O); e.g. Valzano *et al.*, 2001).

SURFACE ROUGHNESS   A rough soil surface has many depressions and barriers, so that excess rainfall can be stored at the surface (depression storage) and is allowed more time to infiltrate. Depression storage can be quite important on freshly ploughed fields (> 20 mm of rainfall), but values for seedbeds are much lower (< 3 mm) (Govers *et al.*, 2000). Furthermore, runoff will start well before all depressions are filled. The trapping of water and sediment in depressions during rain causes rough surfaces to erode at slower rates than do smooth soil surfaces under similar conditions. Also, on rough soil surfaces more runoff energy is dissipated than on smooth soil surfaces, leaving less flow energy to detach and transport particles. Soil surface roughness can be increased or maintained through tillage practices. When tilling on the contour, roughness in a downslope direction is increased (compared to up and down tillage) leading to an increased depression storage and hence a reduced interrill and rill erosion risk. On a uniform slope, the efficiency of contour tillage in reducing the erosion risk depends on ridge height, slope gradient and rainfall intensity (Renard *et al.*, 1997). However, in two-dimensional landscapes where flow divergence and flow convergence occur, tillage along the contour may lead to the concentration of water in hollows, causing increased erosion rates (Takken *et al.*, 2001).

MACROPOROSITY   Many studies have shown the tremendous soil crusting effect on infiltration rates. Whereas a freshly tilled seedbed has infiltration capacities often exceeding 50 mm/h, the infiltration capacity of crusted soils is often only 1–2 mm/h (Cerdan *et al.*, 2002). Hence, management options that reduce the potential for crusting can drastically reduce the soil erosion risk. The main strategy to prevent soil crusting is the maintenance of a good soil structure through the adequate management of soil OM and, if necessary, other soil amendments like gypsum (e.g. Pikul and Zuzel, 1994). With respect to infiltration, specific attention has to

be given to wheel tracks. Several studies have documented that they are often the dominant sources of runoff and sediment on arable land, mainly due to their low infiltration capacity (Auzet *et al.*, 1995; Basher and Ross, 2001). Management techniques to reduce the risk of erosion from wheel tracks include the use of controlled traffic and/or wheel track removers.

ROOTING   EFFECTS   ON   TOPSOIL   COHESION Various studies have shown that (crop) roots can reinforce soil by increasing the shear strength (e.g. Morgan and Rickson, 1995). As a consequence, crop roots are effective in reducing rill erosion rates (Gyssels *et al.*, 2002). The topsoil can be reinforced in this way through crop selection and its resistance to runoff erosion accordingly increased.

In conclusion, it can be stated that rill and interrill erosion can be significantly reduced by appropriate management of soil cover, soil OM, soil structure, surface roughness or root density. The relationships between rill and interrill erosion and controlling factors such as soil cover are non-linear, and indicate that a small increase in, for instance, soil cover or soil OM at low values usually has a relatively large impact on rill and interrill erosion. Further increases generally have a smaller effect. Thus, although it may not be possible to define clear management thresholds below which erosion rates will always be acceptable it is possible to define soil-quality indicators that should be achieved in order to avoid excessive rill and interrill erosion risk. For example, the organic carbon content of the topsoil of silty soils should exceed 1.5% in order to facilitate good water infiltration (Le Bissonnais and Arrouays, 1997). Above this threshold, additional increases of organic carbon are less important. Although exact data are lacking, it can be assumed that this threshold will be lower on clayey soils and higher on more sandy soils. The general, negative exponential relationship between soil cover and erosion rates implies that on all soils, cover by residues and/or rock fragments or living plants should be maintained at or above 30% for sufficient protection against erosion. The structure of dispersive soils can be improved by the use of gypsum and other amendments.

However, the relationship between gypsum, the cations present in the exchange complex, clay mineralogy and soil pH is quite complex. More research is needed to define optimum amendment levels to control soil erosion economically (Norton et al., 1999). Although specific management strategies could be devised to achieve the required soil-quality indicator values to reduce erosion to an acceptable level, it is far more useful to consider these threshold values within the framework of an integrated management strategy that addresses these issues in a consistent way. In our view, management strategies based on a reduced tillage intensity and appropriate residue management are often most appropriate to reduce water and tillage erosion in intensive, mechanized agriculture. Over recent years, various studies have shown that the use of no-tillage or conservation tillage not only provides a better protection of the soil against erosion due to an increased soil cover, but also leads to an increase in WSA and hence better resistance of the soil against water erosion (Arshad et al., 1999; McGregor et al., 1999). However, the introduction of these management strategies is far from easy and the reasons for non-acceptance by farmers are not always properly understood (Napier et al., 2000). Proper understanding of farmers' attitudes and alternative management techniques may therefore be a more crucial element of successful soil conservation than further technological advances.

Gully erosion

Techniques to reduce ephemeral gully erosion rates can be applied at the catchment scale and in the concentrated flow zones of the landscape. At the catchment scale, all management techniques increasing infiltration rates and hence reducing peak runoff discharges also reduce the risk of ephemeral gully erosion. Within the concentrated flow zone, measures can be taken to increase the resistance of the topsoil through increasing its cohesion. The latter can be achieved by for instance compacting the topsoil (Ouvry, 1989) or through root reinforcement (Gyssels et al., 2002). A classical approach would be to establish a grassed waterway in these zones,

which will not only increase the resistance of the soil surface to erosion but will also increase dissipation of flow energy by increasing the surface roughness. Given that farmers in Europe are reluctant to adopt these control measures, Gyssels et al. (2002) suggested multiple sowing of small grains in concentrated flow zones as a means of controlling rill and gully erosion. Multiple sowing refers to drilling more than once in zones of concentrated flow erosion in order to increase the total root mass in this zone. As to bank gullies, the use of geomembranes at the reshaped gully head to protect the soil in the concentrated flow zone and willow staking to increase the shear strength of the gully banks are efficient techniques (Poesen, 1989).

Soil translocation due to tillage

Obviously, the most straightforward strategy to reduce tillage erosion is by reducing the number and intensity of tillage operations on arable land. In this respect, minimum tillage systems have a clear advantage over conventional tillage systems. A first, rapid comparison of various tillage systems can be made by estimating the tillage transport coefficient for each tillage operation, summing the estimates for a complete crop cycle and then calculating an average annual value for different tillage systems (Govers et al., 1994). Typical values for the tillage transport coefficient of various implements can be found in Van Muysen et al. (2000, 2002). However, care should be taken when estimating tillage erosivity: the erosivity of a tillage operation with a given implement greatly depends on the speed and the depth at which it is operated, as well as the soil condition (loose vs. compacted). Soil type appears to be less important.

Van Muysen et al. (2000, 2002) developed nomographs that allow the erosivity of a chisel tillage or a mouldboard tillage operation to be assessed depending on tillage speed, tillage depth, tillage direction and soil condition. These nomographs show that significant reductions in tillage erosivity may also be obtained by modifying tillage techniques within an existing tillage system. The erosivity of up- and downslope mouldboard tillage increases exponentially with tillage depth. Consequently a moderate reduction in tillage depth from 0.3 m to 0.25 m reduces tillage erosion by about 30% (Van Muysen *et al.*, 2002). A change in tillage direction may also be helpful: mouldboard ploughing is much less erosive along the contour than up- and downslope, especially for tillage depths exceeding 0.2 m. In contrast to mouldboard ploughing, chisel tillage erosivity is more dependent on tillage speed than on tillage depth. In this case, a reduction in tillage speed will therefore be more efficient than an equivalent reduction in tillage depth (Van Muysen *et al.*, 2000).

The fact that significant amounts of soil are moved by tillage implies that tillage can also deliberately be used to remedy or alleviate the effects of water erosion and/or previous tillage erosion by consistently moving the soil upslope during tillage operations. Even if tillage translocation occurs consistently in the upslope direction, net soil loss on convex landscape positions may continue. This will especially be true for chisel tillage operations: experimental data show that upslope translocation during mouldboard ploughing is independent of slope gradient so that no net erosion on convexities will occur during upslope mouldboard tillage (Van Muysen *et al.*, 2002). Also, upslope soil movement by tillage will result in the accumulation of soil on ridgetops and at the downslope side of field boundaries: these are locations where water erosion is minimal. Thus, upslope soil movement by tillage will not exactly counteract the downslope movement of soil by water erosion, nor will it completely annihilate the effects of historical tillage erosion.

Wind erosion and dust emission due to tillage

Management strategies to prevent the displacement of soil particles via the atmosphere should not only focus on wind erosion. In most areas where the land is used for agricultural purposes, emission (and subsequent transportation and deposition) of soil particles is also caused by tillage practices, as indicated above. In scarcely populated arid and semiarid regions, wind erosion is the dominant emission process. In areas like Europe, tillage is more important. Therefore, management strategies should focus on both types of emission.

Some examples of how farmers could help reduce the risk of soil emission by wind are as follows:

- Emission of soil particles will only take place when the topsoil is sufficiently dry. This is true for both wind- and tillage-induced erosion. Therefore, any tillage during episodes of dry topsoil conditions should be avoided. It is the fine particle fraction (dust) that is especially vulnerable to emission during tillage.

- The speed of tillage is also very important, especially in the case of dry topsoil. The more quickly tillage operations are executed, the more sediment will be evacuated. When the topsoil is sufficiently wet and the risk of emission is minimum, the speed of tillage is less important (at least, for atmospheric evacuation).

- Because clods are more difficult to erode than individual soil grains, any tillage resulting in a breakdown of small-size clods should be kept to a minimum.

- To avoid avalanching effects during particle transport, and also to reduce the wind speed, the field length of parcels should be kept within reasonable limits.

- Many techniques to reduce water erosion are also effective in reducing wind erosion: choice of an adequate soil cover, ensuring a sufficient amount of OM in the topsoil and keeping the roughness of the soil above a minimum level.

- Because surface crusts lead to an increase of the deflation threshold, and also affect the amounts of horizontal and vertical dust flux, care should be taken to keep such crusts intact until the percentage of cover crop has reached a level high enough to protect the soil from further erosion. In areas that are vulnerable to water erosion, however, crusts normally stimulate overland flow, leading to an increased risk of interrill, rill and gully erosion. This example illustrates the ambivalent effects that soil conservation techniques may exert, depending on the specific type of erosion.

Policy and Erosion

Policy measures can have an important effect on the decisions ultimately taken by farmers with respect to land-use and soil-quality management. The political (or policy) level operates at different hierarchic levels: municipality level, regional level, national level and supranational level. Decisions made at these levels should focus on various aspects: (i) they should guarantee an adequate organization of the landscape (for example, resulting in maximum field sizes); (ii) they should establish adequate legislation with respect to soil quality and soil use; (iii) they should provide information to farmers about how to comply with this legislation; and (iv) they should stimulate farmers to adopt conservation techniques that guarantee the quality of the soil.

Although it will always be the farmer who decides whether or not to adopt soil conservation, policy regulations can be developed to encourage him/her to accept the idea. The general introduction of a soil conservation code, such as the British *Code of Good Agricultural Practice for the Protection of Soil* (MAFF, 1998) or the German *Bundes-Bodenschutzgesetz* (Anonymous, 1998), can be seen as an important step in this context. Financial support remunerating those farmers who apply soil conservation measures, such as is the case in the *Conservation Reserve Program* of the US Department of Agriculture, can dramatically speed up acceptance of soil conservation practices (Young and Osborn, 1990; Mello *et al.*, 2002).

Proper management for erosion prevention may not only prevent erosion but will often also restore the quality of soils that were degraded by erosion. The degree of success of soil-quality restoration will depend on the soil type and the properties most strongly affected by erosion. Although soil properties such as OM content respond relatively

quickly to changes in soil management, it may not be possible to restore properties such as rooting depth or available water capacity without excessive investments. The timely adoption of soil-management strategies in order to prevent irreversible damage by erosion therefore remains a cornerstone of any strategy aimed at maintaining or improving the quality of arable land.

References

Anonymous (1998) Gesetz zum Schutz vor schädlichen Bodenveränderungen und zur Sanierung von Altlasten (Bundes-Bodenschutzgesetz-BBodSchG). *Bundesgesetzblatt* I 1998, 502.

Arshad, M.A., Franzluebbers, A.J. and Azooz, R.H. (1999) Components of surface soil structure under conventional and no-tillage in northwestern Canada. *Soil and Tillage Research* 53, 41–47.

Auzet, A.V., Boiffin, J. and Ludwig, B. (1995) Concentrated flow erosion in cultivated catchments: influence of soil surface state. *Earth Surface Processes and Landforms* 20, 759–767.

Basher, L.R. and Ross, C.W. (2001) Role of wheel tracks in runoff generation and erosion under vegetable production on a clay loam soil at Pukekohe, New Zealand. *Soil and Tillage Research* 62, 117–130.

Beuselinck, L., Govers, G., Steegen, A., Hairsine, P.B. and Poesen, J. (1999) Evaluation of the simple settling theory for predicting sediment deposition by overland flow. *Earth Surface Processes and Landforms* 24, 993–1007.

Beuselinck, L., Steegen, A., Govers, G., Nachtergaele, J., Takken, I. and Poesen, J. (2000) Characteristics of sediment deposits formed by major rainfall events in small agricultural catchments in the Belgian Loam Belt. *Geomorphology* 32, 69–82.

Bochet, E. (1996) Interactions Sol-végétation et Processus d'Érosion Hydrique dans le Micro-environnement de Trois Espèces du Matorral Méditerranéen. PhD thesis, Department of Geography-Geology, Catholic University of Leuven.

Cerdan, O., Souchère, V., Lecomte, V., Couturier, A. and Le Bissonnais, Y. (2002) Incorporating soil surface crusting processes in an expert-based runoff model: sealing and transfer by runoff and erosion related to agricultural management. *Catena* 46, 189–205.

De Ploey, J. (1980) Some field measurements and experimental data on wind-blown sand. In: de Boodt, M. and Gabriëls, D. (eds) *Assessment of Erosion*. John Wiley & Sons, Chichester, UK, pp. 541–552.

Den Hartog, P. (1999) *Year Report Air Quality 1997*. National Institute of Public Health and the Environment, Report No. 725301 001. Bilthoven, The Netherlands.

Dinel, H. and Gregorich, E. (1995) Structural stability status as affected by long-term continuous maize and bluegrass sod treatments. *Biological Agriculture and Horticulture* 12, 237–252.

Duncan, E.R. and Moldenhauer, W.C. (1968) *Controlling Wind Erosion in Iowa*. Cooperative Extension Service, Iowa State University, Ames, Iowa, 6 pp.

Dunne, T., Zhang, W.H. and Aubry, B.F. (1991) Effects of rainfall, vegetation, and microtopography on infiltration and runoff. *Water Resources Research* 27, 2271–2285.

Fennelly, P.F. (1976) The origin and influence of airborne particulates. *American Scientist* 64, 46–56.

Gantzer, C.J. and McCarthy, T.R. (1987) Predicting corn yields on a claypan soil using a soil productivity index. *Transactions of the ASAE* 30, 347–1352.

Goossens, D. and Gross, J. (2002) Similarities and dissimilarities between the dynamics of sand and dust during wind erosion of loamy sandy soil. *Catena* 47, 269–289.

Goossens, D., Gross, J. and Spaan, W. (2001) Aeolian dust dynamics in agricultural land areas in Lower Saxony, Germany. *Earth Surface Processes and Landforms* 26, 701–720.

Govers, G. (1991) Rill erosion on arable land in Central Belgium: rates, controls and predictability. *Catena* 18, 133–155.

Govers, G. and Poesen, J. (1988) Assessment of the interrill and rill contributions to total soil loss from an upland field plot. *Geomorphology* 1, 343–354.

Govers, G., Vandaele, K., Desmet, P., Poesen, J. and Bunte, K. (1994) The role of tillage in soil redistribution on hillslopes. *European Journal of Soil Science* 45, 469–478.

Govers, G., Quine, T.A., Desmet, P.J.J. and Walling, D.E. (1996) The relative contribution of soil tillage and overland flow erosion to soil redistribution on agricultural land. *Earth Surface Processes and Landforms* 21, 929–946.

Govers, G., Lobb, D.A. and Quine, T.A. (1999) Tillage erosion and translocation: emergence of a new paradigm in soil erosion research. *Soil and Tillage Research* 51, 167–175.

Govers, G., Takken, I. and Helming, K. (2000) Soil roughness and overland flow. *Agronomie* 20, 131–146.

Gray, L.C. (1999) Is land being degraded? A multiscale investigation of landscape change in southwestern Burkina Faso. *Land Degradation and Development* 10(4), 329–343.

Gyssels, G., Poesen, J., Nachtergaele, J. and Govers, G. (2002) The impact of sowing density of small grains on rill and ephemeral gully erosion in concentrated flow zones. *Soil and Tillage Research* 64, 189–201.

Karlen, D.L., Wollenhaupt, N.C., Erbach, D.C., Berry, E.C., Swan, J.B., Eash, N.S. and Jordahl, J.L. (1994) Crop residue effects on soil quality following 10-years of no-till corn. *Soil and Tillage Research* 31, 149–167.

Kemper, W. and Koch, E. (1966) Aggregate stability of soils from western United States and Canada. *United States Department of Agriculture Bulletin* 1355, 1–52.

Kosmas, C., Gerontidis, S., Marathianou, M., Detsis, B., Zafiriou, T., Van Muysen, W., Govers, G., Quine, T.A. and Van Oost, K. (2001) The effects of tillage displaced soil on soil properties and wheat biomass. *Soil and Tillage Research* 58, 31–44.

Lal, R., Mokma, D. and Lowery, B. (1999) Relation between soil quality and erosion. In: Lal, R. (ed.) *Soil Quality and Erosion*. Soil and Water Conservation Society, Ankeny, Iowa, pp. 237–258.

Lal, R., Ahmadi, M. and Bajracharya, R.M. (2000) Erosional impacts on soil properties and corn yield on alfisols in central Ohio. *Land Degradation and Development* 11, 575–585.

Langdale, G.W., Box, J.E., Leonard, R.A., Barnett, A.P. and Fleming, W.G. (1979) Corn yield reduction on eroded Southern Piedmont soils. *Journal of Soil and Water Conservation* 34, 226–228.

Le Bissonnais, Y. and Arrouays, D. (1997) Aggregate stability and assessment of soil crustability and erodibility. 2. Application to humic loamy soils with various organic carbon contents. *European Journal of Soil Science* 48, 39–48.

Le Bissonnais, Y., Cros-Cayot, S. and Gascuel-Odoux, C. (2002) Topographic dependence of aggregate stability, overland flow and sediment transport. *Agronomie* 22, 489–501.

Leathers, C.R. (1981) Plant components of desert dust in Arizona and their significance for man. In: Péwé, T.L. (ed.) *Desert Dust: Origin, Characteristics, and Effect on Man. Geological Society of America Special Paper* 186, 191–206.

Littmann, T. (1997) Atmospheric input of dust and nitrogen into the Nizzana sand dune ecosystem, north-western Negev, Israel. *Journal of Arid Environments* 36, 433–457.

Lowery, B., Hart, G.L., Bradford, J.M., Kung, K.J.S. and Huang, C. (1999) Erosion impact on soil quality and properties and model estimates of leaching potential. In: Lal, R. (ed.) *Soil Quality and Erosion*. Soil and Water Conservation Society, Ankeny, Iowa, pp. 75–94.

MAFF (1998) *Code of Good Agricultural Practice for the Protection of Soil*. Ministry of Agriculture, Fisheries and Foods, Welsh Office Agriculture Department, MAFF Publications, London, 66 pp.

McCool, D.K. and George, G.O. (1983) *A Second-Generation Adaptation of the Universal Soil Loss Equation for Pacific Northwest Drylands*. American Society of Agricultural Engineers Paper 83–2066, 20 pp.

McGregor, K.C., Cullum, R.F. and Mutchler, C.K. (1999) Long-term management effects on runoff, erosion, and crop production. *Transactions of the ASAE* 42, 99–105.

Mello, I., Heissenhuber, A. and Kantelhardt, J. (2002) The American Conservation Reserve Program – the chance to reward farmers for services to the environment? *Berichte uber Landwirtschaft* 80, 85–93.

Meyer, L.D., Line, D.E. and Harmon, W.C. (1992) Size characteristics of sediment from agricultural soils. *Journal of Soil and Water Conservation* 47, 107–111.

Morgan, R.P.C. (1995) *Soil Erosion and Conservation*. Longman, Harlow, UK.

Morgan, R.P.C. and Rickson, R.J. (1995) *Slope Stabilization and Erosion Control. A Bioengineering Approach*. Chapman & Hall, Cambridge, 274 pp.

Nachtergaele, J., Poesen, J., Steegen, A., Takken, I., Beuselinck, L., Vandekerckhove, L. and Govers, G. (2001) The value of a physically based model versus an empirical approach in the prediction of ephemeral gully erosion for loess-derived soils. *Geomorphology* 21, 237–252.

Napier, T.L., Tucker, M. and McCarter, S. (2000) Adoption of conservation production systems in three Midwest watersheds. *Journal of Soil and Water Conservation* 55, 123–134.

Norton, L.D., Shainberg, I., Cihacek, L. and Edwards, J.H. (1999) Erosion and soil chemical properties. In: Lal, R. (ed.) *Soil Quality and Erosion*. Soil and Water Conservation Society, Ankeny, Iowa, pp. 39–56.

Offer, Z.Y. and Goossens, D. (2001) Ten years of aeolian dust dynamics in a desert region (Negev desert, Israel): analysis of airborne dust concentration, dust accumulation and the high-magnitude dust events. *Journal of Arid Environments* 47, 211–249.

Offer, Z.Y., Goossens, D. and Shachak, M. (1992) Aeolian deposition of nitrogen to sandy and loessial ecosystems in the Negev Desert. *Journal of Arid Environments* 23, 355–363.

Olson, K.R., Mokma, D.L., Lal, R., Schumacher, T.E. and Lindstrom, M.J. (1999) Erosion impact on crop yield for selected soils of the North Central United States. In: Lal, R. (ed.) *Soil Quality and Erosion*. Soil and Water Conservation Society, Ankeny, Iowa, pp. 259–284.

Oostwoud Wijdenes, D., Poesen, J., Vandekerckhove, L. and de Luna, E. (1997) Chiselling effects on the vertical distribution of rock fragments in the tilled layer of a Mediterranean soil. *Soil and Tillage Research* 44, 55–66.

Ouvry, J.F. (1989) Effet des techniques culturales sur la susceptibilité des terrains à l'érosion par ruisellement concentré. Expérience du Pays de Caux (France). *Cahiers ORSTOM, Série Pédologie* 15, 157–169.

Pennock, D.J., Anderson, D.W. and de Jong, E. (1999) Landscape-scale changes in indicators in soil quality due to cultivation in Saskatchewan, Canada. *Geoderma* 64, 1–19.

Pikul, J.L. and Zuzel, J.F. (1994) Soil crusting and water infiltration affected by long-term tillage and residue management. *Soil Science Society of America Journal* 58(5), 1524–1530.

Poesen, J. (1989) Conditions for gully formation in the Belgian Loam Belt and some ways to control them. *Soil Technology Series* 1, 39–52.

Poesen, J. (1993) Gully typology and gully control measures in the European loess belt. In: Wicherek, S. (ed.) *Farm Land Erosion in Temperate Plains Environment and Hills*. Elsevier Science Publishers, Amsterdam, pp. 221–239.

Poesen, J., Torri, D. and Bunte, K. (1994) Effects of rock fragments on soil erosion by water at different spatial scales: a review. *Catena* 23, 141–166.

Poesen, J., Vandaele, K. and van Wesemael, B. (1996) Contribution of gully erosion to sediment production in cultivated lands and rangelands. *IAHS Publication* 236, 251–266.

Poesen, J.W.A., Verstraeten, G., Soenens, R. and Seynaeve, L. (2001) Soil losses due to harvesting of chicory roots and sugar beet: an underrated geomorphic process? *Catena* 43(1), 35–47.

Pope, C.A., Hill, R.W. and Villegas, G.M. (1999) Particulate air pollution and daily mortality on Utah's Wasatch Front. *Environmental Health Perspectives* 107, 567–573.

Pretty, J.N., Brett, C., Gee, D., Hine, R.E., Mason, C.F., Morison, J.I.L., Raven, H., Rayment, M.D. and van der Bijl, G. (2000) An assessment of the total external costs of UK agriculture. *Agricultural Systems* 65, 113–136.

Pye, K. and Tsoar, H. (1990) *Aeolian Sand and Sand Dunes.* Unwin Hyman, London, 396 pp.

Renard, K.G., Foster, G.R., Weesies, D.K., McCool, D.K. and Yoder, D.C. (1997) *Predicting Soil Erosion by Water: a Guide to Conservation Planning with the Revised Universal Soil Loss Equation (RUSLE).* USDA Agriculture Handbook 703, 404 pp.

Rhoton, F.E. and Lindbo, D.L. (1997) A soil depth approach to soil quality assessment. *Journal of Soil and Water Conservation* 52, 66–72.

Rosenbloom, N.A., Doney, S.C. and Schimel, D.S. (2001) Geomorphic evolution of soil texture and organic matter in eroding landscapes. *Global Biogeochemical Cycles* 15, 365–381.

Savat, J. (1982) Common and uncommon selectivity in the process of fluid transportation: field observations and laboratory experiments on bare surfaces. In: Yaalon, D.H. (ed.) *Aridic Soils and Geomorphic Processes. Catena Supplement* 1, 139–160.

Schulz, D. (1992) Obergrenze für den Dioxingehalt von Ackerböden. *Zeitschrift für Umweltchemie und Ökotoxicologie* 4, 207–209.

Schumacher, T.A., Lindstrom, M.J., Mokma, D.L. and Nelson, W.W. (1994) Corn yield: erosion relationships of representative loess and till soils in the north central United States. *Journal of Soil and Water Conservation* 49, 77–81.

Sharpley, A.N. and Smith, S.J. (1990) Phosphorus transport in agricultural runoff: the role of soil erosion. In: Boardman, J., Foster, I.D.L. and Dearing, J.A. (eds) *Soil Erosion on Agricultural Land.* John Wiley & Sons, Chicester, UK, pp. 351–366.

Sibbesen, E., Andersen, C.E., Andersen, S. and Flensted-Jensen, M. (1985) Soil movement in long-term field experiments. *Plant and Soil* 91, 73–85.

Soil Science Society of America (2002) Glossary of Soil Science Terms. Available at: http://www.soils.org/sssagloss/

Steegen, A., Govers, G., Nachtergaele, J., Takken, I., Beuselinck, L. and Poesen, J. (2000) Sediment export by water from an agricultural catchment in the Loam Belt of central Belgium. *Geomorphology* 33, 25–36.

Steegen, A., Govers, G., Takken, I., Nachtergaele, J., Poesen, J. and Merckx, R. (2001) Factors controlling sediment and phosphorus export from two Belgian agricultural catchments. *Journal of Environmental Quality* 30, 1249–1258.

Sterk, G. (1998) Quantification of aeolian sediment balances from soil particle transport measurements. In: Sivakumar, M.V.K., Zöbisch, M.A., Koala, S. and Maukonen, T. (eds) *Wind Erosion in Africa and West Asia: Problems and Control Strategies. Proceedings of the Expert Group Meeting, 22–25 April 1997, Cairo, Egypt.* ICARDA, Aleppo, Syria, pp. 155–171.

Sterk, G., Herrmann, L. and Bationo, A. (1996) Wind-blown nutrient transport and soil productivity changes in Southwest Niger. *Land Degradation and Development* 7, 325–335.

Sterk, G., Riksen, M. and Goossens, D. (2001) Dryland degradation by wind erosion and its control. *Annals of Arid Zone* 41, 351–367.

Stone, J.R., Gilliam, J.W., Cassel, D.K., Daniels, R.B., Nelson, L.A. and Kleiss, H.J. (1985) Effect of erosion and landscape position on the productivity of Piedmont soils. *Soil Science Society of America Journal* 49, 987–991.

Strauss, P. and Klaghofer, E. (2001) Effects of soil erosion on soil characteristics and productivity. *Bodenkultur* 52, 147–153.

Takken, I., Jetten, V., Govers, G., Nachtergaele, J. and Steegen, A. (2001) The effect of tillage-induced roughness on runoff and erosion patterns. *Geomorphology* 37, 1–14.

Thompson, A.L., Gantzer, C.J. and Anderson, S.H. (1991) Topsoil depth, fertility, water management and weather influences of yield. *Soil Science Society of America Journal* 55, 1085–1091.

Tisdall, J.M. and Oades, J.M. (1982) Organic matter and water-stable aggregates in soils. *Journal of Soil Science* 33, 141–163.

Uri, N.D. and Lewis, J.A. (1999) Agriculture and the dynamics of soil erosion in the United States. *Journal of Sustainable Agriculture* 14, 63–82.

Valzano, F.P., Murphy, B.W. and Greene, R.S.B. (2001) The long-term effects of lime ($CaCO_3$), Gypsum ($CaSO_4.2H_2O$), and tillage on the physical and chemical properties of a sodic red-brown earth. *Australian Journal of Soil Research* 39, 1307–1331.

Van Muysen, W., Govers, G., Van Oost, K. and Van Rompaey, A. (2000) The effect of tillage depth, tilllage speed and soil condition on chisel tillage erosivity. *Journal of Soil and Water Conservation* 55, 354–363.

Van Muysen, W., Govers, G. and Van Oost, K. (2002) Identification of important factors in the process of tillage erosion: the case of mouldboard tillage. *Journal of Soil and Tillage Research* 65, 77–93.

Van Oost, K., Govers, G., Van Muysen, W. and Quine, T.A. (2000) Modeling translocation and dispersion of soil constituents by tillage on sloping land. *Soil Science Society of America Journal* 64, 1733–1739.

Van Oost, K., Van Muysen, W., Govers, G., Hechrath, G., Quine, T.A. and Poesen, J. (2003) Simulation of the redistribution of soil by tillage on complex topographies. *European Journal of Soil Science* 54, 1–14.

Vandaele, K., Poesen, J., Govers, G. and Van Wesemael, B. (1996) Geomorphic threshold conditions for ephemeral gully incision. *Geomorphology* 16, 161–173.

Weesies, G.A., Livingston, S.J., Hosteter, W.D. and Schertz, D.L. (1994) Effect of soil erosion on crop productivity in Indiana: results of a 10 year study. *Journal of Soil and Water Conservation* 49, 597–600.

Whiting, P.J., Bonniwell, E.C. and Matisoff, G. (2001) Depth and areal extent of sheet and rill erosion based on radionuclides in soils and suspended sediment. *Geology* 29, 1131–1134.

Young, C.E. and Osborn, C.T. (1990) Costs and benefits of the conservation-reserve-program. *Journal of Soil and Water Conservation* 45, 370–373.

Zenchelsky, S.T., Delany, A.C. and Pickett, R.A. (1976) The organic component of wind-blown soil aerosol as a function of wind velocity. *Soil Science* 122, 129–132.

Chapter 13

Recyclable Urban and Industrial Waste – Benefits and Problems in Agricultural Use

R. Naidu, M. Megharaj and G. Owens
University of South Australia, Centre for Environmental Risk Assessment and Remediation, Mawson Lakes Boulevard, Mawson Lakes, South Australia 5095, Australia

---

Summary

The current rate of world population growth is paralleled by a global increase in waste production. Increased population growth places stress both on the demand for agricultural produce and also on the finite world resources. If agricultural activities are to remain sustainable, recycling of urban and industrial waste for agricultural use must be considered. Urban and industrial waste is often rich in nutrients and trace elements that make them a valuable renewable resource. However, some public concern exists with such wastes because of the presence of potentially toxic substances. Good management strategies are essential to ensure effective reuse/recyling of urban and industrial wastes.

©CAB International 2004. *Managing Soil Quality: Challenges in Modern Agriculture*
(eds P. Schjønning, S. Elmholt and B.T. Christensen)

Introduction

The disposal of urban and industrial wastes is a challenging and expensive problem confronting environmental regulators, local councils and industries in many countries throughout the world. The production of these wastes is and will continue to be an ongoing problem as long as human civilization persists. Worldwide the management of urban and industrial wastes has received much attention since the late 1970s, when the implications of the presence of potentially toxic substances (PTSs) for environmental and human health were recognized. Toxic substances in wastes include inorganics (e.g. heavy metals and metalloids, nutrients), organics (e.g. polychlorobiphenols (PCBs), polyaromatic hydrocarbons (PAHs), halogenated aliphatics, pesticides, etc.) and pathogens (Henry and Heinke, 1996).

Urban and industrial wastes can have both beneficial and detrimental effects on the soil environment, and this has led to significant division in public opinion on the potential land application of wastes. Problems associated with land-based disposal arise because, in most metropolitan areas, there is no separation between domestic and industrial waste and wastewater. In some situations, urban runoff and storm water also finds its way into sewage works. However, although industrial contamination is a major problem with sewage sludge, domestic sewage alone is by no means completely free of a range of potential contaminants, including for example the heavy metals copper and zinc, various pathogenic organisms and a whole range of organic chemicals. In this chapter, although sewage sludge is the main focus of discussions, other examples illustrating recycling of industrial wastes will also be discussed to demonstrate the potential benefits and problems associated with waste recycling.

Sustained cropping may result in nutrients being depleted from agricultural soils and transferred off site by crop uptake. In the long term, agricultural production therefore becomes unsustainable without the influx of nutrients from external sources. The recognition that the world population will grow from its current 6 billion people to 8.5 billion by 2020 (Kinsman, 2000) is of concern as the volume of urban and industrial wastes produced will also increase. Indeed, Bastian (1997) reported a steady increase in biosolids production in the USA from < 5 Mt in the early 1970s to > 10 Mt in 1990. With finite resources the cycle eventually becomes unsustainable without some form of recycling. The very wastes produced by a growing population are now being recognized as a valuable resource for application to agricultural land, since many sewage sludges are rich in plant nutrients and also have the potential to improve soil structure.

Land application of wet sludge is therefore a method of disposal that is both economical and beneficial because organic matter (OM), nutrients and water in the sludge are recycled back into the land. However, both metals (Cu, Cd, Ni, Cr, Zn and Pb) and organics (dioxins, PCBs and endocrine disruptors), which are PTSs when present in high concentrations, could be persistent in soils for long periods (McGrath, 1987) because losses from the soil by crop uptake and leaching are normally small (Chang *et al.*, 1984; McGrath, 1987; McLaren and Smith, 1996). Hence, these PTSs accumulate in waste-/sludge-treated soils for long periods of time.

In the past, three major constraints have limited land application of sewage sludge: (i) nitrogen in excess of crop needs; (ii) pathogens; and (iii) heavy metals. Recent concerns include a fourth category: endocrine disruptors. Current regulations address the first three constraints (nitrogen, pathogens, heavy metals), but not the fourth (endocrine disruptors). Finding environmentally acceptable, secure and cost-effective outlets for urban and industrial wastes is already a challenging task because of legislative constraints and public sensitivity. In this chapter, the main issues discussed relate to potential contamination of the soil with both inorganic and organic pollutants and the subsequent effects on plant growth, animal and human health and the overall fertility of the soil itself.

Composition of Urban and Industrial Wastes

Although sludges from both urban and industrial sources may include appreciable concentrations of plant nutrients and PTSs, chemical compositions typifying sewage sludges are difficult to ascertain because sewage sludge varies widely depending on source composition. For instance, the concentration of toxic metals in sludges may range from 1 to > 3000 mg/kg dry weight with a medium value of 10 mg/kg (Logan and Chaney, 1983). This indicates the wide scatter in the distribution of these elements in the wastes. The variations in the chemical composition of sludges may be attributed to considerable geographical differences in climate, landform and land use, population distribution, water quality problems and economic status of the country.

Plant nutrients

Sewage sludge is basically an organic waste material, which contains significant amounts of nitrogen (N), phosphorus (P) and other nutrients (Table 13.1) and, therefore, has been recognized as having considerable potential as a fertilizer material and soil conditioner. The typical concentrations (mg/kg) of trace metals in sludges according to USEPA (1985) are shown in Table 13.1.

Potentially toxic substances

Metals

Generally sewage sludges are rich in metals (Table 13.1), which are consequently introduced into the soils when wastes are

Table 13.1. Concentrations of heavy metal(loids) and nutrients in sewage sludge.

	Concentration (mg/kg dry weight)			
	Minimum	Maximum	Medium	USEPA[b]
Heavy metal(loids)[a]				
As	1.1	230	10	4.6
Be				0.31
Cd	1	3,410	10	8.2
Co	11.3	2,490	30	11.6
Cr				230
Cu	84	17,000	800	410
Fe				28,000
Hg	0.6	56	6	1.5
Mn	32	9,870	260	
Mo				9.8
Ni	2	5,300	80	45
Pb	13	26,000	500	248
Se				1.1
Zn	101	49,000	1,700	678
Nutrients[c]				
Total N	21,100	82,100	38,700	
NH_4-N	600	58,000	10,100	
NO_3-N	100	700	400	
Total P	10,600	89,000	29,800	
Na	500	5,300	1,700	
K	900	3,400	1,700	
Ca	9,500	43,300	18,700	
Mg	1,800	6,800	3,700	

[a]Logan and Chaney (1983).
[b]USEPA (1985).
[c]Ross et al. (1991).

disposed on to land. The most common metals found in sludges are cadmium (Cd), chromium (Cr), copper (Cu), nickel (Ni), lead (Pb) and zinc (Zn). However, other metal(loids) including arsenic (As) and mercury (Hg) may also be present in significant amounts (Table 13.1). Many studies have reported that high levels of metals in soils can reduce plant growth, and that the concentrations of some metals in plants can be increased substantially by applications of sludge to the soil. In many countries guidelines for sludge use, based on sludge and soil metal concentrations, have been developed with the aim of preventing metal toxicity to plants and to human and animal consumers of the plants (e.g. Ross *et al.*, 1991; Department of Health (NZ), 1992; McGrath *et al.*, 1994). Metals also pose a serious risk for contamination of the food chain through bioaccumulation in plants in addition to soil contamination and surface water pollution that may arise from runoff. In some countries, application of sludge to forests (McLaren and Smith, 1996) is being used as a means of ensuring that metals do not enter the food chain. The potential for metal build-up in the soil horizon from repeated applications has led to public concern and considerable uproar in many countries. Consequently, innovative techniques that minimize metal content and metal bioavailability are being researched in developing countries. Some of these techniques include, for instance, incineration (Davoudi, 2000), vermicomposting (Masciandaro *et al.*, 2000) and the Safe Sludge Matrix (ADAS, 2000).

Organics

Similar to metals and nutrients, a significant range in the composition and levels of organic contaminants is also expected in wastes given the large variation in the nature of industries and chemicals used in the urban environment. However, although over 100,000 synthetic chemicals are used by industries including chemicals for pest control, only a few such chemicals are found in sewage sludges and even then these substances are generally present only in trace concentrations (USEPA, 1990). High levels of organic contaminants are still occasionally detected in some sewage sludges. Analysis of 100 sewage sludges from the Federal Republic of Germany (Bergs and Lindner, 1997) revealed a range of organic contaminants including polychlorinated dibenzodioxins and furans (PCDD/Fs) (12–347 ng TEq (dioxin toxic equivalents)/kg) and PAHs (300–1600 µg/kg).

Many organic chemicals present in sewage sludges are now also considered to be endocrine disruptors. Mechanistically these compounds simply bind to a hormone receptor and mimic or block normal hormone responses. These chemicals include natural and synthetic oestrogens, pesticides and industrial chemicals such as halogenated aromatic hydrocarbons (HAHs), polycyclic aromatic hydrocarbons (PAHs), phthalates and phenols, antibiotics and certain vitamins. The presence of endocrine disruptors in sewage sludges has raised much concern in countries where adverse impacts of such chemicals on human and wildlife health have been observed. For example, detrimental effects on amphibians, eagles, alligators and seagulls, as well as decreases in male sperm counts and increases in testicular, prostate and breast cancers in humans have been blamed on endocrine disruptive chemicals. The pharmaceutical oestrogen, ethynyloestradiol, and alkylphenol ethoxylates are the potential cause for vitellogenin induction found in fish caged in waste treatment plant effluents in the UK.

Pathogens

The pathogen content of wastes may vary considerably with the nature and origin of the wastes. For example, raw sewage can contain a whole range of pathogenic organisms, including viruses, bacteria, fungi and parasites such as tapeworms and liver flukes. Some of the common pathogens present in wastes include *Salmonella*, *Shizella*, *Mycobacterium*, *Vibrio comma* and *Entamoeba histolytica*. Although sewage treatment processes are designed to eliminate or reduce pathogen numbers they are never 100% effective and substantial numbers of organisms can be present in sewage sludge. There are many reports that indicate a rapid elimination of pathogenic bacteria from sludge-treated soils. However, some pathogenic bacteria may survive for long periods in soils.

The survival or elimination of pathogens is governed by both abiotic (soil texture, moisture, pH, aeration, temperature, OM content) and biotic (microbially mediated changes in soil characteristics such as pH, OM content, production of microbial toxins, enzymatic lysis, predatory protozoa, parasitic bacteria, fungi, phages, inability to compete with native microorganisms in soil) factors. The antagonistic activities of saprophytic microorganisms in the soil are of considerable importance in controlling pathogenic organisms. Although soil has a high buffering capacity in decontaminating waste material, the pathogens that escape the destructive action of soil and survive longer periods, enough to complete their life cycles, could be of public concern. Pathogens can also be recycled to hosts in the process of recycling wastes to land by direct contamination of food crops.

Compared to human and animal pathogens, plant pathogens in sewage sludge have received little attention. *Erwinia amylovora* (fire blight) and *Plasmodiophora brassicae* (clubroot of cabbage) are the two potential plant pathogens that occur in organic wastes and are resistant to decomposing processes. *E. amylovora* is highly pathogenic and can be transmitted by birds or insects and if any orchards are infected, uprooting and burning are the only alternatives for its control. In the case of *P. brassicae*, the spores are highly resistant and can survive for 6–9 years and still be able to infect host plants.

All plant pathogens have the potential to cause large economic losses. However, Bruns *et al.* (1993) have demonstrated that careful management of the composting process can eliminate these pathogens. Thus the quality of many organic wastes can be improved for its acceptability as a soil amendment by optimizing the compost process and determining the appropriate levels for safe application. Although pathogens may impact beneficial microorganisms, information on the interactions between the pathogens and soil microorganisms is surprisingly scant.

Soil-quality Indicator Thresholds

With the increase in land-based applications of sewage sludges and the potential for adverse environmental impact from the presence of PTSs, many countries have introduced legislative measures that dictate threshold concentrations that must not be exceeded with sludge applications to land (Table 13.2). These values recognize a threshold limit for toxic substance loading and nutrients above which excess loading could lead to adverse effects on soil and environmental quality. For most beneficial elements, increasing sludge loading leads to an initial growth response followed by a steady state where there is no change in yield (Fig. 13.1). Loading beyond this level may show a threshold effect before the onset of toxicity. These indicator thresholds differ between countries (see Table 8 in McGrath *et al.*, 1994 for comparison between Europe

Table 13.2. Comparison of Australian biosolid guidelines (maximum permissible metal loadings; mg/kg dry weight) with USEPA-503 (USEPA, 1984) regulations and Dutch guidelines for selected metals.

Metal	USEPA	Dutch[a]	Australian	Landscape forestry
Arsenic	41	60 (100)	20	30
Cadmium	39	10 (40)	3	11
Chromium	3000	500 (1600)	400	500
Copper	1500	200 (1000)	200	750
Lead	300	300 (1200)	200	300
Zinc	2800	1000 (6000)	250	1400

[a]These guidelines are considered to represent moderate soil contamination requiring further investigation; the values in parentheses are considered to represent a serious threat to environmental or human health.

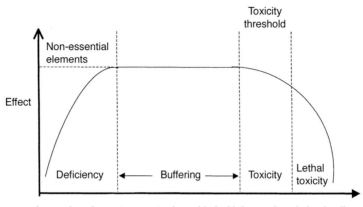

Fig. 13.1. Generalized dose–response curve for soils subjected to sludge loading with vertical broken lines indicating the thresholds for deficiency and toxicity (after McGrath *et al.*, 1994).

and the USA) and depend on the nature of soils including climatic patterns. Unlike the US and European guidelines for sludge loading, Australian biosolid guidelines and the National Water Quality Management Strategy summarizing biosolid management indicates that Australian guidelines are generally based on data obtained elsewhere. Although Australian soils vary significantly in soil type, climatic factors, rainfall and soil temperature, regulatory bodies have placed limited emphasis on establishing data that may be directly relevant to Australian conditions. Indeed, many of the metal contaminants may exist in a number of different ionic forms with varying toxicity to soil and plants. For instance trivalent arsenic [As(III)] is more toxic than pentavalent arsenic [As(V)], and hexavalent chromium [Cr(VI)] is the more toxic chromium species. Although no national environmental guideline is currently available on As(III), some states in Australia have adopted a concentration of 25 mg/kg for Cr(VI). However, legislation is not advanced enough to distinguish between the nature of ionic species that are phytotoxic. Limited studies on metal behaviour in soils by Naidu *et al.* (1996) suggest that metal interactions in soils may vary depending on soil type, thus leading to significant differences in their bioavailability.

Concern about the presence of organic substances in sludge materials arose when biosolids, being given away for use on lawns, gardens and farms (Landrigan *et al.*, 1979) were found to be highly contaminated with PCBs (Chaney and Oliver, 1996). This led to the rapid development of thresholds for trace organics in some countries even though a scientific basis for these values was not well established. However, legislations on threshold values for trace organic contaminants in wastes or soils still do not exist in many countries. In Australia (NEPM, 1999) and also in Germany (Sauerbeck and Leschber, 1992), limits for polyhalogenated compounds have been set. The latest draft of the EU guidelines for disposal of sewage sludge on to agricultural land (EU, 2000) indicates threshold values of 0.8 mg/kg for PCBs, 6 mg/kg for PAHs, 50 mg/kg for nonylphenols and 100 ng TEq/kg for PCDD/Fs. These limits are highly precautionary values for which a toxicological basis does not exist.

In the absence of relevant Australian data, the guidelines are extremely conservative leading to significant difficulties and frustrations by industries and local councils who find management of sludge an expensive and difficult process.

Fertilizer Value of Waste

Sewage sludge has become more attractive as a fertilizer as the price of fertilizers increases, and in many countries a substantial proportion of the sludge produced is applied to land. As mentioned previously in this chapter, 47% of the sludge produced in the UK (1.12 Mt of dry solids) is recycled on to agricultural land (Smith, 1996), and almost all wastes produced in India are recycled (Prasad, 1996).

Long-term field studies of application of diluted raw sewage to agricultural land indicated increased soil fertility due to accumulation of nutrients in the soil and consequently enhanced crop yield of rice (Chakrabarti, 1995). Maize yields have been increased by application of sugarbeet vinasse compost, an agroindustrial waste, in moderate concentration to agricultural land (Madejón et al., 2001).

Many farmers in India have used tannery wastes containing high levels of Cr for crop production. Tannery wastes are initially attractive as a fertilizer substitute because of their high organic content and nutrient load. However, concern over phytotoxicity exists because of the concurrent high concentrations of salt and total Cr. To address these concerns, the impact of tannery waste on crop yield and Cr bioavailability has been studied using a variety of plant types, including vegetables, clover and trees (Naidu et al., 2000; Sara Parwin Banu et al., 2000; Sakthivel et al., 2000). Sara Parwin Banu et al. (2000) observed that although crop yield increased when tannery waste (sludge) was applied to soil up to a total soil Cr concentration of 750 mg/kg, there was a corresponding increase in the uptake of Cr by plants, mainly to the roots. Naidu et al. (2000) also demonstrated severe toxic effects on plant growth when the soil solution concentration exceeded 40 µg Cr/l and that, in addition, sludge amendment also significantly increased the soil pH. Irrigation with effluent sludge also significantly impacted tree growth and, although the three species of trees studied showed no signs of toxicity or other adverse affects from effluent application, all trees exhibited a slight decrease (5–10%) in tree height growth. This reduction in height was attributed to osmotic effects caused by the increased salinity of the soil water (Sakthivel et al., 2000).

Compost and sludge resulted in an overall improvement in soil fertility with no negative impacts when applied to a sandy loam soil of pH 6.8 (Debosz et al., 2002). There is no substantial evidence that persistent organic contaminants pose any detrimental effect to soil quality largely due to their lower water solubility, which makes them less phyto-available than metals to crops. In addition the vast majority of organic compounds are rapidly degraded so that their lifetime in the soil is transient.

Waste Disposal and Soil Quality

The application of sewage sludge to land is of some concern because of the PTSs present within the sludge. As discussed above, PTSs include heavy metals, pesticide residues and other organic compounds, such as PCBs, and a range of pathogenic organisms. Not all sewage sludge will be equally contaminated. For example, those from industrialized cities are likely to have higher metal loadings than those from rural towns. Contamination with organic compounds and pathogens will also vary greatly between sludges. Hence, not all of the issues discussed below are relevant to every situation where sludge is applied to land. Similarly, the exact nature of the sewage treatment process will also affect the nature of the sludge produced and the types of problems likely to be encountered during land disposal. Nevertheless, waste recycling could lead to significant beneficial as well as adverse effects on soil health. Soil quality is frequently used synonymously with soil health (see Karlen et al., Chapter 2, this volume) and is defined as 'the capacity of a soil to promote the growth of (healthy) plants (free from potential hazards to the food chain), protect watersheds by regulating the infiltration and partitioning of precipitation, and prevent water and air pollution by buffering potential pollutants such as agricultural chemicals, organic wastes, and industrial chemicals' (National Research Council, 1993). This suggests that waste reuse can influence soil quality. The following sections present an overview of waste reuse and its potential impact on soil parameters that affect soil quality.

Organic matter and derived properties

The significant levels of OM present in sewage sludge have resulted in a concerted move towards recycling these wastes into agricultural land. For instance, Navas et al. (1999), while investigating the application of biosolids to natural vegetation, found that the increase in plant biomass was directly related to the addition of OM and nutrients present in the waste material. In an earlier study, these investigators reported other benefits for the soil quality, such as a decrease in bulk density and an increase in total porosity (Navas et al., 1998). These results were similar to those reported by many other researchers. Although sewage applications to land have been shown to have a beneficial effect on soil chemical and physical properties (Olness et al., 1998), few studies have reported on the effects of industrial wastes on soil OM content and its subsequent effect on soil physical properties. Zibilske et al. (2000) investigated the effect of multiple applications of paper mill sludge in an agricultural system. They tested five rates of paper sludge applications ranging from 0 to 225 Mg/ha in multiple applications under three management protocols: applied once, applied in alternate years or applied annually. They found a strong relationship between added sludge C and several soil physical properties. Whereas they found either an increase or maintenance of OM with annual or biennial sludge applications, little residual effect of a single application was found after 5 years. Significant increases were observed in soil aggregation and soil moisture-holding properties at higher rates of sludge application and when cumulative C additions reached 225 Mg/ha at lower sludge application rates. Previous studies using pulp sludge had documented that adding pulp and paper-mill sludge to soil can increase soil OM content, moisture-holding capacity and cation exchange capacity (Einspahr et al., 1984) and N and P availability (Brickway, 1983), exchangeable Ca and Mg (Logan and Esmaeilzadeh, 1985) and soil pH (Feagley et al., 1994). Based on these studies, researchers concluded that long-term paper-mill sludge application could be managed to effect positive changes in soil physical properties that were correlated with soil quality. Similar agronomic benefits have been observed after repeated application of sugar-beet vinasse compost (Madejón et al., 2001).

As with the pulp and paper-mill industries, power generation also produces a massive volume of wastes that are rich in OM. Power wastes are produced during the combustion of coal, predominantly for the generation of electricity. During the combustion of coal, fly ash is that portion of the ash stream that is 0.001–0.1 mm in size. Currently, Australia produces in excess of 7.7 Mt of fly ash annually (Beretka and Nelson, 1994); the corresponding figure for the USA is over 60 Mt (Carlson and Adriano, 1993). The physical, mineralogical and chemical properties of fly ash depend on the composition of the parent coal, conditions during coal combustion and the efficiency of emission control devices. In addition, the chemical properties in particular can be affected by storage and handling of the material (Adriano et al., 1980). The potential impact of fly ash on soil quality has been studied by numerous investigators who report improved soil structure (for both fine and coarse-textured soil), improved moisture-holding capacity (Ghodrati et al., 1994), increased soil pH (Warren, 1992) and increased concentrations of most macro- and micronutrients (Adriano et al., 1980; Sims et al., 1995).

Impact on soil biota

There is now considerable evidence that, in the long term, elevated metal and organic concentrations can reduce soil microbial biomass levels, inhibit N_2 fixation by both free-living and symbiotic organisms and reduce certain enzyme activities such as urease and phosphatase (e.g. Tyler, 1981; Baath, 1989; McGrath, 1994). There is increasing evidence that metal pollutants in soil impair the functioning of the soil biota at concentrations well below those where phytotoxicity is first evident (Brookes et al., 1986; McGrath, 1987; Witter, 1993). This is not surprising given that in all countries where sludge is used as an alternative source of nutrients, threshold limits for PTSs were set with the aim of protecting human and animal health and crop yield (McGrath *et al.*, 1994). Some researchers have argued that the limits for soil loadings should be reduced to protect microbiota and important soil microbial processes that protect long-term soil fertility (Chaudri *et al.*, 1993; McGrath, 1993, 1994; Giller *et al.*, 1998). However, many of these studies have been limited to information on processes and bacterial numbers and are not appropriate for analyses of changes in community structures and bacterial diversity. Hence, it is not clear whether PTSs present in the wastes will impact on microbes beneficial for maintaining soil processes. Sandaa *et al.* (2001) studied the influence of long-term heavy metal loading via sludge application on microbial community structure in soils by applying new molecular techniques. Based on the observed change in community structure of sludge-treated soils, they concluded that certain components of the microbial community were no longer present in the contaminated soils and that this could eventually lead to loss of functions associated with the bacterial species studied. Moreover, they reported no dramatic effect of heavy metals when investigating the culturable fraction of the bacterial community, but a severe effect was observed when investigating the structure of the total bacterial community. In contrast, Petersen *et al.* (2003) reported no adverse effect on microbial communities but rather the stimulation of microbial activity, measured in terms of fluorescein diacetate (FDA) hydrolysis and CO_2 evolution rate, when sewage sludge was amended to soil. Sludge application appears not to affect soil fauna. Microarthropod populations were even stimulated following sludge application (Petersen *et al.*, 2003).

Accumulation and degradation of potentially toxic substances

Some countries, including the USA, have long adopted recycling of certain waste

Table 13.3. Maximum permissible content of heavy metals (mg/kg dry solids) in sewage sludge for land application (Paulsrud and Nedland, 1997).

Metal	For use in agriculture and in private gardens and lawns	For use on green areas
Cd	2	5
Pb	80	200
Hg	3	5
Ni	50	80
Zn	800	1500
Cu	650	1000
Cr	100	150

categories as a sustainable management strategy. They recognize that biosolids are beneficial for maintaining and improving soil structure and water-holding capacity (Giroux and Tabi, 1990) and nutrient status of soils (McGrath *et al.*, 1995). For these reasons sludge recycling/reuse in agriculture and land reclamation is being encouraged. However, potential risks from accumulation of heavy metals and organic compounds as well as pathogen contamination must be borne in mind. Hence, many countries have developed criteria for land application of wastes (McGrath *et al.*, 1994). For example, Norwegian regulations for sewage sludge treatment and disposal (Paulsrud and Nedland, 1997) actively promote sludge management practices that provide for the beneficial use of sludge while maintaining or improving environmental quality and protecting human health and aiming to recycle 75% of total sludge production. This means that the sludge must contain only very small amounts of toxic substances when used for agriculture and in green areas. Based on these data, Norway has established the maximum heavy metal concentrations for land application of sewage sludge (Table 13.3). The maximum content of heavy metals in soil before applying any sewage sludge is similar to the lower limits presented in EU Council Directive 86/278 on agricultural use of sludge.

Like Norway, other countries have also established criteria for recycling wastes and there is now a move towards developing strategies for managing contaminants. In North America the main metals of concern are

Zn, Cu and Ni because they commonly occur in sewage sludge and effluents and can be toxic to plants. In Australia, the main metal of concern in biosolids is Cd. This concern has arisen from a marketing viewpoint, because the maximum permissible concentration for Cd in a range of foodstuffs including cereals and agricultural products is 0.1 mg/kg fresh weight (FW), as set by the Australian National Food Authority (NFA, 1994). Other metals of concern in sludge include Pb, Hg, Se and Mo, which, in excess, threaten animals, livestock and wildlife. By contrast with the waterborne sewage systems used in most Western countries, human waste in Asian countries has been collected for thousands of years and spread on crops after little if any treatment (see Chapters 14–25 in Naidu *et al.*, 1996). Therefore, in Asian countries, the problems of contamination with industrial wastes are not as great as in most Western cities where industrial wastes are commingled with domestic wastes and street runoff (Taiganides, 1976). However, in Japan, although human waste was used on land until the end of the Second World War, this practice ceased when chemical fertilizers became abundant at low cost and the high cost of labour prevented the collection of human waste (Takahashi, 1976). Newly constructed sewage treatment works in Japan now produce large amounts of biosolids, much of which are applied to land as fertilizer. As in most Western cities the most serious concern in Japan regarding use of sewage sludge on farmland is the need to enforce limits on heavy metals.

Several studies have found that in the long term, organic contaminants present in sludge do not substantially accumulate in plants. Many field studies also indicate that there is no significant accumulation of nonylphenols, di(2-ethylhexyl)phthalate or alkylbenzene sulphonates (Vikelsoe *et al.*, 1999; Petersen *et al.*, 2003).

A major limitation of the current guidelines is the lack of high quality data, particularly from long-term studies, on which the metal and organic contaminant loadings can be based. Even if data from long-term studies were available, its application to soils varying widely in pH, mineral and OM composition and rainfall would limit its applications, as all of these factors influence the binding of PTSs. It is thus critical that all developed countries (e.g. USA, Europe, Canada and Australia) invest funds in long-term studies with sludge using widely ranging soil types under varying environmental conditions. Such studies can assist countries to establish relevant guideline levels and help predict the long-term impact of PTSs on environmental quality thus enabling more informed and efficient management strategies.

Pathways of Potentially Toxic Substance Transfer

Public concern about contaminants in soils is based on their potential for adverse effects on agriculture, humans, or the environment. The pathways by which PTSs cause harm are principally related to the following exposure scenarios:

1. Sludge–soil–plant–human toxicity
2. Sludge–soil–human toxicity
3. Sludge–soil–animal toxicity
4. Sludge–soil–plant toxicity
5. Sludge–soil–soil biota toxicity
6. Sludge–soil–soil biota–predator toxicity
7. Sludge–soil–airborne particulate–human toxicity
8. Sludge–soil–surface runoff–surface water–human toxicity
9. Sludge–soil–vadose zone–groundwater–human toxicity
10. Sludge–soil–atmosphere vaporization–human toxicity
11. Sludge–soil–plant–animal–human toxicity

For each contaminant, one pathway dominates in causing the highest probability of adverse effects to some environmental receptor. In practice, some of these potential pathways would seem unlikely to cause significant human exposure to certain organic chemicals. O'Connor *et al.* (1991) carried out a detailed review of the bioavailability to plants of sludge-borne toxic organics and concluded that the vast majority of these chemicals in sludges: (i) occurred in sludge-amended soils at low initial concentrations; (ii) were so strongly sorbed in the sludge–soil matrix as to have extremely low bioavailabilities to plants; and (iii) were degraded, or otherwise lost, from the soil during the cropping season, and were accumulated, if at all, at very low concentrations in the edible portion of food crops. Direct contamination of plants is also unlikely since sludge application to standing food crops is not permitted. The implications of such effects on the long-term fertility and sustainability of the soil remain unclear, and several important questions still require answers. In particular we need to understand any changes in metal and organic contaminant bioavailability, which may take place over long periods of time following waste application to the soil. Numerous investigators have hypothesized (Berrow and Burridge, 1980; McBride, 1995) that when the OM added in wastes breaks down, the metals held by it will be released and become more mobile and consequently ecotoxic.

However, it has also been suggested that, with time, metals added to the soil in sludge will react with the soil and 'revert' to less mobile and bioavailable forms (Lewin and Beckett, 1980). Whatever the answers are, the most conservative approach would be to minimize metal inputs to the soil wherever possible.

The most likely pathway for human exposure arises from the ingestion of sludge-contaminated soil by grazing animals. Cattle diets can contain on average 6% soil (Healy, 1968), and sheep can also consume considerable amounts of soil. Organochlorine pesticides in particular are known to accumulate in cattle primarily as a result of soil ingestion (Lindsay, 1982). These chemicals are deposited in the body fat and can appear in meat products and milk. However, where such accumulation has occurred, for example with DDT, it has been related to the past use of pesticides on pastures rather than as a result of sludge application.

The pathogens within sewage sludge can cause diseases if they come in contact with humans. According to Ross et al. (1991), this contact can occur in several ways: by inhaling sludge aerosols or dust while spreading, by eating vegetables or fruits contaminated by sludge, by drinking water contaminated by runoff or percolation from sludge applied to land or by eating meat from livestock infected while grazing on pastures or crops fertilized with sludge. Of the large range of potential hazards, it has been concluded (Ross et al., 1992) that salmonellosis and bovine cysticercosis are the two infectious diseases most likely to affect animals and humans as a result of sludge contamination. Outbreaks of both these diseases in animals in European countries have been associated with surface application of sewage sludge to pastures (Jones, 1986; Nansen and Henriksen, 1986). However, the risks can be minimized by incorporation of sludges into the soil and by using suitable withholding periods before allowing animals to graze sludge-treated land.

Surface runoff and leaching

Although sewage sludge can be a valuable source of plant nutrients (trace elements, N and P) care must be taken to ensure that these nutrients are not allowed to pollute the environment by contaminating surface or groundwater supplies (Smith, 1996; Siddique et al., 2000). Numerous investigators have reported leaching of these nutrients from soils subjected to sludge applications (Gerritse et al., 1982; Melanen et al., 1985; Sloan et al., 1998; Cooke et al., 2001). In a laboratory study, Gove et al. (2002) compared the effect of surface application and sub-surface incorporation of enhanced treated biosolids on the leaching of heavy metals (Zn, Cu, Pb, Ni) and nutrients (N and P) through sand (typic quartzipsamments, %OM = 3.0, pH = 6.5), sandy loam (typic hapludalf, %OM = 4.8, pH = 7.6) and silversand 'repacked semi-structured cores' (0.2 m by 0.1 m diameter). Biosolids were applied at a rate equivalent to 250 kg N/ha/year and followed five 8 h simulated rainfall events (4.9 mm/h). They found that subsurface incorporation increased the risk of P and metal leaching compared with surface applications. In a similar, but long-term field study in India, researchers investigated the migration of toxic heavy metals in soils amended with sewage sludge at Nagpur (Olaniya et al., 1991). Using lysimeters, they showed that, whereas the composition of the effluent (heavy metals and organic content) varied with sludge composition, moisture content, temperature, etc., the interaction between soil and leachate depended largely on the individual characteristics of the soils. Olaniya et al. (1991) found that soils leached with leachates from sludge over a period of 1 year showed considerable movement of both nitrate and heavy metals (Cu, Cr, Ni and Zn) to the subsurface horizon. The high mobility of Zn in these soils was also reflected in the high Zn concentration (20–30 mg/l) of the leachate from the sludge and the low Zn adsorption capacity of the surface soil. Based on the dissolved organics in the leachate samples, Olaniya et al. (1991) concluded that fulvic acid was the predominant carrier of toxic heavy metals, presumably as metal–organic complexes.

One of the major problems with the application of metal-containing sludges and wastes to soils, particularly with repeated applications, is that the metals accumulate in the surface layers of the soil, as there appears to be little movement of heavy metals below the zone of sludge incorporation (Emmerich et al., 1982). It is, however, evident from many previous studies on metal interactions in soils, that solute breakthrough only occurs following saturation of metal binding sites. Thus, repeated application of sludge may lead to the saturation of sorption sites causing metals to move into the subsurface environment. Indeed, Dowdy and Volk (1983), in an extensive review of heavy metal movement, concluded that movement was most likely to occur where heavy applications of sewage sludge are made to sandy, acid, low-OM soils receiving high rainfall or irrigation. Such soils often have low metal sorption capacity. Lamy et al. (1993) reported direct leaching of Cd from a sludge-treated agricultural soil, and Cameron et al. (1994) demonstrated increased leaching of Zn from a sludge-treated forest soil. Although in both studies the concentrations of metals in drainage leachates from the sludge-treated soils remained below drinking water limits, their actual mode of transfer is of concern. Lamy et al. (1993) found increased Cd (average of 3 and 1.5 µg/l, in treated and control plots, respectively) in

leachates containing the highest fractions of soluble OM.

In addition to vertical leaching of metals, the potential for surface transport of metals from sludge-treated soils has also been assessed by a number of investigators who found no significant difference in water quality when sludge was used in agricultural practices under controlled and sludge-amended conditions (McLaren and Smith, 1996). However, significant contamination of surface waters with heavy metals was recorded (Duncomb et al., 1982) when snow melted or heavy rain followed surface application of sludge to grasslands in Minnesota, USA.

To minimize the adverse impact of sludge on soil quality, the aim should be to match nutrient application rates to plant requirements. In the case of N, this may not be easy since a high proportion of the N in sludge will be present in organic forms and will only become available for plant uptake after mineralization to ammonium-N (NH_4^+) or nitrate-N (NO_3^-). Rates of N mineralization in the field are difficult to predict and if NO_3^- in particular is released from the sludge quicker than it is taken up by plants, leaching of excess NO_3^- from the soil may result. However, nitrate is generally found in insignificant amounts and will only accumulate if the sludge has a large potential for ammonification, which is uncommon. Once leached from the soil profile, the NO_3^--N has the potential to contaminate groundwater and ultimately drinking water supplies. In addition the leached NO_3^--N may eventually find its way into open water bodies such as streams and lakes, and cause eutrophication.

Compared to N, P is relatively immobile in soils, and applications of sewage sludge are unlikely to result in leaching of P from the soil profile (Lindo et al., 1993) unless it is subsurface incorporated (Gove et al., 2002). However, with both P and N there is a risk that if sewage sludge is surface applied in unsuitable locations, heavy rain may cause runoff leading to the contamination of nearby surface water. In such cases, transport of the pollutants may not necessarily take place in solution but as suspended solids. Clearly, sludge application management practices must minimize such possibilities, usually by ensuring good buffer zones around bodies of surface water, and by not applying sludge during high-rainfall periods (Ross *et al.*, 1991). Incorporation of sludge into the soil will also minimize pollution of surface waters, although subsurface incorporation may increase the risk of nitrate leaching. O'Brien and Mitsch (1980) reported that surface application of sewage sludge resulted in 50% lower N leaching rate than subsurface incorporation. Therefore, care should be taken to ensure that sludge incorporation is controlled so that the roots of the crop plant can intercept nutrients, thus reducing the potential for nutrient leaching.

Bioamplification in food chains

Although sludge recycling is an attractive environmental alternative for both soil conservation (Navas *et al.*, 1999) and residue disposal (see references in Davoudi, 2000), potential risks from accumulation of PTSs as well as pathogen contamination and subsequent transfer to food (Sandaa *et al.*, 2001) must be borne in mind. In order to minimize such transfer, many countries have introduced new technologies that, along with sludge-loading legislation are expected to minimize the potential transfer of PTSs into food crops. For example, the UK has introduced the Safe Sludge Matrix (ADAS, 2000) in response to consumer concerns about the use of sludge as fertilizers in agriculture. The Safe Food Matrix reconciles the need to beneficially recycle sludge to land with concerns about the safety of this practice, particularly in relation to risks of pathogen transfer to food. In addition to posing a risk to crop quality, PTSs also pose a risk to surfacewater and groundwater quality.

The risk of transfer of organic contaminants to plants via roots is probably extremely low. For example, when sewage sludge was applied to agricultural soils (50–500 t of dry matter/ha) an elevated level of PCBs was observed in the soils but there was no significant transfer of PCBs to plants (Bergs and Lindner, 1997). However, further work needs to be conducted to ascertain the degree of foliar uptake following volatilization.

Volatilization and airborne transfers

Although the majority of wastes are generally non-volatile, some wastes contain volatile metalloids (mercury and selenium) or organic substances (e.g. ammonia, methane, DDT) that may be inadvertently transferred to the atmosphere. Airborne transfer of applied sludge dusts may be a problem if incorporation into soil is poor and weather conditions are unfavourable, e.g. dust storms are common in Australian rural areas. This mode of transfer needs further investigation to establish if it is a significant pathway.

Managing Urban and Industrial Wastes

During the last 30 years, we have witnessed a changing perspective in waste management brought about primarily by the environmental movement in the 1970s. Better public appreciation of the potentially adverse impact of wastes on environmental and human health resulted in the adoption of a new perspective, that wastes, both industrial and urban, should be recovered or disposed of without jeopardizing human health and without using processes or methods that could harm the environment (Tchobanoglous *et al.*, 1993). With the recent ban on ocean-based disposal of wastes, there has been a significant move towards disposal into prescribed landfills. Landfill is still the preferred option in many countries.

However, this mode of management is becoming increasingly expensive and the introduction of new legislation banning certain forms of wastes from being disposed into these cavities, as well as the rising cost of landfill, have led to the slow acceptance of other waste management strategies that encourage a move away from traditional landfill and incineration options to more modern waste prevention/minimization technologies.

In an excellent review on 'Planning for Waste Management', Davoudi (2000) commented that 'while the rhetoric for waste management conceals these conflicts, the tensions are increasingly manifested in the polarisation of debate around recycling vs incineration'.

The author comments that at one end of the spectrum are the environmental groups who are in favour of a moratorium on new incinerators in order to promote establishment of recycling activities, while at the other end are the influential packaging lobby and the fast-growing private sector waste management industry who are promoting the profitable, capital intensive option of incineration with energy recovery (Davoudi, 2000). Davoudi (2000) argues that while under the EU-led regulatory pressures, the intention to move away from landfill was universal; the selection of alternatives was also diverse, reflecting the influences derived from locally specific factors. For instance, in Germany, thermal use of sludge is being promoted based on the total organic carbon (TOC) and loss on ignition of wastes (Dorschel, 1997). Since sewage sludge cannot fulfil the German criteria for dumping (given in Table 13.4) due to its large organic carbon content (45–55%), its dumping has been largely terminated. Thus, a large proportion of Germany's wastes are being thermally treated with 13 incinerating plants already treating wastes throughout the country.

Unlike the UK and Germany, where legislation is well advanced, Europe has many countries that are still implementing legislations and new strategies for waste management. For example, Poland has seen a steady increase in the amount of industrial and municipal wastes since the early 1970s (Grodzinska-Jurczak, 2000): in excess of 4 billion t of wastes are land filled, and this increases by 145 Mt annually.

In order to improve waste management, Poland will: (i) aim to bring waste legislation into compliance with EU standards; (ii) undertake projects involving alternative means of

waste disposal (including recycling technologies that permit processing part of each material introduced to the market); (iii) increase funding for waste management projects; (iv) encourage local authorities to implement principles of sustainable waste management; and (v) promote principles of rational waste management in the society.

As with Poland, other countries have also introduced new measures for waste management with many complying with the EU directive (see, for example, Paulsrud and Nedland, 1997; Melanen *et al.*, 2002). These stringent guidelines have led to a shift in the waste management from landfill to better quality waste, due to a dramatic reduction in the inputs of metals to sewers. This reduction results from: (i) improved trade effluent control imposed by the water undertakings; (ii) changes in the nature of traditional manufacturing industries; and (iii) adoption of cleaner manufacturing technologies. In the UK, quoting Rowlands (1992), Smith (1996) reported 80 and 98% reduction in Cd and Zn concentration, respectively, in sewage sludge from the Nottingham Sewage Treatment Works (STW) between 1962 and 1992, and a similar decrease in Cr was recorded at Coventry STW. Although further decreases in metal contents may be expected, complete elimination of metals from sewage wastes will not be achieved given the relative increase in metal contributions from diffuse sources.

Conclusions and Future Outlook

The high nutrient and OM contents of many sludges make them a cheaper alternative than commercial fertilizers, and result in enhanced soil structure, water-holding capacity and crop productivity when managed correctly. However, if managed incorrectly, sludge application in excess of plant requirements may result in increased nutrient leaching to groundwater and soil subsurfaces. Likewise, the presence of PTSs (such as heavy metals/metalloids) is of minor concern if application rates are sensible (below the binding capacity of the soil). The potential risk of bioaccumulation is low for many persistent organic

Table 13.4. Criteria for dumping wastes (Dorschel, 1997).

	Dump Class I (mass %)[a]	Dump Class II (mass %)
Defined as loss on ignition	≤ 3	≥ 5
Defined as TOC	≤ 1	≥ 3

[a]The organic share of the dry residue of the original substance.

chemicals. Pathogens present in sewage sludge are a potential public health concern, but use of appropriate management protocols should alleviate these risks.

Although it is impossible to cover all potential urban and industrial wastes, we have attempted to highlight some of the issues related to sludge recycling on agricultural soils. Whereas other types of wastes, for example, wastes from the paper pulp and food processing industries are of importance in some areas, many of the problems discussed in this chapter are relevant to all wastes. It is clear that in all cases good management strategies are essential to ensure protective and sustainable reuse, and each chemical needs to be assessed on a case-by-case basis. Many wastes that have the potential to seriously damage the environment and have adverse effects on human health will need continued monitoring after sludge amendment and toxicity assessment should be a vital tool in any ongoing monitoring of waste reuse.

In order to maximize the benefits of waste reuse for soil application as a more acceptable disposal method, and to eliminate the possible risks to public and environmental health, more research should be directed towards: (i) understanding the movement and survival of pathogenic microbes in soil, particularly field studies under different management conditions; (ii) the impact of waste on microbial biodiversity in soil; (iii) the possible role of pathogens in transmitting antibiotic resistance (via genetic exchange, plasmid transfer) to native soil microorganisms; (iv) the presence and effects of oestrogenic substances on soil biota; and (v) developing relevant indicators and methods for assessing soil biological quality.

References

ADAS (2000) *The Safe Sludge Matrix. Guidelines for the Application of Sewage Sludge to Agricultural Land*. Information sheet available from ADAS Gleadthorpe Research Centre, Mansfield.

Adriano, D.C., Page, A.L., Elseewi, A.A., Chang, A.C. and Straughan, I. (1980) Utilisation and disposal of fly ash and other coal residues in terrestrial ecosystems: a review. *Journal of Environmental Quality* 9, 333–334.

Baath, E. (1989) Effects of heavy metals in soils on microbial processes and populations (a review). *Water Air and Soil Pollution* 47, 335–379.

Bastian, R. (1997) The biosolids (sludge), treatment, beneficial use, and disposal situation in the USA. *European Water Pollution Control Journal* 7, 62–79.

Beretka, J. and Nelson, P. (1994) The current state of utilisation of fly ash in Australia. In: *Proceedings of the 2nd International Symposium Ash – a Valuable Resource*. South African Coal Ash Association, Johannesburg, Vol. 1, pp. 51–63.

Bergs, C.-G. and Lindner, K.-H. (1997) Sewage sludge use in the Federal Republic of Germany. *European Water Pollution Control Journal* 2, 47–52.

Berrow, M.L. and Burridge, J.C. (1980) Trace element levels in soils: effects of sewage sludge in inorganic pollution and agriculture. *MAFF Reference Book* 326, HMSO, London, pp. 159–183.

Brickway, D.G. (1983) Forest floor, soil and vegetation response to sludge fertilization in red and white pine plantations. *Soil Science Society of America Journal* 47, 776–784.

Brookes, P.C., McGrath, S.P. and Heijnen, C. (1986) Metal residues in soils previously treated with sewage-sludge and their effects on growth and nitrogen fixation by blue-green algae. *Soil Biology and Biochemistry* 18, 345–353

Bruns, C., Gottschall, R., Zeller, W., Schuler, C. and Vogtmann, H. (1993) Survival rates of plant pathogens during composting of biogenic wastes in commercial composting plants under different decomposing conditions. In: Paoletti, M.G., Foissner, W. and Coleman, D. (eds) *Soil Biota, Nutrient Cycling and Farming Systems*. Lewis, Boca Raton, Florida, pp. 41–52.

Cameron, K.C., McLaren, R.G. and Adams, J.A. (1994) Application of municipal sewage sludge to low fertility forest soils: the fate of nitrogen and heavy metals. *Transactions of 15th World Congress of Soil Science*, Vol. 3a, 467–482.

Carlson, C.L and Adriano, D.C. (1993) Environmental impacts of coal combustion residues. *Journal of Environmental Quality* 22, 227–247.

Chakrabarti, C. (1995) Residual effects of long-term land application of domestic wastewater. *Environmental International* 21, 333–339.

Chaney, R.L. and Oliver, D.P. (1996) Sources, potential adverse effects and remediation of agricultural soil contaminants. In: Naidu, R., Kookana, R.S., Oliver, D.P., Rogers, S. and McLaughlin, M.J. (eds) *Contaminants and the Soil Environment in the Australasia–Pacific*

Region. Kluwer Academic Publishers, Dordrecht, pp. 323–359.

Chang, A.C., Warneke, J.E., Page, A.L. and Lund, L.J. (1984) Accumulation of heavy metals in sludge treated soils. *Journal of Environmental Quality* 13, 87–91.

Chaudri, A.M., McGrath, S.P., Giller, K.E., Rietz, E., Sauerbeck, D.R. (1993) Enumeration of indigeneous *Rhizobium leguminiosarum* biovar *trifolii* in soils previously treated with metal-contaminated sewage sludge. *Soil Biology and Biochemistry* 25, 301–309.

Cooke, C.M., Gove, L., Nicholson, F.A., Cook, H.F. and Beck, A.J. (2001) Effects of drying and com-posting biosolids on the movement of nitrate and phosphate through repacked soil columns under steady-state hydrological conditions. *Chemosphere* 44, 799–806.

Davoudi, S. (2000) Planning for waste management: changing discourses and institutional relation-ships. *Progress in Planning* 53, 165–216.

Debosz, K., Petersen, S.O., Kure, L.K. and Ambus, P. (2002) Evaluating effects of sewage sludge and household compost on soil physical, chemical and microbiological properties. *Applied Soil Ecology* 19, 237–249.

Department of Health (NZ) (1992) *Public Health Guidelines for the Safe Use of Sewage Effluent and Sewage Sludge on Land.* Department of Health, Wellington, New Zealand, 66 pp.

Dorschel, W. (1997) Future prospects for sewage sludge disposal under consideration of new framework legislation. *European Water Pollution Control Journal* 7, 68–73.

Dowdy, R.H. and Volk, V.V. (1983) Movement of heavy metals in soils. In: Nelson, D.W. (ed.) *Chemical Mobility and Reactivity in Soil Systems. Soil Science Society of America Special Publication No 11*, Madison, Wisconsin, pp. 229–240.

Duncomb, D.R., Larson, W.E., Clapp, C.E., Dowdey, R.H., Linden, D.R. and Johnson, W.K. (1982) Effect of liquid wastewater sludge application on crop yield and water quality. *Journal of Water Pollution Control Federation* 54, 1185–1193.

Einspahr, D., Fiscus, M.H. and Gargan, K. (1984) Paper mill sludge as a soil amendment. In: *TAPPI Proceedings, 1984 Environmental Conference.* TAPPI Press, Atlanta, Georgia, pp. 253–257.

Emmerich, W.E., Lund, L.J., Chang, A.C. and Chang, A.L. (1982) Movement of heavy metals in sewage sludge treated soils. *Journal of Environmental Quality* 11, 174–178.

EU (2000) Working draft on sludge, 3rd draft. ENV.E3/LM, Brussels, 27 April 2000.

Feagley, S.E., Vladez, M.S. and Hudnal, W.H. (1994) Bleached primary paper mill sludge effect on bermudagrass grown on a mine soil. *Soil Science* 157, 389–397.

Gerritse, R.G., Vriesema, R., Dalenberg, J.W. and De Roos, H.P. (1982) Effect of sewage sludge on trace element mobility in soils. *Journal of Environmental Quality* 11, 359–364.

Ghodrati, M., Sims, J.T. and Vasilas, B.L. (1994) Evaluation of fly ash as a soil amendment for the Atlantic Coastal Plain, I. Soil hydraulic properties and elemental leaching. *Water, Air, and Soil Pollution* 81, 349–361.

Giller, K.E., Witter, E. and McGrath, S.P. (1998) Toxicity of heavy metals to microorganisms and microbial processes in agricultural soils: a review. *Soil Biology and Biochemistry* 30, 1389–1414.

Giroux, M. and Tabi, M. (1990) Considerations relatives a l'evaluation des sols comme milieu receptuer de boues de stations d'epuration. *Le premier colloque Quebecois sur la valorisation des boues de stations d'epuration municipals 18/19 Sept. Hull, Quebec*, pp. 109–119.

Gove, L., Nicholson, F.A., Cock, H.F. and Beck, A.J. (2002) Comparison of the effect of surface application and subsurface incorporation of enhanced treated biosolids on the leaching of heavy metals and nutrients through sand and sandy loams. *Environmental Technologies* 23, 189–198.

Grodzinska-Jurczak, M. (2000) Ecological education in the Polish educational system. *Environmental Science and Pollution Research* 7, 235–238.

Healy, W.B. (1968) Ingestion of soil by dairy cows. *New Zealand Journal of Agricultural Research* 13, 664–672.

Henry, H. and Heinke, G.W. (1996) *Environmental Science and Engineering*, 2nd edn. Prentice-Hall, Englewood Cliffs, New Jersey.

Jones, P.W. (1986) Sewage sludge as a vector of Salmanellosis. In: Block, J.C., Havelaar, A.H., L'Hermite, P. (eds) *Epidemiological Studies of Risks Associated with the Agricultural Use of Sewage Sludge: Knowledge and Needs.* Elsevier Applied Science, London, pp. 21–33.

Kinsman, J. (2000) Principles for pragmatic environ-mental policy. *Environmental Science and Policy* 3, 55–56.

Lamy, I., Bourgeois, S. and Bermond, A. (1993) Soil cadmium mobility as a consequence of sewage sludge disposal. *Journal of Environmental Quality* 22, 731–737.

Landrigan, P.J., Wilcox, K.R. Jr and Silva, J.R. Jr (1979) Cohort study of Michigan residents exposed to polybrominated biphenyls: epi-demiologic and immunologic findings. *Annals of the New York Academy of Sciences* 320, 284–294.

Lewin, V.H. and Beckett, P.H.T. (1980) Monitoring heavy metal accumulation in agricultural soils treated with sewage sludge. *Effluent water Treatment Journal* 71, 205–208.

Lindo, P.V., Taylor, R.W. and Shuford, J.W. (1993) Accumulation and movement of residual phosphorus in sludge treated decatur silty clay loam soil. *Communications in Soil Science and Plant Analysis* 24, 1805–1816.

Lindsay, D.G. (1982) Effects arising from the presence of persistent organic compounds in sludge. In: Davis, R.D., Hucker, G. and Hermite, L. (eds) *Environmental Effects of Organics and Inorganic Contaminants in Sewage Sludge*. Reidel, Dordrecht, The Netherlands, pp. 19–26.

Logan, T.J. and Chaney, R.L. (1983) Utilisation of waste water and biosolids on land – metals. In: Page, A.L. (ed.) *Proceedings of the 1983 Workshop on Utilisation of Municipal Wastewater and Biosolids on Land*. University of California, Riverside, California, pp. 223–235.

Logan, T.J. and Esmaeilzadeh, H. (1985) Utilizing paper mill sludge: use on cropland. *BioCycle* 26, 52–53.

Madejón, E., López, R., Murillo, J.M. and Cabrere, F. (2001) Agricultural use of three (sugar-beet) vinasse compost: effect on crops and chemical properties of a Cambisol soil in the Guadalquivir river valley (SW Spain). *Agriculture, Ecosystems and Environment* 84, 55–65.

Masciandaro, G., Ceccanti, B. and Garcia, C. (2000) 'In-situ' vermicomposting of biological sludges and impacts on soil quality. *Soil Biology and Biochemistry* 32, 1015–1024.

McBride, M.B. (1995) Toxic metal accumulation from agricultural use of sludge: are USEPA regulations protective? *Journal of Environmental Quality* 24, 5–18.

McGrath, S.P. (1987) Long-term studies of metal transfers following application of sewage sludge. In: Coughtrey, P.J., Martin, M.H. and Unsworth, M.H. (eds) *Pollutant Transport and Fate in Ecosystems*. Blackwell Scientific Publishers, Oxford, pp. 301–317.

McGrath, S.P. (1993) Soil quality in relation to agricultural uses. In: Eijsackers, H.J.P. and Halmers, T. (eds) *Integrated Soil and Sediment Research: a Basis for Proper Protection*. Kluwer Academic Publishers. Dordrecht, The Netherlands, pp. 187–200.

McGrath, S.P. (1994) Effects of heavy metals from sewage sludge on soil microbes in agricultural ecosystems. In: Ross, S.M. (ed.) *Toxic Metals in Soil–Plant Systems*. John Wiley & Sons, Chichester, UK, pp. 247–274.

McGrath, S.P., Chang, A.C., Page, A.L. and Witter, E. (1994) Land application of sewage sludge: scientific perspectives of heavy metal loading limits in Europe and the United States. *National Research Council Environmental Reviews*, Ottawa, pp. 108–118.

McGrath, S.P., Chaudri, A.R. and Giller, K.E. (1995) Long-term effects of metals in sewage sludge on soils, microorganisms and plants. *Journal of Industrial Microbiology* 14, 94–104.

McLaren, R.G. and Smith, C.J. (1996) Issues in the disposal of industrial and urban wastes. In: Naidu, R., Kookana, R.S., Oliver, D.P., Rogers, S. and McLaughlin, M.J. (eds) *Contaminants and the Soil Environment in the Australasia-Pacific Region*. Kluwer Academic Publishers, Dordrecht, The Netherlands, pp. 183–212.

Melanen, M., Jaakola, A., Melkas, M., Ahtianen, M. and Matinvesi, J. (1985) *Leaching Resulting from Land Application of Sewage Sludge and Slurry*. Publications of the Water Research Institute, 61. Veshiallitus- National Board of Waters, Finland.

Melanen, M., Kautto, P., Saarikoski, H., Ilomÿki, M. and Yli-Kauppila, H. (2002) Finnish waste policy – effects and effectiveness. *Resources, Conservation and Recycling* 35, 1–15.

Naidu, R., Kookana, R.S., Oliver, D.P., Rogers, S. and McLaughlin, M.J. (eds) (1996) *Contaminants Under the Soil Environment in the Australasian-Pacific Region*. Kluwer Acadamic Publishers, Dordrecht, The Netherlands.

Naidu, R., Smith, L., Mowat, D. and Kookana, R.S. (2000) Soil–plant transfer of chromium from tannery waste sludge: results from a glasshouse study. In: Naidu, R., Willet, I., Mahimairajah, S., Kookana, R. and Ramasamy, K. (eds) *ACIAR Proceedings No. 88: Towards Better Management of Soils Contaminated with Tannery Waste*, pp. 132–143.

Nansen, P. and Henriksen, S.A.A. (1986) The epidemiology of bovine cysticercosis (*C. bovis*) in relation to sewage and sludge application on farmland. In: Block, J.C., Havelaar, A.H. and L'Hermite, P. (eds) *Epidemiological Studies of Risks Associated with the Agricultural Use of Sewage Sludge: Knowledge and Needs*. Elsevier Applied Science Publishers, London, pp. 21–33.

National Research Council (NRC) (1993) *Soil and Water Quality*. National Academy Press, Washington, DC.

Navas, A., Bermúdez, F. and Machín, J. (1998) Influence of sewage sludge application on physical and chemical properties of Gypsisols. *Geoderma* 87, 123–135.

Navas, A., Machín, J. and Navas, B. (1999) Use of biosolids to restore the natural vegetation cover on degraded soils in the badlands of Zaragoza (NE Spain). *Bioresource Technology* 69, 199–205.

NEPM (1999) *National Environment Protection (Assessment of Site Contamination) Measure.* National Environment Protection Council Service Corporation, Adelaide.

NFA (1994) *Australian Food Standards Code, March 1993.* National Food Authority, Australian Government Publication Service, Canberra, Australia.

O'Brien, P. and Mitsch, W.J. (1980) Root zone nitrogen simulation model for land application of sewage sludges. *Ecological Modelling* 8, 233–257.

O'Conner, G.A., Chaney, R.L. and Ryan, J.A. (1991) Bioavailability to plants of sludge-borne toxic organics. *Reviews of Environmental Contamination and Toxicology* 121, 129–155.

Olaniya, M.S., Bhoyar, R.V and Bhide, A.D. (1991) Effects of solid waste disposal on land. *Indian Journal of Environmental Health* 34, 143–150.

Olness, A., Clapp, C.E., Liu, R. and Palazzo, A.J. (1998) Biosolids and their effects on soil properties. In: Wallace, A. and Terry, R.E. (eds) *Handbook of Soil Conditioners: Substances that Enhance the Physical Properties of Soil.* Marcel Dekker, New York, pp. 141–165.

Paulsrud, B. and Nedland, K.T. (1997) Strategy for land application of sewage sludge in Norway. *Water Science Technology* 36, 283–290.

Petersen, S.O., Henriksen, K., Mortensen, G.K., Krogh, P.H., Brandt, K.K., Sørensen, J., Madsen, T., Petersen, J. and Grøn, C. (2003) Recycling of sewage sludge and household compost to arable land: fate and effects of organic contaminants, and impact on soil fertility. *Soil and Tillage Research* 72, 139–152.

Prasad, R. (1996) Soil contaminants in India: an overview. In: Naidu, R., Kookana, R.S., Oliver, D.P., Rogers, S. and McLaughlin, M.J. (eds) *Contaminants and the Soil Environment in the Australasia-Pacific Region.* Kluwer Academic Publishers, Dordrecht, The Netherlands, pp. 513–542.

Ross, A.D., Lawrie, R.A., Whatmuff, M.S., Keneally, J.P. and Awad, A.S. (1991) *Guidelines for the Use of Sewage Sludge on Agricultural Land.* NSW Agriculture, Sydney, Australia, 17 pp.

Ross, A.D., Lawrie, R.A., Keneally, J.P. and Whatmuff, M.S. (1992) Risk characterisation and management of sewage sludge on agricultural land – implications for the environment and the food chain. *Australian Veterinary Journal* 69, 177–181.

Rowlands, C.L. (1992) Sewage sludge in agriculture: a UK perspective. In: *Water Environment*

Federation 65th Annual Conference and Exposition, New Orleans, 20–24 September, pp. 305–315.

Sakthivel, S., Mahimairajah, S., Divakaran, J., Saravanan, K., Kookana, R.S., Ramasamy, K. and Naidu, R. (2000) Tannery effluent irrigation for tree plantations: preliminary observations from field experiments. In: Naidu, R., Willet, I.R., Mahimairajah, S., Kookana, R. and Ramasamy, K. (eds) *ACIAR Proceedings No. 88: Towards Better Management of Soils Contaminated with Tannery Waste,* pp. 144–150.

Sandaa, R.A., Torsvik, V. and Enger, Ø. (2001) Influence of long-term heavy metal contamination on microbial communities in soil. *Soil Biology and Biochemistry* 33, 287–295.

Sara Parwin Banu, K., Ramesh, P.T., Ramasamy, K., Mahimairajah, S. and Naidu, R. (2000) Is it safe to use tannery chrome sludge for growing vegetables? Results from a glasshouse study. In: Naidu, R., Willet, I.R., Mahimairajah, S., Kookana, R. and Ramasamy, K. (eds) *ACIAR Proceedings No. 88: Towards Better Management of Soils Contaminated with Tannery Waste,* pp. 127–132.

Sauerbeck, D.R. and Leschber, R. (1992) German proposal for acceptable contents of inorganic and organic pollutants in sewage sludge and sludge-amended soils. In: Hall, J.E., Sauerbeck, D.R. and L'Hermite, P. (eds) *Treatment and Use of Sewage Sludge and Liquid Agricultural Wastes: Review of COST 68/681 Programme, 1972–90,* Office for Official Publications of the European Details Communities, Luxembourg.

Siddique, M.T., Robinson, J.S. and Alloway, B.J. (2000) Phosphorus reactions and leaching potential in soils amended with sewage sludge. *Journal of Environmental Quality* 29, 1931–1938.

Sims, J.T., Vasilas, B.L. and Ghadrati, M. (1995) Evaluation of fly ash as a soil amendment for the Atlantic Coastal Plains. II. Soil chemical properties and plant growth. *Journal of Water, Air and Soil Pollution* 81, 363–372.

Sloan, J.J., Dowdy, R.H. and Dolan, M.S. (1998) Recovery of biosolids-applied heavy metals sixteen years after application. *Journal of Environmental Quality* 27, 1312–1317.

Smith, S.R. (1996) *Agricultural Recycling of Sewage Sludge and the Environment.* CAB International, Wallingford, UK, 384 pp.

Taiganides, E.P. (1976) Principles and techniques of animal wastes management and utilization in Asia. *Organic Recycling in Asia. FAO Soils Bulletin* 36, 341–362.

Takahashi, J. (1976) Role of night soil in Japanese agriculture. *Organic Recycling in Asia. FAO Soils Bulletin* 36, 363–364.

Tchobanoglous, G., Theisen, H. and Vigil, S. (1993) *Integrated Solid Waste Management.* McGraw-Hill, New York.

Tyler, G. (1981) Heavy metals in soil biology and biochemistry. In: Paul, E.A. and Ladd, J.N. (eds) *Soil Biochemistry*, Vol. 5. Marcel Dekker, New York, pp. 371–414

USEPA (1984) *Use and Disposal of Municipal Wastewater Sludge.* EPA 625/10–84–003.

USEPA (1985) *Summary of Environmental Profiles and Hazard Indices for Constituents of Municipal Sludge: Methods and Results.* USEPA Office of Water Regulations and Standards, Washington, DC.

USEPA (1990) *National Pesticide Survey: Phase I Report. Report Number PB91–125765.* US Department of Commerce, National Technical Information Service, Springfield, Virginia.

Vikelsoe, J., Thomsen, M., Johansen, E. and Carlen, L. (1999) Phthalates and nonylphenols in soil – a field study of different profiles. *NERI Technical Report No. 268.* Danish Ministry of Environment and Energy.

Warren, C.J. (1992) Some limitations of sluiced fly ash as a liming agent for acidic soil. *Waste Management Research* 10, 317–327.

Witter, E. (1993) Heavy metal concentrations in agricultural soils critical to microorganisms. *National Swedish Environmental Protection Board Report*, pp. 40–79.

Zibilske, L.M., Clapham, W.M. and Rourke, R.V. (2000) Multiple applications of paper mill sludge in an agricultural system: soil effects. *Journal of Environmental Quality* 29, 1975–1981.

Chapter 14

Pesticides in Soil – Benefits and Limitations to Soil Health

M.A. Locke* and R.M. Zablotowicz

USDA-ARS, Southern Weed Science Research Unit, PO Box 350, Stoneville, Mississippi 38776, USA

Summary

Pesticides are important components of many agricultural management systems and their effects on soil and its ability to process them should be included when evaluating soil quality. Pesticides help maintain agricultural productivity by controlling pests, however, management thresholds must be established to minimize potential non-target effects on soil biota and processes. In this chapter, we review: (i) selected examples of pesticide effects on soil biology and ecosystem function; (ii) methodologies to assess these effects; and (iii) conservation management options that may improve the capability of soils to process pesticides, and thereby to function safely and productively. Soil biota typically are resilient to pesticides applied at recommended rates, with only transient disruptions. Conservation management practices that

---

* See Contributors list for present address.

©CAB International 2004. *Managing Soil Quality: Challenges in Modern Agriculture*
(eds P. Schjønning, S. Elmholt and B.T. Christensen)

increase soil organic matter (OM) and promote accumulation of plant residues on the soil surface help the soil to process many pesticides through sorption and degradation. Our review provides information needed to begin comprehensive coordinated initiatives to establish criteria for assessing and managing pesticide impacts on soil quality.

Introduction

The agricultural use of chemical agents to control pests such as weeds, insects, rodents, nematodes, fungi and bacteria has been practised since the latter part of the 19th century, when Bordeaux spray was first used by Millardet as a fungicide to prevent downy mildew on grapes. During the latter half of the 20th century, pesticide development and use became widespread as improved application and production technologies made it more practical and economical to use pesticides as a management tool. Worldwide pesticide use in 1997 was estimated at 2.58 billion kg (Aspelin and Grube, 1999). Weed control with herbicides was the largest single category of pesticide usage (40%), followed by usage of insecticides (26%) and fungicides (9%).

Integrating pesticide use with other technologies such as improved crop varieties, application formulations and farm equipment, spurred on the Green Revolution and the greatest capacity to produce food and fibre in history. With increased pesticide use, questions on potential effects regarding public health and the environment developed. Concerns arose that pesticides would pollute air, soil and water resources, contaminate the food chain and disrupt ecosystem balance. Environmental consequences might also result from pesticide application at rates higher than recommended, accidental spills or long *in situ* residence time in soil. Given these issues, evaluating soil quality in the agricultural context must also consider pesticide interactions with soil and soil biota.

Various philosophies about soil quality have been proposed and discussed at length elsewhere. A basic definition adopted by the Soil Science Society of America and accepted here (see Schjønning *et al.*, Chapter 1, this volume) considers soil quality as: '... the capacity of a specific kind of soil to function, within natural or managed ecosystem boundaries, to sustain plant and animal productivity, maintain or enhance water and air quality, and support human health and habitation'. Little information is available on relationships between pesticide and soil quality, thus, assessing soil quality with respect to pesticide management necessitates: (i) defining factors that may affect pesticide dissipation or activity; (ii) identifying soil characteristics that may be influenced directly or indirectly by pesticide use; and (iii) evaluating how soil management can mitigate unintended effects of pesticides or enhance intended effects of pesticides. According to Larson and Pierce (1994), soil quality in agricultural settings should be assessed on the basis of the soil's ability to serve as a medium for plant growth (i.e. soil productivity), facilitate water flow in the environment and function as an environmental buffer. All three functions relate to pesticides. For example, high concentrations of pesticides in soil may influence processes such as plant growth and the activity and diversity of biotic populations. The ability of a soil to regulate and partition water flow could determine pesticide residence time in soil or the potential for pesticide movement to non-target areas. The capacity of a soil for pesticide sorption and degradation determines its efficacy for buffering their impact. How effectively a soil processes excess or unused pesticides and mitigates detrimental effects to humans and other species helps to determine its value.

This chapter reviews research on pesticides and soil quality in two sections. The first section reviews recent research concerning pesticide effects on soil components and processes related to soil quality. Numerous studies have evaluated pesticide effects on individual aspects, such as soil biota, biochemical activity and nutrient processing.

Many of these studies only examined initial effects of pesticides in laboratory microcosms, rather than in integrated systems over long periods and at larger scales. Soil is not only a medium for agricultural production, but is often viewed as a filter and processor for xenobiotics; and how a soil is managed can determine its ability to function in this capacity. The second section addresses how conservation management practices: (i) may enhance soil's potential to serve as a medium for facilitating the intended function of an applied pesticide; and (ii) may improve the soil's ability to filter and process pesticides.

Pesticide Effects on Soil Components and Processes

Determining assessment criteria

Assessing the effects of pesticides requires an understanding of which soil components will be influenced by a pesticide and the longevity of these effects. The most sensitive (responsive) components need to be identified and threshold pesticide tolerance concentrations determined. Pesticides have many purposes; therefore establishing general soil quality criteria is difficult, if not impossible. The simplest way to begin the assessment is to categorize pesticides based on similarities in purpose of use and properties. General goals of pesticide use in agricultural systems are the prevention, suppression and eradication of pests that reduce production quality and quantity. Agriculturally important herbicides, insecticides and fungicides are this chapter's focus.

One measure of the productivity of agricultural soils is crop yield, and the ability of a soil to sustain yields is a measure of soil quality. Therefore, beneficial aspects of judicious pesticide use in crop production systems must be considered in a discussion of management criteria when assessing pesticide effects on soil quality. Weeds compete with crops for nutrients, water and sunlight, resulting in reduced yields. Sensitive crop species might be used to establish thresholds at which herbicides are impacting growth and production.

Similarly, insects and disease reduce both yield and quality of crop production, and sensitive or beneficial populations might be used in those assessments. It is critical to establish the *minimum* quantity of pesticide necessary to achieve sustainable crop production goals with negligible adverse effects to the soil ecosytem and environment, i.e. the management threshold. The strategy, therefore, would be to adopt management measures (e.g. reduced tillage, site-specific application) that minimize potential non-target effects while achieving the needed productivity.

Exposure to a substance to which an organism is not adapted will probably cause some effect, with certain organisms being inherently more sensitive. Distinction must be made between toxicity to target organisms (pests) and non-target organisms. Target toxicity is the dose required to harm or kill pests of interest. Non-target toxicity is a measure of the degree to which a pesticide can harm or injure beneficial organisms and is usually assessed on a limited number of representative species. Several standard parameters to interpret toxicity have been established, e.g. LD_{50} (lethal dose), LC_{50} (lethal concentration), EC_{50} (effective concentration), that are based on the dose or concentration of a substance that affects or kills 50% of a test population. Parameters such as LOEC (lowest observed effect concentration) and NOEC (no observed effect concentration) provide information on concentrations that result in adverse effects not found in control populations. Whether or not pesticide EC_{50} or LC_{50} values can be used in assessments of soil quality needs systematic, comprehensive evaluation. A simplistic approach for obtaining an estimate of potential toxicity would extract a pesticide from soil and determine whether its concentration falls within published EC_{50}/LC_{50} ranges for key indicator species that inhabit the soil. Measuring pesticide concentrations in soil, however, does not give sufficient information to determine impacts on the total soil ecosystem, i.e. bioavailability.

Limited information exists on standard testing protocols for evaluating lethal or sublethal toxicity in pesticide-treated soil. In aquatic systems, sediment quality assessments have received much attention because contaminated sediment can readily impact water quality (Suter, 1993; Ingersoll, 1995). Partitioning of sediment contaminants between sorbed and interstitial water phases occurs until equilibrium is established. This assumes that sensitive aquatic organisms are more susceptible to solution-phase contaminants than sorbed contaminants. Thus, indices of contaminant availability can be obtained from sediment sorption partition coefficients. One weakness of this approach is that some organisms ingest contaminated sediment particles, thus assimilating the associated pesticide. There are similarities between sediment and soil, and it has been suggested that similar concepts be applied to assess potential toxicity in soil (Suter, 1993). Distinct differences between soil and sediment present some complications in directly relating sediment evaluations to soil. For example, sediments are usually saturated, whereas soils undergo wetting and drying cycles. During drier periods, soil pore space is exposed and pesticides in the vapour phase could impact organisms. Alternatively, pesticides sorbed to dry soil may be more tightly retained, and thus less bioavailable than during moist periods.

Edwards (2002) reviewed some of the currently available methods for assessing effects of pesticides in soil, including: single-species laboratory tests, multispecies tests, integrated soil microcosms and terrestrial model ecosystems. Although much effort has been directed towards harmonizing tests and criteria for assessing non-target effects of pesticides, there are still greatly disparate testing criteria among various countries. Regulatory agencies worldwide have developed a series of diagnostic tests to assess non-target effects, e.g. USEPA (1996) and OECD (2001), for use during registration of agrochemicals. Bioindicators in these tests include plants, earthworms and components of the soil and aquatic biotic community.

Establishing the link from pesticide effects on one or more indicator species to soil ecosystem health is more complicated. Factors such as soil resilience to adverse conditions, pesticide longevity, or indirect effects become important in this assessment. Many pesticides dissipate rapidly from soil. Others may

accumulate, thus becoming more concentrated in soil with repeated use; for example, DDT is a classic case of a long-lived, immobile pesticide. How do we assess long-term impact of these pesticides on the soil ecosystem? When a pesticide indirectly inhibits a biotic process, what is the effect on the total soil ecosystem?

The soil ecosystem is considered relatively resilient to short-term stress and capable of recovery from various perturbations, including pesticide stress. Work by Domsch *et al.* (1983) compared the effect of natural stressors (temperature and drought) on soil biological activities to the effects of pesticides in terms of magnitude and duration. As summarized in Fig. 14.1, these studies show that soil can be inhibited as much as 90% of a given parameter, but if recovery is within pre-stress levels, this effect may be considered tolerable to the soil ecosystem.

Several of the soil biotic parameters selected for a discussion of pesticide effects in this chapter include microbial populations, fauna, microbial biomass, enzyme activity, nitrogen fixation and mycorrhizas. Other factors more indirectly influenced by pesticides include nutrient availability and carbon turnover. Although evaluation of most of these parameters was not designed to assess soil quality, a review of individual pesticide effects gives an indication of which parameters are potential candidates for use in these assessments.

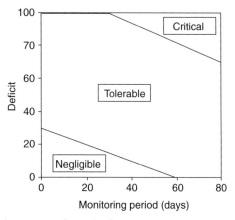

Fig. 14.1. Relationship between monitoring period and deficit vs. the ecological significance of injuries (adapted from Domsch *et al.*, 1983).

Soil microbial populations

Most modern pesticides were developed to target specific pests and even individual enzymes of a particular pest. However, effects on non-target soil microbial populations have been observed. Specific enzymes associated with these non-target microorganisms may be sensitive to pesticides. For instance, the mode of action for sulphonylurea herbicides is inhibition of the enzyme acetolactate synthetase, a component of the biosynthetic pathway of branched amino acids (leucine and isoleucine). This pathway is present in many bacteria; for example, certain pseudomonads are inhibited by sulphonylurea herbicides (Boldt and Jacobsen, 1998).

The effects of several pesticides on microorganisms in studies ranging from *in vitro* culture studies to field assessments are summarized in Table 14.1. *In vitro* studies may be useful for determining mechanisms of inhibition or toxicity for specific pesticides; however, these effects may be reduced or not observed in field soils where sorption and degradation affect pesticide bioavailability. When considering the effects of specific pesticides on soil microorganisms, caution must be used to discern effects of the active ingredient versus the formulated product. In this respect, effects of long-term normal field applications of two 2,4-D formulations were assessed on soil microbial populations (Narain Rai, 1992). Twenty-five weeks after application, total bacteria, fungi and actinomycete populations were reduced over 50% by applying 2,4-D as the *iso*-octyl ester versus the dimethylamine salt. Since herbicides are used to kill vegetation, a microbial response might be attributed to either direct toxic effects of the compound or to indirect effects such as vegetation loss limiting carbon substrate to soil microflora. Nevertheless, numerous studies observed no significant negative effects of normal herbicide applications on culturable soil microorganisms, e.g. Seifert *et al.* (2001).

Fungicides and insecticides may have a greater effect on soil microbial populations than herbicides. Application of the fungicide captan at normal field application rates (2–10 kg/ha) significantly reduced numbers of culturable fungi in four soils over a 30-day study, with a clear dose-dependent response

Table 14.1. Non-target effects of pesticides on soil microorganisms.

Pesticide	Organism	Effects	Reference
2,4-D-*iso*-octyl ester (H)[a]	Culturable soil bacteria, fungi and actinomycetes	Reduced soil populations	Narain Rai (1992)
Bromopropylate (I)[a]	*Azospirillum brasilense*	No effect on growth or N_2 fixation	Gomez *et al.* (1998)
Captan (F)[a]	Culturable soil bacteria, fungi and actinomycetes	Reduced soil populations	Martinez-Toledo *et al.* (1998)
Diazinon (I)	*Azospirillum brasilense*	No effect on growth or N_2 fixation	Gomez *et al.* (1999)
Fenamiphos (I)	Algae and cyanobacteria	None	Megharaj *et al.* (1999)
Fenpropimorph (F)	Actinomycetes, *Pseudomonas* sp., active fungi	Active fungi reduced, no effect on others	Thirup *et al.* (2001)
Glyphosate (H)	*Bradyrhizobium japonicum*	Inhibition, death	Moorman *et al.* (1992)
Imazaquin (H)	Culturable soil bacteria and fungi	No effect	Seifert *et al.* (2001)
Metsulphuron methyl (H)	*Pseudomonas* sp.	Growth inhibition	Boldt and Jacobsen (1998)
Methidathion (I)	*Azospirillum brasilense*	Reduced nitrogen fixation and ATP content	Gomez *et al.* (1998)
Simazine (H)	*Azotobacter chroococcum*	No effect on growth, high concentrations increased N_2 fixation	Martinez-Toledo *et al.* (1991)

[a]F, fungicide; H, herbicide; I, Insecticide.

as a tool for assessing soil quality. Interactions of tillage intensity and pesticides on diversity of microarthropod communities were evaluated at three sites over a 6-month period (Cortet *et al.*, 2002a). These studies indicated that greater microarthropod diversity was maintained under minimum tillage compared to deep tillage, but effects attributed to herbicides were negligible. Litterbag protocols assessed the effects of herbicides (atrazine, alachlor) and insecticides (fipronil and carbofuran, 0.2 and 0.6 kg/ha) on microarthropods (Cortet *et al.*, 2002b). Fiprinol had the greatest toxicity to microarthropods, significantly reducing numbers of *Oribatida* (about 50% after 61 and 102 days). Minimal effects of the other pesticides tested were observed. Martikainen *et al.* (1998) observed that dimethoate insecticide (tenfold field rate) reduced microarthropod populations with the greatest effect on the collembolans, but acari were reduced only when both benomyl fungicide (tenfold field rate) and dimethoate were applied. Although populations of collembolans recovered, species composition changed. Neither pesticide, alone or combined, had an effect on nematodes and enchytraeids.

Effects of long-term benomyl application in a tall grass prairie were assessed on nematode populations (Smith *et al.*, 2000). Benomyl had no significant effect on herbivores, but significantly reduced certain fungal feeders (Tylenchidae) by 13% and predatory nematodes (Dorylaimidae) by 33%. Protozoa may also be affected by various soil-applied herbicides. For example, Ekelund (1999) demonstrated that various groups of protozoa are affected by field application rates of the fungicide fenpropimorph.

Soil enzymatic activity

Metabolic capabilities of soils are often characterized by quantifying the activities of various hydrolytic and oxidoreductive enzymes. Pesticide effects on soil enzymatic activity were extensively reviewed by Schäffer (1993). Soil enzymatic activities are useful indicators of soil health and have been used to determine whether adverse effects of a management practice affect soil biochemical functions. A dehydrogenase enzyme assay is used to assess pesticide soil toxicity in German pesticide registration (Anderson *et al.*, 1992).

Enzyme status in soil, extracellular and bound to clay or humic acids, or as a component of viable organisms, can determine how pesticides affect enzymatic activities. Some pesticides may inactivate an enzyme by competitive binding at an active site, thus blocking the enzyme from acting on the test substrate. Other enzymes, e.g. dehydrogenase, measure oxidative electron transfer during carbon substrate utilization and are reflective of total biotic activity in soil.

Pesticide effects on enzyme activities depend on soil conditions and the pesticide application rate. Omar and Abdel-Sater (2001) observed that bromoxynil (herbicide) and profenophos (insecticide) inhibited cellulase activity when applied at 1 and 5× field application rates. Bromoxynil inhibited cellulase activity by over 30% and 20% after 4 and 10 weeks, respectively. Bromoxynil and profenophos transiently increased acid phosphatase activity at field rates but reduced activity by about 30% at the higher rate for 6 weeks. In a series of studies evaluating sulphonylurea herbicide effects on enzyme activities, conflicting results were observed. Sulphonylurea herbicides applied at normal field rates had no detrimental effects on dehydrogenase or fluorescein diacetate (FDA) hydrolytic activity (Perucci *et al.*, 1999), but FDA activity was reduced in other cases (Perucci *et al.*, 2000). Some negative effects on these enzymes were observed at rimsulfuron rates 10–100× higher than field rates, but effects were only slight and transitory (Perucci *et al.*, 1999).

Several soil enzymes play a role in facilitating soil nutrient availability. Urease is an important enzyme in nitrogen transformations. Urea is often applied as a nitrogen source, and urease hydrolyses urea to ammonium. Substituted phenylurea (monuron, diuron, linuron) herbicides were shown to inhibit urea hydrolysis (10–30%) in soil (Cervelli *et al.*, 1976). These herbicides behaved both as competitive and noncompetitive inhibitors of soil urease activity.

Nitrogen fixation

Although agriculture depends heavily on fertilizer nitrogen, nitrogen fixation by free-living, associative and symbiotic bacteria is the major input of usable nitrogen into global ecosystems. Under USEPA guidelines for pesticide registration, detailed protocols are given to assess pesticide effects on symbiotic nitrogen fixation (see USEPA, 1996).

Several studies have addressed the effects of specific pesticides on non-symbiotic nitrogen fixation. Effects of several insecticides on *Azospirillium brasilense* were evaluated *in vitro* (Gomez *et al.*, 1998, 1999). Bromopropylate and diazinon had no effect on microbial growth or nitrogen fixation, but methidathion and profenophos reduced N_2 fixation in synthetic media. The fungicide captan (3.5–10 kg/ha) significantly reduced populations of aerobic nitrogen-fixing bacteria and nitrogen-fixation activity (40–80%) in four soils over a 30-day incubation with inhibition being highly dose-dependent (Martinez-Toledo *et al.*, 1998).

Symbiotic nitrogen fixation by legumes can provide 30–70% of the nitrogen requirement for a crop. Fungicides and herbicides may either directly affect colonization of rhizobial symbionts, or indirectly influence performance of either the plant or the nitrogen-fixing bacteria *in planta* (Moorman, 1989). Fungicides to control seedling diseases have potential for inhibiting rhizobial inoculant establishment. Commercial application rates of carboxin, captan, pentachloronitrobenzene and thiram to seed reduced the survival of *Bradyrhizobium japonicum* and *Rhizobium phaseoli* (Curly and Burton, 1975; Graham *et al.*, 1980).

Glyphosate-resistant transgenic crops have revolutionized weed control and the acceptance of conservation management practices in North America. The basis of resistance is insertion of an insensitive 5-enolpyruvylshikimate-3-phosphate synthetase (EPSP) gene from an *Agrobacterium* strain allowing expression of a functional shikimic acid pathway, hence glyphosate tolerance. The soybean symbiont *B. japonicum*, however, possesses glyphosate-sensitive EPSP, thus *B. japonicum* growth is inhibited by glyphosate, ultimately resulting in death at concentrations exceeding 5 mM (Moorman *et al.*, 1992; Hernandez *et al.*, 1999). Nodule mass accumulation of glyphosate-resistant soybeans was significantly but inconsistently inhibited by field rates of glyphosate under greenhouse conditions (Reddy *et al.*, 2000; King *et al.*, 2001) and reduced by 21–28% in a 2-year field study (Reddy and Zablotowicz, 2003). Hernandez *et al.* (1999) demonstrated 10–30% inhibition of *B. japonicum* nitrogen fixation by glyphosate under laboratory conditions using bacteroids from glyphosate-treated plants or by treating bacteroids with glyphosate. Inhibition of nitrogen-fixation activity corresponded to glyphosate sensitivity of the *B. japonicum* strain *in vitro*. Nitrogen fixation in glyphosate-resistant soybeans was most severe under moisture stress, and reductions in soybean yield due to glyphosate application were observed in the field during drought (King *et al.*, 2001). The magnitude of inhibition of nitrogen fixation in soybeans due to glyphosate application has not been critically ascertained under field conditions. However, even a small reduction in nitrogen-fixation potential may have long-term effects on sustainable soil nitrogen pools, considering the widespread adoption of glyphosate-resistant technology.

Mycorrhizas

Symbiosis of mycorrhizal fungi with plants is another mutualistic plant–microbial association. Mycorrhizal roots have an altered morphology that enhances nutrient and water uptake. Fumigants used as nematicides and fungicides can profoundly influence mycorrhizal establishment. O'Bannon and Nemec (1978) showed that chloropicrin and methyl bromide completely inhibited *Glomus mossae* and *G. fasiculatum* on citrus, whereas citrus infection by these fungi was not affected by either 1,3-dichloropropene or ethylene dibromide. Cotton mycorrhizal infection was stimulated by nematicides 1,3-dichloropropane and dibromochloropropane in soils highly productive for cotton (Bird *et al.*, 1974).

Herbicide effects on mycorrhizal infection of citrus were evaluated in greenhouse and field studies (Nemec and Tucker, 1983). Bromacil, diuron or trifluralin had no effect on citrus growth or *Glomus etunicatum* chlamydospores in soil. Only high rates of simazine reduced citrus infection by *G. etunicatum*, indicating minor effects of herbicides on mycorrhizal symbiosis. Benomyl reduced mycorrhizal colonization more than 75% in 7-year field trials on a tallgrass prairie, indicating that long-term fungicide application may drastically alter beneficial plant–microbial interactions with significant implications for ecosystem sustainability (Smith *et al.*, 2000).

Carbon and nitrogen mineralization

Assessing pesticide effects on carbon and nitrogen mineralization is a standardized component of testing pesticides for non-target effects in the registration process by the USEPA (1996) and the OECD (2001). Understanding the effects on both processes is important in understanding pesticide interactions in soil and their role in supporting plant growth and overall ecosystem health (Edwards, 2002).

Chen *et al.* (2001) compared the effects of benomyl, captan and chlorothalonil on soil-nitrogen dynamics in laboratory incubations with or without additions of organic materials. Both nitrogen mineralization and nitrification rates were influenced by all fungicides, with captan eliciting the greatest influence on mineralization rates. Captan increased soil NH_4-N, whereas benomyl or chlorothalonil had little impact. Martinez-Toledo *et al.* (1998) showed that captan (2–10 kg/ha) inhibited nitrifying bacteria in four soils (50–90%) during a 30-day study. Applying bensulfuron at normal field rates had no effect on nitrification (Gigliotti *et al.*, 1998), whereas cinosulfuron transiently inhibited nitrification after 1 week, but had no effect at 4 weeks (Allievi and Gigliotti, 2001).

Examining endogenous respiration by monitoring cumulative CO_2 evolution in pesticide-treated soil is one approach to evaluating pesticide effects on mineralization. Under USEPA pesticide registration guidelines, pesticide effects on soil respiration are determined 5 and 28 days after application. Alternatively, short-term monitoring of the respiration of exogenous added substrates, e.g. substrate-induced respiration (SIR) as described by Anderson and Domsch (1978), measures active microbial biomass and can provide relative bacterial and fungal biomass when used with appropriate inhibitors. Assessments using SIR are a component of certain OECD guidelines, e.g. Germany (Anderson *et al.*, 1992). SIR techniques were used in studies by Engelen *et al.* (1998) to demonstrate that dinoseb had a greater effect on microbial activity than metamitron.

Altering soil conditions may influence the proportion of microorganisms that are sensitive to pesticides. In one study, when soils were amended with soybean residue, respiration in metribuzin-treated soil was lower than in soil without metribuzin (Locke and Harper, 1991a). However, in soils without soybean residue, respiration did not differ between metribuzin-treated and non-treated soils.

Selection of tests for assessing pesticide effects

A review of pesticide effects on various organisms and processes in soil provides an indication of how pesticides may affect soil quality. Standard tests of a soil's quality need to be simple, relatively inexpensive and repeatable. We propose several assays that fit these criteria (Table 14.2). Overall, diagnostic tests that assess pesticide effects on microbial biomass, SIR, dehydrogenase activity and nitrogen mineralization, are easy to perform and provide essential information on overall soil microbial activity. DNA-based microbial community analysis can ascertain minute changes in diversity, but does not consider physiological function in soil. Toxicity of pesticides to soil fauna such as earthworms can provide information on a group of organisms involved in processes such as OM decomposition, soil structure formation, and also on

Table 14.2. Proposed criteria for assessing pesticide effects on soil quality and ability of soil to process pesticides or buffer effects.

Criterion	Method
Pesticide effects	
Microbial biomass	Chloroform fumigation Substrate-induced respiration
Soil enzyme activity	Dehydrogenase Fluorescein diacetate hydrolysis
Soil fauna	Earthworm survival
Mineralizable N	Ammonification Nitrification
Processing ability	
Texture	
Sorption	Batch
Soil organic matter	
Pesticide degradation	2,4-D ring or carboxy label mineralization
pH	
Microbial biomass	Chloroform fumigation Substrate-induced respiration
Soil enzyme activity	Dehydrogenase Fluorescein diacetate hydrolysis

the possible entry of pesticides into the food chain of terrestrial species. An integrated soil microcosm approach (Burrows and Edwards, 2002) assessing microbial activity, nematodes, earthworms, plant productivity and carbon and nitrogen transformations in one system offers much promise.

Conservation Management Practices that Influence Soil Quality Factors Related to Pesticides

The impact of individual conservation management practices on soil quality will determine the long-term profitability and sustainability of a given management system. Understanding interactions among biological, chemical and physical soil properties that relate to pesticide bioavailability and dissipation is necessary to identify appropriate soil management strategies to maintain or improve soil health. Mechanisms associated with pesticide persistence in soil include degradation, sorption, movement and volatilization, and the extent to which farm managers can influence these mechanisms will determine the persistence of pesticide residues and their effects. Key soil factors related to pesticide dissipation that can be influenced by management include plant residue accumulation, quantity and character of soil OM, and soil microflora, chemistry and structure (Locke and Bryson, 1997).

Pesticide transport to groundwater or surface water bodies has major implications relevant to soil quality. International concerns about potential pesticide contamination of drinking water and impacts on human health and the environment have led to increased government regulation. The US Food Quality Protection Act of 1996 (FQPA) provides for major changes in pesticide regulation with respect to assessments of non-occupational exposure from drinking water, residential and dietary sources. FQPA gives the USEPA more latitude in promoting programmes such as Integrated Pest Management and use of alternative management practices that may reduce the risk of pesticide exposure. Similar initiatives are being adopted in the European Economic Community and elsewhere.

These regulatory trends, along with a depressed world agricultural market, have motivated efforts to develop low-cost, environmentally compatible management systems for agriculture. Conservation management strategies that substantially improve soil characteristics while improving the environment include reduced tillage, the use of cover crops or organic amendments, and practices of crop and herbicide rotations (Locke et al., 2002a). These conservation measures are gaining more widespread acceptance as farmers recognize the benefits that accrue from not only preserving soil, but also improving it.

Soil characteristics influenced by conservation management

Reducing tillage quantity and intensity modifies most factors that could potentially affect pesticide persistence and bioavailability in soil (Locke and Bryson, 1997). For many cropping systems, when soil disturbance is reduced or eliminated, a narrow organic soil layer develops at the surface (e.g. Doran, 1980; Edwards *et al.*, 1992). Reduced tillage imparts several changes in the soil surface that gradually diminish with increasing depth (Rhoton, 2000). A crust containing mixtures of soil and plant residues usually develops at the soil surface, forming a protective seal conserving soil moisture and promoting pesticide degradation. However, soil crust formation may enhance pesticide sorption to clays and plant materials, thus minimizing bioavailability. Both the size and diversity of biotic populations increase in the soil surface under reduced tillage (Doran, 1980; Reddy *et al.*, 1995a; Lupwayi *et al.*, 1998; Cortet *et al.*, 2002a), but as organic substrate declines with soil depth, biotic activity diminishes (Zablotowicz *et al.*, 2000). Reduced tillage alters soil structural properties. Increased microbial activity and soil OM promote the formation of macroaggregates, cemented together with microbial exudates (Locke and Bryson, 1997). Conservation tillage soils develop more defined and stable structures, resulting from increased proportions of water-stable organo-clay microaggregates (Rhoton, 2000) and the OM trapped within microaggregates (Bossuyt *et al.*, 2002). Under reduced tillage, there is greater potential for preferential macropore flow in soils because of tunnels formed by faunal activity and voids formed from *in situ* plant residue decomposition.

Cover crops provide erosion control, green manures and weed inhibition. Organic amendments are used as mulch for weed control or moisture retention, for adding nutrients and for accelerated degradation of xenobiotics. Examples of organic amendments are animal manures, plant residues or manufacturing by-products or waste. Rotation of crops may increase soil OM, but this effect is dependent on the cropping sequence (Edwards *et al.*, 1992). If a crop produces abundant biomass (e.g. maize, sorghum), resulting changes in soil OM may be significant. Cover crop residues and organic amendments also stimulate soil biota and enhance soil OM (Reddy *et al.*, 1997a,b; Wardle *et al.*, 1999; Locke *et al.*, 2002b). Microbial activities and

populations in hairy vetch and ryegrass cover crop residues were six- and 100-fold greater, respectively, than in underlying soil (Zablotowicz *et al.*, 1998a), whereas ryegrass and poultry litter amendments enhanced soil bacterial and fungal populations (Wagner *et al.*, 1995).

Pesticide dissipation and conservation management practices

Pesticides dissipate from soil via several processes, including degradation, sorption or binding, leaching, movement in surface runoff or volatilization. These processes are interrelated, making it difficult to assess each independently. We will address individual processes in a logical sequence, each building on the next, in the order: (i) pesticide sorption; (ii) pesticide mobility; (iii) initial pesticide transformation; and (iv) transformation of pesticide metabolites, and sequestration, binding and bioavailability of pesticide metabolites or residues.

Pesticide sorption

Pesticide sorption in soil is a dissipation mechanism that is significantly influenced by soil OM (Locke and Bryson, 1997). Pesticide sorption kinetics are often described as biphasic, characterized by the pesticide achieving rapid initial equilibrium distribution into soil components followed by a more gradual sorption phase (e.g. Locke, 1992). The rapid initial phase is believed to involve more accessible, exposed sites on the surface of soil particles and microaggregates. The more gradual phase follows as the pesticide permeates the soil complex to more restricted sorption sites. Pesticide desorption from these restricted sites is usually a slow process resulting from release from physical entrapment within the soil matrix, slow diffusion from restricted sites along tortuous pathways, or greater affinity for soil components than for the bulk solution (Locke, 1992). Within the soil matrix, an increased density of functional sorption sites could lessen the attraction of pesticides for more aqueous phases.

Bulk, electronic and hydrophobic functional areas within pesticide structures drive reactions between pesticides and sorbents (Reddy and Locke, 1994). In turn, soil OM contains numerous sites that react with pesticide functional groups. For many pesticides, sorption and soil carbon content are positively correlated (e.g. Locke *et al.*, 2002c). Since quantity and composition of soil OM play such an important role in pesticide sorption to soil, management strategies that alter characteristics and levels of soil organic components should be considered in pesticide management. As reviewed above, management practices such as reduced tillage, cover crops and animal waste applications can positively influence soil OM characteristics and distribution in soil.

Several studies have assessed herbicide sorption in soils where management practices increased soil OM (e.g. Locke, 1992; Locke *et al.*, 1995; Reddy *et al.*, 1997a; Gaston and Locke, 2000). A consistent trend for many herbicides (alachlor, cyanazine, fluometuron, acifluorfen) was that sorption was greater in surface soils managed through reduced tillage, with a cover crop, or where an organic amendment had been made. This phenomenon was attributed to greater quantities of soil OM accumulation. Comparisons of sorption to soil at lower depths where soil OM was equivalent between tillage treatments indicated little difference (Gaston and Locke, 2000; Zablotowicz *et al.*, 2000).

Plant residue characteristics can affect pesticide sorption, and crop residue management practices can influence pesticide sorbent properties. Some soil OM constituents have a greater affinity for pesticides than do others. Greater proportions of soil OM as humic material were important for oxyfluorfen and metribuzin sorption (Stearman *et al.*, 1989). Reduced-tillage soil had a greater proportion of soil carbon as humin, whereas tilled soil had more humic and fulvic acids. Age or weathered condition of plant residues in soil, or on the soil surface, can also affect pesticide sorption. In comparisons of herbicide sorption to plant residues in various stages of decomposition, more herbicide was sorbed on aged material (Dao, 1991; Reddy *et al.*, 1995b).

Pesticide mobility

Three primary processes are involved in the transfer of pesticides within and from soil: (i) movement through pores or matrices within the soil profile; (ii) transport in surface runoff; and (iii) volatilization. Mobility of a pesticide in soil is closely linked to its sorption affinity for soil components. The stronger the affinity for soil and its constituents, the less likely it will detach and move.

Avenues for pesticide movement through soil include both macropore and micropore flow. Macropore flow involves large pore sizes and consists of channels among soil aggregates, cracks caused by shrink–swell processes, voids left by decayed roots, or tunnels created by fauna (Locke and Bryson, 1997). Macropore flow is important for drainage when soil is saturated. Since large hydraulic conductivities are associated with macropore flow, dissolved pesticides moving through macropores have little interaction with reactive pore surfaces, thus facilitating rapid pesticide leaching to greater soil depths, i.e. preferential flow. Aggregation and channel development in reduced tillage soils potentially provide paths for enhanced pesticide movement when flow is saturated, but results from field studies have been mixed (Hall *et al.*, 1991; Isensee and Sadeghi, 1995). It appears that the most important factors determining pesticide leaching via preferential flow are timing, amount and intensity of rainfall following pesticide application (Isensee and Sadeghi, 1995).

Micropore flow involves inter- and intraparticle diffusion through small pores. Rate of pesticide diffusion through micropores is dependent on the affinity of the pesticide for sorbents on micropore surfaces and moisture gradients within the soil. Interparticle movement is micropore flow between soil particles, whereas intraparticle diffusion is flow within particle matrices and clay lattices. The more interaction a pesticide has with active sorption sites, the greater the likelihood of inhibiting further movement through the profile. Thus, higher concentrations of OM in the surface layer of reduced tillage soils should increase sorption and inhibit downward and lateral movement.

Surface runoff is the movement of water, sediment and chemicals over the soil surface. Runoff occurs when saturated soil cannot take in any more water during precipitation and rainfall is too intense for soil to accommodate water entry. Pesticide movement in runoff occurs when pesticides applied to foliage or plant residues are washed off during precipitation or pesticides sorbed to soil particles move with sediment or are desorbed into runoff water. Plant residues on the soil surface generally reduce volume of runoff, hence lessening pesticide loss. Pesticide applied to the surface of reduced tillage soils may be sorbed by the plant residues or the soil OM layer, thus further movement into the soil may be inhibited.

In soil, a pesticide can either degrade *in situ* by microbial or chemical processes, volatilize or photodegrade, but as long as pesticide residues remain at the surface, they are vulnerable to loss in runoff. The affinity of individual pesticides for soil constituents will determine the extent to which this occurs. The longer pesticides remain in soil or on plant debris, the less likely that they will desorb, because they transfer to restricted sites during the more gradual phase of sorption. Moderate precipitation shortly after pesticide application may be sufficient to move a pesticide into the soil, but perhaps not to the extent of leaching via channelling. When cover crops or fallow-season weeds provide dense foliage covering the soil surface, pesticides applied to soil will be intercepted and retained to some degree (Reddy and Locke, 1996). The quality or condition of the plant residues (Dao, 1991; Reddy et al., 1995b, 1997a) may determine the degree to which pesticides are retained in plant debris or are removed as leachate or runoff.

Volatilization is a major mechanism for pesticide dissipation. Factors regulating pesticide volatilization include soil moisture content, temperature, physicochemical pesticide and formulation properties and pesticide-soil attractions. Pesticide volatilization from soil involves a two-step processes: (i) the evaporation of pesticide molecules; and (ii) dispersion of the resulting vapour into the atmosphere (Taylor and Spencer, 1990). During evaporation, pesticide transforms from a liquid or solid phase to a gas. Further pesticide dissipation in the gas phase depends on air turbulence and diffusion. Reduced-tillage soils may be more insulated from extremes in temperatures, and wetter because of the surface plant residues. Cooler temperatures and lower evaporation rates may also decrease the potential for pesticide loss by volatilization (Gish et al., 1995).

Initial pesticide transformation

Initial modification to pesticide structure can occur via chemical or biological activity. The transformed compound (metabolite) is typically a product of hydrolysis, reductive or oxidative processes and may retain the basic structure of the parent compound, but possess altered biological and physicochemical properties. Hydrolysis is an important initial reaction of pesticide metabolism, since many pesticides have susceptible moieties (e.g. amides, carbamates and ester linkages). Many soil microorganisms produce hydrolytic enzymes, but hydrolysis also results from chemical processes, including low redox potential, soil acidity and interactions with soil OM. Nitroaromatic and aromatic pesticides (dinitroaniline and nitrodiphenylether herbicides; the fungicide pentachloronitrobenzene (PCNB); the insecticide parathion) are all susceptible to microbial metabolism by oxidative and reductive processes, depending on the characteristics of microbial populations, chemical structure and soil environmental factors (Zablotowicz et al., 1998b). Other microbial enzymes (halidohydrolases, dehydrohalogenases and glutathione S-transferases) cleave the halogen–carbon bond present in many pesticides. Many pesticides, such as carbamates, chloroacetamides, phenylureas and triazines, additionally contain N-alkyl groups, making them susceptible to microbially mediated N-dealkylation.

The effects of conservation management on initial pesticide degradation, i.e. disappearance of parent pesticide, have been evaluated with varied results. Levanon et al. (1994) observed an enhanced rate of atrazine degradation in reduced-tillage soil, whereas more rapid degradation of bentazon (Gaston et al., 1996a) and acifluorfen (Gaston and Locke,

2000) occurred in conventional-tillage soil relative to no-tillage soil. The enhanced metolachlor degradation in soil from an untilled vegetative buffer area as compared to degradation in soil from an adjacent tilled field, was attributable to higher microbial activities (Staddon et al., 2001). Fluometuron degraded more rapidly in surface conventional-tillage soil even though microbial activities were higher in no-tillage soil (Zablotowicz et al., 2000). Increased fluometuron degradation in conventional tillage resulted from lessened sorption (greater vulnerability to degradation) and better distribution of plant residues and soil due to mixing during tillage. Wagner et al. (1996) found that some soils with a history of bentazon use were better able to degrade it under no-tillage than conventional-tillage management, whereas in other soils there was negligible tillage effect. Still other research showed little or no effect of tillage on initial pesticide degradation, e.g. alachlor (Locke et al., 1996), metribuzin (Locke and Harper, 1991b), fluometuron (Locke et al., 1995) and chlorimuron (Reddy et al., 1995a).

Pesticides applied to live cover crops or decomposing plant residues may be intercepted and retained by the plant tissue, which may also delay pesticide degradation. Zablotowicz et al. (1998a) determined that although desiccated cover-crop residues were more biologically active than associated underlying soils, 2,4-D mineralization was slower in plant residues than in surface soil. Fluometuron degradation in ryegrass residues was as rapid as in soil, but required high moisture conditions. Sorption to cover-crop material may provide a degree of protection from degradation. Eventually, pesticides retained by plant residues will either degrade in place or elute to soil below. However, soils associated with cover crops can provide an environment conducive to pesticide metabolism. Enhanced 2,4-D degradation in surface soil from a cereal rye cover crop as compared to fallow soil was attributed to elevated populations of 2,4-D degraders in the cover-crop soil (Bottomley et al., 1999). Laboratory results, however, indicated that the half-life of fluometuron in surface soil from cover-crop areas was approximately 22 days longer than that from areas without a cover crop (Brown et al., 1994).

Amending soil with plant materials or manure can accelerate pesticide degradation via a process called biostimulation, used to detoxify contaminated soils (Zablotowicz et al., 1998b). Adding poultry litter, maize meal or ryegrass to soil enhanced cyanazine and fluometuron degradation (Wagner and Zablotowicz, 1997a), an effect attributed to greater microbial activity in amended soils. Fluometuron degradation was stimulated in soils amended with rice, hairy vetch or ryegrass in another study (Wagner and Zablotowicz, 1997b). In contrast to other studies, metribuzin degradation did not differ between soil amended and not amended with soybean plant residues, although microbial respiration was higher (Locke and Harper, 1991a).

How important are soil characteristics and management to the *initial* degradation of a pesticide? For many pesticides applied to soil, the initial chemical structure may be necessary for it to function successfully as a pesticide, and metabolites may exhibit less or no activity. Ideally, if pesticide efficiency is to be achieved, a balance must be struck between the factors of pesticide bioactivity, sorption and vulnerability to degradation. If pesticides can function as intended under conservation management systems for the same duration as in conventional systems, then the lack of difference in the initial degradation of the parent pesticide structure is a positive outcome for environmental quality.

Transformation and binding of pesticide metabolites or residues

Although generalized conclusions cannot be made about initial pesticide degradation in conservation-tillage soils, distinct trends were observed concerning the dynamics of pesticide metabolites in these systems. Pesticide metabolites possess a myriad of characteristics, and their fate is ultimately determined by metabolite physicochemical properties and soil conditions. Metabolites that are mobile or readily vulnerable to degradation may be transient, whereas less mobile pesticides may have a long residence in soil. Some metabolites are more polar than the parent pesticide, e.g. sulphonic and oxanilic acids associated with chloroacetamide metabolism.

The characteristics of metabolites found in soil may help to explain how soil management influenced pesticide transformation. In several studies, conventional-tillage soils tended to have a larger proportion of applied pesticide accumulated as polar metabolites than reduced-tillage soils. For example, in two of three soils, recovery of one polar metabolite was greater in conventional tillage than in no-tillage soils, 63 days after chlorimuron application (Reddy et al., 1995a). More polar metabolites of metolachlor accumulated in a tilled-field soil than in an untilled vegetative buffer soil (Staddon et al., 2001). Alachlor metabolites oxanilic and sulphonic acids were present in greater quantities in conventional-tillage soil than in no-tillage soil (Locke et al., 1996). Polar metribuzin metabolites were more abundant in conventional-tillage and non-amended soils than in no-tillage soil or soil amended with plant residues (Locke and Harper, 1991a,b). These studies suggest that greater accumulation of polar metabolites in conventional-tillage soils resulted from lower microbial degradation activity, compared with reduced-tillage soils. However, in soils amended with ryegrass residues, more hydroxycyanazine was recovered than in non-amended soil (Wagner and Zablotowicz, 1997a). Greater proportions of polar metabolites in conventional-tillage soils may have implications relevant to subsequent movement of pesticide residues. Some polar metabolites are more easily desorbed into the aqueous solution phase (e.g. cyanazine amide, chloroacid cyanazine; Reddy et al., 1997b), thus making them more vulnerable to loss in runoff or through leaching. Other polar metabolites, such as hydroxycyanazine acid or hydroxyatrazine have greater affinity for soil humic components, making them less mobile (Lerch et al., 1997; Reddy et al., 1997b).

Some metabolites are more non-polar than their parent pesticides, resulting in potential for greater sorption to soil (e.g. aminoacifluorfen; Locke et al., 1997). More non-polar metolachlor metabolites were extracted from untilled vegetative buffer soil than from tilled soil after 13 days of incubation (Staddon et al., 2001). Lower quantities of non-polar alachlor metabolites were extracted from either conventional- or no-tillage soils

than polar metabolites, probably because they were more tightly sorbed (Locke et al., 1996).

Mineralization is the endpoint of pesticide metabolism, where pesticide carbon is completely degraded to CO_2. Pesticide residues that are mineralized would therefore have to be labile, accessible to microorganisms and have low recalcitrance. The impact of management on mineralization can vary with each pesticide, and no clear pattern emerges. Mineralization of alachlor residues in no-tillage soil was greater than in conventional-tillage soil, perhaps corresponding to the decline in polar metabolites (Locke et al., 1996). In another study, greater initial mineralization of atrazine residues was attributed to higher microbial activity in no-tillage soils (Levanon et al., 1994). However, lower mineralization of metribuzin and chlorimuron residues was observed in no-tillage or soybean-residue-amended soils (Locke and Harper, 1991a,b; Reddy et al., 1995a).

The proportion of a pesticide that cannot be extracted from soil by solvents is termed non-extractable or bound pesticide residue. It is extremely difficult to identify non-extractable pesticide residues, but possible components include: (i) strongly sorbed parent pesticide; (ii) strongly sorbed metabolites or fragments of the degraded pesticide molecule; (iii) pesticides or metabolites polymerized into humic material via oxidative coupling reactions catalysed by various enzymes; and (iv) physical sequestration of pesticides or metabolites. Physical sequestration occurs when pesticide residues become entrapped within soil microaggregates or within organo-clay matrices or lattices from which they are removed only with great difficulty. In undisturbed soil, pockets of aggregates may be protected from weathering and remain entrapped. Channels of water move around rather than through these aggregates, and thereby reduce the potential for release of bound pesticide residues. If pesticide residues are sequestered within soil aggregates, release may occur if the soil is disrupted, as with tillage.

In several pesticide fate studies involving soils from conservation-managed areas, we observed that surface soils from reduced tillage systems contained a greater proportion of applied pesticide as non-extractable residues, e.g. metribuzin (Locke and Harper, 1991a,b), alachlor (Locke *et al.*, 1996), sulfentrazone (Reddy and Locke, 1998) and metolachlor (Staddon *et al.*, 2001), but there are exceptions, e.g. fluometuron (Zablotowicz *et al.*, 2000), bentazon (Gaston *et al.*, 1996a,b; Wagner *et al.*, 1996) and chlorimuron (Reddy *et al.*, 1995b). In some cases, the process is reversible; for example formerly non-extractable hydroxyatrazine residues can be extracted, e.g. after liming soils (Kells *et al.*, 1980).

Sorbed or bound pesticides or metabolites may lose substantial activity and toxicity, i.e. become less bioavailable (Alexander, 1994). For example, sorption of herbicides to soil may reduce toxicity to weeds. Gaston *et al.* (2001) observed a positive correlation between soil OM content and weed control activity of fluometuron, implying that as soil OM increases, herbicide bioactivity decreases. Crop residues can intercept herbicide applied to the soil surface (e.g. Locke *et al.*, 2002b), perhaps reducing its bioavailability for controlling weeds. Reduced weed control with a cover crop, however, might be offset by the cover crop inhibiting weed growth by shading or allelopathy (Locke *et al.*, 2002a). If sorption of a pesticide prevents or hinders degradation, it could have prolonged *in situ* residence. The balance between maintaining sufficient bioavailability for adequate weed management and reducing herbicide longevity in soil becomes critical when crops are rotated. If crop rotation is practised, it is necessary that herbicides applied for weed control in the first crop not interfere with management and production of the subsequent crop.

Selection of tests for assessing pesticide processing ability in soil

Changes in the soil environment caused by increased carbon accumulation have impacts on pesticide fate. Management activities such as adding manures or plant material, using cover crops and reducing tillage promote accumulation of carbon-rich material in the soil surface, resulting in increases in biotic activity and changes to soil structure. Our discussion of effects of conservation soil and crop management on pesticide dissipation provides a background for selecting methods to assess the ability of soils to process pesticides. Several protocols proposed in Table 14.2 should provide an indication as to whether a pesticide is bioavailable (i.e. active or residual) and a given soil's intrinsic ability for pesticide dissipation.

Conclusions

As pesticides are an extremely diverse group of chemicals; it is impossible to thoroughly explore the ramifications of all pesticides on soil and environmental quality in this limited space. We have introduced pesticides and emphasized their importance to crop production, while recognizing environmental and human health concerns. The management threshold for pesticides was defined as the *minimum* quantity of pesticide necessary to achieve sustainable crop-production goals with negligible adverse effects to the soil ecosystem and environment. Microbial and faunal populations are generally tolerant of pesticides, exhibiting only minor transient perturbations when recommended rates are used. Likewise, major biotic processes such as enzyme activity, respiration and carbon and nitrogen transformations are minimally impacted. In many cases, crop and soil management practices that increase soil OM and plant residues impart attributes to soil that can impede pesticide movement and enhance degradation, while not hindering pesticide efficacy.

Satisfactory and generally applicable measures of pesticide effects and management in relationship to soil quality do not exist, although some regulatory organizations have made substantial progress. This chapter provides a background to facilitate the establishment of comprehensive strategies for determining acceptable protocols for these assessments. Questions for discussion include:

- How do we measure or assess soil quality with regard to pesticides?
- How do we determine the most meaningful indicators?
- What can we do to ameliorate negative effects and enhance positive attributes of pesticide use?

- How will we evaluate measured side effects?
- Can we establish management thresholds below which adverse pesticide effects are negligible?

Based on the current state of knowledge, we summarized a group of diagnostic tests that may be performed to assess pesticide impact on soil health and the ability of a soil to process pesticides or limit negative effects. Issues of experimental scale (*in vitro* to watershed) and duration of assessment are challenges. Differences in regulatory approaches and philosophies may preclude establishing international standards, but within regions or nations, efforts to develop strategies can be initiated using the following logical steps.

- Use the information reviewed here as a basis for further debate on pesticide contributions to soil quality issues.
- Establish testing criteria for measuring pesticide impacts on soil quality.
- Develop and adopt standard protocols for cost–benefit analysis and risk management assessments.
- Plan, coordinate and implement a series of multilocation, holistic studies to provide appropriate databases for refinement and modelling.

References

Alexander, M.A. (1994) *Biodegradation and Bioremediation*. Academic Press, San Diego, California, 302 pp.

Allievi, L. and Gigliotti, C. (2001) Response of the bacteria and fungi of two soils to the sulfonylurea herbicide cinosulfuron. *Journal of Environmental Science and Health Part B – Pesticides, Food Contaminants and Agricultural Wastes* 36, 161–175.

Anderson, J.P.E. and Domsch, K.H. (1978) A physiological method for the quantitative measurement of microbial biomass in soil. *Soil Biology and Biochemistry* 10, 215–221.

Anderson, J.P.E., Castle, D., Ehle, H., Eichler, D., Laerman, H.-T., Maas, G. and Malkomes, H.-P. (1992) *Guideline for the Official Testing of Plant Protection Products*, 2nd edn. Part VI, 1–1 Effects on the activity of soil microflora. Biologische Bundesanstalt für Land- und Forstwirtsshaft, Braunschweig, Germany.

Aspelin, A.L. and Grube, A.H. (1999) *Pesticide Industry Sales and Usage: 1996 and 1997 market estimates*. Office of Prevention, Pesticide, and Toxic Substances, United States Environmental Protection Agency, 733-R-99-001.

Bird, L.S., Rich, J.R. and Glover, S.U. (1974) Increased endomycorrhizae of cotton roots in soil treated with nemacides. *Phytopathology* 64, 48–51.

Bjørnlund, L., Ekelund, F., Christensen, S., Jacobsen, C.S., Krogh, P.H. and Johnsen, K. (2000) Interactions between saprophytic fungi, bacteria and protozoa on decomposing wheat roots in soil influenced by the fungicide fenpropimorph (Corbel®): a field study. *Soil Biology and Biochemistry* 32, 967–975.

Boldt, T.S. and Jacobsen, C.S. (1998) Different toxic effects of the sulfonyl urea herbicides metsulfuron methyl, chlorsulfuron, and thifensulfuron methyl on fluorescent pseudomonads isolated from soil. *FEMS Microbiology Letters* 161, 29–35.

Bossuyt, H., Six, J. and Hendrix, P.F. (2002) Aggregate-protected carbon in no-tillage and conventional tillage agroecosystems using carbon-14 labeled plant residue. *Soil Science Society of America Journal* 66, 1965–1973.

Bottomley, P.J., Sawyer, T.E., Boersma, L., Dick, R.P. and Hemphill, D.D. (1999) Winter cover crop enhances 2,4-D mineralization potential of surface and subsurface soil. *Soil Biology and Biochemistry* 31, 849–857.

Brown, B.A., Hayes, R.M., Tyler, D.D. and Mueller, T.C. (1994) Effect of tillage and cover crop on fluometuron adsorption and degradation under controlled conditions. *Weed Science* 44, 171–175.

Burrows, L.A. and Edwards, C.A. (2002) The use of integrated soil microcosms to predict effects of pesticides on soil ecosystems. *European Journal of Soil Biology* 38, 245–249.

Busse, M.D., Ratcliff, A.W., Shestak, C.J. and Powers, R.F. (2001) Glyphosate toxicity and the long-term vegetation control on soil microbial communities. *Soil Biology and Biochemistry* 33, 1777–1789.

Cervelli, S., Nannipieri, P., Giovanni, G. and Perna, A. (1976) Relationship between substituted urea herbicides and soil urease activity. *Weed Research* 16, 365–368.

Chen, S.-K., Edwards, C.A. and Subler, S. (2001) Effects of the fungicides, benomyl, captan, and chlorothalonil on soil microbial activity and nitrogen dynamics in laboratory incubations. *Soil Biology and Biochemistry* 33, 1971–1980.

Cortet, J., Ronce, D., Poinsot-Balaguer, N., Beaufreton, C., Chabert, A., Viaux, P. and de Fonseca, J.P.C. (2002a) Impacts of different agricultural practices on the biodiversity of microarthropod communities in arable crop systems. *European Journal of Soil Biology* 38, 239–244.

Cortet, J., Gillon, D., Joffre, R., Ourcival, J.-M. and Poinsot-Balaguer, N. (2002b) Effects of pesticides on organic matter recycling and microoarthropods in a maize field: use and discussion of the litterbag methodology. *European Journal of Soil Biology* 38, 261–265.

Crecchio, C., Curci, M., Pizzigallo, M.D.R., Ricciuti, P. and Ruggiero, P. (2001) Molecular approaches to investigate herbicide-induced bacterial changes in soil microcosms. *Biology and Fertility of Soils* 33, 460–466.

Curly, R.L. and Burton, J.C. (1975) Compatability of *Rhizobium japonicum* with chemical seed protectants. *Agronomy Journal* 67, 807–808.

Dalby, P.R., Baker, G.H. and Smith, S.E. (1995) Glyphosate, 2,4-DB and dimethoate: effects on earthworm survival and growth. *Soil Biology and Biochemistry* 27, 1661–1662.

Dao, T.H. (1991) Field decay of wheat straw and its effects on metribuzin and S-ethyl metribuzin sorption and elution from crop residues. *Journal of Environmental Quality* 20, 203–208.

Domsch, K.H., Jagnow, G. and Anderson, T.-H. (1983) An ecological concept for assessing the side effects of agrochemicals on soil microorganisms. *Residue Reviews* 86, 65–105.

Doran, J.W. (1980) Soil microbial and biochemical changes associated with reduced tillage. *Soil Science Society of America Journal* 44, 765-771.

Edwards, C.A. (2002) Assessing the effects of environmental pollutants on soil organisms, communities, processes, and ecosystems. *European Journal of Soil Biology* 38, 225–231.

Edwards, C.A. and Bohlen, P.J. (1992) The effects of toxic chemicals on earthworms. *Reviews of Environmental Contamination and Toxicology* 125, 23–99.

Edwards, J.H., Wood, C.W., Thurlow, D.L. and Ruf, M.E. (1992) Tillage and crop rotation effects on fertility status of a Hapludult soil. *Soil Science Society of America Journal* 56, 1577–1582.

Ekelund, F. (1999) The impact of the fungicide fenpropimorph (Corbel) on bacteriovorous and fungivorous protozoa in soil. *Journal of Applied Ecology* 36, 233–243.

El Fantroussi, S., Verschuere, L., Verstraete, W. and Top, E.M. (1999) Effect of phenylurea herbicides on soil microbial communities estimated by analysis of 16S rRNA gene fingerprints and community-level physiological profiles.

Applied and Environmental Microbiology 65, 982–988.

Engelen, B., Meinken, K., von Wintzgerode, F., Heuer, H., Malkomes, H.-P. and Backhaus, H. (1998) Monitoring of a pesticide treatment on bacterial soil communities by metabolic and genetic fingerprinting in addition to conventional testing procedures. *Applied Environmental Microbiology* 64, 2814–2821.

Gaston, L.A. and Locke, M.A. (2000) Acifluorfen sorption, degradation, and mobility in a Mississippi Delta soil. *Soil Science Society of America Journal* 64, 112–121.

Gaston, L.A., Locke, M.A. and Zablotowicz, R.M. (1996a) Sorption and degradation of bentazon in conventional- and no-till Dundee soil. *Journal of Environmental Quality* 25, 120–126.

Gaston, L.A., Locke, M.A., Wagner, S.C., Zablotowicz, R.M. and Reddy, K.N. (1996b) Sorption of bentazon and degradation products in two Mississippi soils. *Weed Science* 44, 678–682.

Gaston, L.A., Locke, M.A., Zablotowicz, R.M. and Reddy, K.N. (2001) Spatial variability of soil properties and weed populations in the Mississippi Delta. *Soil Science Society of America Journal* 65, 449–459.

Gennari, M., Fournier, J.C., Dughera, R. and Negre, M. (1998) A study on the mineralization of C-14 acifluorfen in soil: comparison between treated and untreated soils. *Fresenius Environmental Bulletin* 7, 338–344.

Gigliotti, C., Allievi, L., Salardi, C., Ferrari, F. and Farini, A. (1998) Microbial ecotoxicity and persistence in soil of the herbicide bensulfuron-methyl. *Journal of Environmental Science and Health Part B – Pesticides, Food Contaminants and Agricultural Wastes* 33, 381–398.

Gish, T.J., Sadeghi, A. and Wienhold, B.J. (1995) Volatilization of alachlor and atrazine as influenced by surface litter. *Chemosphere* 31, 2971–2982.

Gomez, F., Salmeron, V., Rodelas, B., Martinez-Toledo, M.V. and Gonzalez-Lopez, J. (1998) Response of *Azospirillum brasilense* to the pesticides bromopropylate and methidathion on chemically defined media and dialyzed-soil media. *Ecotoxicology* 7, 43–47.

Gomez, F., Martinez-Toledo, M.V., Salmeron, V., Rodelas, B. and Gonzalez-Lopez, J. (1999) Influence of the insecticides profenofos and diazinon on the microbial activities of *Azospirillum brasilense*. *Chemosphere* 39, 945–957.

Graham, P.H., Ocambo, G., Ruiz, L.D. and Duque, A. (1980) Survival of *Rhizobium phaseoli* in contact with chemical seed protectants. *Agronomy Journal* 72, 625–627.

Hall, J.K., Mumma, R.O. and Watts, D.W. (1991) Leaching and runoff losses of herbicides in a tilled and untilled field. *Agricultural Ecosystem and Environment* 37, 303–314.

Haney, R.L., Senseman, S.A., Hons, F.M. and Zuberer, D.A. (2000) Effect of glyphosate on soil microbial activity and biomass. *Weed Science* 48, 89–93.

Haney, R.L., Senseman, S.A. and Hons, F.M. (2002) Effect of Roundup Ultra on microbial activity and biomass from selected soils. *Journal of Environmental Quality* 31, 730–735.

Hart, M.R. and Brookes, P.C. (1996) Soil microbial biomass and mineralization of soil organic matter after 19 years of cumulative application of pesticides. *Soil Biology and Biochemistry* 28, 1641–1649.

Hernandez, A., Garcia-Plazzola, J.I. and Becerril, J.M. (1999) Glyphosate effects on phenolic metabolism of nodulated soybean (*Glycine max* L.Merr.) *Journal of Agricultural and Food Chemistry* 47, 2920–2925.

Ingersoll, C.G. (1995) Sediment tests. In: Rand, G.M. (ed.) *Fundamentals of Aquatic Toxicology.* 2nd edn. Taylor and Francis, Washington, DC, pp. 231–255.

Isensee, A.R. and Sadeghi, A.M. (1995) Long-term effect of tillage and rainfall on herbicide leaching to shallow groundwater. *Chemosphere* 30, 671–685.

Johnsen, K., Jacobsen, C.S., Torsvik, V. and Sorensen, J. (2001) Pesticide effects on bacterial diversity in agricultural soils – a review. *Biology and Fertility of Soils* 33, 443–453.

Katayama, A., Funassaka, K. and Fujie, K. (2001) Changes in respiratory quinone profile of a soil treated with pesticides. *Biology and Fertility of Soils* 33, 454–459.

Kells, J.J., Rieck, C.E., Blevins, R.L. and Muir, W.M. (1980) Atrazine dissipation as affected by surface pH and tillage. *Weed Science* 28, 101–104.

King, A.C., Purcell, L.C. and Vories, E.D. (2001) Plant growth and nitrogenase activity of glyphosate-tolerant soybean in response to glyphosate applications. *Agronomy Journal* 93, 179–186.

Larson, W.E. and Pierce, F.J. (1994) The dynamics of soil quality as a measure of sustainable management. In: Doran, J.W. (ed.) *Defining Soil Quality for a Sustainable Environment. Soil Science Society of America Special Publication* No. 35, Madison, Wisconsin, pp. 37–51.

Lerch, R.N., Thurman, E.M. and Kruger, E.L. (1997) Mixed-mode sorption of hydroxylated atrazine degradation products in soil: a mechanism for bound residue. *Environmental Science and Technology* 31, 1539–1546.

Levanon, D., Meisinger, J.J., Codling, E.E. and Starr, J.L. (1994) Impact of tillage on microbial activity and the fate of pesticides in the upper soil. *Water Air and Soil Pollution* 72, 179–189.

Locke, M.A. (1992) Sorption–desorption kinetics of alachlor in surface soil from two soybean tillage systems. *Journal of Environmental Quality* 21, 558–566.

Locke, M.A. and Bryson, C.T. (1997) Herbicide–soil interactions in reduced tillage and plant residue management systems. *Weed Science* 45, 307–320.

Locke, M.A. and Harper, S.S. (1991a) Metribuzin degradation in soil: I Effects of soybean residue, metribuzin level, and soil depth. *Pesticide Science* 31, 221–237.

Locke, M.A. and Harper, S.S. (1991b) Metribuzin degradation in soil: II Effects of tillage. *Pesticide Science* 31, 239–247.

Locke, M.A., Zablotowicz, R.M. and Gaston, L.A. (1995) Fluometuron interactions in crop residue-managed soils. In: Kingery, W.L. and Buehring, N. (eds) *Conservation Farming: a Focus on Water Quality.* Mississippi Agriculture Forestry Experiment Station Special Bulletin 88–7, pp. 55–58.

Locke, M.A., Gaston, L.A. and Zablotowicz, R.M. (1996) Alachlor biotransformation and sorption in soil from two soybean tillage systems. *Journal of Agricultural and Food Chemistry* 44, 1128–1134.

Locke, M.A., Gaston, L.A. and Zablotowicz, R.M. (1997) Acifluorfen sorption and sorption kinetics in soil. *Journal of Agricultural and Food Chemistry* 45, 286–293.

Locke, M.A., Reddy, K.N. and Zablotowicz, R.M. (2002a) Weed management in conservation production systems. *Weed Biology and Management* 2, 123–132.

Locke, M.A., Zablotowicz, R.M., Steinriede, R.W. and Dabney, S.M. (2002b) Conservation management practices in Mississippi Delta agriculture: implications for crop production and environmental quality. In: van Santen, E. (ed.) *Making Conservation Tillage Conventional: Building a Future on 25 Years of Research.* Alabama Agricultural Experiment Station and Auburn University Special Report No. 1, pp. 320–326.

Locke, M.A., Zablotowicz, R.M. and Gaston, L.A. (2002c) Environmental fate of fluometuron in a Mississippi Delta lake watershed. In: Arthur, E.L., Clay, V.C. and Barefoot, A.C. (eds) *Terrestrial Field Dissipation Studies: Purpose, Design, and Interpretation.* American Chemical Society Symposium Series 842, American Chemical Society, Washington, DC, pp. 206–225.

Lupwayi, N.Z., Rice, W.A. and Clayton, G.W. (1998) Soil microbial diversity and community structure under wheat as influenced by tillage and crop rotation. *Soil Biology and Biochemistry* 13, 1733–1741.

Martikainen, E., Haimi, J. and Ahtianen, J. (1998) Effects of dimethoate and benomyl on soil organisms and soil processes – a microcosm study. *Applied Soil Ecology* 9, 381–387.

Martinez-Toledo, M.V., Salmeron, V. and Gonzalez-Lopez, J. (1991) Effect of simazine on the biological activity of *Azotobacter chroococcum*. *Soil Science* 151, 459–467.

Martinez-Toledo, M.V., Salmeron, V., Rodelas, B., Pozo, C. and Gonzalez-Lopez, J. (1998) Effects of the fungicide captan on some functional groups of soil microflora. *Applied Soil Ecology* 7, 245–255.

Megharaj, M., Singelton, I., Kookana, R. and Naidu, R. (1999) Persistence and effects of fenamiphos on native algal populations and enzymatic activities in soil. *Soil Biology and Biochemistry* 31, 1549–1553.

Moorman, T.B. (1989) A review of pesticide effects on microorganisms and soil fertility. *Journal of Production Agriculture* 2, 14–22.

Moorman, T.B., Becerrill, J.M., Lydon, J.M. and Duke, S.O. (1992) Production of hydroxybenzoic acids by *Bradyrhizobium japonicum* strains after treatment with glyphosate. *Journal of Agricultural Food Chemistry* 40, 289–293.

Narain Rai, J.P. (1992) Effect of long-term 2,4-D application on soil microbial populations. *Biology and Fertility of Soils* 13, 187–191.

Nemec, S. and Tucker, D. (1983) Effects of herbicides on endomycorrhizal fungi in Florida citrus (*Citrus* sp.). *Weed Science* 31, 427–431.

O'Bannon, J.H. and Nemec, S. (1978) Influence of soil pesticides on vesicular arbuscular mycorrhizae in citrus. *Nematropica* 8, 56–61.

OECD (2001) OECD Dossier Guidance: Section 6 Ecotoxicological Studies and Risk Assessment. Available at: http://www.oecd.org/findDocument/0,2350,en_2649_34383_1_119820_1_1_1,00.htm

Omar, S.A. and Abdel-Sater, M.A. (2001) Microbial populations and enzyme activities in soil treated with pesticides. *Water, Air and Soil Pollution* 127, 49–63.

Perucci, P. and Scarponi, L. (1996) Side effects of rimsulfuron on the microbial biomass of a clay-loam soil. *Journal of Environmental Quality* 25, 610–613.

Perucci, P., Vischetti, C. and Battistoni, F. (1999) Rimsulfuron in a silty clay loam soil: effects upon microbiological and biochemical properties under varying microcosm conditions. *Soil Biology and Biochemistry* 31, 195–204.

Perucci, P., Dumontet, S., Bufo, S.A., Mazzatura, A. and Casucci, C. (2000) Effects of organic amendment and herbicide treatment on soil microbial biomass. *Biology and Fertility of Soils* 32, 17–23.

Racke, K.D. and Coats, J.R. (eds) (1990) *Enhanced Biodegradation of Pesticides in the Environment*. American Chemical Society Symposium Series 426, American Chemical Society, Washington, DC, 326 pp.

Reddy, K.N. and Locke, M.A. (1994) Prediction of soil sorption K_{oc} of herbicides using semi-empirical molecular properties. *Weed Science* 42, 453–461.

Reddy, K.N. and Locke, M.A. (1996) Imazaquin spray retention, foliar washoff, and runoff losses under simulated rainfall. *Pesticide Science* 48, 179–187.

Reddy, K.N. and Locke, M.A. (1998) Sulfentrazone sorption, desorption, and mineralization in soils from two tillage systems. *Weed Science* 46, 494–500.

Reddy, K.N. and Zablotowicz, R.M. (2003) Glyphosate-resistant soybean response to various salts of glyphosate and glyphosate accumulation in soybean nodules. *Weed Science* 51, 494–502.

Reddy, K.N., Zablotowicz, R.M. and Locke, M.A. (1995a) Chlorimuron adsorption, desorption, and degradation in soils from conventional tillage and no-tillage systems. *Journal of Environmental Quality* 24, 760–767.

Reddy, K.N., Locke, M.A., Wagner, S.C., Zablotowicz, R.M., Gaston, L.A. and Smeda, R.J. (1995b) Chlorimuron ethyl sorption and desorption kinetics in soils and herbicide-desiccated cover crop residues. *Journal of Agricultural Food Chemistry* 43, 2752–2757.

Reddy, K.N., Locke, M.A. and Gaston, L.A. (1997a) Tillage and cover crop effects on cyanazine adsorption and desorption kinetics. *Soil Science* 162, 501–509.

Reddy, K.N., Locke, M.A. and Zablotowicz, R.M. (1997b) Soil type and tillage effects on sorption of cyanazine and degradation products. *Weed Science* 45, 727–732.

Reddy, K.N., Hoagland, R.E. and Zablotowicz, R.M. (2000) Effect of glyphosate on growth, chlorophyll content and nodulation in glyphosate-resistant soybeans (*Glycine max*) varieties. *Journal of New Seeds* 2, 37–52.

Rhoton, F.E. (2000) Influence of time on soil response to no-till practices. *Soil Science Society of America Journal* 64, 700–709.

Schäffer, A. (1993) Pesticide effects on enzyme activities in the soil ecosystem. In: Bollag, J.-M. and Stotzky, G. (eds) *Soil Biochemistry*. Marcel Dekker, New York, pp. 273–340.

Seifert, S., Shaw, D.R., Zablotowicz, R.M., Wesley, R.A. and Kingery, W.L. (2001) Effect of tillage on microbial characteristics and herbicide degradation. *Weed Science* 49, 685–693.

Smith, M.D., Hartnett, D.C. and Rice, C.W. (2000) Effects of long-term fungicide applications on microbial properties in tallgrass prairie soil. *Soil Biology and Biochemistry* 32, 935–946.

Staddon, W.J., Locke, M.A. and Zablotowicz, R.M. (2001) Microbiological characteristics of a vegetative bufferstrip soil and degradation and sorption of metolachlor. *Soil Science Society of America Journal* 65, 1136–1142.

Stearman, G.K., Lewis, R.J., Tortorelli, L.J. and Tyler, D.D. (1989) Herbicide reactivity of soil organic matter fractions in no-tilled and tilled cotton. *Soil Science Society of America Journal* 53, 1690–1694.

Suter, G.W. (1993) *Ecological Risk Assessment*. CRC Press, Boca Raton, Florida, pp. 156–159.

Taylor, A.W. and Spencer, W.F. (1990) Volatilization and vapor transport processes. In: Cheng, H.H. (ed.) *Pesticides in the Soil Environment*. Soil Science Society of America Book Series 2, Madison, Wisconsin, pp. 213–270.

Thirup, L., Johnsen, K., Torsvik, V., Spliid, N.H. and Jacobsen, C.S. (2001) Effects of fenpropimorph on bacteria and fungi during the decomposition of barley roots. *Soil Biology and Biochemistry* 33, 1517–1524.

Tiedje, J.M. and Zhou, Z. (1996) Analysis of non-culturable bacteria. In: Hall, G.S. (ed.) *Methods for the Examination of Organismal Diversity in Soils and Sediments*. CAB International, Wallingford, UK, pp. 53–66.

USEPA (1996) Ecological Effects Test Guidelines. Available at: http://www.epa.gov/opptsfrs/ OPPTS_Harmonized/850_Ecological_ Effects_Test_Guidelines/index.html

Wagner, S.C. and Zablotowicz, R.M. (1997a) Effect of organic amendments on the bioremediation of cyanazine and fluometuron in soil. *Journal*

of Environmental Science and Health Part B – Pesticides, Food Contaminants and Agricultural Wastes 32, 37–54.

Wagner, S.C. and Zablotowicz, R.M. (1997b) Utilization of plant material for remediation of herbicide-contaminated soils. In: Kruger, E.L., Anderson, T.A. and Coats, J.R. (eds) *Phytoremediation of Soil and Water Contaminants*. American Chemical Society Symposium Series 664, American Chemical Society, Washington, DC, pp. 65–76.

Wagner, S.C., Zablotowicz, R.M., Locke, M.A., Smeda, R.J. and Bryson, C.T. (1995) Influence of herbicide-desiccated cover crops on biological soil quality in the Mississippi Delta. In: Kingery, W.L. and Buehring, N. (eds) *Conservation Farming: a Focus on Water Quality*. Mississippi Agricultural Forestry Experiment Station Special Bulletin 88–7, pp. 86–89.

Wagner, S.C., Zablotowicz, R.M., Gaston, L.A., Locke, M.A. and Kinsella, J. (1996) Biodegradation of bentazon in soil, effects of tillage and bentazon history. *Journal of Agricultural and Food Chemistry* 44, 1593–1598.

Wardle, D.A., Yeates, G.W., Nicholson, K.S., Bonner, K.I. and Watson, R.N. (1999) Response of soil microbial biomass dynamics, activity, and plant litter decomposition to agricultural intensification over a seven-year period. *Soil Biology and Biochemistry* 31, 1707–1720.

Zablotowicz, R.M., Locke, M.A. and Smeda, R.J. (1998a) Degradation of the herbicides 2,4-D and fluometuron in herbicide-desiccated cover crop residues. *Chemosphere* 37, 87–101.

Zablotowicz, R.M., Hoagland, R.E. and Locke, M.A. (1998b) Biostimulation: enhancement of co-metabolic pathways in pesticide-contaminated soils. In: Kearney, P.C. and Roberts, T.R. (eds) *Pesticide Remediation in Soils and Water*. John Wiley & Sons, Chichester, UK, pp. 217–250.

Zablotowicz, R.M., Locke, M.A., Gaston, L.A. and Bryson, C.T. (2000) Interactions of tillage and soil depth on fluometuron degradation in a Dundee silt loam soil. *Soil Tillage Research* 57, 61–68.

Chapter 15
Systems Approaches for Improving Soil Quality

M.R. Carter,[1] S.S. Andrews[2] and L.E. Drinkwater[3]

[1]Agriculture and Agri-Food Canada, Crops and Livestock Research Centre,
440 University Avenue, Charlottetown, Prince Edward Island C1A 4N6, Canada;
[2]USDA-NRCS, Soil Quality Institute, 2150 Pammel Drive, Ames, Iowa 50011, USA;
[3]Cornell University, Department of Horticulture, Plant Science Building,
Ithaca, New York 14853, USA

Summary

The complexity of agricultural systems and the wide array of possible management inputs require that soil quality be evaluated at individual management or single function level as well as at the agroecosystem level. A systems approach to soil quality assessment involves a synthesis of agriculture and ecology. It allows the identification and characterization of indirect and emergent properties and their effects on soil quality attributes. Emergent properties are defined as the functional interaction of system components that are not observable from those smaller units of organization characteristic of reductionist approaches. At each level of organization in an agricultural system, properties emerge that were not present at the level below. Emergent system qualities influence nutrient, energy, tertiary structure, water regulation, and biotic community dynamics and processes in soil, and thus can improve or build soil quality. The objective of this chapter is to evaluate the improvement of soil quality from an agricultural system viewpoint, using information from two agricultural systems (conservation tillage and organic farming). Specific objectives are to evaluate the advantages and disadvantages of these systems with respect to their ability to build soil quality.

©CAB International 2004. *Managing Soil Quality: Challenges in Modern Agriculture*
(eds P. Schjønning, S. Elmholt and B.T. Christensen)

Introduction

From the beginning of agriculture there has been an innate interest in soil and land quality. Although precise definitions of soil quality and soil health can be variable and controversial, the central idea that soil serves various functions in agroecosystems is well established (Ellert *et al.*, 1997). Present concerns with soil degradation and sustainable soil management in both extensive and intensive agroecosystems have emphasized the need for ongoing improvement of soil and land quality (FAO, 1983; Doran and Parkin, 1994; Lal, 1998).

Soil quality is an evolving term (Warkentin, 1995). The basic definitions of soil *quality*, within an agricultural context, are as follows: *fitness for a specific purpose* (Carter *et al.*, 1997); *suitability for chosen uses* (Warkentin, 1995); and *capacity to function* (Karlen *et al.*, 1998). These premises are reflected in early attempts to classify soil suitability and land capability (Pierce, 1996; Carter *et al.*, 1997; Singer and Ewing, 1999). In comparison, the concept of soil *health* is concerned with the *efficiency* or *good condition* of the functions and processes that compose soil or ecosystems. Emphasis here is placed on the continued capacity of a soil to function as a vital living system (Doran *et al.*, 1996). At the agricultural system level, both *quality* and *health* tend to be used synonymously (see Karlen *et al.*, Chapter 2, this volume). An important feature of soil quality is the differentiation between inherent and dynamic soil properties (Carter *et al.*, 1997). Generally, the need for improved and sustainable soil management places the emphasis on managing and improving dynamic soil quality properties.

In the context of soil management, a systems approach involves a synthesis of agriculture and ecology. It places an emphasis on the integrated function or holistic approach to obtain an ecosystem perspective (Drinkwater, 2002). An important feature of agroecosystems is that through the functional interaction of system components, as a result of multiple varied and interacting relations, new properties or system qualities emerge at the system level (Gliessman, 2000). Soil quality itself can be viewed as an emergent property of a managed system. Emergent properties in an agricultural system include, for example: internal nutrient cycling and energy flow leading to nutrient storage and carbon retention; stability of soil structure and improved watershed functioning (e.g. reduced erosion and runoff); and mutualistic interactions that regulate pest populations, and suppress pathogens and plant disease. Many of these emergent system qualities build soil quality by improving dynamic soil quality properties. Altieri (2002) noted that use of agroecological technologies to diversify the agricultural system can optimize agroecosystem processes and maintain or improve soil quality and productivity. Broad-scale indices need to be identified to assess functioning at the system level, such as net primary production and biological productivity, stable levels of soil organic matter (OM), and biogeochemical mass balance and internal nutrient cycling (Ellert *et al.*, 1997).

The general objective of this chapter is to focus on the agricultural management system and its influence on soil quality. The ability or propensity of a system to build or improve soil quality is described, using information from two agricultural systems (conservation tillage and organic farming). Specific objectives are to evaluate both agricultural systems and agroecological approaches to soil quality management, and to outline the advantages and disadvantages of selected agricultural systems with respect to emergent qualities and the production, environment, and health aspects of soil quality.

Systems Approaches

Systems approaches are based on the concepts of systems theory. Systems theory has been described as a bridge between pure mathematics and the empirical sciences (Boulding, 1956). Others have outlined ways that systems thinking can be used as a framework to balance generality and detail (Rountree, 1977) or descriptive and mechanistic science (Bouma, 1997). Agriculture is just one of many disciplines in which general systems theory and systems approaches are used.

General systems concepts in an agricultural context

Spedding (1988) defined a system as a group of interacting components, operating together for a common purpose, capable of reacting as a whole to external stimuli within a specified boundary. Stephens and Hess (1999) emphasized that a system boundary is not predetermined; it is tied to the relevant spatial or temporal scale of the research question or management unit. A system could be a plant, a field, a farm, a watershed or a region. Policy makers are likely to choose spatially larger boundaries, whereas most agricultural researchers choose much smaller scales.

A closed system is one that has no inputs or outputs, i.e. no matter, energy or information transfers across the system boundary (Stephens and Hess, 1999). Open systems allow such transfers 'horizontally' among subsystems, 'vertically' between higher and lower level systems, or both (Hart, 1984). In vertical transfers, the inputs and outputs of one system are components of its suprasystem (or higher level system of which it is a component). For example, nutrient and sediment losses from farm field or agroecosystem contribute to degradation of water quality in the watershed system. A related example of horizontal transfer would be nutrients and sediments lost from one field that accumulate in a neighbouring field – both equal subsystem components of possibly the same farm. Modern agricultural systems are, by definition, open systems (Odum, 1984), with, at a minimum, inputs of nutrients and management, and outputs of harvestable crops. Further, the desired outputs from agroecosystems (crop products) cannot be sustained without human inputs, therefore humans are an integral part of agroecosystems. Considering humans as a component within open agroecosystems introduces management decision making and its external influences (or controllers) into agricultural research.

In addition to being open or closed, systems can also be classified as 'hard' or 'soft'. The distinction between the two lies primarily in whether or not humans are included within the system boundary. Soft systems, which include purposeful human activity, are driven by reasons (as opposed to causes) and have negotiated, common goals (Röling, 1997). In addition to this, systems also exhibit hierarchical organization (Allen and Starr, 1982; Hart, 1984; O'Neill et al., 1986). While there has been much debate about whether hierarchies are a human construct or a true phenomenon (Allen and Starr, 1982), the ability to organize our thinking into spatio-temporal units has clear benefits for research and management, and could serve as a central paradigm for a general theory of agricultural systems (Hart, 1984).

Often systems are considered nested within other systems (Allen and Starr, 1982). Figure 15.1 illustrates the nested quality of a hypothetical agroecosystem along with the concepts of 'open', 'hard' and 'soft'. Connectedness, or transfer of materials, energy or information, between and among open systems and subsystems is an important component of this hierarchical organization (Allen and Starr, 1982), contributing to many system properties or management outcomes (connections or subsystem interactions are shown in the figure as arrows). Ellert et al. (1997) argue that understanding of these interactions is a prerequisite to effective agroecosystem and soil quality management.

Application of agricultural systems theory to soil quality

The central underlying assumption of a systems approach to agriculture is that since agroecosystems are complex, then the interrelationships among environmental conditions, management and biological processes are important in determining outcomes such as yield, pest pressure and environmental impact (Drinkwater, 2002). Using systems theory to organize thinking about this complexity can be beneficial for both agricultural research and management.

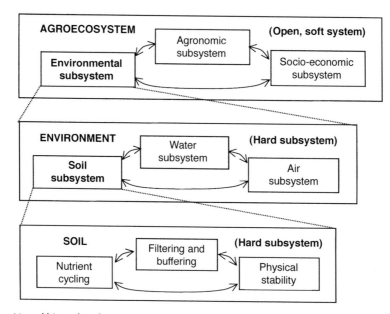

Fig. 15.1. Nested hierarchy of agroecosystem components, including the soil system and its component functions, encompassing soil quality (adapted from Andrews *et al.*, 2002).

Adding to system complexity are issues of scale, which can be addressed through hierchical systems constructs. Temporal and spatial scales usually coincide, with spatially larger processes often requiring longer time periods than spatially smaller ones. Ellert *et al.* (1997) defined the relationships between spatial and temporal scales for a variety of soil systems, subsystems and components, illustrating this proportionality between space and time for soils. As a result of this relationship, different management approaches or practices may require different levels of systems analysis for assessment of sustainability or quality. For example, soil biophysical processes are often defined at the field level, whereas rotational cropping might be assessed at a field or farm level using a time-scale at least equivalent to the rotation length. Filter-strip systems probably need to be assessed at a watershed or regional level at a time-scale long enough to allow species establishment and account for precipitation variability. Microeconomics would be properly addressed at the farm level. A watershed- or regional-level analysis would be appropriate to examine macroeconomic sustainability (Lowrance, 1990).

Hoosbeek and Bouma (1998) emphasized the need for the use of different methodologies for land quality assessment depending on scale, and Karlen *et al.* (1998) listed various potential soil-quality indicators according to spatial scale of interest. These calls for scale-dependent methods underscore the role of systems approaches as a necessary component for designing soil assessment strategies. Systems approaches can also be used to place reductionist research into a broader context at the whole farm or agroecosystem level, like pieces of a puzzle, bringing research results closer to application.

Hart (1984) states that a hierarchical systems approach to study agriculture allows for greater integration of biological and social sciences. Ellert *et al.* (1997) compare differences in soil quality research when using an ecosystem versus a reductionist perspective. In part, they conclude that using a systems view: (i) places soil within a larger ecosystem; (ii) recognizes a broad array of support services or soil functions (beyond crop production); (iii) incorporates humans as internal controllers; (iv) allows for multiple management goals including production, conservation and aesthetics; and (v) uses integrative science to identify possible pathways to sustainability. These benefits are needed to make soil quality research results applicable in a management context.

Agroecology and Agroecosystems

Agroecology is a subdiscipline of ecology that applies the principles of ecosystems to agriculture. Tansley (1935) first used the term 'ecosystem' to define a set of interacting biotic and abiotic (physical and chemical) components. The concept was popularized by Odum (1953), who recognized the need to examine a range of spatiotemporal scales. In agroecology, an agroecosystem can be broadly defined, spatially and temporally, to incorporate both the ecological and the social components of the (soft) system or defined more narrowly to include only environmental interactions, such as crop–weed competition (Altieri, 1995).

Application of ecological theory to soil quality

There are strong parallels between the agroecological approach and the agricultural systems approach and their application to agriculture. However, there are several concepts emanating from ecological theory that may benefit the study of soil quality management in agricultural systems.

Structure and function

One dominant theme in ecology and agroecology is the identification and interaction of system structure and function. This theme is found across scales: it applies to communities (groups of interacting populations) (e.g. Bongers, 1990), ecosystems (Drinkwater et al., 1998) and landscapes (e.g. Lowrance and Groffman, 1987). To ecologists, the term structure can have various meanings. The most prevalent definition applies to the organization of a food web: a purely biological function. In this case structure refers to the biological species that interact with one another through predator–prey, mutualistic, competitive and other relationships to form the complex cycle of energy flow through a system.

Based on the above definition for the term structure, the function versus structure question refers mainly to species population size or diversity versus its activity and interactions (i.e. ecosystem services). Soil quality assessment frequently attempts to estimate soil performance of essential functions or services (e.g. Doran and Parkin, 1994; Carter et al., 1997; Andrews and Carroll, 2001).

Having diverse structure may lead to increased functioning. Having many species in a system means that each species occupies its own niche space (sensu Hutchinson, 1959); this could mean that different system services are performed. However, there can be considerable niche overlap, which could result in functional redundancy. For example, given two species of nematodes, one could be a predator and one a fungal feeder or they could both be fungal feeders. In the former case, a greater number of functions are being performed but in terms of structure (species richness, a type of biodiversity) the two examples are equal. Beare et al. (1995) argue that the spatiotemporal heterogeneity of interactions (creating a hierarchy of scales of controlling factors) in the soil food web leads to increased functioning such as biogeochemical cycling.

In addition to species diversity, other types of system structure (or diversity) can be defined. The most obvious structural component is physical structure (below or above ground). This could refer to plant structures (e.g. root or shoot architecture) that would be closely related to plant species diversity but have implications for competition for light, water or nutrients among them, or provide refugia for predatory insects (e.g. Letourneau, 1990). In soils, physical structure has implications not only for species composition (e.g. plant access to nutrients and water or microbial protection from predators) but also for numerous other system functions such as water partitioning and filtering or buffering. Again, the system effects of structure depend on the scale of interest. Many management practices impact soil physical structure, e.g. tillage, organic amendments or traffic pattern. Structure (or diversity) can also be defined temporally (Gliessman, 2000). For instance, growing cover crops or using crop rotations increases temporal diversity, with implications for disease resistance, erosion losses, carbon sequestration and insect pest management. Table 15.1 outlines the influence of management practices on soil quality function.

Stability and disturbance

Over time, the structure and function of a healthy agroecosystem should remain relatively stable (oscillating around a stable equilibrium), even in the face of disturbance. If a stress or disturbance does alter the sustainably managed agroecosystem, it should be able to bounce back quickly. Stability in this context has two components: resistance – the ability of the ecosystem to continue to function without change when stressed by disturbance; and resilience – the ability of the ecosystem to recover after disturbance (Pimm, 1984; Herrick and Wander, 1998; Seybold et al., 1999). Because agroecosystems have reduced structural and functional diversity compared with natural systems, they may have less resilience than natural systems (Gliessman, 2000). One hypothesis supporting this claim is that structural (food-web) diversity leads to functional redundancy, such that several populations may perform similar functions in the system. If one of these redundant species is extirpated, another will quickly fill the void, hence leading to functional resiliency.

Complexity and emergent properties

Increased structural and functional diversity impart complexity to the system. Complexity may lead to system stability (through enhanced resistance and resilience). However, the relationship between complexity and stability can be complex (Pimm, 1984). The outcome of the relationship depends on which factor in the system is being disturbed. In addition, disturbances must be defined for various spatial (and temporal) scales, making generalizations difficult.

There is strong recognition across disciplines of the need for spatial approaches to resource management to deal with system complexity (e.g. Christensen et al., 1996; Herrick et al., 2002). Applying hierarchy theory to agroecosystems, one can identify unique phenomena with increasing scale, not predictable from lower levels of organization (Odum, 1984). These unique phenomena, or emergent properties, are an important property of complex systems arising from the functional interaction of system components. They are the source of the phrase, systems are 'greater than the sum of their parts' (Odum, 1953; Rountree, 1977; Stephens and Hess, 1999).

Some argue that emergent properties are a function of incomplete understanding of the lower level components (Pomeroy et al., 1988). Nevertheless, interactions within and between subsystems continue to result in properties not predictable from examination of the lower level alone. The phenomenon of 'overyielding' (i.e. productivity of a diverse

Table 15.1. General effects of management practices on soil functions.

Management practices	Productivity	Nutrient cycling	Water and solute flow	Biodiversity and habitat	Filtering and buffering	Structural support	Physical stability
Tillage	X	X	X	X	X	X	X
Organic amendments	X	X	X	X	X	X	X
Residue cover	X	X	X	X			X
Rotations	X	X		X			
Fertilizer	X	X		X	X		
Pesticide	X			X	X		
Irrigation	X	X	X	X			X
Drainage	X	X	X	X			
Filter strips	X				X		
Timing applications	X	X	X				

system greater than that of a less complex system) seen in many intercropping systems (Vandermeer, 1990) is an example of an emergent property. Because overall function and viability of a system emerge from the interactions of component systems (Bossel, 2001), understanding the principle of emergent properties is essential for agroecosystem management.

Sustainability and soil quality as emergent properties

In the last two decades, interest in the concept of sustainable agriculture has grown considerably (Jackson, 1980; Rodale, 1983). Although there is no one definition of sustainable agriculture, most incorporate three main components: ecological (e.g. spatial and temporal relations, diversity, stability and resilience); economic (i.e. resource distribution and allocation); and social (e.g. equity, access, stewardship and institutions) (Miller and Wali, 1995). However, sustainability is not a component of the agroecosystem *per se* but rather a property (or goal) that stems from appropriate interactions within the system. Hence, sustainability is an emergent property of a system (*sensu* Röling, 1997).

Within the cognitive context laid out by Schjønning *et al.* (Chapter 1, this volume), the individual farmer's goal is typically maximization of yield or economic net revenue. As a soft system, however, agroecosystem sustainability is seen as emerging from many interacting components (or possible management goals), including environmental (soil, air and water quality) and social (e.g. quality of life) as well as economic (i.e. net revenue) subsystems. Site specificity of sustainable practices due to differences in soil types, climatic factors and a myriad of economic and social dimensions add to the significant challenges facing farm managers. The challenge to scientists is to provide understanding of alternatives at a variety of scales: spatial scales from plant–soil interactions to implications for regional and global agriculture (Miller and Wali, 1995), and temporal scales

from short-term nutrient cycling to long-term interactions. Due to difficulties in defining sustainability (i.e. the lack of a cohesive societal cognitive context), indicators to assess sustainability are difficult to identify and interpret. For instance, environmental stewardship goals often wane with decreasing land tenure (a societal construct).

Similar to agroecosystem sustainability, soil quality is an emergent property at a lower level of agroecosystem organization: the soil system. The concept of soil quality is related to all three of areas of sustainability, with its emphases on productivity (an economic goal), environment and human health (a social goal). Emerging soil quality arises from the interaction of the chemical, physical and biological properties and processes that comprise the soil system (*sensu* Karlen *et al.*, 1998). It is this synergy that makes soil quality measurement so challenging and leads us to the current emphasis on manipulating management systems rather than individual indicators to assess soil quality. Thus, using practices that enhance or maintain soil quality may be a more attainable goal than trying to enhance sustainability, however closely the two concepts may be related.

Agricultural Management Systems that Build Soil Quality

Both Doran *et al.* (1996) and Lal (1998) stressed the importance of holistic management to optimize the multiple functions of soil and thereby promote and maintain soil quality. Building and/or restoring soil quality may involve a range of measures, adopted at the field, farm or watershed level, to optimize resource conservation. Integrative approaches to land use, such as conservation tillage and organic farming, have noted that management inputs (e.g. crop-residue level) and system diversity (e.g. crop rotation) strongly influence dynamic soil quality properties (Carter, 1994; Drinkwater *et al.*, 1998). Table 15.2 lists ecological attributes and emergent properties of various agricultural management systems that can improve and build soil quality.

Table 15.2.   Comparison of ecological factors and corresponding examples of agricultural management systems with emerging qualities at the system level that build soil quality and health.

Ecological factors affecting system stability and function	Selected system management strategies to build soil quality/health	Emerging qualities at the system level
Decrease disturbance • Chemical • Biological • Physical	Environmental integrity/quality • Minimize use of pesticides • Promote beneficial allelopathies • Use conservation tillage • Minimize soil erosion	• Beneficial mutualisms and interactions • Tertiary soil structure • Efficient energy use • Internal nutrient cycling • Internal management of pests and diseases • Improved water storage and regulation • Enhanced retention of organic C and N
Increase diversity/complexity • Species • Habitat • Trophic group	Diversify agricultural system • Rotations • Intercropping • Use of legumes and perennials • Field margins • Organic management	
Increase nutrient/energy flux • Organic matter • Nutrients	Conserve organic matter/nutrients • Rotation • Cover crops • Residue management • Organic amendments • Nutrient management	

Conservation tillage

Tillage operations modify soil structure and distribute energy-rich organic substances into the soil. Thus, the type and degree of tillage inputs can have a major influence on soil properties and processes, and thereby modify soil quality. Conservation tillage is an umbrella term covering a very wide range of diverse tillage practices that have, as a common characteristic, the potential to reduce soil and water loss relative to some form of conventional tillage (Mannering and Fenster, 1983). A well-accepted operational definition of conservation tillage is tillage, or a tillage and planting combination, that retains a 30% or greater cover of crop residue on the soil surface.

Management tools commonly used to achieve the above operational definition for conservation tillage are: (i) non-inversion tillage (usually implies replacement of a mouldboard plough with a chisel plough or cultivator); (ii) tillage depth confined to < 15 cm (deeper tillage may be retained in the row for row crops); and (iii) number of tillage passes minimized. The major outcome of these management options (relative to some conventional system that employs full-soil-profile tillage) is to provide some degree of permanent soil cover (i.e. 30% or more residue in the non-crop period), to increase the OM content and structural stability of the soil surface depth (< 15 cm) over time and, in many cases (i.e. depending on a reduction in vehicular traffic), to improve the soil structural form (pore size and continuity) of the lower non-tilled soil depth (> 15 cm). Soil stratification, which mainly involves enrichment of the soil surface with OM, is the dominant management outcome of conservation tillage (Franzluebbers, 2002). Although dependent on soil type and climate (Carter, 1994), soil stratification can impact on several system qualities such as nutrient storage and soil aggregation, improved water regulation at the soil surface and throughout the soil profile, and beneficial mutualistic interactions. In general, conservation tillage is yield neutral although yield variations, in comparison with conventional tillage, are possible due to soil and climatic interactions (e.g. limiting soil density and wet conditions) or under extreme forms of conservation tillage (e.g. no-tillage).

Emergent soil structural and nutrient quality

Soil OM is a relatively transient component of the soil that controls many chemical, physical and biological properties, and consequently influences a wide range of soil functions (see Dick and Gregorich, Chapter 7, this volume). Relative enrichment of the surface soil with OM results in an increase in microbial activity and a concomitant increase in the size and stability of soil aggregates (Carter and Stewart, 1996). Stratification of OM and biologically active soil carbon and nitrogen pools has implications for storage and release of soil nutrients (Franzluebbers, 2002). Increased biological activity at the soil surface, especially an increase in the soil microbial biomass, can directly influence soil quality functions such as plant growth, regulation of energy and soil structural stability (Carter *et al.*, 1999).

Although not confined to conservation tillage systems, the above development and formation of soil aggregation, in relation to OM turnover, is considered to be an emergent structural feature (Christensen, 2001). Table 15.3 illustrates the increasing soil structural and functional complexity that can emerge under OM enrichment, especially the emergence of the tertiary structure that can control many soil and biological processes. However, under conservation tillage, the improved tertiary structure is not associated with OM stratification alone. Evidence suggests that an interaction or combination of OM and reduction in tillage passes (i.e. reduced mechanical disturbance) serve to positively influence soil tertiary structure (Kay and Angers, 1999;

Bissonette *et al.*, 2001). These two components interact to provide improvement in tertiary structure.

The above scenario can have important implications for root growth and development, the way soil holds and partitions water, and flow of both water and air (Carter, 2002). Although variable with soil texture, OM enrichment and enhanced aggregate stability, in combination with minimum tillage, can reduce the propensity for soil compactibility and enhance soil resistance to deformation (Kay and Angers, 1999). Based on these emergent properties, the management options associated with conservation tillage would help improve both the resistance and resilience of the system.

Emergent water regulation features

Non-point or diffuse pollution is the water pollution associated with land-use activities. Surface runoff losses are related to low infiltration rates, the volume of sediment lost and the concentration of nutrients and chemicals in the runoff. Thus, practices that affect these factors will address agricultural non-point source pollution (see Locke and Zablotowicz, Chapter 14, this volume). Generally, control of soil erosion will reduce the loss of pesticides that are strongly adsorbed to soil particles, whereas control of runoff water volume can reduce the loss of moderate to weakly adsorbed pesticides.

Table 15.3. Emerging levels of soil structural and functional complexity in agricultural systems (after Christensen, 2001).

Soil structural entities	Soil functional features
Primary structure (scale in μm–mm)	
Organomineral complexes	Modification of microenvironment
Uncomplexed organic matter	Surface reactivity
	Chemical stabilization of organic matter
Secondary structure (scale in mm–cm)	
Aggregated organomineral complexes	Physical protection of organic matter
Uncomplexed organic matter	Soil porosity and aeration
Fine roots	Microfaunal habitat
Fungal hyphae	Water retention
Tertiary structure (scale in cm–m)	
Intact soil *in situ*	Macrofauna and bioturbation
Macropores	Pore continuity and preferential flow
Macro roots	Structural form and density

Table 15.4. Influence of conservation tillage on runoff and leaching parameters, compared to some form of conventional tillage, in a wide range of North American studies. After Carter (1998). Based on 40 studies on eight soil texture classes from sandy loam to clay, 1991–1996.

Measurement	No. of studies	Decrease (% of studies)	No effect (% of studies)	Increase (% of studies)
Runoff				
Runoff volume	16	38	38	25
Sediment in runoff	20	100	0	0
Phosphorus in runoff	13	69	8	23
Nitrogen in runoff	13	69	15	15
Pesticides in runoff	7	71	29	0
Leaching				
Leaching volume	5	20	0	80
Nitrogen leached	13	36	27	36
Pesticide leached	11	8	8	85

For many soil types, emerging soil and system properties under conservation-tillage and crop-residue systems can have an impact on water regulation in the soil profile (Langdale *et al.*, 1994). Usually surface soils under conservation tillage can better resist the harmful impact of raindrops and have a greater capacity to accept infiltrating water than their conventional-tilled counterparts. A continuous pore network down the soil profile can develop in loam to clay soils, which can have implications for rapid movement or leaching of nutrients and pesticides (Carter, 1994).

Carter (1998) summarized a wide range of mainly long-term field or farm experiments with conservation tillage with respect to both runoff and leaching (Table 15.4). Generally, the volume of runoff was not greatly influenced by conservation tillage. In some cases this was related to the presence of soil horizons or layers with poor permeability (Langdale *et al.*, 1994) or the development of surface crusts, which impeded the rate of water infiltration. Studies that characterized sedimentation showed a consistent decrease in mass of soil sediments in the runoff under conservation tillage, compared to conventional tillage. Of those studies that evaluated other pollutants, about 70% indicated that conservation tillage decreased the phosphorus (P), nitrogen (N) and pesticide content of the runoff. In the studies that measured leaching losses, both leaching volume and pesticide content of the leachate were increased by conservation tillage in 80% of the studies, whereas

the amount of N leached was not strongly related to tillage system. Generally, most of the studies that measured pesticide loss in runoff or leachate indicated that only minor amounts were lost (Table 15.4).

Overall, the data from the above studies indicated that conservation tillage is a useful management option for control or reduction of sedimentation, but less effective for decrease of P movement. Generally, P resides in both inorganic and organic forms in the soil, and the latter is related to increasing levels of OM. In the presence of crop residue, a large proportion of the organic P would be in the particulate organic form (i.e. partially decomposed crop residue). Furthermore, the activity and size of the soil microbial biomass is increased by additions of crop residue. This tends to increase or maintain the level of labile organic P at the soil surface under conservation systems, which may – depending on concentration level – present an environmental risk (Kleinman *et al.*, 2000).

Emergent mutualistic interactions

Recent studies have emphasized that crop resistance or tolerance to diseases and pests can be related to soil quality or optimum soil properties (Altieri, 2002). Conservation-tillage practices can have a varied effect on mutualistic interactions among soil organisms (Bockus and Shroyer, 1998). Beneficial interferences (mutualisms) are seen as emergent system qualities (Gliessman, 2000). In this context, soil health can be characterized as the efficiency with which soil functional processes are able to support beneficial mutualistic interactions (Sturz and Christie, 2003). Enhanced soil biological activity, energy flow and nutrient cycling, associated with adequate levels of soil OM, are considered the key processes that influence and encourage such interactions.

Soils with a biological propensity to reduce the harmful effect of soilborne pathogens are termed disease suppressive. Sturz and Christie (2003) have outlined the various beneficial microbial mechanisms involved, along with management strategies that can both enhance and manipulate these beneficial mutualistic interactions. For example, selection of complementary rotation crops and tillage systems that retain crop residues near the soil surface are conducive to a build-up of beneficial microbial communities over time (Altieri, 2002). Although the relationship between crop management and microbial diversity, and its consequences for soil health, is unclear, there is some consensus that species diversity is beneficial for agricultural systems subject to ongoing perturbations (see Brussaard et al., Chapter 9, this volume). As described by Sturz and Christie (2003), managed soil microbial communities in agroecosystems may benefit from moderate perturbation stress and result in the emergence of a beneficial microflora over time.

Emerging beneficial mutualistic interactions that provide incidences of plant disease suppression have been recorded for several conservation tillage systems. Sturz et al. (1997) showed that elevated populations of saprophytic microflora associated with incorporated crop residues in the surface soil, under conservation tillage for cereal rotations (barley–soybean), were related to an inhibition of Rhizoctonia solani activity. This decreased disease expression was not associated with a reduction per se in disease-causing microbes in the rhizosphere, but rather the development of a relatively extensive and balanced community of compatible disease antagonists that could effectively outcompete the pathogens for space and energy substrate in the root zone (Sturz et al., 1997). Similar results on plant disease suppression have been found for root crops (e.g. Solanum

tuberosum L.) under conservation-tillage comparisons (Carter and Sanderson, 2001). Peters et al. (2003) demonstrated that a range of plant pathogens (e.g. R. solani, Fusarium spp., Helminthosporium solani and Phytophthora erythroseptica) could be significantly reduced by a combination of crop rotation and conservation tillage.

Challenges in conservation-tillage systems

The potential emergent properties associated with conservation-tillage systems, as outlined above, may be impeded under certain circumstances. Adoption of extreme versions of conservation practices, such as no-tillage, on certain soil types can result in less than optimum soil aeration associated with excessive soil compaction. This, in turn, can lead to the adverse build-up or activity of deleterious microbes, resulting in plant root disease (Allmaras et al., 1988) or adversely influencing soil structure (Kay and Angers, 1999). Overall, a good quality soil physical environment is an important indicator for root health under conservation tillage (Sturz et al., 1997). In some instances, disease reduction under conservation tillage is associated with changes in environmental conditions and is thus site or region specific.

Generally, the emerging soil quality condition is related to a combination of both conservation-tillage and optimum crop-rotation practices. Conservation tillage needs to be combined with beneficial crop rotations to control plant diseases (Bockus and Shroyer, 1998). Such combinations address all the ecological factors listed in Table 15.2.

Organic cropping systems

Legal definitions of organic agriculture emphasize the elimination of synthetic fertilizers and pesticides and replacement with organic soil amendments and biological pesticides (National Organic Program, 2002). Although the avoidance of synthetic inputs is a key characteristic of organic agriculture, organic management is not simply a matter of substituting *organic* inputs for conventional inputs. The distinguishing feature of organic agriculture has always been the application of a holistic, systems approach to crop production (Howard, 1945; Balfour, 1948). This systems view of food production and the underlying assumptions and goals that go with it are the basis for identifying the management tools used in organic agriculture.

Organic agriculture is a biologically intensive agricultural system that aims to optimize biological processes that are both directly and indirectly important for maintaining crop productivity (Organic Farming Research Foundation, 2002). Organic agriculture has a unique relationship to the concept of soil health because the proponents and practitioners have always assumed that maintaining healthy soil is crucial to the production of healthy, nutritious food (Howard, 1945; Balfour, 1948). Soil health is seen as the foundation that supports the entire agroecosystem. Recognition of the central role played by soil OM in maintaining soil health is a key principle guiding soil-management strategies. The practices used in organic farming systems reflect the assumption that optimizing soil health will increase plant vigour and contribute to prevention of pest problems (Coleman, 1989).

Structure and function of organically managed systems

In agroecosystems, management plays a large role in determining structure. A major advantage of organic food production systems is the potential for improvements in ecosystem functions that may reduce the need for agrochemical inputs and the resulting environmental consequences. Many management practices used by organic farmers are known to enhance soil OM and promote biological nutrient cycling. There are four common management tools used in organic cropping systems that have a significant impact on soil function and thus on soil health: (i) organic residues other than senescent crop residues are added on a regular basis (green manures, animal manures, composts); (ii) crop species are diversified in space and time; (iii) active plant growth in the soil is maintained as much as possible (through relay cropping, intercropping and cover crops); and (iv) pesticides are not applied.

The studies cited in this chapter generally classified farms/treatments as organic when plant nutrients were supplied through the use of leguminous green manures and/or organic soil amendments and no synthetic fertilizers and/or pesticides (including herbicides) were used. Conventional farms/treatments generally used synthetic fertilizers and/or pesticides. In reality, farming practices fall along a continuum rather than into discrete groups, and use of these management designations does not imply that some overlap in terms of specific practices did not occur, particularly in on-farm studies.

Emergent aspects of nutrient and energy flow ecosystem functions

The extensive literature comparing organic and conventional cropping systems demonstrates conclusively that organic management usually leads to clear improvements in soil quality, particularly in those properties related to soil organic carbon (SOC) pools (e.g. Wander et al., 1994; Drinkwater et al., 1995, 1998; Reganold et al., 2001). Studies of actual farms as well as long-term systems experiments indicate that organically managed systems do have increased levels of SOC (Drinkwater et al., 1998; Robertson et al., 2000). These increases in SOC are most pronounced in the labile fractions such as particulate OM (Wander et al., 1994). Decomposer biomass and secondary consumer populations have also been shown to increase (Reganold et al., 1993; Scow et al., 1994; Wander et al., 1994), reflecting the increased role of decomposers in determining the availability of major nutrients. As a result, the overall capacity of the soil to supply nutrients such as N and P through mineralization is generally increased (Fig. 15.2; Wander et al., 1994; Drinkwater et al., 1995). These increases in C pools and biological activity impact chemical and physical soil properties. Organically managed soils have greater aggregate stability (Colla et al., 2000; Schjønning et al., 2002), increased infiltration rates, and increased water-holding capacity compared to soils under conventional management. These

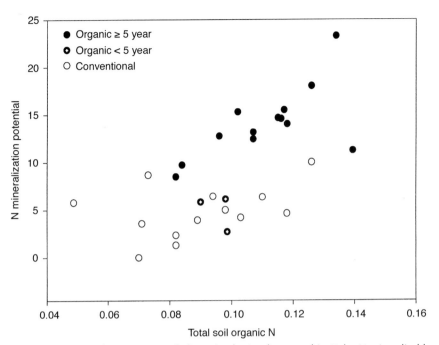

Fig. 15.2. Nitrogen mineralization potential (determined using the anaerobic, 7-day N-mineralizable potential incubation method; Drinkwater et al., 1996) relative to total soil N among organically and conventionally managed farm sites. Potential N mineralization is proportionately greater in soils under organic management indicating increased pools of labile organic N relative to total organic N (modified from Drinkwater et al., 1995).

changes are often significant enough to impact plant–water relations (Petersen et al., 1999; Colla et al., 2000).

Organically managed systems increase organic C and N storage partially through achieving a more favourable balance of nutrient inputs and exports compared to conventionally managed systems (Drinkwater et al., 1998; Clark et al., 1999; Poudel et al., 2001). Mass balance calculations in long-term studies indicate that the conversion of N inputs into organic forms that can be retained in the soil was greater in organically managed systems compared to conventional systems (Clark et al., 1998; Drinkwater et al., 1998; Poudel et al., 2001). Despite a significant N surplus, the conventional systems receiving inorganic N did not show a net accrual of soil N. These increases in SOC pools and living biomass do not coincide with net increases in inputs of organic residues or N (Drinkwater et al., 1998). The increased retention of soil C and N can be attributed to multiple aspects of organic systems including: (i) the use of

organic amendments as N sources (Harris et al., 1994); (ii) expanding the time frame of C fixation and N uptake (McCracken et al., 1994); (iii) increased rotational diversity (Tisdall and Oades, 1979; Drinkwater, 1999; Drinkwater and Puget, 2001; Puget and Drinkwater, 2001); and (iv) management-induced differences in microbial community structure and function (cf. Anderson and Domsch, 1990; van Bruggen and Semenov, 2000).

Emergent properties related to biodiversity

A few studies of intact organic systems have provided evidence that these systems do seem to promote plant health and reduce the incidence of plant pathogens (Vereijken, 1989; Workneh et al., 1993). An on-farm comparison of organic and conventional tomato (*Lycopersicon esculentum* L.) systems found that organically managed soils had reduced levels of root disease compared to conventionally managed soils (Workneh et al., 1993). Subsequent experiments confirmed that the organically managed soils inhibited growth of the root pathogen, and that this was at least partially due to the presence of antagonistic microorganisms (Workneh and van Bruggen, 1994). Many practices used in organically managed systems have been linked to reductions in plant diseases that are mediated through a variety of mechanisms involving the microbial community, including use of organic fertility sources (Hoitink and Boehm, 1999; Bulluck and Ristaino, 2002), incorporation of green manures (Abawi and Widmer, 2000) and increased rotational diversity (Cook and Baker, 1983). When these practices are used in isolation, however, they do not always result in reductions of plant diseases (Cook and Baker, 1983). For example, some cover crops have been shown to increase disease problems in conventional systems (Abawi and Widmer, 2000).

Crop productivity

Although some cases of reduced yields on a per area basis have been reported in organically managed systems, competitive yields have been found in numerous on-farm and experiment-station comparisons of organic and conventional cropping systems (e.g. Andow and Hidaka, 1989; Clark *et al.*, 1998; Colla *et al.*, 2000; Reganold *et al.*, 2001). Often, when yields are reduced under organic compared to conventional management, it is because the organically managed system has not yet reached an equilibrium state: it is still in transition from conditions developed under previous management (Liebhardt *et al.*, 1989). In other cases, clear problems that can be remedied are identified (Hannukkala *et al.*, 1990). When yields are viewed relative to other resources, organic systems generally outperform conventional systems. For example, organic production systems generally have higher yields per unit of fossil fuel energy used (Mäder *et al.*, 2002). From a soft-system perspective, the price premiums available for organic crops increase the efficiency of cash input:output ratios (i.e. net revenues), making organic systems viable even with some yield reductions.

Considering that in most industrialized countries only a very small effort has focused on improving these systems (Lipson, 1997) it is likely that, with additional research, the management of biological processes in the systems can be improved. It is also likely that cultivars that are better adapted to organic production systems can be developed. With an increased research effort, yields comparable to conventional systems on a per hectare basis may be achieved for most crops.

Cascading effects of soil quality on other agroecosystem properties

The effects of soil quality on the weed community and arthropod pest dynamics in agroecosystems are not well understood. Certainly there are differences in weed and insect pest dynamics on organic farms that cannot be adequately explained by current knowledge of agroecosystems. Changes in the soil environment and microbial community may contribute to reductions in weed pressure (Gallandt *et al.*, 1999) and arthropod damage (Phelan *et al.*, 1995) in organic farming systems. However, with the exception of a few selected topics, data are sparse, probably due to the complex nature of these interactions. Clearly, the soil environment impacts weed community dynamics through a variety of mechanisms, both direct and indirect (Liebman and Gallandt, 1997). For instance, the reduced nitrate concentrations that occur under organic management regimes have been linked to reductions in weed germination and competitive ability (Dyck and Liebman, 1995; Dyck *et al.*, 1995). In addition to soil quality, a complex set of factors related to soil microbial and faunal community composition also influence weed vigour and population dynamics in these systems (DiTomaso, 1995; Liebman and Gallandt, 1997; Gallandt *et al.*, 1999). While opportunities do exist to supplement other non-chemical weed-management strategies by fostering weed-suppressive soil environments (Gallandt *et al.*, 1999) the extent to which soil quality can assist in weed management has not been fully explored.

Soil quality effects on arthropod pests are less direct in that they stem from soil–plant interactions that, in turn, influence population dynamics of plant-eating insects. There is evidence that plant tissue composition differs under conventional and organic management (Letourneau *et al.*, 1996). In a greenhouse study, Phelan *et al.* (1995) found that the European corn borer (*Ostrinia nubilalis*) preferentially laid eggs on maize (*Zea mays* L.) plants growing in conventionally managed soils. The intriguing herbivore preference observed in this experiment may have been due to the mineral content of the leaves, but the mechanism, as well as the significance of this finding at the field scale, remains unclear. Changes in soil nutrient availability alter plant tissue composition and have been linked to changes in the growth and reproductive rates of herbaceous arthropods (Letourneau, 1997; Ritchie, 2000). However, the response of plant-feeding arthropods to changes in host quality are quite varied (Brodbeck and Strong, 1987; Letourneau, 1997) and are also influenced by other factors such as climate and the composition of the aboveground biotic community (Strauss, 1987; Stiling and Rossi, 1997; Ritchie, 2000). Although the influence of soil management on plant quality and the connection between plant-tissue composition and herbivore populations have both been documented in agricultural systems, the role of soil–plant–herbivore interaction in agroecosystems remains unclear. For example, the so-called *nitrogen limitation hypothesis* predicts that foliage N can be an important factor affecting herbivore populations and has been studied extensively in natural as well as agricultural ecosystems (Letourneau, 1997). In agriculture, the expectation is that reduced N availability will lead to lower rates of herbivore damage. This hypothesis is clearly supported in greenhouse studies where large-scale ecosystem processes are absent; however, it is not supported in most field-scale experiments (Letourneau, 1997). Thus, even a fairly simple case involving changes in a single nutrient appears to involve numerous interacting factors. The role of soil quality in plant insect resistance is an area in need of more research.

Resistance and resilience

Very few studies have explicitly addressed hypotheses about either resistance or resilience in agroecosystems. One study did examine microbial-community function in response to a wet–dry cycle using organically and conventionally managed soils (Lundquist *et al.*, 1999a). The response of the microbial community to soil drying and re-wetting was similar. Relative changes in microbial biomass and respiration did not differ between soils from the two management systems (Lundquist *et al.*, 1999b). Organic systems may be able to maintain yields during times of increased environmental stress. Yields in non-irrigated cropping systems are often greater in organic compared to conventional systems under drought conditions (Petersen *et al.*, 1999). Colla *et al.* (2000) found that soil structural changes were significant enough to alter infiltration and water-use dynamics to the extent that irrigation in the organic cropping system could potentially be reduced.

Organic management and human and animal health

Determining the impact of organic management on human and animal health through its impacts on soil quality is a truly difficult question. To the extent that organic systems are successful at eliminating the use of agrochemicals that have been linked to environmental degradation and identified as threats to human health, one can say that organic systems will contribute positively to human health (Clancy, 1997). Although differences in nutritional value of foods produced under organic and conventional systems have been documented, they are often inconsistent (Woese *et al.*, 1997) and the methodology differs across studies so that cause–effect relationships cannot be identified (Clancy, 1997).

Challenges in organic management systems

Although soil quality is generally improved by implementation of organic production systems, there are some aspects of organic management that are probably detrimental to soil health. First, because herbicides are not used, tillage intensity tends to be greater in organic than in conventional systems (cf. Wander *et al.*, 1994). Organic management *per se*, does not rule out the use of no-till or direct-drill planting as long as supplementary tillage combined with use of mulches is used for weed control. Conservation or alternating tillage practices need to be developed for organic systems (Drinkwater *et al.*, 2000). Secondly, although overall nitrate leaching losses are usually reduced in organically managed systems, it is clear that, under certain circumstances, incorporation of a large biomass of leguminous residues will result in significant nitrate leaching (Drinkwater *et al.*, 2000). Improved methods for managing low C:N residues, particularly green manures, need to be developed. Finally, many organic systems, particularly perennial crops such as stone fruit production, are dependent on applications of heavy metals as fungicides (i.e. Cu). Over time these heavy metals will accumulate in the soil and may reach levels that are detrimental to soil organisms (see Brussaard *et al.*, Chapter 9, this volume) and plants, and may also become environmental contaminants.

Conclusions

Both the agricultural systems and agroecological approach place soil within a larger ecosystem and allow for multiple management goals and integrative science. These approaches allow the translation of general ecological perspectives and principles into practical soil-management decisions, and provide a potential to improve or build soil quality. An understanding of ecosystem interactions, such as the relation between biodiversity and agroecosystem function, is seen as a prerequisite to effective soil quality management. Agroecosystems are driven by human reason and negotiated goals, and therefore farm management decisions are a central focus. In most cases, soil quality must be built into the farming system using best-management scenarios.

A systems approach emphasized the issue of both hierarchical organization and scale. Assessment of both conservation tillage and organic farming for improving soil quality must take into consideration a time-scale, as many of the benefits for soil quality are not immediate. Further, the needed employment of crop-rotation practices in conservation tillage and organic systems emphasizes the importance of hierarchical organization. The ecological concepts of structure and function have significant implications for soil quality improvement, specifically via species diversity. Interaction of biological species (especially through mutualisms) and the various structural components impart complexity to the agricultural system. This in turn has implications for stability, especially the components of resistance and resilience, in agricultural systems subject to ongoing perturbations.

Integrative approaches to land use and management inputs can manipulate and build dynamic soil quality properties. Management strategies or tools that maintain environmental integrity, diversify the farming system and conserve OM and nutrients allow beneficial system qualities to emerge over time. These emerging system qualities arise from the interactions of the various properties and processes that comprise the soil–plant system. As illustrated by conservation-tillage and organic cropping systems case studies, emerging system qualities such as beneficial mutualisms, tertiary soil structure, internal nutrient cycling and retention, improved water storage and enhanced retention of organic C and N are common features of improved farming systems that lead to a concomitant improvement of soil quality.

References

Abawi, G.S. and Widmer, T.L. (2000) Impact of soil health management practices on soilborne pathogens, nematodes and root diseases of vegetable crops. *Applied Soil Ecology* 15, 37–47.

Allen, T.F.H. and Starr, T.B. (1982) *Hierarchy: Perspectives for Ecologial Complexity.* University of Chicago Press, Chicago, Illinois, 310 pp.

Allmaras, R.R., Kraft, J.M. and Miller, D.E. (1988) Effects of soil compaction and incorporated crop residue on root health. *Annual Review of Phytopathology* 26, 219–243.

Altieri, M.A. (1995) *Agroecology: the Science of Sustainable Agriculture.* Westview Press, Boulder, Colorado, 433 pp.

Altieri, M.A. (2002) Agroecology: the science of natural resource management for poor farmers in marginal environments. *Agriculture, Ecosystems and Environment* 93, 1–24.

Anderson, T.H. and Domsch, K.H. (1990) Application of eco-physiological quotients (qCO_2 and qD) on microbial biomasses from soils of different cropping histories. *Soil Biology and Biochemistry* 22, 251–255.

Andow, D.A. and Hidaka, K. (1989) Experimental natural history of sustainable agriculture: syndromes of production. *Agriculture, Ecosystems and Environment* 27, 447–462.

Andrews, S.S. and Carroll, C.R. (2001) Designing a soil quality assessment tool for sustainable agroecosystem management. *Ecological Applications* 11(6), 1573–1585.

Andrews, S.S., Karlen, D.L. and Mitchell, J.P. (2002) A comparison of soil quality indexing methods for vegetable production systems in Northern California. *Agriculture, Ecosystems and Environment* 90, 25–45.

Balfour, E. (1948) *The Living Soil*. Faber and Faber, London.

Beare, M.H., Coleman, D.C., Crossley, D.A. Jr, Hendrix, P.F. and Odum, E.P. (1995) A hierarchical approach to evaluating the significance of soil biodiversity to biogeochemical cycling. *Plant and Soil* 170, 5–22.

Bissonnette, N., Angers, D.A., Simard, R.R. and Lafond, J. (2001) Interactive effects of management practices on water-stable aggregation and organic matter of a Humic Gleysol. *Canadian Journal of Soil Science* 81, 545–551.

Bockus, W.W. and Shroyer, J.P. (1998) The impact of reduced tillage on soilbourne plant pathogens. *Annual Review of Phytopathology* 36, 485–500.

Bongers, T. (1990) The maturity index: an ecological measure of environmental disturbance based on nematode species composition. *Oecologia* 83, 14–19.

Bossel, H. (2001) Assessing viability and sustainability: a systems-based approach for deriving comprehensive indicator sets. *Conservation Ecology* 5(2), 12 [online]. Available at: http://www.consecol.org/vol5/iss2/art12/

Boulding, K.E. (1956) General systems theory – the skeleton of science and management. *Management Science* 21, 197–208.

Bouma, J. (1997) The land use systems approach to planning sustainable land management at several scales. *Special Issue: Geo-Information for Sustainable Land Management. ITC Journal* 3/4, 237–242.

Brodbeck, B. and Strong, D.R. (1987) Amino acid nutrition of herbivorous insects and stress to host plants. In: Barbosa, P. and Schultz, J. (eds) *Insect Outbreaks*. Academic Press, New York, pp. 347–364.

Bulluck, L.R. and Ristaino, J.B. (2002) Effect of synthetic and organic soil fertility amendments on southern blight, soil microbial communities, and yield of processing tomatoes. *Phytopathology* 92, 181–189.

Carter, M.R. (ed.) (1994) *Conservation Tillage in Temperate Agroecosystems*. CRC Press, Boca Raton, Florida, 390 pp.

Carter, M.R. (1998) Conservation tillage practices and diffuse pollution. In: Petchey, A.M., D'Arcy, B.J. and Frost, C.A. (eds) *Diffuse Pollution and Agriculture II*. Scottish Agricultural College, Edinburgh, pp. 51–60.

Carter, M.R. (2002) Soil quality for sustainable land management: organic matter and aggregation interactions that maintain soil functions. *Agronomy Journal* 94, 38–47.

Carter, M.R. and Sanderson, J.B. (2001) Influence of conservation tillage and rotation length on potato productivity, tuber disease and soil quality parameters on a fine sandy loam in eastern Canada. *Soil and Tillage Research* 63, 1–13.

Carter, M.R. and Stewart, B.A. (eds) (1996) *Structure and Organic Matter Storage in Agricultural Soils*. CRC Press, Boca Raton, Florida, 477 pp.

Carter, M.R., Gregorich, E.G., Anderson, D.W., Doran, J.W., Janzen, H.H. and Pierce, F.J. (1997) Concepts of soil quality and their significance. In: Gregorich, E.G. and Carter, M.R. (eds) *Soil Quality for Crop Production and Ecosystem Health*. Elsevier, Amsterdam, pp. 1–19.

Carter, M.R., Gregorich, E.G., Angers, D.A., Beare, M.H., Sparling, G.P., Wardle, D.A. and Voroney, R.P. (1999) Interpretation of microbial biomass measurements for soil quality assessment in humid temperate regions. *Canadian Journal of Soil Science* 79, 507–520.

Christensen, B.T. (2001) Physical fractionation of soil and structural and functional complexity in organic matter turnover. *European Journal of Soil Science* 52, 345–353.

Christensen, N.L., Bartuska, A.M., Brown, J.H., Carpenter, S.C., D'Antonio, C., Francis, R., Franklin, J.F., MacMahon, J.A., Noss, R.F., Parsons, D.J., Peterson, C.H., Turner, M.G. and Woodmansee, R.G. (1996) The report of the Ecological Society of America committee on the scientific basis for ecosystem management. *Ecological Applications* 6(3), 665–691.

Clancy, K. (1997) Research and policy issues related to the nutritional quality of alternatively produced crops. In: Lockeretz, W. (ed.) *Proceedings of Agricultural Production and Nutrition.* Tufts University, Medford, Massachusetts, pp. 53–58.

Clark, M.S., Horwath, W.R., Shennan, C. and Scow, K.M. (1998) Changes in soil chemical properties resulting from organic and low-input farming practices. *Agronomy Journal* 90, 662–671.

Clark, M.S., Horwath, W.R., Shennan, C., Scow, K.M., Lanini, W.T. and Ferris, H. (1999) Nitrogen, weeds and water as yield-limiting factors in conventional, low-input, and organic tomato systems. *Agriculture, Ecosystem and Environment* 73, 257–270.

Coleman, E. (1989) *The New Organic Grower.* Chelsea Green Press, Chelsea, Vermont.

Colla, G., Mitchell, J.P., Joyce, B.A., Huyet, L.M., Wallender, W.W., Temple, S.R., Hsiao, T.C. and Poudel, D.D. (2000) Soil physical properties and tomato yield and quality in alternative cropping systems. *Agronomy Journal* 92, 924–932.

Cook, R.J. and Baker, K.F. (1983) *The Nature and Practice of Biological Control of Plant Pathogens.* American Phytopathological Society, St Paul, Minnesota, 539 pp.

DiTomaso, J.M. (1995) Approaches for improving crop competitiveness through manipulation of fertilizer regimes. *Weed Science* 43, 491–497.

Doran, J.W. and Parkin, T.B. (1994) Defining and assessing soil quality. In: Doran, J.W., Coleman, D.C., Bezedick, D.F. and Stewart, B.A. (eds) *Defining Soil Quality for a Sustainable Environment.* Special Publication No. 35, American Society of Agronomy, Madison, Wisconsin, pp. 3–21.

Doran, J.W., Sarrantanio, M. and Liebig, M.A. (1996) Soil health and sustainability. *Advances in Agrology* 56, 1–54.

Drinkwater, L.E. (1999) Using plant species composition to restore soil quality and ecosystem function. In: Olesen, J.E., Eltun, R., Gooding, M.J., Jensen, E.S. and Kopke, U. (eds) *Designing and Testing Crop Rotations for Organic Farming.* DARCOF Report No. 1, Danish Research Centre for Organic Farming, Tjele, Denmark, pp. 37–46.

Drinkwater, L.E. (2002) Cropping systems research: reconsidering agricultural experimental approaches. *HortTechnology* 12, 355–361.

Drinkwater, L.E. and Puget, P. (2001) Fate of root-derived carbon in annual cropping systems. *Proceedings of the 86th Annual Meeting of the Ecological Society of America.* Madison, Wisconsin, USA.

Drinkwater, L.E., Workneh, F., Letourneau, D.K., van Bruggen, A.H.C. and Shennan, C. (1995) Fundamental differences in organic and conventional agroecosystems in California. *Ecological Applications* 5, 1098–1112.

Drinkwater, L.E., Cambardella, C.A. and Rice, C.W. (1996) Potentially mineralizable N as an indicator of active soil N. In: Doran, J.W. and Jones, A.J. (eds) *Handbook of Methods for Assessment of Soil Quality.* Soil Science Society of America, Madison, Wisconsin.

Drinkwater, L.E., Wagoner, P. and Sarrantonio, M. (1998) Legume-based cropping systems have reduced carbon and nitrogen losses. *Nature* 396, 262–265.

Drinkwater, L.E., Janke, R.R. and Longnecker, L. (2000) Effects of reduced tillage intensities on nitrogen dynamics and crop productivity in legume-based cropping systems. *Plant and Soil* 227, 99–113.

Dyck, E. and Liebman, M. (1995) Crop–weed interference as influenced by a leguminous or synthetic fertilizer nitrogen source: II. Rotation experiments with crimson clover, field corn, and lambsquarters. *Agriculture, Ecosystems and Environment* 56, 109–120.

Dyck, E., Liebman, M. and Erich, M.S. (1995) Crop–weed interference as influenced by a leguminous or synthetic fertilizer nitrogen source: I. Double cropping experiments with crimson clover, sweet corn, and lambsquarters. *Agriculture, Ecosystems and Environment* 56, 93–108.

Ellert, B.H., Clapperton, M.J. and Anderson, D.W. (1997) An ecosystem perspective of soil quality. In: Gregorich, E.G. and Carter, M.R. (eds) *Soil Quality for Crop Production and Ecosystem Health.* Elsevier, Amsterdam, pp. 115–141.

FAO (1983) *Guidelines: Land Evaluation for Rainfed Agriculture.* Food and Agriculture Organization, Rome, 237 pp.

Franzluebbers, A.J. (2002) Soil organic matter stratification ratio as an indicator of soil quality. *Soil and Tillage Research* 66, 95–106.

Gallandt, E.R., Liebman, M., Huggins, D.R. and Buhler, D.D. (1999) Improving soil quality: implications for weed management. *Special issue. Expanding the context of weed management. Journal of Crop Production* 2, 95–121.

Gliessman, S.R. (2000) *Agroecology.* Lewis Publishers, Boca Raton, Florida, 357 pp.

Hannukkala, A.O., Korva, J. and Tapio, E. (1990) Conventional and organic cropping systems at Suitia. I. Experimental design and summaries. *Journal of Agricultural Science in Finland* 62, 295–307.

Harris, G.H., Hesterman, O.B., Paul, E.A., Peters, S.E. and Janke, R.R. (1994) Fate of legume and fertilizer nitrogen-15 in a long-term cropping systems experiment. *Agronomy Journal* 86, 910–915.

Hart, R.D. (1984) The effect of interlevel hierarchical system communication on agricultural system input–output relationships. *Options* 1, 111–124.

Herrick, J.E. and Wander, M.M. (1998) Relationships between soil organic carbon and soil quality in cropped and rangeland soils: the importance of distribution, composition and soil biological activity. In: Lal, R., Kimble, J.M., Follet, R.F. and Stewart, B.A. (eds) *Soil Processes and the Carbon Cycle*. CRC Press, Boca Raton, Florida, pp. 405–426.

Herrick, J.E., Brown, J.R., Tugel, A.J., Shaver, P.L. and Havstad, K.M. (2002) Application of soil quality to monitoring and management: paradigms from rangeland ecology. *Agronomy Journal* 94, 3–11.

Hoitink, H.A.J. and Boehm, M.J. (1999) Biocontrol within the context of soil microbial communities: a substrate-dependent phenomenon. *Annual Review of Phytopathology* 37, 427–446.

Hoosbeek, M.R. and Bouma, J. (1998) Obtaining soil and land quality indicators using research chains and geostatistical methods. *Nutrient Cycling in Agroecosystems* 50, 35–50.

Howard, A. (1945) *Farming and Gardening for Health or Disease*. Faber and Faber, London.

Hutchinson, G.E. (1959) Homage to santa rosalia or why are there so many kinds of animals? *The American Naturalist* 93, 145–159.

Jackson, W. (1980) *New Roots for Agriculture*. Friends of the Earth, San Francisco, California, 155 pp.

Karlen, D.L., Gardner, J.C. and Rosek, M.J. (1998) A soil quality framework for evaluating the impact of crop production. *Journal of Production Agriculture* 11(1), 56–60.

Kay, B.D. and Angers, D.A. (1999) Soil structure. In: Sumner, M. (ed.) *The Handbook of Soil Science*. CRC Press, Boca Raton, Florida, pp. 229–276.

Kleinman, P.J.A., Bryant, R.B., Reid, W.S., Sharpley, A.N. and Pimentel, D. (2000) Using soil phosphorus behavior to identify environmental thresholds. *Soil Science* 165, 943–950.

Lal, R. (1998) Soil erosion impact on agronomic productivity and environmental quality. *Critical Reviews in Plant Science* 17, 319–464.

Langdale, G.W., Alberts, E.E., Bruce, R.R., Edwards, W.M. and McGregor, K.C. (1994) Concepts of residue management: infiltration, runoff and erosion. In: Hatfield, J.L. and Stewart, B.A. (eds) *Crop Residue Management*. Lewis, Boca Raton, Florida, pp. 109–123.

Letourneau, D.K. (1990) Abundance patterns of leafhopper enemies in pure and mixed stands. *Environmental Entomology* 19(3), 505–509.

Letourneau, D.K. (1997) Plant–arthropod interactions in agroecosystems. In: Jackson, L.E. (ed.) *Ecology in Agriculture*. Academic Press, New York, pp. 239–291.

Letourneau, D.K., Drinkwater, L.E. and Shennan, C. (1996) Soil management effects on crop quality and insect damage in commercial organic and conventional tomato fields. *Agriculture, Ecosystems and Environment* 57, 179–187.

Liebhardt, W.C., Andrews, R.W., Culik, M.N., Harwood, R.R., Janke, R.R., Radke, J.K. and Rieger-Schwartz, S.L. (1989) Crop production during conversion from conventional to low-input methods. *Agronomy Journal* 81, 150–159.

Liebman, M. and Gallandt, E.R. (1997) Many little hammers: ecological management of crop–weed interactions. In: Jackson, L.E. (ed.) *Ecology in Agriculture*. Academic Press, New York, pp. 291–343.

Lipson, M. (1997) *Searching for the 'O-Word': an Analysis of the USDA Current Research Information System (CRIS) for Pertinence to Organic Farming*. Organic Farming Research Foundation, Santa Cruz, California.

Lowrance, R. (1990) Research approaches for ecological sustainability. *Journal of Soil and Water Conservation* 45, 51–54.

Lowrance, R. and Groffman, P.M. (1987) Impacts of low and high input agriculture on landscape structure and function. *American Journal of Alternative Agriculture* 2(4), 175–183.

Lundquist, E.J., Jackson, L.E., Scow, K.M. and Hsu, C. (1999a) Changes in microbial biomass and community composition, and soil carbon and nitrogen pools after incorporation of rye into three California agricultural soils. *Soil Biology and Biochemistry* 31, 221–236.

Lundquist, E.J., Scow, K.M., Jackson, L.E., Uesugi, S.L. and Johnson, C.R. (1999b) Rapid response of soil microbial communities from conventional, low input, and organic farming systems to a wet/dry cycle. *Soil Biology and Biochemistry* 31, 1661–1675.

Mäder, P., Fliessbach, A., Dubois, D., Gunst, L., Fried, P. and Niggli, U. (2002) Soil fertility and biodiversity in organic farming. *Science* 296, 1694–1697.

Mannering, J.V. and Fenster, C.R. (1983) What is conservation tillage? *Journal of Soil and Water Conservation* 38, 140–143.

McCracken, D.V., Smith, M.S., Grove, J.H., MacKown, C.T. and Blevins, R.L. (1994) Nitrate leaching as influenced by cover cropping and nitrogen source. *Soil Science Society of America Journal* 58, 1476–1483.

Miller, F.P. and Wali, M.K. (1995) Soils, land use and sustainable agriculture: a review. *Canadian Journal of Soil Science* 75, 413–422.

National Organic Program (2002) United States Department of Agriculture, Organic Foods Standards and Labels. Available at: http://www.ams.usda.gov/nop/NOP/NOPhome.html

Odum, E.P. (1953) *Fundamentals of Ecology*. Saunders, Philadelphia, Pennsylvania, 546 pp.

Odum, E.P. (1984) Properties of agroecosystems. In: Lowrance, R., Stinner, B.R. and House, G.J. (eds) *Agricultural Ecosystems*. John Wiley & Sons, New York, pp. 5–11.

O'Neill, R.V., DeAngelis, D.L., Waide, J.B. and Allen, T.F.H. (1986) *A Hierarchical Concept of Ecosystems*. Princeton University Press, Princeton, New Jersey, 253 pp.

Organic Farming Research Foundation (2002) General Information. Available at: http://www.ofrf.org/general/index.html

Peters, R.D., Sturz, A.V., Carter, M.R. and Sanderson, J.B. (2003) Developing disease-suppressive soils through crop rotation and tillage management practices. *Soil and Tillage Research* 72, 181–192.

Petersen, C., Drinkwater, L.E. and Wagoner, P. (1999) *The Farming Systems Trial: the First Fifteen Years*. Rodale Institute, Kutztown, Pennsylvania.

Phelan, P.L., Mason, J.F. and Stinner, B.R. (1995) Soil fertility management and host preference by European corn borer, *Ostrinia nubilalis* (Huebner), on *Zea mays* L.: a comparison of organic and conventional chemical farming. *Agriculture, Ecosystems and Environment* 56, 1–8.

Pierce, F.J. (1996) Land management: the purpose for soil quality assessment. In: MacEwan, R.J. and Carter, M.R. (eds) *Proceedings of Symposium Soil Quality for Land Management: Science, Practice, and Policy*. University of Ballarat, Ballarat, Victoria, Australia, pp. 53–58.

Pimm, S.L. (1984) The complexity and stability of ecosystems. *Nature* 307, 230–237.

Pomeroy, L.R., Hargrove, E.C. and Alberts, J.J. (1988) The ecosystem perspective. In: Pomeroy, L.R. and Alberts, J.J. (eds) *Concepts of Ecosystem Ecology: a Comparative View*. Springer-Verlag, New York, pp. 1–17.

Poudel, D.D., Horwath, W.R., Mitchell, J.P. and Temple, S.R. (2001) Impacts of cropping systems on soil nitrogen storage and loss. *Agricultural Systems* 68, 253–268.

Puget, P. and Drinkwater, L.E. (2001) Short-term dynamics of root and shoot-derived carbon from a leguminous green manure. *Soil Science Society of America Journal* 65, 771–779.

Reganold, J.P., Palmer, A.S., Lockhart, J.C. and Macgregor, A.N. (1993) Soil quality and financial performance of biodynamic and conventional farms in New Zealand. *Science* 260, 344–349.

Reganold, J.P., Glover, J.D., Andrews, P.K. and Hinman, H.R. (2001) Sustainability of three apple production systems. *Nature* 410, 926–930.

Ritchie, M.E. (2000) Nitrogen limitation and trophic vs. abiotic influences on insect herbivores in a temperate grassland. *Ecology* 81, 1601–1612.

Robertson, G.P., Paul, E.A. and Harwood, R.R. (2000) Greenhouse gases in intensive agriculture: contributions of individual gases to the radiative forcing of the atmosphere. *Science* 289, 1922–1925.

Rodale, R. (1983) Breaking new ground: the search for a sustainable agriculture. *The Futurist* 1(1), 15–20.

Röling, N. (1997) The soft side of land: socio-economic sustainability of land use systems. *Special Issue: Geo-Information for Sustainable Land Management. ITC Journal* 3/4, 248–262.

Rountree, J.H. (1977) Systems thinking – some fundamental aspects. *Agricultural Systems* 2, 247–254.

Schjønning, P., Elmholt, S., Munkholm, L.J. and Debosz, K. (2002) Soil quality aspects of humid sandy loams as influenced by organic and conventional long-term management. *Agriculture, Ecosystems and Environment* 88, 195–214.

Scow, K.M., Somasco, O., Gunapala, N., Lau, S., Venette, R., Ferris, H., Miller, R. and Shennan, C. (1994) Transition from conventional to low-input agriculture changes soil fertility and biology. *California Agriculture* 48, 20–26.

Seybold, C.A., Herrick, J.E. and Brejda, J.J. (1999) Soil resilience: a fundamental component of soil quality. *Soil Science* 164(4), 224–234.

Singer, M.J. and Ewing, S. (1999) Soil quality. In: Sumner, M. (ed.) *The Handbook of Soil Science*. CRC Press, Boca Raton, Florida, pp. 271–298.

Spedding, C.R.W. (1988) *An Introduction to Agricultural Systems*. Elsevier Applied Science, New York, 189 pp.

Stephens, W. and Hess, T. (1999) Systems approaches to water management research. *Agricultural Water Management* 40, 3–13.

Stiling, P. and Rossi, A.M. (1997) Experimental manipulations of top down and bottom up factors in a tri-trophic system. *Ecology* 78, 1602–1606.

Strauss, S.Y. (1987) Direct and indirect effects of host plant fertilization on an insect community. *Ecology* 68, 670–678.

Sturz, A.V. and Christie, B.R. (2003) Beneficial microbial allelopathies in the root zone: the management of soil quality and plant disease with rhizobacteria. *Soil and Tillage Research* 72, 107–123.

Sturz, A.V., Carter, M.R. and Johnston, H.W. (1997) A review of plant disease, pathogen interactions and microbial antagonism under conservation tillage in temperate humid agriculture. *Soil and Tillage Research* 41, 169–189.

Tansley, A.G. (1935) The use and abuse of vegetational concepts and terms. *Ecology* 16, 284–307.

Tisdall, J.M. and Oades, J.M. (1979) Stabilization of soil aggregates by the root system of ryegrass. *Australian Journal of Soil Research* 17, 429–441.

van Bruggen, A.H.C. and Semenov, A.M. (2000) In search of biological indicators for soil health and disease suppression. *Applied Soil Ecology* 15, 13–24.

Vandermeer, J.H. (1990) Intercropping. In: Carroll, C.R., Vandermeer, J.H. and Rosset, P.M. (eds) *Agroecology*. McGraw-Hill, St Louis, Missouri, pp. 481–516.

Vereijken, P. (1989) Experimental systems of integrated and organic wheat production. *Agricultural Systems* 30, 187–197.

Wander, M.M., Traina, S.J., Stinner, R.B. and Peters, S.E. (1994) The effects of organic and conventional management on biologically-active soil organic matter pools. *Soil Science Society of America Journal* 58, 1130–1139.

Warkentin, B.P. (1995) The changing concept of soil quality. *Journal of Soil and Water Conservation* 50, 226–228.

Woese, K., Lange, D., Boess, C. and Bogl, K.W. (1997) A comparison of organically and conventionally grown foods – results of a review of the relevant literature. *Journal of the Science of Food and Agriculture* 74, 281–293.

Workneh, F. and van Bruggen, A.H.C. (1994) Suppression of corky root of tomatoes in soil from organic farms associated with soil microbial activity and nitrogen status of soil and tomato tissue. *Phytopathology* 84, 688–694.

Workneh, F., van Bruggen, A.H.C., Drinkwater, L.E. and Shennan, C. (1993) Relationships between soil variables and corky root and *Phytophthora* root rot of tomatoes in organic and conventional farms. *Phytopathology* 83, 581–589.

Chapter 16

Implementing Soil Quality Knowledge in Land-use Planning

J. Bouma

Wageningen University and Research Center, Environmental Sciences Group, PO Box 47, NL-6700 AA Wageningen, The Netherlands

Summary

Land-use negotiation rather than land-use planning tends to become the standard in our modern network society, where stakeholders, policy makers and scientists in interdisciplinary teams are engaged in joint learning experiences. Soil scientists are, as yet, not well prepared to take an active part in these teams, which is due to both attitude and lack of skills. When handled from the viewpoint of *reflexive objectivity*, the soil quality concept can be quite helpful to focus soils information input into the team discussions. Moreover, this can be strengthened by also illustrating the role of soil management to reach certain qualities because discussions in the team are centred on future demands and not on current conditions alone. Soil qualities can be defined for different soil functions but soil quality in general can best be expressed by defining a *window of opportunities* for every type of soil to be defined by field observations and monitoring *and* simulation modelling. A plea is made to emphasize field work through studying effects of past management on soil properties in any given type of soil, after defining their locations on published soil maps. Soil scientists are eminently qualified to play a central role in land negotiation teams if they possess the proper set of tools and a mind open to joint learning.

Introduction

Land-use planning has a different meaning in different countries. Rigid, centralized planning of land use has not worked well in communist countries in the past and is now certainly not part of modern governmental repertoires. Even though it is easy to criticize the shortcomings of the centralized planning of the past, finding ways by which the alternative free-for-all approach can be avoided is far from easy. Of course, some argue that 'the market' will solve all problems and that any interference by government in this process will unnecessarily complicate manners. But such voices are criticized, as the existence of common interests is increasingly recognized together with the realization that these common interests are not necessarily dealt with adequately by commercial outfits operating on 'the market'. Taking these considerations into account, the term 'land-use negotiation' has therefore been proposed as an alternative to 'land-use planning'. The term suggests that actual land-use patterns result from negotiations between policy makers and individuals or organizations (the 'stakeholders') that have interests of various kinds into whatever land-use patterns develop in future. The term 'negotiation' suggests interaction, but the manner in which interaction finds its most effective form and the various roles being played by the participants in the process are still the object of much debate. We will discuss modern developments in land-use negotiation and the role of non-governmental organizations (NGOs), policy makers and scientists. Emphasis will be on the latter category with particular attention for soil scientists taking part in the land-use negotiation process (e.g. Bouma, 2001b) and the manner in which the soil quality concept can play a role here.

The term land-use planning is usually associated with larger spatial scales relating to counties, provinces and larger areas that are linked to administrative units of various kinds. Strictly speaking, however, any farmer is also engaged in land-use planning, but then for his own farm only where his decision-making powers are increasingly restricted by environmental laws and regulations. Even though the larger spatial scales are emphasized in this chapter, reference is also made to the farm level because many issues are comparable at both scales and our rural areas are primarily occupied by farmers making their independent decisions and, thereby, decisively affect land-use patterns.

The notion of *reflexive objectivity* is appropriate for the different spatial scales because, increasingly, the input of scientists is ignored when they play the role of providing clear-cut answers to problems they defined themselves (see also Fig. 1.1, Chapter 1). Rather, the scientist of today operates in societies with well-educated citizens and farmers cherishing norms and values, priorities and goals, which need to be taken into account one way or the other to arrive at conclusions that are broadly supported. And, obviously, only conclusions that are broadly supported have at least a chance of being followed up in the real world. This does not imply, however, that scientists should only meekly respond to any question being asked by stakeholders. By contributing their expertise they can not only refine and focus these questions but they can also show that sometimes questions can partly be answered by existing knowledge, whereas specific new research is needed to fill any knowledge gap. Increasingly, scientists have a function to explain, to help negotiate and to be an intermediary in many other ways, as they help to bridge contrasting viewpoints. They often find themselves in positions where they can – at a certain distance and from an independent viewpoint – formulate trade-offs between conflicting interests, thus neutralizing conflicts and contributing to the solution of problems. Such activities go far beyond consideration of agricultural land-use aspects and should, rather, focus on multifunctionality, also considering land use for nature, recreation, building and water storage, just to mention a few non-agricultural forms of land-use (e.g. Tait, 2001).

Within this context, the soil quality concept is introduced in this chapter as a means of entering soils into the overall land-use negotiation process. No one can argue against the necessity that discussions on land use should somehow incorporate a discussion on soil. Unfortunately, the type of soil information that is prevalent when soil scientists get together is often difficult to interpret in the much broader context of land-use negotiation. In fact we will argue in this chapter that soil information may not be considered at all, let alone soil quality. Soil scientists are not sufficiently aware of this as they are, like many colleagues in other disciplines, rather inward looking. Substantial attention must therefore be paid here to the manner in which soil information is considered in the land-negotiation process, so as to avoid a virtual debate.

Emphasis in this book on *management thresholds* rather than soil-quality *indicator thresholds* is important and wise because land-use discussions always have the implication that 'something is to be done somewhere', and linking management to objectives to be achieved provides a more concrete, practical and attractive framework than one in which more theoretical threshold values are discussed.

Finally, we realize that land use should be properly addressed in a realistic socio-economic context. The traditional approach of land evaluation where soil suitabilities were defined for a wide range of land uses is quite irrelevant for high-capital land development. Building houses, factories and roads requires such large investments that some additional expense for overcoming local soil limitations becomes insignificant. However, agricultural and recreational forms of land use, including nature building, present a different picture. Certainly, here also 'anything can be done anywhere' in a technical sense, but when striving for sustainable land-use systems, it is advisable to consider soil processes at a given location as being part of natural processes in the surrounding larger ecosystem. Straying

too far away from these natural processes implies taking significant ecological risks, thereby limiting the potential sustainability of the system being designed.

An attempt is therefore made in this chapter to approach soils from a quality perspective and to use this approach to define characteristic 'windows of opportunity' with characteristic *management thresholds* for each soil within a landscape context.

The Land-use Negotiation Process

Land-use negotiation is focused on future land use and therefore part of 'future research', which is a popular research topic with a large literature that is beyond the scope of this text. However, reference can be made to Castells (1996) who defined the network society and to Rescher (1998) who reviews theories of forecasting. Different approaches to future research relating to land use have been described by Healy *et al.* (1997), Bouma *et al.* (2000) and Bouma (2001a). One general observation needs to be made when describing future research. Most often, series of exploratory land-use scenarios are described for the future reflecting demands by society and by stakeholders. Rather than defining one ideal scenario, to be followed by implementation measures, sets of scenarios are produced as input for discussions. From these sets, selections can be made. With the availability of information and communication technologies (ICT), it becomes increasingly easy to produce glossy maps, and we see a proliferation in The Netherlands and elsewhere of future studies with relatively little follow-up. Governments have a 4-year time span, at least when they complete their term, and it is much more attractive for a new minister to initiate broad scenario studies for the future that tickle the imagination, rather than to proceed to work out the nitty-gritty details of earlier studies. A recent study in The Netherlands on future research and environmental policy focused for the first time on the question of whether long-term future studies are being implemented in the context of short-term policy (T&O, 2001). The study concludes on the basis of four specific case studies that this implementation is highly problematic now, and a large number of recommendations are made to clarify the objectives of future research and to facilitate the interaction process between future researchers, policy makers and stakeholders. In this chapter, when discussing the use of the soil-quality concept in land-use negotiation, such interaction processes are of paramount importance.

The challenge will be to present soil information in the right form at the right time to the right people. A number of recommendations of T&O (2001) are relevant in the context of this chapter. Some of them may appear rather obvious at first sight, but the case studies reveal that lack of acceptance of future land-use studies is often the result of simple oversights or poor timing and conceptualization. Some of the major recommendations include:

- Try to define common objectives of land-use studies that are accepted by both stakeholders and policy makers. All too often not enough time is available for a true negotiation process that results in common objectives, and political deadlines result in immature plans that, not surprisingly, fall apart before implementation can take place.

- Try to understand the type of reasoning that is being followed by the parties involved. Often, the same terminology may have different meanings for different groups, and perceptions of needs and limitations may differ significantly. If such differences are not openly discussed and analysed before plans are made, the chances are that yet another 'dialogue of the deaf' is being initiated.

- Separate what we *want* from what we *know* and what we are *able to do*. Often these points of departure are mixed up, thus creating much confusion. Ask all participants to make this analysis, which can clarify discussions that often get bogged down because 'wishful thinking' of various types is tempting when dealing with future developments that can only be speculated about.

- Stimulate creativity and generation of wild ideas and try to improve the

learning ability of participants and organizations, if necessary by introducing techniques of social learning. The ideal is to create joint learning processes by not stifling creativity with bureaucratic procedures.

- Try to arrive at broadly shared visions for the future that trigger the imagination and that may lead to powerful coalitions within the group of participants. An important ingredient of such visions may be the result of 'win-win' situations in which what appeared to be conflicts at first turn out to be potential opportunities seen by all.

The Role of Soils during Land-use Negotiation

As indicated in the introduction, soil aspects – at least in The Netherlands – play a very minor role when land use is discussed in land-use negotiations at the national and regional levels. In the most recent Dutch government report on spatial planning (VROM, 2001) areas of the country are delineated by red and green lines. The red lines encircle areas where building is allowed, whereas the green lines delineate nature areas. The remaining 70% of the country is described in terms of 'balance areas', where various forms of agriculture represent the most prominent form of land use (even though its contribution to the Gross National Product (GNP) is less than 5%). There is no relation to soil conditions. Areas are identified on the basis of locations of and distances between current settlements and nature areas. Not only does the report provide very little information about desirable future trends of land use in the 'balance areas', it does not emphasize any relationship to soil conditions either. This is surprising because the 'balance areas' cover the whole range of soil conditions in The Netherlands, from clays and sands to peats. Clearly, soil scientists have not been effective in demonstrating the significant impact of different soil conditions on natural vegetation and on the potential of different forms of agriculture.

Also, environmental regulations ignore the differences in soil conditions. For example, the important nitrogen guideline for nitrogen fertilization in the countries of the European Union of 1991 (EC, 1991) mentions a critical fertilization rate of organic manure but provides no specifics in terms of soil conditions, which are obviously quite diverse within the EU.

Fortunately, all is not lost. Some years ago the so-called metropolitan debate was started for the western part of The Netherlands (HMD, 1998). Here, the land-use system was stratified into four conceptual 'layers', each with characteristically different dynamics, resulting from human impact. The first layer consisted of the soil and water system (dynamic cycle approx. 100 years), the second layer of mobility and infrastructure (cycle approx. 50 years), the third layer of the municipal system (approx. 25 years) and the fourth layer of the social system (cycle quite variable from 5 to 15 years). The entire system is characterized by superimposing actions and effects in the different layers, thus allowing a stratified and transparent approach to the analysis of the complex land-use system. The implicit suggestion of the four-layer approach was that the dynamics of the lower layers should affect the dynamics of the higher ones. Though intriguing from a conceptual point of view, this suggestion has not yet resulted in clear results. Still, the layer concept provides an excellent opportunity for soil and water scientists to more specifically define their roles in this particular framework and in the land-use negotiation process.

Soils play a prominent role when defining new management systems at farm level in the context of precision agriculture (Bouma et al., 1999; van Alphen, 2002). Rather than implicitly considering a farmer's field to be homogeneous, soil differences are expressed by a detailed soil survey and models are run on a real-time basis to keep track of water and nutrient fluxes during the growing season using real weather data. Next, the application of agrochemicals is fine-tuned to differences within the field in space and time (that are different each year) using ICT technology. Such input is valuable in discussions with farmers who easily see the benefits, thereby facilitating the negotiation process.

Emphasizing Soil Quality, But Only in a Broader Context

The discussion so far indicates that much needs to be done before soil input can play a significant and – in our vision – proper role in land-use negotiations. The current procedures for presenting and discussing soil science expertise need a drastic overhaul. Emphasizing soil quality aspects rather than soil information in general would appear to be quite attractive because general information is increasingly difficult to see in our digital age, which is characterized by a torrent of easily accessible data and information. Information in general terms tends to disappear, but attention to definition of *indicators* that facilitate communication can lead to a certain impact. For example, economists use many indicators, such as the GNP. Even though it involves gross generalizations, it has proved quite effective in ranking different countries and in judging different economic policies. Corresponding examples can be given for the social sciences, where the average age of citizens or birth rates are effective indicators to judge and guide public policy. Soil quality could have the same indicative function for soil science. But before defining soil quality in the next section, the role that the soil quality concept can play in a land-negotiation context needs to be analysed.

One problem of the soil quality concept at this time, at least in our view, is the fact that mainly soil scientists and an occasional agronomist take part in discussions about soil quality. This leads to long discussions about definitions and shopping lists of possible indicators, without having much impact. The soil quality discussion is, in our view, bound to evaporate without substantial results if we do not seriously engage policy makers and fellow scientists beyond our own fields in our discourse.

When discussing land negotiation earlier in this chapter, five success factors for the negotiation proces were presented. How do these relate to the soil quality concept?

Common objectives

A unifying principle and common objective for the land-use negotiation team is the concept of sustainable land use, emphasizing productivity, security, protection, viability and acceptability (e.g. Smyth and Dumanski, 1993; Bouma et al., 1998b). The link of soil science with agronomy provides the keys to expressing the productivity function in both a descriptive and a quantitative manner, the latter using simulation modelling. This procedure is particularly attractive for expressing production risks by simulating time series of years (e.g. Bouma and Jones, 2001; Kropff et al., 2001). The involvement of soil chemists, biologists and physicists allows the development of health, security and protection aspects relating to soil behaviour. Such contributions by soil scientists function as building blocks for team-wide discussions on viability and acceptability that involve many social and economic considerations.

A focus on sustainable land use in general terms is rather broad and leaves many degrees of freedom for research. Increasingly, however, environmental laws are introduced, which have characteristic *threshold values* that function as clear guiding posts for land-use practices. A focus on management practices that aim to reach the environmental *thresholds* is an excellent way to define common objectives. Many researchers are not yet inclined to follow this road and prefer to apply their own methods, models or expert systems rather independently, which means that much environmental research, as reported in the literature, is, as yet, not as effective as it might be (e.g. Bouma et al., 2002).

Reasoning by team members

Soil scientists often have a rather exclusive agronomic or environmental focus when discussing soils. Achieving high agricultural productivity of high quality with minimal losses of agrochemicals to the environment is seen by many as the ultimate objective. It is, indeed, quite an achievement when such results are obtained but some policy makers have a quite different focus. For example, some economists wonder whether there should still be agriculture in The Netherlands. Why not import food and use the expensive and precious soil resource in the most densely populated country in the world for building large houses with gardens and for substantially extending nature areas? What used to be our most productive grain-producing area with clay soils in the northern part of the country as little as 30 years ago is now partly being taken out of production and flooded. The lakes, thus created, are to be surrounded by large villas for commuters from large cities in Germany. This has come as a great shock to those still adhering to the agricultural thinking of only a few decades ago. However, we live in an urbanizing society and cannot afford to be unaware of urban thinking when charting our disciplinary course. The soil scientist can play a role in this broad discussion on future land-use patterns in the country by pointing out the relative advantages of using certain soils for agriculture, even at a time when World Trade Organization (WTO) agreements are bound to lead to free trade. They can also demonstrate that our landscape was formed over the centuries by farmers struggling with nature. This resulted in characteristically different landscapes in our sandy, clayey and peaty areas, and these landscapes are part of our cultural heritage. Clearly, soil scientists cannot decide by themselves where land-use changes should be made and where not, but they can add essential elements to the debate provided that they can package their message in an accessible manner.

What is described here for conditions in The Netherlands also largely applies to other western European countries, although sometimes in terms of different expressions of the same basic phenomenon.

To want, to know and to be able to: creativity and joint learning

The immense impact of technology on agricultural development has led to a certain degree of euphoria: 'anything can be done anywhere'. This is true, but often much less attractive when all associated economic and environmental costs are considered. The latter are often not part of the economic balance book. When dealing with land-use negotiations, the soil scientist is in a unique position to point out what is known about soil behaviour and what the costs and associated effects of certain plans will be. This does not, as in the past, take the form of defining strict soil suitabilities for different forms of land use but simply provides the data, allowing the negotiation team to draw conclusions. Housing developments and roads are still being built at locations where costs to improve soil conditions are high beforehand, and they may be located at poor positions on the watershed. Sometimes more attractive alternatives can be found using soil expertise. Nature areas are still planned in locations where soil and hydrological conditions are not likely to result in attractive biodiverse conditions. Also here, alternatives based on soil and landscape knowledge may provide attractive alternatives.

The above discussion could suggest that soil scientists should particularly focus on providing data to allow better decisions by others (*to know*). Nothing is farther from the truth. Unfortunately, however, soil scientists have not been at the frontier where truly innovative ideas are generated. Aspects of '*to want*' need more attention and some wild ideas would be welcome. For example, the concept of precision agriculture has been generated by agronomists and technicians and not by soil scientists, even though the dynamic behaviour of soils during the growing season is the core of the concept (e.g. Bouma *et al.*, 1999; van Alphen, 2002). The aspect of '*to be able to*' is better represented within the profession by looking at different management procedures that have been developed, for example to improve soil structure, the organic matter content and soil and water conservation. The linkage between soil quality and soil management, as proposed in this book, is therefore a very good one.

Broad visions for the future

There are complaints that the quality of air and water are receiving increasing attention while few appear to be concerned about soil quality. Part of this may be caused by the fact that soils are 'invisible', while air and water can be seen, tasted and smelled. Still, when sketching the outlines of a sustainable future world where humans use natural resources only to the extent that unacceptable depletion is avoided, a focus on soils provides a good starting point. Again, as stated for the common objectives, a focus on sustainable management of natural resources provides the core of an attractive and comprehensive vision for the future when dealing with land negotiation. This vision focuses on both the sustainable production of food for nine billion people and on maintaining the soil as a vital part of ecosystems at different spatial-scale levels. Broad visions for the future must also include growing crops as sources for energy and as basic materials for industry, as our current resources are being depleted rapidly. Which soils are most suitable to address these different challenges and how can international trade-offs be arranged to achieve optimal uses of our soil resources at the world level? Currently, economic trade-offs are discussed in the context of the WTO arrangement. Within 10 years I expect we will need a World Soil Agreement (WSA), where a binding treaty on optimal use of the soil resource at a world level will be realized.

The Management Factor

Returning to the issue of land-use negotiations, attention is being paid to different options for future land use. Being able to list soil qualities for current conditions can be helpful when making selections for certain activities at certain locations. Still, it is more interesting to see how soil management can help to achieve certain objectives in the future and whether or not different forms of management can play a role here. Numerous experiments have been reported in the literature on the effects of different forms of land management on soil properties and its performance. They usually involve replicate experiments on plots where a given treatment is being varied systematically. This relates to the entire range of management options, which will not be discussed further here as reference is made to the literature. Here, attention is given to a supplementary approach studying soil conditions on operating farms in a given soil type and relating soil performance and soil properties to management practices by statistical analyses. For example, Droogers and Bouma (1997) identified three distinctly different *phenoforms* of a given soil series (the *genoform*), resulting from high-tech arable farming, organic farming and permanent use as meadow. They showed by simulation modelling that nitrogen transformations were characteristically different in the three phenoforms. But management thresholds for soil tillage were also different. The higher organic matter content of the organic-farming phenoform resulted in a higher soil-moisture-supply capacity, which increased soil quality. At the same time these higher moisture contents made the soil more vulnerable to compaction (Droogers *et al.*, 1996). The example illustrates that the entire production system should be considered when defining soil qualities for biomass production, including trade-offs that can be made when conflicting trends arise. Tillage is a means towards a purpose, not a purpose in itself. Biomass production, its quality and effects on the environment were the logical main focal points in this study. Using a published soil map to find the proper locations, Pulleman *et al.* (2000) visited 40 farmers to investigate the effects of management on soil properties in the same soil type, which qualified as prime agricultural land. Pulleman *et al.* (2000) were able to derive a regression equation predicting the soil organic matter (SOM) content as a function of previous management history in six different, successive periods from 1940 to 2000, as follows:

$$SOM = 20.7 + 29.7C_1 + 7.5\,C_v + 7.5\,M_{iv} \qquad (16.1)$$

where C_1 and C_v are the crop type in periods 1 and v, respectively (grass = 1 and arable = 0)

and M_{iv} is the management type in period iv (organic = 1 and conventional = 0).

Sonneveld *et al.* (2002) did the same for a prominent sandy soil in the north of the country. They also included landscape characteristics, which were associated with the groundwater classes distinguished as follows:

$$\text{C-organic (\%)} = 3.40 - 1.54 \times \text{Maize} + 0.19 \times \text{Old} + 0.55 \times \text{GWC} \qquad (16.2)$$

where Maize = 1 for continuous maize cropping and 0 otherwise; Old = 1 for old grassland and 0 otherwise; and GWC is groundwater class, type Vb = 1 and VI = 0. GWC defines the average highest and lowest groundwater level during the growing season.

These equations allow estimates of the organic matter content of surface soil as a function of soil management practices in the past. In turn, simulation techniques can translate such conditions in important soil functions as discussed and can define *management thresholds*. This type of field observation also allows assessments of *soil resilience* when studying the reversal of adverse effects of certain management practices (e.g. ploughing or driving under wet conditions) as a time series at different locations within the same soil type.

In addition to expressing soil quality as a function of management the argument can also be turned around and soil management emphasized as a means to protect soil quality. Bouma and Droogers (1998) demonstrated the occurrence of different soil quality indexes for a given soil type when subjected to certain types of management.

Windows of Opportunities Rather Than Fixed Assessments

Soil functioning

Even though a universal definition of soil quality can be given (Doran *et al.*, 1994), it is not possible to define universal *indicators* as well. Soils have different functions, as defined by Blum (1997), and each function will require different indicators, which is no problem as long as it remains clear which function is being addressed.

The definition of soil quality has the key phrase: the capacity of a specific kind of soil to function. Functioning can be interpreted for each soil type in terms of water regimes and the associated physical and chemical fluxes that define the habitat for crops, flora and fauna and its mechanical properties. As suggested by Bouma (2001b), soil classification may be used to define 'specific types of soil' as a means of stratification, realizing that some soils that are taxonomically different function identically, whereas soils identical from a taxonomic point of view sometimes function differently. The first problem can be solved by lumping soils together. The second problem is more difficult to solve, but appears to be the exception rather than the rule. The functioning of soils can be characterized by measurement, by using an increasing number of new *in situ* sensors. In addition, simulation models for water and solute fluxes and crop growth are by now well tested and provide a reliable tool to define soil functioning over extended periods of time, reflecting temporal variability due to weather conditions (e.g. Bouma and Jones, 2001; Kropff *et al.*, 2001). A key problem here is the availability of basic data to feed the models. Pedotransfer functions, which use available soil data to predict parameters needed for modelling, have been defined succesfully, certainly for soil physical data (e.g. Bouma, 1997).

Soil functioning can be characterized at different levels of detail. This is a very important point to be made because disciplinary scientists tend to focus on detailed studies with a high data demand. This is not necessarily a problem for single studies, but when statements have to be made about areas of land with a wide variety of different soils, problems can be severe. Rather than estimating parameters from standard tables, we would advocate the use of simpler models with a lower data demand, or even only qualitative assessments instead. In fact, soil functioning should be defined at different levels of detail if only to allow effective interaction with other disciplines in the negotiation team. Too many interdisciplinary projects collapse because scientists of the different disciplines can only express themselves in a highly detailed manner that is inaccessible to other disciplines. Bouma and Droogers (1999) presented and discussed five different procedures of increasing complexity to define the soil moisture supply capacity. This has a clear function in the discussion process during land-use negotiations: one starts with the simplest (and cheapest) method, but its limitations may convince the client that a more complex (and more expensive) method would yield better results, etc. This step-by-step process may ultimately result in the use of sophisticated methods after all, but this would most likely not happen if these methods were presented right away (Bouma, 2001b).

Quality indicators for different functions

Many quality indicators have been defined for the various soil functions as described by Blum (1997). On the other hand, soil quality as a source of raw materials, such as sand or gravel, can be based on geological assessments, and the traditional suitabilities of soil survey interpretations can perform their function here. On the other hand, production of biomass is more difficult to assess and can be characterized by simulating potential productions and by comparing them with water-limited yields, thus developing a quality indicator with values between 0 and 100 (e.g. Bouma *et al.*, 1998a; Bouma, 2002). Bouma and Droogers (1998) combined the production function with the transforming function of nitrogen to nitrate and suggested a selection procedure for an indicator (again between 0 and 100) that also expressed the level of risks involved. The value of the selected indicator expresses the risk that the user is willing to accept. This indicator turned out to be significantly different for the same soil type as a function of previous management. Listings of soil characteristics, such as bulk densities and organic matter contents, cannot act as distinguishing features for soil qualities as they have no direct, but only an indirect, relation with soil functioning. Clearly, soil quality in terms of a *gene reserve for plant and animal organisms* needs more attention, as does the quality *cultural heritage* (Blum, 1997). In principle, though, soil qualities can be defined for the different functions and they can sometimes be logically combined when, for example, agricultural production is tied to environmental quality, as was done for nitrogen fertilization by Bouma and Droogers (1998). This was so because nitrate pollution of groundwater was the most important environmental issue of the day. Even though procedures for estimating soil qualities can be defined, standardization of procedures is urgently needed to avoid anarchy (see Karlen *et al.* Chapter 2, this volume for considerations on this topic).

regimes and biomass production, allows input of exploratory results into the debate and serves to sketch and explore the limits of the *window of opportunity*.

The crucial position of the soil within any ecosystem requires an active and creative role of soil scientists in teams struggling with land-use negotiations. Modern techniques allow the soil scientist to present many data on soil behaviour at different levels of detail, including expected effects of management. Thus the soil scientist can perhaps play a crucial role in changing the tone of future discussions from a strictly economic to a more environmental focus.

The window of opportunity as a general measure of soil quality

Field studies were crucial for the two case studies by Pulleman *et al.* (2000) and Sonneveld *et al.* (2002) reviewed above. Available soil maps were used to identify locations where a certain type of soil occurred. Observation of actual soil conditions and interviews with farmers, focused on their management practices, were a rich source of information that turned out to be crucial for defining effects of soil management on soil qualities. A large number of real-life experiments are out there in the field, waiting to be explored. We believe that each particular type of soil has a characteristic *window of opportunity* as a function of management (see Bouma, 1994). Defining the width and height of that *window* for the six functions of Blum (1997) can serve as a true measure for soil quality for any given type of soil. Soil quality indicators can be defined for each function, as explored above, and for combinations of functions as needed. Whether or not this will be needed will depend on the discussions during the land-negotiation process. However, the soil scientist who has succeeded in being a participant in such discussions (i.e. he or she is being taken seriously by the other partners) should also be ready to launch his or her repertoire in a proactive manner when appropriate. Having basic data available to run simulation models for solute

Considering the policy level: impact of rules and regulations

Any discussion on implementation of soil-quality knowledge in land-use planning should acknowledge the guiding and sometimes suffocating effects of various rules and regulations that are increasingly important in many countries. Within the EU, such regulations originate in Brussels to be enforced in the different countries of the Union. If they focus on objectives to be achieved within a broad framework, much room is left for local creativity in trying to achieve the set objectives. Activities described in this chapter will then have a purpose. If, however, these regulations emphasize the means to reach certain objectives rather than the actual objectives, then there is a problem. For example, the nitrate guideline was introduced by the European Commission in 1991. A maximum fertilization rate of 170 kg nitrogen, derived from organic manure, was allowed per hectare to avoid excessive and harmful nitrate leaching into groundwater. But groundwater quality was the real concern, so it would have been much better to focus the regulations on groundwater quality directly, for instance in terms of a maximum nitrate concentration of 50 mg nitrate per litre. The 170 kg of manure per hectare is likely to have quite a different effect on groundwater quality in different climate zones and in different soils, quite apart from following different types of land use.

Also, farmers found that manure quality and associated nitrification could be significantly manipulated by changing the feeding of their cattle. They knew that leaching could be restricted by applying the manure at different times of the year taking into account the effects of weather conditions. None of this fitted within the rules, which focused solely on *total* manure quantity, and this provided the yardstick by which farmers were to be judged and fined. This rigidity has been highly demotivating for many innovative farmers and it is therefore important to make rules and regulations that allow a high degree of local decision making within a context and objectives provided by the authorities. We believe this to be very important because creative and innovative management of our soils can only be achieved when land users are motivated by acknowledgement of their professional qualities.

The Spatial Scale Problem

The above discussion focused on the level of the individual soil, which is considered to be representative for corresponding areas of land shown on soil maps with defined spatial scales. Each type of soil has 'stories' to tell, which are characteristic for and define its *window of opportunity*. The soil quality aspect, however, must not only be considered at the level of the individual soil, but also at larger geographic scales where decisions on land use are being made in administrative entities such as farms, communities, provinces, countries and, in fact, the entire world. Even at the level of a farmer's field, different soil types may occur, and in the past a farmer used to be forced to decide on uniform management practices that were less than optimal when considering the separate soils. Now, however, the use of ICT technology in the context of precision agriculture allows differential treatment within a field, as discussed above. Questions in larger geographical areas are much broader and focus on alternative forms of land use involving different stakeholders and policy makers and requiring a different negotiation process. In all cases, however,

different soils occur in the area and their properties are likely to play a role in the discussions, be it that different 'stories' from the 'window', with its associated 'qualities', apply to different types of multifunctional land use (Tait, 2001). Agriculture will always be considered within this context because it still occupies major areas of land in all European countries. It should be considered in its most modern form, to be taken seriously in the discussions.

Conclusions

1. Top-down land-use planning does not work in a modern network society where land-use negotiations among stakeholders, policy makers and scientists represent a much more realistic concept. The soil science profession is not very successful so far in playing a role in such negotiations, and drastic changes are needed in the research and communication approach of the profession before it can play its proper role.
2. The success of the negotiation process depends in general on the ability of soil scientists to operate from the viewpoint of *reflexive objectivity* and to consider a number of issues when claiming their role in the negotiation team. These include the need to: (i) define common objectives; (ii) understand the reasoning and background of other disciplines; (iii) separate what one wants from what one knows and what one is able to do; (iv) stimulate creativity and an atmosphere of joint learning; and (v) develop a joint, broad vision for the future.
3. A focus on soil quality rather than on soil properties in general is attractive for input into the negotiation process. Quality indicators for six different soil functions (Blum, 1997) can be defined and – depending on demand – sometimes indicators can be combined, as was illustrated for biomass production and environmental quality. Still, negotiations are likely to benefit further from emphasis on management procedures that lead to certain soil qualities, and consider the associated *thresholds* rather than soil-quality indicators as such.

4. A single, comprehensive soil-quality indicator for any given soil is difficult to define because different soil functions are associated with quite different demands. Still, actual and potential functioning of a given soil (its overall quality) can be defined by a characteristic *window of opportunity* determined by data obtained by monitoring and dynamic simulation. The latter technique, in particular, offers unique opportunities to explore effects of innovative forms of soil management.

5. The negotiation process requires good contacts between stakeholders and policy makers and mutual respect and trust. Rather than only providing theoretical or abstract data on soil behaviour, field observations should be made for any given type of soil reflecting the effect of different forms of past management. Existing soil maps, showing where certain soils occur in the field, are an as yet unexplored source of information as is illustrated. There is a wealth of information that remains to be explored.

References

Blum, W.E.H. (1997) Basic concepts: degradation, resilience and rehabilitation. In: Lal, R., Blum, W.H.M., Valentine, C. and Stewart, B.A. (eds) *Methods for Assessment of Soil degradation*. CRC Press, Boca Raton, Florida, pp. 1–16.

Bouma, J. (1994) Sustainable land use as a future focus of pedology. *Soil Science Society of America Journal* 58, 645–646.

Bouma, J. (1997) Long-term characterization: monitoring and modeling. In: Lal, R., Blum, W.H., Valentin, C. and Stewart, B.A. (eds) *Methods for Assessment of Soil Degradation*. CRC Press, Boca Raton, Florida, pp. 337–358.

Bouma, J. (2001a) The role of soil science in the land use negotiation process. *Soil Use and Management* 17, 1–6.

Bouma, J. (2001b) The new role of soil science in a network society. *Soil Science* 166, 874–879.

Bouma, J. (2002) Land quality indicators of sustainable land management accross scales. *Agriculture, Ecosystems and Environment* 88, 129–136.

Bouma, J. and Droogers, P. (1998) A procedure to derive land quality indicators for sustainable agricultural production. *Geoderma* 85, 103–110.

Bouma, J. and Droogers, P. (1999) Comparing different methods for estimating the soil moisture supply capacity of a soil series subjected to different types of management. *Geoderma* 92, 185–197.

Bouma, J. and Jones, J.W. (2001) An international collaborative network for agricultural systems applications (ICASA). *Agricultural Systems* 70, 355–368.

Bouma, J., Batjes, N. and Groot, J.J.H. (1998a) Exploring soil quality effects on world food supply. *Geoderma* 86, 43–61.

Bouma, J., Finke, P.A., Hoosbeek, M.A. and Breeuwsma, A. (1998b) Soil and water quality at different scales: concepts, challenges, conclusions and recommendations. *Nutrient Cycling in Agroecosystems* 50, 5–11.

Bouma, J., Stoorvogel, J., van Alphen, B.J. and Booltink, H.W.J. (1999) Pedology, precision agriculture and the changing paradigms of agricultural research. *Soil Science Society of America Journal* 63, 343–348.

Bouma, J., Jansen, H.G.P., Kuyvenhoven, A., van Ittersum, M.K. and Bouman, B.A.M. (2000) Introduction. In: Bouman, B.A.M., Jansen, H.G.P., Schipper, R.A., Hengsdijk, H. and Nieuwenhuyse, A. (eds) *Tools for Land Use Analysis on Different Scales*. Kluwer Academic Publishers, Dordrecht, pp. 1–7.

Bouma, J., van Alphen, B.J. and Stoorvogel, J.J. (2002) Fine tuning water quality regulations in agriculture to soil differences. *Environmental Science and Policy* 5, 113–120.

Castells, M. (1996) The information age: economy, society and culture: 'The rise of the network society'. Oxford University Press, Oxford, 380 pp.

Doran, J.W., Coleman, D.C., Bezdicek, D.F. and Stewart, B.A. (1994) *Defining Soil Quality for a Sustainable Environment*. SSSA Special Publication 35, Soil Science Society of America, Madison, Wisconsin, 244 pp.

Droogers, P. and Bouma, J. (1997) Soil survey input in exploratory modeling of sustainable soil management practices. *Soil Science Society of America Journal* 61, 1704–1710.

Droogers, P., Fermont, A. and Bouma, J. (1996) Effects of ecological soil management on workability and trafficability of a loamy soil in the Netherlands. *Geoderma* 73, 131–145.

European Commission (1991) Council directive 91/676/EEC concerning the protection of waters against pollution caused by nitrates from agricultural sources. *Official Journal* L 375, 31/12/1991: 1–8.

Healy, P., Khakee, A., Motte, A. and Needham, B. (1997) *Making Strategic Spatial Plans*. University College of London Press, London.

HMD (The Metropolitan Debate) (1998) *Laagland (The Low Country)*. HRM-design. Amsterdam, 107 pp. [in Dutch].

Kropff, M.J., Bouma, J. and Jones, J.W. (2001) Systems approaches for the design of sustainable agricultural systems. *Agricultural Systems* 70, 369–393.

Pulleman, M.M., Bouma, J., van Essen, E.A. and Meyles, E.W. (2000) Soil organic matter content as a function of different land use history. *Soil Science Society of America Journal* 64, 689–694.

Rescher, N. (1998) *Predicting the Future: an Introduction to the Theory of Forecasting.* State University of New York, Albany, New York, 320 pp.

Smyth, A.J. and Dumanski, J. (1993) *FESLM: an International Framework for Evaluating Sustainable Land Management.* World Resources Reports 73. Land and Water Development Division, FAO, Rome, 77 pp.

Sonneveld, M.P.W., Veldkamp, T. and Bouma, J. (2002) Refining soil survey information for a Dutch soil series using land use history. *Soil Use and Management* 18, 157–163.

Tait, J. (2001) Science, governance and multi-functionality of European agriculture. *Outlook on Agriculture* 30, 91–95.

T&O (2001) *Toekomstonderzoek en Strategisch Omgevingsbeleid (Future Research and Strategic Environmental Policy).* Final Report. Scientific Council for Government Policy, The Hague, The Netherlands, 358 pp.

van Alphen, B.J. (2002) A case study on precision nitrogen management in Dutch arable farming. *Nutrient Cycling in Agro-Ecosystems* 62, 151–161.

VROM (Ministry of Housing, Spatial Planning and Environment) (2001) *Een Wereld en een Wil (A World and a Desire).* The Fourth National Environmental Policy Plan. The Hague, The Netherlands, 410 pp.

Chapter 17

Soil Quality in Industrialized and Developing Countries – Similarities and Differences

R. Lal

The Ohio State University, Carbon Management and Sequestration Center, Columbus, Ohio 43210, USA

Summary

The world population of 6 billion in 2000 will increase to 9.32 billion by 2050, with 8 billion, or 86% of the total population, living in the developing countries where soil and water resources are limited and already under great stress due to historic land misuse and soil mismanagement. Thus, soil quality and its management are more important now than ever before, especially in developing countries that are characterized by high risks of soil degradation, predominantly resource-poor and small landholders, and weak institutions. The demand on soil resources is increasing, especially for new and conflicting soil functions: enhancing food security, improving water quality, disposing urban and industrial wastes and mitigating climate change. Although some of the concepts of soil quality have been understood for a long time, new developments include identification of key soil-quality indicators and their threshold values in relation to specific soil functions. Whereas the key soil indicators are similar in developed and developing countries, the difference lies in threshold/critical values and the rate of change with change in land use and management. It is important to strengthen the database on key soil-quality indicators and their threshold values for soil organic carbon (SOC) concentration, aggregation, pH, available water capacity

(AWC), NPK storage, cation exchange capacity (CEC), microbial biomass carbon and biomass of earthworms and termites. There is a need for a paradigm shift towards developing an interdisciplinary team with a participatory approach involving farmers/land managers. The objective is to link scientific information with farmers' knowledge, and translate data into usable practices. New challenges of the 21st century warrant development of a high-quality, original, innovative, demand-driven and problem-solving research programme in developing countries.

Introduction

The fact that the world population of 6.2 billion is increasing at the rate of 1.8% per annum and may reach 8 billion by 2025 and 9.32 billion by 2050 (Litvin, 1998) is not as serious a challenge as is the reality of its uneven geographical distribution and growth. The seriousness of the challenge lies in the fact that most of the future increase in population will occur in developing countries, where about 1 billion people are already underfed and malnourished, and soil and water resources are under great stress. Developing countries that will experience the largest absolute growth in population by 2050 relative to 2000 include India (575 million), China (300 million), Nigeria (200 million), Pakistan (200 million) and Ethiopia (140 million) (Fisher and Helig, 1997). Whereas the global demand for food may double by 2030 compared to that in 1990, the relative increase may be 2.5–3 times in developing countries and as much as five times in countries of sub-Saharan Africa (SSA) (Daily et al., 1998; Greenland et al., 1998). Penning de Vries et al. (1995a,b) observed the following regions that have a serious threat of food security: southern Asia with either medium or high population growth but with high external input systems of agricultural production; southern and South-east Asia with medium population growth; and Central America, western Africa, South-east and southern Asia with high population growth but low external input systems of agricultural production. However, Penning de Vries and colleagues did not consider 'soil quality' as a variable in their study. Yet, the adverse impacts of soil degradation (e.g. erosion, compaction, salinization, acidification) on current and future production cannot be ignored.

The objective of this chapter is to describe the importance of soil quality management and enhancement in achieving food security and agricultural sustainability in the developing countries. Although a considerable knowledge of soil-quality indicators has been accrued for soils of the temperate/developed regions (Doran and Parkin, 1994, 1996; Carter et al., 1997), the challenge lies in transferring and adapting this knowledge towards achieving sustainable use of the soil and water resources of the developing countries.

Agricultural Sustainability

In the context of developing countries, agricultural sustainability implies 'a nonnegative trend in productivity and soil/water quality over time'. There are three important aspects of sustainability: space, time and dimension (Herdt and Steiner, 1995). The spatial aspect refers to the scale of its assessment, which may be at molecular/cellular, plant, crop, farming system, regional or global scale. There is a strong link among scales, and weakness at any level can jeopardize the sustainability at others. The temporal aspect implies that quantification at any one scale is related to another over time, and the process is dynamic and highly time dependent. The minimum time-span to assess sustainability of a system at the farm or larger scale is 25 years, or a generation. The dimensional aspect of sustainability refers to four distinct but interrelated processes – biophysical, economic, social and environmental. To be sustainable, an agricultural system must be biophysically achievable, economically viable, socially/ethnically/politically acceptable and environmentally compatible. The biophysical dimension refers to the agronomic output or yield (e.g. Mg/ha/year, Mg/ha/input of energy), the economic dimension to the profitability (net return), the social dimension to the ability of the system to support a farming community in the context of its ethnic culture and traditions, and the environmental dimension to water and air quality (Lal, 1998).

With high and increasing demographic pressure in developing countries, coupled with scarcity of soil and water resources, sustainable agriculture is not synonymous with 'low-input', 'organic' or 'alternative' agriculture as is sometimes implied in developed countries. In some cases, a low-input system may be acceptable for a short time, but in others it may not be acceptable at all. In some cases, application of organic amendments may be profitable, but in other cases the bulk required and the logistics of procurement and transport may make it either unfeasible or unprofitable. In some cases, pest incidence may be reduced through crop rotations and other biological measures, but in others it may not.

With the aim of enhancing production on a continuous basis in densely populated developing countries, sustainable agriculture is the 'most economic-cum-efficient harnessing of solar energy in the form of agricultural products without degrading soil and environment quality'. The overall goal is to achieve food security while restoring soil resources, improving water quality, mitigating climate change and preserving soil/natural resources for long-term use rather than exploiting them for short-term gains. The strategy is to optimize output while minimizing input, restore degraded soils while enhancing efficiency of input and enhance quality of water resources while using chemicals judiciously and discriminately.

The developing countries have no alternative to 'agricultural intensification'. This means using soil resources as per their capability and according to methods which produce a positive soil balance in nutrients (N, P, K, S, Zn), organic matter (OM) pool and maintain a favourable balance in water resources. Such systems can be identified by combining traditional knowledge with scientific information, and harnessing the potential of soil resources to their fullest.

Soil Quality

Soil quality, the capacity of soil to perform specific functions of interest to humans, has been of interest to humankind ever since the dawn of civilization and settled agriculture (see Schjønning et al., Chapter 1, and Karlen et al., Chapter 2, this volume). Some ancient civilizations (e.g. those in the valleys of the Nile, Indus and Yangtze) thrived for millennia because soil quality was maintained by the alluvial processes of the specific rivers that supported these cultures. However, many ancient cultures perished because they could not maintain or enhance the quality of their soil resources. Decline in soil quality by accelerated erosion and the attendant degradation toppled many ancient civilizations by washing/blowing away the very foundation on which they developed (Lowdermilk, 1939; Eckholm, 1976; Olson, 1981). Whereas the concern of soil quality has challenged humankind for at least 10,000 years, the definition and the basic concepts remain a work in progress and keep evolving with every generation. A soil scientist of Moorish Spain, Ibn-Al-Awam, wrote several volumes during the 12th century on agricultural issues. The book 'Kitab al-Felhah' or 'Book of Agriculture' was translated into Spanish in 1802. The book was brought to public attention in 1802 in the 'Encyclopedia of Islam' (1760–1777). In the book, the author writes: 'The first step in science of agriculture is the recognition of soils and of how to distinguish that which is of good quality and that which is of inferior quality.' The best of all soils, according to the author, are the alluvium of river valleys 'because of the mud with which they are mixed, for the running water brings sediments removed from the surface of the soil along with dead leaves and manure' (Banqueri, 1802).

There are several definitions of soil quality (Doran and Parkin, 1994, 1996; Carter et al., 1997; Johnson et al., 1997). Schjønning et al. (Chapter 1, this volume) adopt the definition by the Soil Science Society of America (SSSA, 1995), which states that 'it is the capacity of a specific kind of soil to function within natural or managed ecosystem boundaries, to sustain plant and animal productivity, maintain or enhance water and air quality, and support human health and habitation'. However, soil functions keep changing with time, and are different in developing compared with developed countries. That being the case, soil-quality indicators and/or their threshold values may also differ in developing *vis-à-vis* developed countries. Although the definition of soil quality may be universal, the fact that its application is soil/society specific needs to be recognized for the concept to be useful in addressing the problems of resource-poor farmers in developing countries.

Soil Functions Relevant to Developing Countries

Basic human needs since the dawn of human civilizations (e.g. food, feed, fodder, fibre and fuel) have been met by five identified soil functions. These are:

1. Medium for plant growth;
2. Repository for gene pool;
3. Archives for planetary/human history;
4. Foundation for engineering structures; and
5. Raw material for industry.

Soil management focused on meeting basic needs of humanity through these functions. Serious food shortages of the 1950s and 1960s were addressed by successful applications of the so-called Green Revolution technologies (Hazel and Ramaswamy, 1991) to enable soils to meet the first function (as a medium for plant growth). Such technologies enhanced production per unit soil area. For example, from 1950 to 2000, the number of people fed by a single US farmer increased from 19 to 129 (Bond, 2000). The effect of Green Revolution technology was especially spectacular in South Asia. Food grain production in India quadrupled over the second half of the 20th century from 50 million Mg to 200 million Mg (Swaminathan, 2000), and there was a linear relationship between energy input (e.g. tillage, fertilizers, pesticides and irrigation) and crop yield (Panesar, 1996).

Plant breeders developed short-stem (dwarf) varieties that responded to high-energy input. Both yield gains per hectare and total grain production increased substantially with increase in energy input (Pimentel *et al.*, 2002).

While food production increased substantially in developed countries and in South and South-east Asia, agricultural production stagnated in SSA. Further, yields of wheat (*Triticum aestivum*) and rice (*Oryza sativa*) have started to decline even in the highly productive areas of South Asia (Pimentel, 1996). Subsidy of N fertilizers and their excessive use have mined P, K and other soil nutrients (Tandon, 1992). Because of the nutrient imbalance, the yield of wheat per unit input of nitrogenous fertilizer has been progressively declining (Twyford, 1994). The gap between food production and demand in South Asia and SSA is widening because of the imbalance between people and soil quality. Soil resources in the developing countries are not only limited in extent but are also prone to severe degradation caused by anthropogenic perturbations of the delicate balance between soil, environment and people. Consequently, there is a need to revisit soil functions of interest to humans and the definition and underlying concept of soil quality.

In the context of the 21st century, soil quality must meet the needs of four basic functions of interest to humans.

1. Maximize long-term productivity per unit input of the non-renewable resources.
2. Minimize risks of environmental pollution, especially with regard to quality of water and air.
3. Moderate fluctuations in components of the water and energy budget due to change in land use and land cover.
4. Proxy for interpretations of past global changes.

Whereas the need for food production is widely recognized, the importance of soil in mitigating the rate of enrichment of atmospheric CO_2 and improving water quality is not. The drastic increase in atmospheric concentration of CO_2 is attributed to fossil fuel combustion and land-use change. The latter includes deforestation, biomass burning, conversion from natural to managed ecosystems and soil cultivation (IPCC, 2001). The emission source with regards to land-use change is primarily deforestation in developing countries. Houghton *et al.* (1987) estimated C emission from tropical deforestation at 1.7 Pg C/year (Pg = petagram = 1 billion tonnes). Soil erosion, ploughing and predominantly low-input agriculture in developing countries are also major sources of emission of greenhouse gases (GHGs). Lal (2002b) estimated that accelerated soil erosion in the tropics causes emission of 0.2–0.5 Pg C/year. Therefore, soils and ecosystems of the developing countries have been and presently are principal sources of CO_2 and other GHGs, and soils of the tropics may have contributed 17–39 Pg of C as CO_2 (Lal, 2002b).

Depletion of the SOC pool leads to decline in soil quality and agronomic productivity. The adverse impact of declining soil quality on productivity is strikingly apparent in developing countries where resource-poor farmers cannot afford the inputs needed to restore soil quality and enhance/replace the depleted SOC pool. Indeed, soil resources of the developing countries are critical in ensuring food security and mitigating climate change.

Soil Resources of Developing Countries

Predominant soils of the tropics include Oxisols, Ultisols and Alfisols in the humid regions and Alfisols, Aridisols, Inceptisols, Entisols and Vertisols in the dry regions

(Table 17.1). Some of these soils have severe constraints towards meeting the functions to address the issues of the 21st century. Soil-related constraints are especially severe in arid and semi-arid regions. Important constraints include nutrient depletion and low soil fertility, Al toxicity and P deficiency, physical degradation and accelerated soil erosion. Over and above the inherent constraints, there are also severe problems of human-induced soil degradation. Soil-nutrient depletion continues to be a major problem in Africa. During the early 1990s, average annual nutrient loss on arable land in Africa was estimated at 22 kg N/ha, 2.5 kg P/ha, and 15 kg K/ha (Stoorvogel *et al.*, 1993; Smaling and Oenema, 1998). Oldeman *et al.* (1991) estimated the extent of soil degradation in Africa to be 65% on agricultural lands, 31% on permanent pastures and 19% on woodlands and forests. Serious soil degradation affected 19% of the total land area. Soil degradation is also a serious problem in South Asia (FAO/UNDP, 1994), where 218 Mha are affected by different processes of which 137 Mha are affected by a moderate-plus level of severity. The problem of soil degradation in Asia extends to the entire Asia-Pacific region (Scherr, 1999), where 31% of the total land area (535 Mha) is affected by different processes of soil degradation. Latin America and the Caribbean regions are equally prone to human-induced soil degradation (Oldeman *et al.*, 1991).

Desertification, i.e. land degradation in dry/arid regions, is also a severe problem

Table 17.1. Soil resources of the tropics and soil-related constraints to enhancing soil quality.

Soil order	Land area[a] (Mha)	% of the world area	Soil-related constraints
Alfisol	800	46.2	Weak soil structure, crusting, compaction, erosion by water
Aridisols	900	36.3	Drought, wind erosion, desertification, secondary salinization
Entisols	400	36.7	Erosion, nutrient depletion, low soil organic matter
Inceptisols	400	34.2	Erosion, low soil organic matter, nutrient imbalance
Mollisols	50	4.4	Few if any limitations
Oxisols	1100	98.2	Nutrient imbalance, toxicity of Al, P fixation, acidification
Ultisols	550	75.3	Nutrient imbalance, acidification, P fixation
Vertisols	100	43.5	Massive structure, poor tilth, drought stress, water erosion
Highlands	600	21.4	Severe erosion, terrain deformation
Total	4900	37.5	

[a]Land area for different orders is from Van Wambeke (1991).

in developing countries (Dregne and Chou, 1992; Le Houerou, 1994) but especially so in Central Asia (Babaeva, 1999; Kharin *et al.*, 1999). Irrigated cotton (*Gossypium hirsutum*) and wheat production has brought about a severe ecological disaster in the region. As a result of the excessive and inefficient irrigation, the Aral Sea has decreased in volume by 54%, sea level has fallen by 25%, salinity has increased from 10 to 27 g/l, the shoreline has receded by 20 Mha, and about 75 Tg (million tons) of salt blows out of it annually on to the surrounding land (Kaser and Mehrotra, 1996). Secondary salinization is a serious problem throughout the region (World Bank, 1993a,b,c).

There are several causes of the widespread problem of human-induced degradation in developing countries. In addition to the biophysical factors (e.g. fragile soils, harsh climate), the human dimensions have also played an important role in exacerbating the extent and severity of degradative processes (Steiner, 1996). Important among these are high demographic pressure, ill-defined rights of ownership and use, and inappropriate subsidies and fiscal policies. Another important reason is 'taking soil resources for granted', as is evidenced by human greed, short-sightedness, poor planning and cutting corners for quick economic returns. These exploitative policies have mined the soil resources for long periods of time without any investment in soil restoration and rehabilitation. The case of severe soil degradation in developing countries indicates that the biophysical process of soil degradation is driven by social, economic, ethnic and political forces.

Recent Developments in Identifying Indicators of Soil Quality

While recognizing some controversy about the basic concepts in soil quality (Sojka and Upchurch, 1999), considerable progress has been made in the 1990s in identifying the indicators of soil quality, with particular reference to agronomic production. Developments in soil-quality indicators for sustainable management of plant nutrients have been highlighted in this volume for cations and soil acidity in Chapter 3 by Johnston, nitrogen in Chapter 4 by Christensen, phosphorus in Chapter 5 by Condron, sustainable management of potassium in Chapter 6 by Askegaard *et al.*, maintenance of soil OM in Chapter 7 by Dick and Gregorich, soil biodiversity and its maintenance in Chapter 8 by Alabouvette *et al.*, and biological soil quality in Chapter 9 by Brussard *et al.* Similar to soil chemical and biological qualities, equally impressive advances have been made in identifying determinants of soil physical quality especially with regard to subsoil compaction (Chapter 10 by Van den Akker and Schjønning), soil structure and its management (Chapter 11 by Kay and Munkholm) and soil erosion and its management (Chapter 12 by Govers *et al.*).

A summary of the key indicators of soil quality is outlined in Table 17.2. Interactive effects among soil-quality indicators outlined in Fig. 17.1 show that SOC concentration is the predominant parameter that affects other determinants of soil physical, chemical and biological qualities, and quality and quantity of plant products (Tiessen *et al.*, 1994). In addition to SOC concentration, other key indicators include bulk density, available water capacity, microbial biomass carbon, biomass of earthworms and termites and soil structure. A similar list was compiled by Lal (1994) with regard to soils and ecosystems of the tropics. These reports highlight the importance of soil structure, especially with regard to the widespread problems of crusting, compaction, erosion and other physical degradative processes. Although compiling a generic list of key indicators is necessary, it is equally important to determine threshold values (critical limits) of these parameters, which may differ among soils, land uses and specific functions of interest.

Knowledge Gained in Developed Countries and its Relevance to Developing Countries

Basic soil processes and underlying mechanisms are similar in tropical and temperate regions. Therefore, key soil-quality indicators as listed in Table 17.2 and Fig. 17.1 are also similar (Lal, 1994, 2000a,b; Bouma *et al.*, 1998; Dumanski and Pieri, 2000; Bouma, 2002). The principal difference lies in: (i) the reference values or baseline characteristics; (ii) threshold values or the critical limits; (iii) the rate of change in these parameters because of anthropogenic perturbations or the susceptibility to human-induced soil degradation; (iv) the magnitude of impact of soil degradation on biomass production and the environment; and (v) high spatial variability in soils of the tropics. The issue of soil variability, as was also pointed out by Sojka and Upchurch (1999), needs to be addressed. There are 5 million series and 10 million phases of series in the tropics (Eswaran *et al.*, 1992). In addition, the variability may be at the macro-scale and the micro-scale. Macro-variability refers to variability that persists on a large scale and results in distinct geographic patterns and

Table 17.2. Recent developments in identifying indicators of soil quality.

Issue	Soil properties
Minimum data set	Aggregate stability, bulk density, soil organic carbon concentration, pH, CEC, total N, P and S, potentially mineralizable N and microbial biomass, clay concentration
Soil nitrogen	Nitrogen use efficiency and integrated nutrient management
Soil phosphorus	Environmental threshold levels of soil P
Soil acidity and cations	Al^{3+}, Mn^{2+} toxicity, threshold soil pH and cations
Soil potassium	Critical soil K threshold values, maintaining positive K balance
Soil organic matter	A key indicator of soil quality and environment moderation capacity
Soil biological quality	Microbial biomass and activity, earthworm/termite biomass
Subsoil compaction	Soil strength
Soil structure	Threshold values of organic carbon concentration
Erosion	Available water capacity, effective rooting depth, soil organic carbon concentration

Fig. 17.1. Key indicators of soil quality and its dynamics. NPS labels nitrogen, phosphorus and sulphur, and CEC is cation exhange capacity.

taxonomic groupings as reported by Eswaran *et al.* (1992). In contrast, micro-variability refers to local and often recurrent variations at a small scale. These variations may be permanent or persist through several cropping cycles and affect soil quality as reflected in crop yield (Moorman and Kang, 1978). Effects of soil variability on soil quality (and thus agronomic yield) are generally more pronounced under low-input farming as practised by resource-poor farmers of the developing countries. It is the knowledge and understanding of the causes and processes of soil micro-variability that is needed to develop specific soil-quality indicators. In this regard, it is important to use modern technology for soil survey (e.g. spatial databases, ground-penetrating radar, video image analysis).

The differential response of soils in developing compared to developed countries is attributed to differences in temperature and moisture regimes, clay minerals, leaching and rate of erosion. For example, SOC concentration is the key soil quality index for both temperate and tropical ecoregions. The difference lies in the rate of its change due to change in land use and in the threshold values. The rate of decline in SOC following conversion from natural to agricultural ecosystems is extremely rapid, and its impact on soil structure and accelerated erosion is drastic in tropical compared with temperate ecosystems. Some studies have shown that the rate of mineralization of OM content in the tropics may be four times greater than that in the temperate regions (Jenkins and Ayanaba, 1977). These observations are supported by the classical work of Hans Jenny on the relation between annual temperature and OM content on a transect from north to south in the USA (Jenny, 1930, 1949, 1950, 1961). Another confounding factor is the effect of clay type and amount on the retardation of OM decomposition in soils. High-activity clays (2:1 type) have greater surface areas than low-activity clays. The slow rate of decomposition in soils with predominantly high-activity clays may be due to adsorption and a relatively high proportion of micropores (< 1 μm) in which SOC is inaccessible to microbial attack (Greenland *et al.*, 1992). Protection may also occur due to interlamellar adsorption by the expanding lattice of high-activity clays (Allison, 1973). Consequently, cultivated soils in the tropics have lower levels of OM than similar soils in the temperate latitudes. Mohr (1922) noted that at temperatures above 30°C, the rate of decomposition of OM by anaerobic organisms becomes sufficiently rapid so that even poor drainage does not necessarily lead to accumulation of OM. Mohr's observations were confirmed by the data of Neue and Scharpenseel (1987) following the decomposition of ^{14}C-labelled rice (*Oryza sativa*) straw under continuously flooded conditions. Similar observations were made for paddy soils in Asia by Kyuma (1985).

Differences in the dynamics of key soil indicators are primarily due to differences in micro- and mesoclimate, especially high temperatures throughout the year and intense rains. The high rate of chemical reaction (e.g. decomposition of OM) is in direct response to high temperatures. High risks of soil erosion are due to intense rains and weak soil structure. Thus, the high rate of depletion of SOC following conversion from natural to managed ecosystems is due to a high rate of mineralization and preferential removal of the light fraction in water runoff and eroded sediment. Similarly, low use efficiencies of applied nutrients are due to high losses (by leaching, erosion and volatilization) and low net ecosystem/primary productivity.

There are also differences in social, economic and policy factors (Table 17.3) Farmers in developing countries are understandably concerned with maximizing rather than optimizing yields, optimizing rather than minimizing the use of off-farm input, increasing household income rather than maximizing profit, ensuring adequate supply of water rather than reducing non-point source pollution, providing food for the family in the immediate future rather than long-term sustainability, and addressing concerns of the family (food, health, schooling) rather than of community at large (Table 17.3). Any strategy of technology transfer to developing countries must consider these issues of concern to resource-poor farmers.

It is because of these differences in social, economic and political factors that sustainability/soil-quality indices used in developed

(Table 17.4) and developing countries (Table 17.5) are different even though key soil parameters (Table 17.2) are similar. Indices commonly used in developed countries (Table 17.4) involve total-factor or natural-resource productivity, and energy or C use efficiency.

In contrast, common indices of soil quality and sustainability in developing countries are land-oriented because predominant production systems are resource-based rather than science-based. Understanding these differences is important for successfully

Table 17.3. Issues of soil quality in developed and developing countries.

Developed countries	Developing countries
1. Optimizing crop yields per unit input	1. Maximizing crop yields per unit area, time
2. Minimizing input of chemicals and energy	2. Optimizing the use of off-farm input
3. Maximizing farm profit	3. Increasing household income, including off-farm employment
4. Reducing risks of pollution/eutrophication of surface and ground waters	4. Ensuring adequate supply of water for human and animal consumption
5. Sustaining productivity on a long-term basis	5. Providing food for the family before the next harvest
6. Addressing issues of regional, national and global importance (e.g. global climate change)	6. Addressing concerns of the family

Table 17.4. Sustainability/soil quality indices used in developed countries.

Index	Reference
1. Total factor productivity (TFP) $= P/\sum_{i=1}^{n}(R_iC_i)$ where P is total production, R_i is resource and C_i is the cost of specific input	Herdt (1993)
2. Total natural resource productivity (TNRP) $= \text{TFP}/\Delta S$ where ΔS is change in soil quality	Herdt and Steiner (1995)
3. Index of sustainability $I_s = f(P_i \times S_i \times W_i \times C_i)t$ where I_s is the index of sustainability, P_i is the change in production, S_i is the alteration in soil properties, W_i is the change in water resources and quality, C_i is modification in climatic factor and t is time	Lal (1993)
4. Energy use efficiency = energy output : energy input	Pimentel et al. (2002)
5. Carbon use efficiency $= [\text{NPP}/\Sigma(C_i)]_t$, where NPP is the net primary production measure in C, and C_i is carbon equivalent of all input	Lal (2002b)
6. Soil quality index $= S_q = f(\text{SOC, CEC, pH, AWC, WSA})$ where SOC is soil organic carbon, CEC is cation exchange capacity, AWC is available water capacity and WSA is water-stable aggregates	Lal (1993); Doran and Parkin (1994, 1996); Gregorich and Carter (1997)
7. Soil quality renewal or Sr $= \int_{0}^{t}(S_n - S_d + I_m)dt$, where S_n is renewal rate, S_d is degradation rate, I_m is input and t is time	Lal (1997)

Table 17.5. Soil quality, land use and sustainability indices used in developing countries.

Index[a]	Reference
1. Land use factor, $L = (C + F)/C$	Okigbo (1978)
2. Land equivalent ratio, LER $= \sum_{i=1}^{n}(Y_iY_m)$	Willey and Osiru (1972)
3. Area × Time Equivalency Ratio, ATER $= \dfrac{1}{t}\left(\sum_{i=1}^{n}\left(\dfrac{d \cdot Y_i}{Y_m}\right)\right)$	Hiebsch and McCollum (1987)

[a]L, land use factor; C, cropping period; F, fallow period; Y_i, yield of component crops in intercrop; Y_m, yield of component crops in monoculture; d, growth period of crop in days; t, time in days for which the field remained occupied or the growth period of the longest-duration crop.

transferring/adapting the technological innovations in soil quality enhancement and management. Resource-poor and small landholders of the developing countries cannot afford the inputs needed to maintain/enhance the value of key soil indicators above the threshold values. Although farmers realize the importance of returning crop residue to the soil, applying farmyard manure to recycle the nutrients and using fertilizers in a balanced way, they lack the resources to implement the recommended management practices. Crop residues and cattle dung are needed as fodder and fuel, and P fertilizer (no subsidies) is more expensive than subsidized N fertilizer.

Linking Food Security in Developing Countries with Emerging Concepts in Soil Quality

Because of the high demographic pressure and the overriding concerns of food security, using soil quality concepts for enhancing crop yield will remain a higher priority in developing countries than maintaining/improving the environment. The key issue is 'how to maintain soil quality under growing and conflicting modes of land use, non-availability or prohibitive cost of essential input, and urgent need for enhancing food production'. The competing land uses in densely populated regions of the developing world are: crop land, grazing land, plantations, restoration of degraded soils, urbanization, nature conservancy, C sequestration and waste disposal. Thus, there is a strong need to adopt a holistic approach to soil quality management (Fig. 17.2). The holistic approach must consider possible impacts of interactions of the pedosphere with the biosphere, hydrosphere, atmosphere and lithosphere as confounded by anthropogenic perturbations.

Food security will remain a primary concern in developing countries during the first half of the 21st century and beyond. The overall goal of soil-quality enhancement is to achieve high yields that are constantly increasing on a long-term basis per unit: (i) land area; (ii) time; (iii) input of irrigation water, NPK and lime; and (iv) input of energy for tillage, other farm operations and pesticides. Because of growing environmental concerns, high yields must also be obtained per

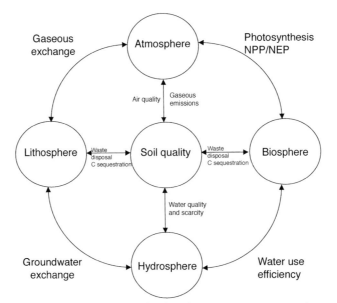

Fig. 17.2. A holistic approach to management of soil quality. NPP, net primary production; NEP, net ecosystem productivity.

unit output of: (i) GHGs (CO_2, CH_4, N_2O), and (ii) water pollutants (nitrates, phosphates, pesticides) (Lal, 1999). There have been drastic changes in storage of terrestrial carbon (Houghton, 1995, 1999), and the atmospheric concentration of CO_2 has steadily increased (Prentice et al., 2001; Schimel et al., 2001).

Next in priority to food security is the availability of high-quality water. At least 1 billion people in the developing countries do not have access to hygienically clean water (Litvin, 1998). There is a severe shortage of renewable fresh water both for agricultural use and human/animal consumption. The shortage is being exacerbated by a rapid rise in population. Gardner-Outlaw and Engelman (1997) reported that at least 30 populous countries will face a severe water shortage by 2025. Most populous among these are India, Nigeria, Egypt, Ethiopia and Iran.

Soil quality is also closely linked to air quality and especially to the atmospheric enrichment of GHGs. In fact, soil degradation or decline in soil quality is a principal contributor to the increasing concentration of GHGs in the atmosphere. Waste disposal and urbanization are two other predominant issues with regards to soil quality and functions that cannot be ignored. Waste disposal, an important environmental liability, has to be addressed in relation to an important soil function in rapidly industrializing countries.

The potential of soil C sequestration is high in developing compared with developed countries because soils are depleted of their SOC pool. The relatively high magnitude of SOC depletion in soils of the developing countries is attributed to predominantly low-input agriculture (especially in SSA) and the widespread problem of erosion and other soil-degradation processes. The potential of soil C sequestration is 0.9–1.9 Pg C/year for desertification control (Lal, 2001), 0.2–0.4 Pg C/year in the WANA (West Asia–North Africa) region (Lal, 2002a) and 0.3–0.5 Pg C/year for soils of the tropics (Lal, 2002b). This vast potential can be realized through application of the concepts of soil quality. Adoption of recommended practices can lead to soil C sequestration (IPCC, 2000).

Challenges of Technology Transfer in Developing Countries

There are four challenges with regard to the transfer of technologies for soil-quality management in developing countries (Lal, 2000a,b). One deals with the development of a database in soil-resources characterization to facilitate validation/adaptation of known/proven technologies. The second deals with the development of a methodological protocol for characterizing soil quality and assessing the impact of land use and management on soil quality parameters. The third, and perhaps the most important challenge, is to define critical limits or threshold values of key soil-quality indicators for the highly variable soils of the tropics in the developing countries. The fourth challenge is to link scientific information with traditional knowledge.

Soil quality database

It is important to develop a comprehensive national soil database at appropriate scales that land managers, decision makers and extension agents can use. The desired database may comprise digital maps and interpretative information on key soil-quality indicators as outlined in Table 17.2. The database must be developed to quantitatively assess soil quality for different functions (e.g. crop land, grazing land, forest land, waste disposal, C sequestration), evaluate water/nutrient use efficiencies, assess structural changes in relation to transport of pollutants into surface and ground water, and evaluate changes in pools and fluxes of SOC in relation to emission of GHGs into the atmosphere. The soil database needs to be strengthened with reference to the following processes.

Soil physical processes

Important soil physical parameters to be assessed include: soil aggregation, available water capacity, texture, bulk density, infiltration rate and rooting depth. A quantitative assessment of these parameters is needed to predict biomass productivity, SOC dynamics, transport processes of water and solutes, gaseous exchange and emission of GHGs into the atmosphere, and fate of C and pollutants transported by erosional processes.

Soil chemical processes

Important soil chemical parameters are pools and fluxes of principal nutrients (e.g. N, P, K, S, Zn, Mo), threshold values of N and P to obtain the desired yield, soil reaction in relation to ionic concentration, and toxic levels of Al^{3+}, Mn^{2+}, etc. In addition to their impact on biomass production, knowledge of the dynamics of these parameters is needed to assess the eutrophication/pollution of natural waters and emission of N_2O, NO_x into the atmosphere.

Soil biological processes

The importance of soil biological processes in soil quality is widely recognized, especially for the developing countries. A quantitative knowledge of soil biological processes is important with regard to: (i) availability of nutrients to plants (Christensen, Chapter 4, this volume); (ii) atmospheric enrichment of GHGs and strategies to mitigate it; (iii) recycling of urban and industrial waste (Naidu et al., Chapter 13, this volume); (iv) biodegradation of pollutants; (v) biological control of rhizosphere pests (Alabouvette et al., Chapter 8, this volume); (vi) improving quality of groundwater especially with regard to waterborne pathogens; (vii) strengthening recycling mechanisms of nutrients; (viii) discovering new products from microbial processes; and (ix) reducing risks of GHG emissions.

Methodological protocol

There is a need to develop and standardize methodological protocol to assess soil quality in relation to specific soil functions, especially biomass production (Lal, 1994). A series of steps needed to develop the necessary database include the following: (i) assess the baseline (reference point) of key soil properties under natural ecosystems; (ii) establish threshold values or critical limits of key soil parameters with reference to basic processes to be addressed; (iii) conduct an objective constraint analysis with regard to specific issues (e.g. erosion, compaction, nutrient depletion, decline in SOC pool); (iv) evaluate the impact of land-use change and soil-management practices on key soil properties, and assess the suitability of a given land use/management; and (v) re-evaluate soil functions and land-use objectives in view of the database thus established.

Developing threshold values of soil-quality indicators

Threshold values or critical limits of key soil properties must take into consideration the micro-variability and be based on crop response to management practices that alter soil quality. Of necessity, such information must be based on long-term soil management experiments conducted on benchmark soils in principal agroecoregions. The strategy is to establish pedotransfer functions based on the cause–effect relationship, and yield-response functions for key soil properties. Although a considerable amount of research information on soil management in the tropics is available, there have been few attempts to establish generic relationships between soil properties (soil quality) and agronomic yield. A wealth of information exists on fertilizer response to yield of principal crops, but often without consideration of the relevant soil properties. Even less information is available with regard to crop response to soil physical properties (e.g. available water capacity, rooting depth, soil compaction, soil temperature, etc.).

Lal (1994) provided quantitative information on threshold levels of key properties for soils of the tropics. These values included: (i) soil strength properties; (ii) soil mechanical properties; (iii) porosity and available water capacity; (iv) water transmission properties; (v) soil organic carbon content; and (vi) chemical and nutritional properties. Lal also proposed a methodology for combining threshold levels of soil indicators into an index or several indices to assess sustainability of a specific management system. Possible statistical approaches to developing a soil quality index include: (i) parametric methods; (ii) geostatistical procedures; (iii) linear or stepwise regression techniques; (iv) principal component analyses; (v) pedotransfer functions; and (vi) modelling.

The data from long-term experiments are needed to assess temporal changes in soil-quality indicators in relation to land use. Functional relationships between soil-quality indicators and crop yield must be assessed over time. Knowledge about such temporal trends is necessary to assess change in actual/ potential productivity due to change in soil-quality indicators over time. Aune and Lal (1997) used this approach and identified threshold values of soil-quality indicators for Oxisols, Ultisols and Alfisols of the tropics.

Linking scientific information with farmer knowledge

There has been a lack of or a slow rate of adopting improved technology in developing countries, but especially so in SSA. A wealth of research information of relevance to small landholders and resource-poor farmers is available, but is awaiting adoption. There are several reasons for the lack or slow rate of adoption. Important among these is the top-down approach that did not involve farmers in the decision-making process. It is important to develop a participatory approach that involves farmers/land managers in the planning process. The research programme must be demand-driven, and address issues of community concern.

It is also important to integrate scientific information with traditional knowledge in order to translate theory into practice by developing 'management tools' or 'decision support systems'. Although many farmers in the developing countries are aware of the benefits of returning OM to the soil, practical issues are serious constraints to translating this knowledge into practice. Palm *et al.* (2001)

observed that the principal considerations by farmers in the developing countries that limit the use of organic materials on land are: (i) immediate rather than long-term production goals; (ii) lack of the organic materials needed to apply at the appropriate time and at the desired rate; and (iii) labour constraints, etc. Given the choice between subsidized chemical fertilizers with quick response and the use of bulky manures with slow and long-term gains, farmers will choose chemical fertilizers. Therefore, addressing farmers' concerns through a participatory approach is an important strategy.

Above all, there is a strong need to adopt a positive and a holistic approach to soil quality enhancement in the developing countries. Reversing soil degradative processes requires a holistic approach by an interdisciplinary team involving both biophysical and social scientists. The interdisciplinary approach is critical because of a close link between soil quality, economic progress and environment quality.

Conclusions

The concept of soil quality in relation to sustainability is as old as the history of settled agriculture. However, new scientific knowledge in measurement and monitoring has enabled us to identify key soil properties, their threshold values and changes in soil-quality parameters under different land uses and management systems. New information is also available on the interdependence between pedosphere, biosphere, atmosphere and lithosphere, and the impact of anthropogenic perturbations on ecosystem functions.

Key soil-quality indicators are similar in developed (temperate) and developing countries. These are SOC concentration, bulk density, clay content and type, aggregation, available water capacity, CEC, NPK reserves, biomass carbon and biomass of earthworms and termites. The SOC concentration, and its dynamics in relation to land-use change and soil management, is an important indicator of soil quality. Soil-specific research information is needed to establish baseline or reference values, threshold values or critical limits, and the rate of change of these parameters under diverse land uses and soil management.

There is a need for a paradigm shift in responding to new challenges of the 21st century, in reaching out to other disciplines in addressing issues of global concern, and in the transfer of technology to land managers and small landholders. Adopting a participatory or bottom-up approach is necessary to link scientific information with farmer knowledge and to translate information into practice.

Two basic strategies for achieving agricultural sustainability in developing countries are: (i) preventing soil degradation, and (ii) enhancing soil quality. Although management options for both strategies may be similar, enhancing soil quality necessitates information on threshold levels of soil properties in relation to specific soil functions (e.g. crop yield). Identification of threshold levels is a researchable priority with regard to specific land use, management system and expected output. There is also a close relationship between soil resilience and threshold values, which needs to be determined for restoring degraded soils.

References

Allison, F.E. (1973) *Soil Organic Matter and Its Role in Crop Production.* Elsevier, Amsterdam, 637 pp.

Aune, J.B. and Lal, R. (1997) Agricultural productivity in the tropics and critical limits of properties of Oxisols, Ultisols and Alfisols. *Tropical Agriculture (Trinidad)* 74, 96–103.

Babaeva, T.A. (1999) The mapping of desertification processes. In: Babaev, A.G. (ed.) *Desert Problems and Desertification in Central Asia.* Springer, Berlin, pp. 89–100.

Banqueri, J.A. (1802) *Libro de Agricultura: Su Autor El Doctor Excelente Abu Zacaria Lahia Aben Mohamed Ben Ahmed Ebu El Awam.* Sevillano, Madrid, Vol. I, Vol. II.

Bond, C.S. (2000) Politics, misinformation and biotechnology. *Science* 287, 1201.

Bouma, J. (2002) Land quality indicators of sustainable land management across scales. *Agricultural Ecosystems and Environment* 88, 129–136.

Bouma, J., Batjes, N.H. and Groot, J.J.R. (1998) Exploring land quality effects on world food supply. *Geoderma* 86, 43–59.

Carter, M.R., Gregorich, E.G., Anderson, D.W., Doran, J.W., Janzen, H.H. and Pierce, F.J. (1997) Concepts of soil quality and their significance. In: Gregorich, E.G. and Carter, M.R. (eds) *Soil Quality for Crop Production and Ecosystem Health.* Elsevier, Amsterdam, 1–19.

Daily, C., Dasgupta, P., Bolin, B., Crosson, P., Guerry, J.D., Ehrlich, P., Folke, C., Jansson, A.M., Jansson, B.-O., Kautsky, N., Kinzig, A., Levin, S., Mäler, K.-G., Pistrup-Anderson, P., Sinisealco, D. and Walker, B. (1998) Food production, population growth and the environment. *Science* 281, 1291–1292.

Doran, J.W. and Parkin, T.B. (1994) Defining and assessing soil quality. In: Doran, J.W., Coleman, D.C., Bezdicek, D.F. and Stewart, B.A. (eds) *Defining Soil Quality for a Sustainable Environment.* Soil Science Society of America Special Publication No. 35, Madison, Wisconsin, pp. 3–21.

Doran, J.W. and Parkin, T.B. (1996) Quantitative indicators of soil quality. In: Doran, J.W. and Jones, A.J. (eds) *Methods for Assessing Soil Quality.* Soil Science Society of America Special Publication No. 49, Madison, Wisconsin, pp. 25–37.

Dregne, H.E. and Chou, N.T. (1992) Global desertification: dimensions and costs. In: Dregne, H.E. (ed) *Degradation and Restoration of Arid Lands.* Texas Tech University, Lubbock, Texas, pp. 249–282.

Dumanski, J. and Pieri, C. (2000) Land quality indicators: research plan. *Agricultural Ecosystems and Environment* 81, 93–102.

Eckholm, E.P. (1976) *Losing Ground.* Norton, New York, 223 pp.

Eswaran, H., Kimble, J., Cook, T. and Beinrothm, F. (1992) Soil diversity in the tropics: implications for agricultural development. In: Lal, R. and Sanchez, P.A. (eds) *Myths and Science of Soils of the Tropics.* Soil Science Society of America Special Publication No. 29, Madison, Wisconsin, pp. 1–16.

FAO/UNDP (1994) *Land Degradation in South Asia: Its Severity, Causes and Effects Upon the People.* World Soil Resources Reports 78, FAO, Rome, 100 pp.

Fisher, G. and Helig, G.K. (1997) Population momentum and the demand on land and water resources. *Philosophical Transactions of the Royal Society (London), Series (B)* 352, 869–889.

Gardner-Outlaw, T. and Engelman, R. (1997) *Sustaining Water, Easing Scarcity.* Population Action International, Washington, DC, 20 pp.

Greenland, D.J., Wild, A. and Adams, D. (1992) Organic matter dynamics in soils of the tropics: from myths to complex reality. In: Lal, R. and

Sanchez, P.A. (eds) *Myths and Science of Soils of the Tropics.* Soil Science Society of America Special Publication No. 29, Madison, Wisconsin, pp. 17–33.

Greenland, D.J., Gregory, P.J. and Nye, P.H. (1998) Land resources and constraints to crop production. In: Waterlow, D.J., Armstrong, L.F. and Riley, R. (eds) *Feeding a World Population of More Than Eight Billion People.* Oxford University Press, New York, pp. 39–51.

Gregorich, E.G. and Carter, M.R. (eds) (1997) *Soil Quality for Crop Production and Ecosystem Health.* Elsevier, Amsterdam, 448 pp.

Hazel, P.B.R. and Ramaswamy, C. (1991) *The Green Revolution Reconsidered: the Impact of High Yielding Rice Varieties in South India.* Johns Hopkins University Press, Baltimore, 286 pp.

Herdt, R.W. (1993) Measuring sustainability using long-term experiments. *Proceedings of a Conference held at Rothamsted Experimental Station, 29–30 April, 1993.* Rothamsted Experimental Station, UK, pp. 3–13.

Herdt, R.W. and Steiner, R.A. (1995) Agricultural sustainability: concepts and conundrums. In: Barnett, V., Payne, R. and Steiner, R. (eds) *Agricultural Sustainability: Economic, Environmental and Statistical Considerations.* John Wiley & Sons, Chichester, UK, pp. 1–13.

Hiebsch, C.K. and McCollum, R.E. (1987) Area × time equivalency ratio: a method for evaluating the productivity of intercrops. *Agronomy Journal* 79, 15–22.

Houghton, R.A. (1995) Changes in storage of terrestrial carbon since 1850. In: Lal, R., Kimble, J.M., Levine, E. and Stewart, B.A. (eds) *Soils and Global Change.* CRC/Lewis, Boca Raton, Florida, pp. 45–65.

Houghton, R.A. (1999) The annual net flux of carbon to the atmosphere from changes in land use from 1850 to 1990. *Tellus* 50B, 298–313.

Houghton, R.A., Boone, R.D., Fruci, J.R., Hobbie, J.E., Melillo, J.M., Palm, C.A., Peterson, B.J., Shaver, G.R. and Woodwell, G.M. (1987) The flux of carbon from terrestrial ecosystems to the atmosphere in 1980s due to change in land use: geographical distribution of the global flux. *Tellus* 39B, 122–139.

IPCC (2000) *Land Use, Land Use Change and Forestry.* Special Report of the Intergovernmental Panel on Climate Change. Cambridge University Press, Cambridge.

IPCC (2001) *Climate Change 2001. The Scientific Basis.* Special Report of the Intergovernment Panel on Climate Change. Cambridge University Press, Cambridge.

Jenkins, D.S. and Ayanaba, A. (1977) Decomposition of carbon-14 labeled plant material under tropical conditions. *Soil Science Society of America Journal* 41, 912–916.

Jenny, H. (1930) *A Study of the Influence of Climate upon the Nitrogen and Organic Matter Content of the Soil.* Missouri Agricultural Experiment Station Research Bulletin No. 152.

Jenny, H. (1949) Comparative study of decomposition rates of organic matter in temperate and tropical regions. *Soil Science* 68, 419–432.

Jenny, H. (1950) Causes of high nitrogen and organic matter content of certain tropical forest soils. *Soil Science* 69, 63–69.

Jenny, H. (1961) *Comparison of Soil Nitrogen and Carbon in Tropical and Temperate Regions as Observed in India and America.* Missouri Agricultural Experiment Station Bulletin No. 765.

Johnson, D.L., Ambrose, S.H., Bassett, T.J., Bowen, M.L., Crummey, D.E., Isaacson, J.S., Johnson, D.N., Lamb, P., Saul, M. and Winter-Nelson, A.E. (1997) Meaning of environmental terms. *Journal of Environmental Quality* 26, 581–589.

Kaser, M. and Mehrotra, S. (1996) The Central Asian economies after independence. In: Allison, R. (ed) *Challenges for the Former Soviet South.* The Royal Institute of International Affairs, London, pp. 217–305.

Kharin, N.G., Tateishi, R. and Harahsheh, H. (1999) *Degradation of Drylands of Asia.* Chiba University, Japan, 81 pp.

Kyuma, K. (1985) Fundamental characteristics of wetland soils. In: IRRI (ed.) *Wetland Soils: Characterization, Classification and Utilization.* International Rice Research Institute, Los Banos, Philippines, pp. 191–206.

Lal, R. (1993) Tillage effects on soil degradation, soil resilience, soil quality and sustainability. *Soil and Tillage Research* 27, 1–8.

Lal, R. (1994) *Methods and Guidelines for Assessing Sustainable Use of Soil and Water Resources in the Tropics.* SCS Technical Monograph No. 21, Soil Management Support Services, Washington, DC, 78 pp.

Lal, R. (1997) Degradation and resilience of soils. *Philosophical Transactions of the Royal Society of London (B)* 352, 997–1010.

Lal, R. (1998) Soil quality and agricultural sustainability. In: Lal, R. (ed.) *Soil Quality and Agricultural Sustainability.* Ann Arbor Press, Chelsea, Michigan, pp. 3–12.

Lal, R. (1999) Soil management and restoration for carbon sequestration to mitigate the accelerated greenhouse effect. *Progress in Environmental Science* 1, 307–326.

Lal, R. (2000a) Soil management in developing countries. *Soil Science* 165, 57–72.

Lal, R. (2000b) Physical management of soils of the tropics: priorities for the 21st century. *Soil Science* 165, 191–207.

Lal, R. (2001) Potential of desertification control to sequester carbon and mitigate the greenhouse effect. *Climate Change* 51, 35–72.

Lal, R. (2002a) Carbon sequestration in dryland ecosystems of West Asia and North Africa. *Land Degradation and Development* 13, 45–59.

Lal, R. (2002b) The potential of soils of the tropics to sequester carbon and mitigate the greenhouse effect. *Advances in Agronomy* 76, 1–30.

Le Houerou, H.N. (1994) *Zero-order Draft on Climate Change, Drought and Desertification*. IPCC, Geneva, 22 pp.

Litvin, D. (1998) Dirt poor. *Economist*, 21 March, 3–16.

Lowdermilk, W.C. (1939) *Conquest of the Land through 7000 Years*. USDA-ARS Bulletin 99, Washington, DC, 30 pp.

Mohr, E.C.J. (1922) *Die Grund van Java en Sumatra*. J.H. de Bussy, Amsterdam.

Moormann, F.R. and Kang, B.T. (1978) Micro-variability of soils in the tropics and its agronomic implications with special reference to West Africa. In: *Diversity of Soils in the Tropics*. ASA Special Publication No. 34, Madison, Wisconsin, pp. 29–43.

Neue, U. and Scharpenseel, H.W. (1987) Decomposition pattern of ^{14}C labelled rice straw in aerobic and submerged rice soils of the Philippines. *Science of the Total Environment* 62, 431–434.

Okigbo, B.N. (1978) *Cropping Systems and Related Research in Africa*. AAAS Occasional Publication Series OT-1, Addis Ababa, Ethiopia, 81 pp.

Oldeman, L.R., Hakkeling, R.T.A. and Sombroek, W.G. (1991) *World Map of the Status of Human-induced Soil Degradation: an Explanatory Note*. ISRIC, Wageningen, The Netherlands.

Olson, G.W. (1981) Archaeology: lessons on future soil use. *Journal of Soil and Water Conservation* 36, 261–264.

Palm, C.A., Giller, K.E., Mofongoya, P.L. and Swift, M.J. (2001) Management of organic matter in the tropics: translating theory into practice. *Nutrient Cycling in Agroecosytems* 61, 63–75.

Panesar, B.S. (1996) Changing patterns of energy use in agriculture. In: Kansal, B.D., Dhaliwal, G.S. and Bajwa, M.S. (eds) *Agriculture and Environment*. National Agricultural Technical Information Center, PAU, Ludhiana, India, 26, pp. 228–239.

Penning de Vries, F.W.T., Van Keulen, H. and Rabbinge, R. (1995a) Natural resources and the limits of food production. In: Bouma, J. *et al.*

(eds) *Ecoregional Approaches for Sustainable Land Use and Food Production*. Kluwer Academic, Dordrecht, The Netherlands, pp. 65–87.

Penning de Vries, F.W.T., Van Keulen, H. and Luyten, J.C. (1995b) The role of soil science in estimating global food security in 2040. In: Wagnet, R.J., Bouma, J. and Hutson, J.L. (eds) *The Role of Soil Science in Interdisciplinary Research*. Soil Science Society of America Special Publication No. 45, Madison, Wisconsin, pp. 17–37.

Pimentel, D. (1996) Green revolution agriculture and chemical hazards. *Science of the Total Environment* 188 (Suppl.), S86–S98.

Pimentel, D., Doughty, R., Carothers, C., Lamberson, S., Bora, N. and Lee, K. (2002) Energy inputs in crop production in developing and developed countries. In: Lal, R., Hansen, D.O., Uphoff, N. and Slack, S. (eds) *Food Security and Environment Quality in the Developing World*. Lewis Publishers, Boca Raton, Florida, pp. 129–151.

Prentice, I.C., Farquhar, G.D., Fasham, M.J.R. *et al.* (2001) The carbon cycle and atmospheric carbon dioxide. In: Houghton, J.T., Ding, Y., Griggs, D.J., Noguer, M., van der Linden, P.J. and Xiaosu, D. (eds) *Climate Change 2001. The Scientific Basis*. Special Report of the Inter-governmental Panel on Climate Change. Cambridge University Press, Cambridge, pp. 183–237.

Scherr, S.J. (1999) *Soil Degradation: a Threat to Developing Country Food Security by 2020*. Food, Agriculture and the Environment Discussion Paper No. 27, IFPRI, Washington, DC, 63 pp.

Schimel, D.S., House, J.I., Hibbard, K.A. *et al.* (2001) Recent patterns and mechanisms of carbon exchange by terrestrial ecosystems. *Nature* 414, 169–172.

Smaling, E.M.A. and Oenema, O. (1998) Estimating nutrient balance in agroecosystems at different spatial scales. In: Lal, R., Blum, W.H., Valentin, C. and Stewart, B.A. (eds) *Methods for Assessment of Soil Degradation*. CRC, Boca Raton, Florida, pp. 229–251.

Soil Science Society of America (1995) Statement on Soil Quality. *Agronomy News*, June.

Sojka, R.E. and Upchurch, D.R. (1999) Reservations regarding the soil quality concept. *Soil Science Society of America Journal* 63, 1039–1054.

Steiner, K.G. (1996) *Causes of Soil Degradation and Development Approaches to Sustainable Soil Management*. Margraf Verlag, Germany, 50 pp.

Stoorvogel, J.J., Smaling, E.M.A. and Janssen, B.H. (1993) Calculating soil nutrient balances in

Africa at different scales. I. Supra-national scale. *Fertilizer Resource* 35, 227–335.

Swaminathan, M.S. (2000) Science in response to basic human needs. *Science* 287, 425.

Tandon, H.L.S. (1992) *Assessment of Soil Nutrient Depletion.* Paper presented at FADINAP Regional Seminar 'Fertilization and Environment', 7–11 September, Chiang Mai, Thailand.

Tiessen, H., Cuevas, E. and Chacaon, P. (1994) The role of soil organic matter in sustaining soil fertility. *Nature* 371, 783–785.

Twyford, I. (1994) *Fertilizer Use and Crop Yields.* Paper presented to 4th National Congress of the Soil Science Society of Pakistan, Islamabad, 1992.

Van Wambeke, A. (1991) *Soils of the Tropics: Properties and Appraisal.* McGraw Hill, New York, 343 pp.

Willey, R.W. and Osiru, D.S.O. (1972) Studies on mixtures of maize and beans with particular reference to plant population. *Journal of Agricultural Science.* 79, 519–529.

World Bank (1993a) *Kazakhstan: the Transition to a Market Economy.* World Bank, Washington, DC, 87 pp.

World Bank (1993b) *Uzbekistan: an Agenda for Economic Reform.* World Bank, Washington, DC.

World Bank (1993c) *Kyrgyzstan: Social Protection in a Reforming Economy.* World Bank, Washington, DC.

Chapter 18
Soil Quality Management – Synthesis

P. Schjønning, S. Elmholt and B.T. Christensen

*Danish Institute of Agricultural Sciences, Department of Agroecology,
PO Box 50, DK-8830 Tjele, Denmark*

Summary

There is an urgent need to analyse the term soil quality in more detail. Most frequently, it is used to describe soil attributes. The term should be used for this purpose only when related to sustainability concerns: (i) soil productivity; (ii) impact on the environment; and (iii) effect on human health. The soil quality concept has been adopted mainly as a technical framework for grading soil and evaluating management effects. We advocate more focus on soil quality as a cognitive concept associated with sustainability. Grading of soils by indicators is difficult across soil types, climates and cropping systems. The indexing of soil-quality indicators introduces a significant loss of information on the complex agroecosystem. The contributions to the present book show that it is generally not possible to identify simple thresholds for sustainable farming at either the soil-quality indicator level or the management level. More adequate means of communicating management prescriptions to farmers and other decision makers

©CAB International 2004. *Managing Soil Quality: Challenges in Modern Agriculture*
(eds P. Schjønning, S. Elmholt and B.T. Christensen)

in society are needed. Decisions should be based on knowledge of management effects on individual soil types as well as on stochastic and mechanistic models simulating soil processes and functions. Teams of researchers and stakeholders are needed in a joined effort towards optimal decisions. This calls for well-educated farmers and consultants. Scientific contributions towards sustainable farming should be based on combinations of on-farm system studies and analyses of specific soil functions. The main purpose of on-farm system studies is to develop that particular system rather than to compare it with other systems. System studies may enable identification of emergent properties, which are important in system development and in identifying gaps in our understanding of the soil ecosystem. There is an urgent need to communicate scientific results with explicit expressions of sustainability priorities. This is to facilitate transfer of results from agricultural research and to encourage scientists to cope with the precautionary principle, adduced by environmentalists and other concerned citizens.

Introduction

One of the main ideas behind this book was to address major challenges in modern agriculture in the context of sustainability as it becomes manifest in the soil quality concept. The authors were encouraged to address their subject with reference to the terms and concepts discussed in Chapter 1. Here we will not reiterate conclusions reached for each specific topic. Rather we will draw attention to statements and conclusions of general relevance to the development of sustainable agriculture and to the advancement of communication in science and society.

One key observation is the blurred semantics of the soil quality term. We found that the term has quite different meanings to different persons. We regard it as most urgent that the semantics of the soil quality term be discussed thoroughly within the scientific community. This chapter also addresses the term sustainability as exemplified by contributions to the book, and reflects on the authors' attempts to identify management thresholds. Further, we consider scientific approaches in the framework of the cognitive context (cf. Chapter 1), i.e. reflections on the role of science in society. Finally, we comment on the implementation and dissemination of results obtained in soil quality studies.

Soil Quality Semantics

Soil quality – concept or attribute

Numerous publications addressing 'soil quality' have appeared over the last decade.

Most papers assign specific soil attributes to the term (e.g. organic matter (OM) content, structural stability, and microbial activity). Only a few consider the soil quality *concept* in terms of: (i) productivity; (ii) environment impact; and (iii) human health (see Fig. 1.2). Although (some of) these concerns are implicitly included in the attributes (indicators) chosen to express soil quality, the soil quality term is often used unreflectively for a vast number of soil characteristics. Carter (2002) noted that some studies are purely descriptive, whereas others apply a functional approach by relating soil attributes to soil fitness for a specific use. Hence, papers reporting on soil attributes are not necessarily addressing soil quality as a concept.

Carter *et al.* (Chapter 15) conclude that soil quality should be considered as an emergent property of the agricultural system. Emergent properties are defined as functional interactions of system components that cannot be observed in studies of smaller units of organization. Using this approach, soil quality can be regarded as an objective state of the soil; soil quality may be addressed by measuring soil attributes. However, since humans are part of an agricultural system (Carter *et al.*, Chapter 15), soil quality incorporates an expression of sustainability. We find that this dualism of the 'soil quality' term is a significant barrier in communicating the soil quality concept.

Figure 18.1 illustrates the dualism for the soil quality as well as for the sustainability term. Most soil quality studies may be assigned to 'quadrant IV', i.e. studies addressing specific soil properties with no explicit consideration of the concerns set up in the soil

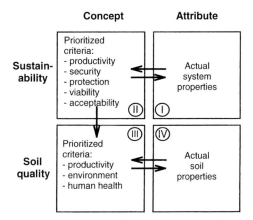

Concept   **Attribute**

Fig. 18.1. Relationship between *concepts* and *attributes* for the sustainability and soil quality terms. Double arrows indicate that actual attributes are evaluated by comparing with set goals. The single arrow indicates that the soil quality criteria are sustainability criteria set up for soil. The five sustainability criteria are those suggested by Smyth and Dumanski (1993), and the soil quality criteria are those focused in the SSSA definition (Allan *et al.*, 1995). Consult text for details.

quality concept ('quadrant III'). Such studies are relevant, but we advocate that the term soil quality is used only when soil properties are evaluated in the framework of the three soil quality concerns. Terms like soil properties, functions and processes should be preferred in studies addressing specific soil attributes.

Figure 18.1 also illustrates how the soil quality concerns are expressions of sustainability ('quadrant II'). These may include concerns beyond soil quality. By analogy with soil quality attributes (Herdt and Steiner, 1995), the system fulfilment of sustainability criteria should also be evaluated (Fig. 18.1, 'quadrant I', and Fig. 1.5).

Soil quality – cognitive or technical concept

The soil quality term emerged in North America (Alexander, 1971; Warkentin and Fletcher, 1977). Despite an early, intense discussion in the USA (e.g. Allan *et al.*, 1995; Karlen *et al.*, 1997) and governmental support for an institute addressing the soil quality concept (Anonymous, 1996), the relevance

and impact of the concept are currently being disputed (Sojka and Upchurch, 1999; Karlen *et al.*, 2001; Letey *et al.*, 2003; Sojka *et al.*, 2003). It appears that this dispute is fuelled largely by the lack of a clear objective for the use of the soil quality concept. To us, soil quality is a cognitive concept that incorporates sustainability. We concur that sustainability criteria are explicitly quantified in the soil quality concept due to the mere fact that values and goals are in play in science even if we do not recognize this (Chapter 1). However, scientists addressing the soil quality concept often focus on the identification of indicators and, especially, the indexing of soil-quality indicators (e.g. Doran and Jones, 1996; Karlen *et al.*, Chapter 2), considering soil quality as a technical concept (a tool for presentation of information) rather than as a cognitive concept.

Sojka *et al.* (2003) illustrated the shortcomings of indexing in expressing complex properties and functions of soils. We agree with this criticism as detailed in the section 'Indexing of soil quality knowledge' below. However, we disagree with Sojka *et al.* (2003) when stating that (soil) science is an activity independent of values and goals in society (Chapter 1). Science is a human activity and as such is influenced by personal and societal contexts and priorities. Clearly, there is a most urgent need to discuss the content and the purpose of the soil quality concept (Fig. 18.2).

Sustainability

Sustainability is a concept that reflects societal priorities (Chapter 1). The term has been adopted in common language worldwide since the release of the 'Brundtland Report' (WCED, 1987). To some extent this has turned out to be a problem because it has become a 'buzz' word with no clear meaning. As stated by Herdt and Steiner (1995), it is invoked by 'presidents, priests and plebeians'! When one person includes the term 'sustainable' in a dialogue, there is no 'guarantee' that the implicit values and goals are apprehended by the other person in that dialogue. They may refer to different personal or societal priorities. For example, the statement 'we

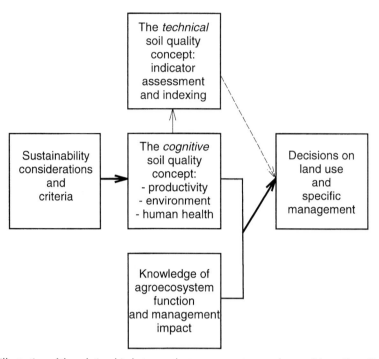

Fig. 18.2. Illustration of the relationship between the two perceptions and uses of the soil quality concept. Soil quality may be regarded as a cognitive concept to be combined with knowledge of soil function to guide management (components linked by boldface arrow), or it may be taken as a technical concept (a tool) for assessing soil properties. The broken arrow indicates our viewpoint that this link to management guidance is weak and inadequate. The thin line between the cognitive and technical concept boxes indicates that the latter has been derived from the former in principle, but that this link has been toned down when using the technical concept.

have introduced sustainable agriculture in our country' has no meaning if the criteria for sustainability are not defined explicitly. This problem becomes evident especially when values and politics change rapidly.

In Chapter 1, we adopted the five sustainability criteria suggested by Smyth and Dumanski (1993) and later promoted by Bouma *et al.* (1998) in the context of soil quality. Sustainable management of agricultural land should simultaneously: (i) maintain or enhance production and services; (ii) reduce the level of production risk; (iii) protect the potential of natural resources and prevent degradation of soil and water quality; (iv) be economically viable; and (v) be socially acceptable. Below, we give an example where these concerns have been addressed.

Askegaard *et al.* (Chapter 6) note that K reserves are easily accessible and long lasting,

and that K lost from agriculture represents no threat to the environment (part of criterion (iii) above). Further, there appear to be no potential health hazards from a balanced use of K (criterion (ii) above). Nevertheless, crop production under reduced K input is a key issue in the presentation. Askegaard *et al.* recall the suggestions made by the Brundtland Commission (WCED, 1987) and base their contribution on the initial part of criterion (iii) that agricultural management should protect the potential of the natural resources, thereby demonstrating an awareness of criterion (v) on 'acceptability'. The general increase in organic farming avoiding the use of mined fertilizers has increased funding of research into low-K-input farming. Their focus on sustained productivity in such low-K-input systems is based on the sustainability criteria (i) and (iv) above.

The regulation of soil acidity also requires application of mined minerals. Although naturally occurring deposits of calcitic and magnesian limestones are a non-renewable resource, there are vast deposits of both (Johnston, Chapter 3). Hence, in principle we have the same situation as for K. However, the use of limestone is accepted by most alternative farming movements.

These examples reflect how the societal and intentional dimensions of the cognitive context (cf. Fig. 1.1) affect the research agenda. However, the research work involved in specific studies emerging from different research agendas should meet the same universally recognized quality standards (the observational dimension of the cognitive context, Fig. 1.1).

Soil Quality at Work

Sojka and Upchurch (1999) and subsequently Sojka et al. (2003) pleaded for 'quality soil management' rather than 'soil quality management' and argued against the soil quality concept and the compiling of universal indicators for soil attributes across soil types. With the present book project, we intended to evaluate the potential of regulating soil quality through management rather than assessing quality indicators. Sojka et al. (2003) also focused on management throughout their paper, the difference being that we attempt to focus management within the cognitive concept of soil quality discussed above. We find that there is an urgent need to explicitly consider sustainability in relation to soil science, and that the soil quality concept constitutes a useful framework for this endeavour.

We recognize that the indicator assessment approach can be ineffective in developing sustainable management systems. Accordingly, in Chapter 1 the term 'management threshold' was introduced as the most severe disturbance any management may accomplish without inducing significant changes towards unsustainable conditions. Figure 1.4 gives a comparison between the 'indicator threshold approach' and the 'management threshold approach'. Authors were encouraged to identify management thresholds for their specific subjects. For some issues, the focus is almost exclusively on a lower threshold (e.g. potassium), whereas for some a lower as well as an upper limit has been indicated (e.g. phosphorus).

Soil chemistry and plant nutrition

The regulation of soil acidity is fundamental to any agricultural use of soils. Johnston (Chapter 3) reviews this area, including soil type and cultivar differences in optimal pH. A management threshold may be defined as the rate of liming needed to maintain pH at a prescribed level. According to data presented in Chapter 3, this threshold increases with increasing equilibrium soil pH values. For a neutral soil (pH = 7), it amounts to about 1000 kg $CaCO_3$/ha/year (Table 3.4).

The nitrogen cycle in arable soil is complex. Christensen (Chapter 4) considers that natural ecosystems generally exhibit a high degree of temporal and spatial coincidence (termed synchrony and synlocation, respectively) between N release and root uptake potentials within the soil. This causes the transfer of N across system borders to be small compared to the amount of N cycled within the ecosystem. In such systems, the N cycle can be considered as relatively tight. The challenge in managed ecosystems is to optimize synchrony and synlocation for all N sources in the system, i.e. to optimize the use of management tools influencing the N cycle. As these are highly dependent on soil type and crop sequence, no simple management threshold may be given. The section 'Communication and Implementation' below discusses the important impact that this has on how to best optimize N management.

The chapter on soil P reveals that quite different challenges prevail in different agricultural systems (Condron, Chapter 5). Many soils in developed countries have for years received more P than is taken up by crops. Although soil P is normally regarded as tied up in organomineral particles, soils with a large surplus of P may lose significant amounts of P to the environment. An Olsen-P level of 60 mg P/kg soil has been suggested as a 'change point' above which leaching of P in drainage water is triggered (Heckrath et al., 1995). Although it appears that this critical threshold varies considerably among soil types (McDowell et al., 2001), Condron (Chapter 5) concludes that soil P test values and the degree of P saturation can be used to predict the risk of P loss by surface or subsurface transfer. A management threshold may then be defined as the rate of P application that prevents soil P (soil-quality indicator) to rise above a specific level. Condron suggests that implementation would require P budgets at the farm as well as the regional level (variation in animal stocking density). The identification of 'critical source areas' from a combination of the hydrology and P status of soils can be one way of introducing mitigation measures, but the effectiveness of such approaches needs to be evaluated (Condron, Chapter 5). Interestingly, management options to reduce P loss from soils also include tillage strategies because conservation tillage reduces surface runoff of water carrying dissolved and colloidal P.

A shift to low-P-input systems may have no short-term effect on crop production if the soil has received surplus P for decades or centuries (Condron, Chapter 5). However, for soils inherently low in P or grown for a long period with reduced input, P availability will rely on well-developed root systems and associated mycorrhizal fungi, and it appears that no specific lower management threshold may be given for the present.

Although K is one of the most abundant nutrient elements in soil, only a small proportion may become available to growing crops. With reduced inputs of mined K, a greater efficiency in K management is required but no simple management threshold can be identified (Askegaard et al., Chapter 6). The quantity and availability of soil K varies greatly with soil type and past K management, and Askegaard et al. suggest that in managing low-K-input systems the aim should be to minimize losses by leaching on sandy soils and maximize the use of soil K reserves in clayey soils.

Soil organic matter and biodiversity

Soil OM is a key attribute of soil quality. Carter (2002) drew attention to the fact that its quantification is not standardized and suggested protocols for post-sampling handling of soil: it is often not clear if a measured soil OM content includes litter, crop residues, or root material. Kay and Munkholm (Chapter 11) identify three caveats in their discussion of management effects on soil OM. One addresses the risk of confounding several management tools in the interpretation of change in soil OM. Another focuses on the assumption that the relation between management impact and soil OM change is unambiguous. That is, no hysteresis in the relation between OM and the direction of change in the management tool is expected. Thirdly, general relationships between soil OM and management are often limited by the lack of data for long-term effects, although existing long-term experiments have been a unique source of information on the effects of various managements on specific soil properties (Christensen and Johnston, 1997).

Given the above, it seems logical that Dick and Gregorich (Chapter 7) use 'bold strokes of the brush' in their review of management effects on soil OM. They suggest a relative listing of major benchmark management systems that affect soil OM levels (Table 7.5). The list provided in Table 7.5 does not pretend to point out critical thresholds. According to Dick and Gregorich, it is useful in assessing the change of direction in soil OM content introduced by a change in management. In conclusion, management thresholds are difficult to establish for soil OM. This is also true for soil OM used as a soil-quality indicator. Further studies are needed for better understanding of the soil OM capacity factors (Dick and Gregorich, Chapter 7) as well as the relationship between soil OM and specific soil processes and functions (Carter, 2002). At present, critical limits are mainly established by consensus.

Chapters 8 and 9 address soil biodiversity and its importance to agroecosystem functioning. Brussaard et al. (Chapter 9) conclude that there is no evidence so far that agricultural practice has irreversibly changed the soil community, as also exemplified by Locke and Zablotowicz (Chapter 14). Generally, soil biota are only transiently disturbed by pesticides applied at recommended rates. Brussaard *et al.* advocate that lack of evidence is mainly due to lack of research. Methodological problems may contribute to this. Thus Alabouvette *et al.* (Chapter 8) elucidate how recent advances in molecular methodology will increase our understanding of genetic and functional biodiversity among populations of soilborne pathogens and antagonists. This may improve our understanding of 'disease suppression', i.e. some soils being suppressive towards certain soilborne diseases. This phenomenon is quite relevant to the soil quality concept, seeing pathogens as components of soil biodiversity, and with disease resulting from disturbances of the balance among functional groups in soil. Disease suppression is also discussed by Carter *et al.* (Chapter 15), who use this phenomenon as an example of an agriculturally important 'emergent soil property'.

Alabouvette *et al.* (Chapter 8) use soilborne populations of the genus *Fusarium* to exemplify the relevance of soil biodiversity studies. This is in line with Brussaard *et al.* (Chapter 9), who conclude that changes in diversity within suitable taxonomic groups (e.g. earthworms, nematodes) can be used as indicators of management-induced effects on, for example, OM and nutrient cycling. A causal relationship between soil biodiversity *beyond* the level of taxonomic groups and ecosystem functioning and stability does not seem to exist. Our knowledge at this level of diversity is not yet sufficiently complete and quantitative to be of practical value for management.

Some biological indicators other than biodiversity seem sensitive to different management practices. Brussaard *et al.* (Chapter 9) point to microbial and earthworm biomass and abundance as indicators of soil aggregate stability and to various measurements of microbial activity as indicators of the physiological state of the microbial biomass. Locke and Zablotowicz (Chapter 14) also suggest microbial biomass and earthworm survival

as being valuable soil attributes in the assessment of pesticide side effects. According to Brussaard *et al.* (Chapter 9) the main advantage of biological indicators – including biodiversity – is that they provide early warnings of long-term changes in OM, nutrient status and soil structure.

Carter *et al.* (Chapter 15) conclude that, although the understanding of the relationship between biodiversity and agroecosystem function is a prerequisite for effective soil quality management, the relationships between crop management and microbial diversity, and their consequences for soil health, remain unclear. Certainly, relationships are not straightforward and the contributions to this book document that we are far from finding simple tools to be of use in management decisions. Neither soil-quality indicator thresholds nor management thresholds for soil biota and soil OM have yet been identified. Best management practices have to be based on more complex expressions of soil functions and properties. Without common standards against which to judge soil-quality indicators, and with little evidence on the reversibility of changes in soil biodiversity, we find that the directions of desirable changes remain ambiguous and need to be examined much more thoroughly.

Soil physical form

Although Van den Akker and Schjønning (Chapter 10) take a clear attitude on the extent of subsoil compaction that may be considered sustainable, they have difficulties in arriving at simple management thresholds to secure their goal: no subsoil compaction. This is due to a lack of data on soil strength across a wide range of soil types and water contents. Unambiguous thresholds for the carrying capacity of soils are also difficult to obtain due to differences in wheel characteristics. Empirically derived maxima of allowable wheel load (20–30 kN: Håkansson and Petelkau, 1994) even seem to be too high if compared to predicted carrying capacities for a number of Dutch soils (11–28 kN, Table 10.1). An upper management threshold or 'rule of thumb' of 20–30 kN wheel load should thus be respected.

Chapter 10 calls for a better characterization of soil strength as well as for limited wheel loads. Most recent achievements further indicate a need for an in-depth evaluation of the basic models used for prediction of stress distribution below wheels as well as the mechanisms and criteria for soil failure (Trautner, 2003). The results emphasize the key importance of an equal distribution of the wheel load on the soil surface in contact with the wheel. The new results urgently call for further studies on compaction of undisturbed field soil.

Kay and Munkholm (Chapter 11) emphasize the balance between soil structure degradation and regeneration. Combinations of management practices are required in which the extent of degradation of soil structure caused by one practice should be balanced or exceeded by the extent of regeneration by other practices. As compared to pasture soils and forage cropping systems, continuous cash crop systems may reduce soil OM (e.g. Yang and Kay, 2001). This in turn may significantly impact soil tilth (e.g. Munkholm et al., 2001; Schjønning et al., 2002). Tillage puts a direct stress on soil and can cause significant degradation of soil tilth (e.g. Watts et al., 1996; Watts and Dexter, 1997). The question is which level of soil OM and which intensity of tillage will create unsustainable tilth characteristics.

Soil structure is significantly stabilized by OM. However, Kay and Munkholm (Chapter 11) conclude that no universal threshold value of soil OM and hence no specific management threshold can be given. Also Carter (2002) attempted to identify critical threshold soil OM values for soil functions. However, based on several investigations, he concluded that critical levels for total soil OM would be soil or site specific, related to a single soil process or function, and based on a range rather than a fixed value. Soil OM may have different qualities in its interaction with the mineral particles of soil (e.g. Christensen, 2001), which render it difficult to address the important soil OM soil quality attribute. It thus appears that the content of total soil OM may be of less value as an indicator of soil quality.

Soil type and climate would be expected to interact with tillage intensity. Hence, Kay and Munkholm (Chapter 11) propose that threshold levels for tillage intensity are identified using crop models and pedotransfer functions allowing for soil type differences. A key issue regarding tillage is the vicious circle that may be triggered in an intensively tilled soil: a poorly structured soil may require greater tillage intensity than soils with inherently better structure (e.g. Larney et al., 1988); however, a large input of energy may further decrease soil workability and hence call for even greater tillage intensity (Kay and Munkholm, Chapter 11). A study by Munkholm and Schjønning (2003) showed little resilience of soil surface structure within a 6-month summer period after intensive tillage. Further studies are urgently needed to identify threshold inputs of energy in order not to trigger the vicious circle.

Govers et al. (Chapter 12) give a thorough review of erosion processes and hence demonstrate that our knowledge of the mechanisms causing soil erosion is rather detailed. They also document that erosion may seriously affect important soil functions and services, including a sustained productivity. Throughout the world, conservation management has often been implemented to meet the threat of soil erosion. Govers et al. list some specific attributes known to control erosion. Soil cover, structural stability, surface roughness and water infiltrability are mentioned as key properties decisive for the extent of water erosion. According to Govers et al., integrated approaches like conservation tillage should be preferred for securing appropriate soil conditions to meet the erosion threat.

Soil contaminants

The spatial separation of densely populated urban areas from agricultural land complicates recycling of nutrients and OM produced in larger towns and cities. The challenge is related to the geographical distances, but the main barrier is the content of contaminants in the waste material. This is accentuated by the heterogeneity of the waste. Sewage sludge is often a mixture of industrial and household waste material. However, recycling of the plant nutrients and the OM in waste material is a key issue in the search for a more sustainable development (cf. the five criteria of sustainability addressed in the section 'Sustainability').

Naidu et al. (Chapter 13) evaluate the potential hazards in recycling waste material to agricultural land. The concerns are linked primarily to contents of potentially toxic substances (PTSs) and pathogens in sludge. The PTSs comprise organic chemicals and heavy metals. Naidu et al. conclude that the risk of bioaccumulation is low for many persistent organic contaminants but they recommend a case-by-case approach. They further conclude that appropriate management protocols may alleviate the risks presented by the pathogens. Leaching of nutrients in sludge may also be controlled if managed correctly.

Heavy metals are retained and accumulate in soil with repeated additions, and changes in the bioavailability of these contaminants in soil should be given more emphasis. In many countries, guidelines for sludge use, based on sludge and soil metal concentrations, have been developed with the aim of preventing metal toxicity to plants and to human and animal consumers of the plants (Tables 13.3 and 13.4). Many studies have shown that high levels of metals in soils can reduce plant growth (Naidu et al., Chapter 13), and Brussaard et al. (Chapter 9) found that very high loadings of Cu may be detrimental to soil nematodes and microbes. Pig production constitutes another source of Cu because Cu contained in feed concentrates ends up in the manure. A worst-case scenario indicates that for Danish conditions, a critical level of Cu in soil (40 mg Cu/kg soil) may be reached by 110 years of continuous application of pig slurry to soils (Poulsen, 1998).

The high productivity of modern agriculture is partly achieved by chemicals controlling weeds, pests and diseases. Many pesticides are toxic to non-target organisms and may leach to the aquatic environment or end up as residues in crops for human consumption. In other words, pesticides influence all three aspects of the soil quality concept. Locke and Zablotowicz (Chapter 14) provide a comprehensive review of recent research on pesticide side effects. Interestingly, Locke and Zablotowicz define the management threshold for pesticides as the *minimum* quantity of pesticide necessary to achieve sustainable crop production goals with negligible adverse effects to the soil ecosystem and environment. 'Minimum' indicates that the threshold is identified as related to production goals (at low doses, effects of pesticide on crop production increase with increase in dose). Similarly, a *maximum* threshold for pesticides may be identified as the maximum dose at which critical effects will not appear. Fortunately, for a number of pesticides, the positive effects on crop growth are effective at doses giving tolerable side effects (Locke and Zablotowicz, Chapter 14). However, in cases where the identified maximum tolerable for some pesticide turns out to be *lower* than the minimum identified from sustainable crop production goals, a conflict will arise. It is then a political decision as to which sustainability criteria should be met: appropriate crop production or minimized pesticide side effects.

Locke and Zablotowicz (Chapter 14) conclude that the soil biota are generally quite resilient to pesticides applied at recommended rates, with only transient disturbances (see also section above on 'Soil organic matter and biodiversity'). The authors touch upon the need to evaluate the significance of measured side effects in relation to effects caused by fluctuations in, for example, soil moisture or temperature. They draw attention to the concept of Domsch et al. (1983), in which side effects on microorganisms and microbial processes are characterized as 'negligible', 'tolerable' or 'critical' on the basis of resilience to natural stress and disturbance. We find it crucial that the significance of measured effects is evaluated before management thresholds above which side effects are unacceptable can be identified. For example, it is notable that laboratory tests are often designed as dose–response experiments with concentrations that are unrealistic compared to standard farming practice. Based on their review, Locke and Zablotowicz identify a selection of diagnostic tests that may be performed to assess pesticide impact on soil health. They mention that the issue of experimental scale is a specific challenge. Side effects measured in the laboratory may differ considerably from effects obtained in the field because the field environment modifies pesticide distribution and behaviour (Elmholt, 1992). Under field conditions, the direct effects of chemicals applied at one site are often translated into indirect effects elsewhere (Levin and Kimball, 1984; Elmholt, 1991). In their section on conservation management, Locke and Zablotowicz exemplify the importance of the field environment, showing that increases in soil OM and accumulation of plant residues on the soil surface promote pesticide sorption and degradation.

Scientific Approaches in Agricultural Research

Discipline-oriented, reductionistic research

Any scientific approach involves some degree of 'reduction' of the research object. The reduction typically takes place when a part of the system is examined under controlled conditions. The strength of such experiments is the ability to reveal specific mechanisms that act in the soil system. For example, *in vitro* studies may elucidate mechanisms of inhibition or toxicity for specific pesticides (Locke and Zablotowicz, Chapter 14). In Fig. 18.3, this type of research is depicted with a high degree of reduction, but yielding an understanding of mechanisms (causal relations).

Discipline-oriented research may take place in the laboratory and in the field. Changing specific management procedures may yield quantitative information on important soil functions. Mechanistic models may be developed from a combination of descriptive measurements of system behaviour and studies of management effects in order to identify sustainable management procedures.

Studies of specific soil functions and individual soil properties cannot explain interactions with properties that are kept constant in the 'reduced' experiment but fluctuate in the 'real world'. Discipline-oriented research may therefore misinterpret the results

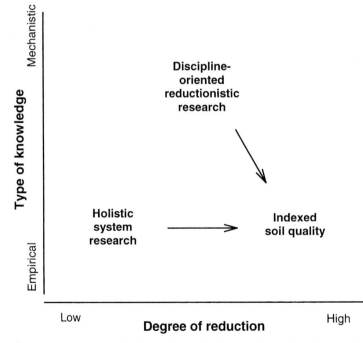

Fig. 18.3. Different research approaches depicted as related to the degree of 'reduction' (abscissa) and the type of knowledge gained (ordinate). The arrows indicate what happens when indexing the information gained by the two types of scientific approaches.

obtained. In the example above, the *in vitro* effects of the pesticide may not be observed in field soils, where sorption and degradation affect pesticide bioavailability (Locke and Zablotowicz, Chapter 14). However, the *reflexive objectivity*, i.e. a high awareness about the experimental conditions and the limitations imposed on the results (the intentional context in Fig. 1.1), may help to avoid erroneous conclusions.

Holistic research

Bezdicek and DePhelps (1994) and, more recently, Sturz and Christie (2002) pleaded for integrated, system-level research in order to develop sustainable farming systems. The strength of holistic system research lies in the low degree of reduction of the research object. Although any scientific study includes some degree of reduction of the 'real world', experiments with the whole agricultural system provide unique information on the integrated function and the direction of change at the system level of function. However, information obtained in this way provides little understanding of causal relationships within the system. Holistic research may thus be depicted in the lower left-hand 'corner' of Fig. 18.3, indicating a small reduction of system complexity, and a rather empirical output of the studies. Vandermeer (1995) emphasized the problems with the multidimensional complexity of systems ecology. Lal (1991) described holistic, integrated, interdisciplinary or systems-oriented approaches as those that can be talked about but not easily translated into practical research. Ellert *et al.* (1997) recognized all these problems, but pleaded for the advantages of having a look at 'the big picture'.

Holistic versus reductionistic research

Carter *et al.* (Chapter 15) introduce systems concepts in an agricultural context, and conclude that humans are an integral part of agroecosystems ('soft' systems). Hence, decisions on management and the basis for these (research results, general knowledge and preferences) are parts of the system. This is in accordance with the discussion on reflexive objectivity in Chapter 1; values and goals prevailing in society as well as preferences of the individual scientist are integrated in the research. The discussion of two sample systems (conservation tillage and organic farming) in Chapter 15 is introduced by reflections on the basic assumptions and goals for each system, but Carter *et al.* realize that farming practices fall along a continuum rather than into discrete groups. This is an important recognition when attempting to compare systems. Emergent properties are important in studies on the effects of soil management, but a system characteristic should not be considered as an emergent property if it is in reality caused by a specific management tool. Analysing and developing agricultural systems may be more relevant than comparing a given system with a 'conventional' system (Lockeretz, 1987; Raupp, 1993; Schjønning *et al.*, 2002).

The complex interactions among ecosystem components and management procedures discussed in Chapters 8 and 9 may be taken as a strong argument for an increased focus on the concept of emergent properties (Carter *et al.*, Chapter 15). However, Alabouvette *et al.* (Chapter 8) showed that OM in some situations favours the antagonistic fungus *Trichoderma*, whereas in others general microbial activity is favoured at the expense of *Trichoderma*. In this case, holistic system studies do not advance our understanding of the mechanisms involved. Mathematical models may organize data and shed more light on the interactions (Alabouvette *et al.*, Chapter 8), but we see concurrent studies of mechanisms under controlled conditions and observations of system performance as a better option.

Bouma (1997) similarly pleaded for an integration of reductionistic and holistic research rather than a shift from one to the other. The perception that much (reductionistic) research is unfocused and inward-looking, and that interactions among users and researchers will solve all problems, imposes a threat to fundamental research in soil science, including the development of the profession (Bouma, 1997). Thus, Bouma suggested a holistic–reductionistic–holistic approach in soil and environmental sciences. This would imply an initial phase with intensive interaction with users and colleagues in order to identify the key issues and the research needed. Specific, discipline-oriented studies would then identify relevant processes and causal relations in the system studied. Subsequently, the research results should be fed back to the larger group involved in the project (Bouma, 1997).

In Chapter 16, Bouma states that top-down land-use planning does not work in a modern network society, and he envisages that profound changes are needed in the research and communication of soil science before it can play its proper role. Bouma (2001b) regarded a squeeze between soil scientists on the one hand and concerned citizens and policy makers on the other not only an interesting option for the future but a necessity for the survival of soil science.

Indexing of soil quality knowledge

Indexing is a way of condensing information and may be regarded as a 'reduction' of system complexity. Indices are typically calculated from specific soil-quality indicators. The soil quality literature repeatedly emphasizes the need for indexing in order to create general and quantitative measures of soil quality (e.g. Andrews *et al.*, 2002a,b). Karlen *et al.* (Chapter 2) claim that indicator selection and indexing are important for soil quality assessments to be understood and used by land managers of soil quality. In contrast, Sojka *et al.* (2003) argued that any index must be deconstructed back to the original components in order to guide management. We fully support that soil managers are provided a full range of specific readouts rather than a red/green warning light (the index value).

Karlen and Stott (1994) suggested that soil-quality indicators were indexed by multi-objective analysis principles used in systems engineering. The procedure normalizes index values (ranging 0–1) by scoring functions and has been used to identify management effects in a number of studies (e.g.

Karlen *et al.*, 1994a,b; Hussain *et al.*, 1999). Karlen *et al.* (Chapter 2) demonstrate a way of scoring P values in soil, but given the complex nature of P dynamics in soil (Condron, Chapter 5), it is obvious that even non-linear indexing includes a significant loss of information.

The process of indexing incorporates intentional values and goals as well as general knowledge about the specific soil characteristic. Sojka *et al.* (2003) focused on the danger of integrating values in scientific studies. We agree with this warning when speaking of indexed information because the values are 'hidden'. Letey *et al.* (2003) demonstrated how this may produce biased conclusions when comparing farming systems. However, we acknowledge that values and goals are also in play in science, and suggest that the communication of research results relies on a combination of 'objective' statements and explicit reflections on the sustainability criteria (cost/benefit) behind the research.

Indexed soil quality information may provide an even higher degree of reduction than traditional research, which normally delivers results in absolute numbers (Fig. 18.3). The indexed information may further be classified as rather empirical because an index necessarily includes salient assumptions. Indices derived from reductionistic research reflect more mechanistic knowledge than those derived from holistic indicators. However, this more subtle aspect is not indicated in Fig. 18.3. The issues raised here become crucial when indexing includes calculations of a general index that incorporates a number of specific indices, which are weighted differently based on subjective/ arbitrary criteria (Andrews *et al.*, 2002b). Munasinghe and Shearer (1995) were also concerned about the problem of 'embedded' values when discussing the possibility of expressing sustainability in one single index.

Communication and Implementation

The local/national perspective

This book deals with the management of soil quality. The primary focus is identification of sustainable management procedures – not identification of soil-type-independent indicators and their reference values. This has implications for the type of information obtained and for how it is implemented in soil management. If universal thresholds are identified for a soil property and/or management tool, these may be communicated directly to decision makers and farmers for implementation. An example is the management threshold of approximately 1000 kg $CaCO_3$/ha/year for keeping soil acidity at a prescribed level (Johnston, Chapter 3). However, for most issues addressed in this book, the interaction between soil and management is less simple. The actions taken to approach sustainable farming systems may depend on soil type, climate, crop rotation and other physical and societal conditions framing agricultural activities. We need to understand the functions and processes in the soil and the impact of different management tools. One example is the important issue regarding biodiversity in soil. The reviews provided by Alabouvette et al. (Chapter 8) and Brussaard et al. (Chapter 9) clearly indicate that the complex information has to be communicated to stakeholders and decision makers with due attention to subtle details.

National and international regulations set rules that frame management options of importance for protecting the environment. For example, the European Commission in 1991 introduced a nitrate guideline in the form of a maximum application of 170 kg N/ha in organic manure (Bouma, Chapter 16). Obviously, this management threshold is much too rigid, given the complex dynamics of N in agricultural soils (Christensen, Chapter 4). Hence, Bouma (Chapter 16) pleads for regulations acknowledging professional qualities at the local level (the farmer). Based on research results, agricultural advisors and farmers are aware of management procedures of key importance to the fate of N in soil. These include the timing of manure application, the decisive influence of soil type and weather, etc. A successful reduction in the impact of agriculture on the environment requires well-educated advisors and farmers and a proper dissemination of research results at the local level. This example also shows that, in some circumstances, management has to be based more on soil-quality indicator thresholds (the soil water nitrate concentration) than on management thresholds (amount of N in organic manure).

We conclude that decisions should preferably be made at the local level. This can be facilitated by a high educational level of farmers and soil managers in general. However, scientists have to be involved in creating tools for organizing the large amount of information. The interaction of stakeholders and researchers in research programmes may be an important vehicle for joint learning. The suggestion is based on an analysis of the modern 'network' society, where policy makers and various stakeholders (who are increasingly members of non-governmental organizations) address the problems encountered (Bouma, 2001a; Bouma, Chapter 16). Bouma et al. (1998) advocated a 'research chain'. This would imply methodical steps in a process of identifying, selecting, resolving and presenting the soil quality problem and the knowledge gained. The 'chain' approach should be performed by an interdisciplinary group of researchers and stakeholders. A step-by-step increase in complexity of succeeding research chains may be a way of further improving the quality of decisions on land use (Bouma, 2001b). At the same time, this will optimize the focus and the contribution from all branches of science, from fundamental, to strategic to applied research. Please consult Bouma's papers and Chapter 16 of this volume for a detailed explanation of these ideas.

The term 'windows of opportunities' should be regarded as a tool for decision makers coping with land-use problems (Bouma, 1994). Each particular type of soil has a characteristic 'window of opportunity' as a function of management (Bouma, Chapter 16). Teams of stakeholders and researchers may use the procedures discussed above for identifying the limits of that window as a true measure of soil quality for any given type of soil. Droogers and Bouma (1997) put the concept to work by applying computer simulation techniques to demonstrate the probability of nitrate leaching for contrasting management systems.

The international perspective

The general concern regarding the protection of non-renewable resources has brought about conventions on, for example, biodiversity, climate change and desertification. Yet, no similar convention has been approved for soil. Several groups (including the International Union of Soil Scientists) have joined to promote a soil protection effort leading to a draft proposal for 'Convention on Sustainable Use of Soils' (Tutzing Project, 1998). Yaalon and Arnold (2000) advocated the participation of soil scientists as experts in this activity in order to bridge the gap between the various regional or international 'top-down' proposals and the many local 'bottom-up' efforts already active. Bouma (Chapter 16) foresees the need for a 'World Soil Agreement' as a binding treaty on optimal use of the soil resources on a world level. However, Wynen (2002) examined the possibility of creating a UN Convention on soil quality and concluded that the most realistic and appropriate approach would be a 'Code of Conduct' for soil management. Please consult Chapter 2 for a somewhat more detailed introduction to these activities.

For developing countries, sustainable agriculture implies 'the most economic-cum-efficient harnessing of solar energy in the form of agricultural products without degrading soil and environment quality' (Lal, Chapter 17), a statement in accordance with the SSSA definition of soil quality (the three concerns, Fig. 1.2). However, the main emphasis is on the 'economic-cum-efficient' productivity, since societal priorities clearly differ from those in developed countries. The developing countries have no alternative to agricultural intensification (Lal, Chapter 17).

Although the concept of soil quality may be universal, its application remains soil/society specific, and this needs to be recognized for the concept to be useful in addressing the problems of resource-poor farmers in developing countries (Lal, Chapter 17). Accordingly, Lal stated that the high and increasing demographic pressure in developing countries, coupled with scarcity of soil and water resources, make sustainable agriculture

not synonymous with 'low-input', 'organic' or 'alternative' agriculture.

Within this clearly different framework of sustainability priorities, Lal (Chapter 17) found that, to a large extent, the same soil attributes may be used as soil-quality indicators. However, due to soil type and climate differences, specific research information is needed to establish threshold values and the rate of change of these parameters under diverse land uses and soil management. The major emphasis should be on transfer of technology to land managers and small landholders: 'adopting a participatory or bottom-up approach is necessary to link scientific information with farmer knowledge, and to translate information to practice' (Lal, Chapter 17). Generally, the emphasis should be on solving problems through a demand-driven agenda where stakeholders are active partners in planning and implementing the agenda (Bouma, Chapter 16; Lal, Chapter 17).

Values in Science and the Precautionary Principle

Chapter 1 considered science and society. We adapted the concept of the *reflexive objectivity* (Alrøe and Kristensen, 2002) as a means of describing how values and goals are linked to science, and found that any research activity arises in a social context. The values and goals of each individual researcher will influence his/her selection of research topics and even the implications drawn from the research work.

Anderson (1994) examined biodiversity in managed ecosystems and recognized the influence of societal priorities and moral concerns. He considered loss of species a normal evolutionary process, but raised the question 'what constitutes the *acceptable* rate of extinctions?' A highly disturbed agricultural soil may show reduced biodiversity (e.g. Crossley *et al.*, 1992; Brussaard *et al.*, Chapter 9) and biodiversity has been linked to ecosystem function and stability (e.g. Tilman and Downing, 1994; Brussaard *et al.*, Chapter 9). But how much diversity is needed within a given functional group of organisms to

maintain a specific function in the face of perturbations (Pankhurst *et al.*, 1997)? Pankhurst
et al. (1997) found that the adoption of conservation management practices and organic
agriculture are attempts to return to a more
natural, less-disturbed ecosystem, but questioned whether such soils are healthier than
those of conventional agriculture. Can the
world afford the production penalty that can
accompany conservation management? In
contrast, Brussaard *et al.* (Chapter 9) use the
same background for recommending reduction of 'sources of ecosystem stress, such as
pesticides and fertilizers . . ., which affect the
community diversity of soil organisms'. In a
comment to Pankhurst *et al.* (1997), Brussaard
et al. (Chapter 9) ask 'can the world afford
not to accept the production penalty?' (from
conservation agriculture). Different views on
the significance of biodiversity of agricultural
soils and different opinions on the need
to sustain biodiversity in management
prescriptions clearly exist.

There are no clear-cut conclusions to be
drawn regarding the influence of biodiversity
on soil functioning as also exemplified by
the confronting views of Pankhurst *et al.*
(1997) and Brussaard *et al.* (Chapter 9). What is
important is the fact that research is influenced by the priorities and prevailing ideas
in society, and researchers have to explicitly
state their priorities and reasons for a given
research activity. We also need to identify and
reveal the value-laden criteria that may enter
into our conclusions, such as sustainability or
precaution in one or another sense. This is crucial for fruitful, critical communication among
scientists as well as for decision making by
stakeholders in the surrounding society.

The precautionary principle is a cul-
turally framed concept that takes its cue
from changing social conceptions about the
appropriate roles of science (O'Riordan and
Cameron, 1994). The concept is related to and
interacts with the sustainability concept. The
basic issues of the precautionary principle are:
(i) thoughtful action in advance of scientific
proof; (ii) leaving ecological space; (iii) care
in management; (iv) shifting the burden of
proof; and, finally, (v) balancing the basis of
proportionality. It is beyond the scope of this
book to go into detail with this principle. Issue

(i) is rather difficult to combine with natural
sciences. 'Thoughtful action in advance of
scientific proof' in the context of agricultural
ecosystems means that management decisions
should be based on a 'burden of evidence'
when 'hard' data are not available (O'Riordan
et al., 2001).

The precautionary principle remains a
challenge to the scientific community. This
principle will increasingly face and provoke
scientists, because concerned citizens, non-
governmental organizations and engaged
political movements become more and more
important in setting the research agenda
in society. Considerations on precautionary
actions will inevitably appear on the agenda of
teams dealing with land management, where
any decision has to be based on evaluations
of benefits, risks and costs. Hence, scientists
will increasingly have to deliver data including the probabilities of occurrence rather than
fixed values. The scientific community should
face this inevitable challenge by explicitly
considering the sustainability concept and on
this background interact with stakeholders in
scientifically founded implementation of the
precautionary principle.

Conclusions and Perspectives

Agricultural scientists participate in solving
challenges related to a sufficient food supply
of high quality and a concurrent protection of
the environment. The soil quality concept as
expressed through the SSSA-suggested definition (Fig. 1.2; Allan *et al.*, 1995) appears to
lend an adequate framework for addressing
these challenges. However, there are a number of barriers associated with insufficient
communication among scientists. One crucial
precondition for a fruitful participation of soil
science in solving the pressing problems is
the recognition of the impact of values and
goals that prevails in society. Only scientific
results backed by explicit expressions of
sustainability criteria and the related cost–
benefit analyses will allow stakeholders and
decision makers to lay down guidelines
for development in agriculture. We find that
the concept of reflexive objectivity is useful

in promoting an increased awareness of science–society interactions.

The soil science community needs to address the concept of soil quality in more detail. It appears that different conceptual interpretations create barriers to a fruitful use of the concept. We strongly advocate the term to be used primarily as a cognitive concept. This means that the soil quality concept should express the sustainability criteria prevalent and decisive for scientific work with soil and agricultural management. We dissuade the focus on indexing soil quality attributes and rather encourage the identification of management thresholds securing the prescribed sustainability of the system.

Most management issues addressed in this book are highly complex. Key terms are tightening nutrient cycles, identifying soil strength thresholds and OM management optimal for specific functions. The challenges further include the identification of thresholds for specific contaminants applied to soils. We conclude that it is generally not possible to identify simple management thresholds. Existing knowledge on processes and management effects may be implemented in soil management by the use of stochastic and mechanistic simulation models. Recognizing the complexity of the soil ecosystem, dissemination of soil quality knowledge may be hampered by the use of indexed information. Stakeholders should be active partners in implementing the scientific knowledge in management strategies.

We see a need for a dual approach in agricultural science. This includes research on specific structures, mechanisms and processes to reveal causal relationships as well as on-farm, holistic investigations of system performance. The identification of emergent system properties should be used to define research needs and to develop the specific system. Development of systems should be preferred, whereas comparisons may obscure specific management effects and emergent system properties.

We foresee an increased focus on the precautionary principle in interactions among scientists and decision makers. This accentuates the need for scientists to accustom themselves to reflexive objectivity and explicit considerations on sustainability, as these are considered crucial for the even stronger challenge of addressing the precautionary principle. The profession of (soil) science will only survive if scientists realize that they have to participate in teams involved with implementation of research results. The presentation of scientific results in a transparent framework of sustainability considerations will enable a more clear distinction between science *per se* and the values and goals set by society.

References

Alexander, M. (1971) Agriculture's responsibility in establishing soil quality criteria. In: *Environmental Improvement – Agriculture's Challenge in the Seventies*. National Academy of Sciences, Washington, DC, pp. 66–71.

Allan, D.L., Adriano, D.C., Bezdicek, D.F., Cline, R.G., Coleman, D.C., Doran, J.W., Haberen, J., Harris, R.G., Juo, A.S.R., Mausbach, M.J., Peterson, G.A., Schuman, G.E., Singer, M.J. and Karlen, D.L. (1995) SSSA Statement on soil quality. In: *Agronomy News*, June, ASA, Madison, Wisconsin, p.7.

Alrøe, H.F. and Kristensen, E.S. (2002) Towards a systemic research methodology in agriculture. Rethinking the role of values in science. *Agriculture and Human Values* 19, 3–23.

Anderson, J.M. (1994) Functional attributes of biodiversity in land use systems. In: Greenland, D.J. and Szabolcs, I. (eds) *Soil Resilience and Sustainable Land Use*. CAB International, Wallingford, UK, pp. 267–290.

Andrews, S.S., Karlen, D.L. and Mitchell, J.P. (2002a) A comparison of soil quality indexing methods for vegetable production systems in northern California. *Agriculture Ecosystems and Environment* 90, 25–45.

Andrews, S.S., Mitchell, J.P., Mancinelli, R., Karlen, D.L., Hartz, T.K., Horwath, W.R., Pettygrove, G.S., Scow, K.M. and Munk, D.S. (2002b) On-farm assessment of soil quality in California's Central Valley. *Agronomy Journal* 94, 12–23.

Anonymous (ed.) (1996) *The Soil Quality Concept*. The Soil Quality Institute, Natural Resources Conservation Service, United States Department of Agriculture, Washington, DC, 89 pp.

Bezdicek, D.F. and DePhelps, C. (1994) Innovative approaches for integrated research and educational programs. *American Journal of Alternative Agriculture* 9, 3–8.

Bouma, J. (1994) Sustainable land use as a future focus for pedology. *Soil Science Society of America Journal* 58, 645–646.

Bouma, J. (1997) Soil environmental quality: a European perspective. *Journal of Environmental Quality* 26, 26–31.

Bouma, J. (2001a) The role of soil science in the land use negotiation process. *Soil Use and Management* 17, 1–6.

Bouma, J. (2001b) The new role of soil science in a network society. *Soil Science* 166, 874–879.

Bouma, J., Finke, P.A., Hoosbeek, M.R. and Breeuwsma, A. (1998) Soil and water quality at different scales: concepts, challenges, conclusions and recommendations. *Nutrient Cycling in Agroecosystems* 50, 5–11.

Carter, M.R. (2002) Soil quality for sustainable land management: organic matter and aggregation interactions that maintain soil functions. *Agronomy Journal* 94, 38–47.

Christensen, B.T. (2001) Physical fractionation of soil and structural and functional complexity in organic matter turnover. *European Journal of Soil Science* 52, 345–353.

Christensen, B.T. and Johnston, A.E. (1997) Soil organic matter and soil quality – lessons learned from long-term experiments at Askov and Rothamsted. In: Gregorich, E.G. and Carter, M.R. (eds) *Soil Quality for Crop Production and Ecosystem Health*. Developments in Soil Science 25, Elsevier, Amsterdam, pp. 399–430.

Crossley, D.A., Mueller, B.R. and Perdue, J.C. (1992) Biodiversity of micro-arthropods in agricultural soils: relations to processes. *Agriculture, Ecosystems and Environment* 40, 37–46.

Domsch, K.H., Jagnow, G. and Anderson, T.-H. (1983) An ecological concept for the assessment of side-effects of agrochemicals on soil microorganisms. *Residue Reviews* 86, 65–105.

Doran, J.W. and Jones, A.J. (eds) (1996) *Methods for Assessing Soil Quality*. Soil Science Society of America Special Publication No. 49.

Droogers, P. and Bouma, J. (1997) Soil survey input in explanatory modeling of sustainable soil management practices. *Soil Science Society of America Journal* 61, 1704–1710.

Ellert, B.H., Clapperton, M.J. and Anderson, D.W. (1997) An ecosystem perspective of soil quality. In: Gregorich, E.G. and Carter, M.R. (eds) *Soil Quality for Crop Production and Ecosystem Health*. Developments in Soil Science 25, Elsevier, Amsterdam, pp. 115–141.

Elmholt, S. (1991) Side effects of propiconazole, (Tilt 250 EC) on non-target soil fungi in a field trial compared with natural stress effects. *Microbial Ecology* 22, 99–108.

Elmholt, S. (1992) Effect of propiconazole on substrate amended soil respiration following laboratory and field application. *Pesticide Science* 34, 139–146.

Håkansson, I. and Petelkau, H. (1994) Benefits of limited axle load. In: Soane, B.D. and van Ouwerkerk, C. (eds) *Soil Compaction in Crop Production*. Developments in Agricultural Engineering 11, Elsevier, Amsterdam, pp. 479–499.

Heckrath, G., Brookes, P.C., Poulton, P.R. and Goulding, K.W.T. (1995) Phosphorus leaching from soils containing different P concentrations in the Broadbalk experiment. *Journal of Environmental Quality* 24, 904–910.

Herdt, R.W. and Steiner, R.A. (1995) Agricultural sustainability: concepts and conundrums. In: Barnett, V., Payne, R. and Steiner, R. (eds) *Agricultural Sustainability: Economic, Environmental and Statistical Considerations*. John Wiley & Sons, Chichester, UK, pp. 1–13.

Hussain, I., Olson, K.R., Wander, M.M. and Karlen, D.L. (1999) Adaptation of soil quality indices and application to three tillage systems in southern Illinois. *Soil and Tillage Research* 50, 237–249.

Karlen, D.L. and Stott, D.E. (1994) A framework for evaluating physical and chemical indicators of soil quality. In: Doran, J.W., Coleman, D.C., Bezdicek, D.F. and Stewart, B.A. (eds) *Defining Soil Quality for a Sustainable Environment*. Soil Science Society of America Special Publication No. 35, Soil Science Society of America, Madison, Wisconsin, pp. 53–72.

Karlen, D.L., Wollenhaupt, N.C., Erbach, D.C., Berry, E.C., Swan, J.B., Eash, N.S. and Jordahl, J.L. (1994a) Crop residue effects on soil quality following 10-years of no-till corn. *Soil and Tillage Research* 31, 149–167.

Karlen, D.L., Wollenhaupt, N.C., Erbach, D.C., Berry, E.C., Swan, J.B., Eash, N.S. and Jordahl, J.L. (1994b) Long-term tillage effects on soil quality. *Soil and Tillage Research* 32, 313–327.

Karlen, D.L., Mausbach, M.J., Doran, J.W., Cline, R.G., Harris, R.F. and Schuman, G.E. (1997) Soil quality: a concept, definition, and framework for evaluation. *Soil Science Society of America Journal* 61, 4–10.

Karlen, D.L., Andrews, S.S. and Doran, J.W. (2001) Soil quality: current concepts and applications. *Advances in Agronomy* 74, 1–40.

Lal, R. (1991) Preface to the book: Soil Management for Sustainability. In: Lal, R. and Pierce, F.J. (eds) *Soil Management for Sustainability*. Soil and Water Conservation Society, Ankeny, Iowa, pp. xi–xii.

Larney, F.J., Fortune, R.A. and Collins, J.F. (1988) Intrinsic soil physical parameters influencing intensity of cultivation procedures for sugar beet seedbed preparation. *Soil and Tillage Research* 12, 253–267.

Letey, J., Sojka, R.E., Upchurch, D.R., Cassel, D.K., Olson, K., Payne, B., Petrie, S., Price, G., Reginato, R.J., Scott, H.D., Smethurst, P. and Triplett, G. (2003) Deficiencies in the soil quality concept and its application. *Journal of Soil and Water Conservation* 58, 180–187.

Levin, S.A. and Kimball, K.D. (1984) New perspectives in ecotoxicology. *Environmental Management* 8, 375–442.

Lockeretz, W. (1987) Establishing the proper role for on-farm research. *American Journal of Alternative Agriculture* 2, 132–136.

McDowell, R.W., Sharpley, A.N., Brookes, P.C. and Poulton, P.R. (2001) Relationship between soil test phosphorus and phosphorus release to solution. *Soil Science* 166, 137–149.

Munasinghe, M. and Shearer, W. (1995) An introduction to the definition and measurement of biogeophysical sustainability. In: Munasinghe, M. and Shearer, W. (eds) *Defining and Measuring Sustainability. The Biogeophysical Foundations.* World Bank, Washington, DC, pp. xvii–xxxii.

Munkholm, L.J. and Schjønning, P. (2003) Structural vulnerability of a sandy loam exposed to intensive tillage and traffic in wet conditions. *Soil and Tillage Research* (in press).

Munkholm, L.J., Schjønning, P. and Petersen, C.T. (2001) Soil mechanical behaviour of sandy loams in a temperate climate: case-studies on long-term effects of fertilization and crop rotation. *Soil Use and Management* 17, 269–277.

O'Riordan, T. and Cameron, J. (eds) (1994) *Interpreting the Precautionary Principle.* Cameron May, London.

O'Riordan, T., Cameron, J. and Jordan, A. (eds) (2001) *Reinterpreting the Precautionary Principle.* Cameron May, London.

Pankhurst, C.E., Doube, B.M. and Gupta, V.V.S.R. (1997) Biological indicators of soil health: synthesis. In: Pankhurst, C.E., Doube, B.M. and Gupta, V.V.S.R. (eds) *Biological Indicators of Soil Health.* CAB International, Wallingford, UK, pp. 419–435.

Poulsen, H.D. (1998) Zinc and copper as feed additives, growth factors or unwanted environmental factors. *Journal of Animal and Feed Sciences* 7, 135–142.

Raupp, J. (1993) Einige Gedanken und Leitlinien zur Forschung im ökologischen Landbau. *Ökologie und Landbau* 21, 9–12.

Schjønning, P., Elmholt, S., Munkholm, L.J. and Debosz, K. (2002) Soil quality aspects of humid sandy loams as influenced by organic and conventional long-term management. *Agriculture, Ecosystems and Environment* 88, 195–214.

Smyth, A.J. and Dumanski, J. (1993) *FESLM: an International Framework for Evaluating Sustainable Land Management.* World Resources Reports 73. Land and Water Development Division, FAO, Rome, 77 pp.

Sojka, R.E. and Upchurch, D.R. (1999) Reservations regarding the soil quality concept. *Soil Science Society of America Journal* 63, 1039–1054.

Sojka, R.E., Upchurch, D.R. and Borlaug, N.E. (2003) Quality soil management or soil quality management: performance versus semantics. *Advances in Agronomy* 79, 1–68.

Sturz, A.V. and Christie, B.R. (2003) Rationale for an holistic approach to soil quality and crop health. *Soil and Tillage Research* 72, 105–106.

Tilman, D. and Downing, J.A. (1994) Biodiversity and stability in grasslands. *Nature* 367, 363–365.

Trautner, A. (2003) On soil behaviour during field traffic. Ph.D. thesis, Swedish University of Agricultural Sciences, Uppsala, Sweden.

Tutzing Project (1998) Preserving Soils for Life. Proposal for a 'Convention on Sustainable Use of Soils' (Soil Convention). Oekom Verlag, Munich.

Vandermeer, J. (1995) The ecological basis of alternative agriculture. *Annual Review of Ecology and Systematics* 26, 201–224.

Warkentin, B.P. and Fletcher, H.F. (1977) Soil quality for intensive agriculture. In: *Proceedings of the International Seminar on Soil Environment and Fertility Management in Intensive Agriculture.* Society of Science Soil and Manure, National Institute of Agricultural Science, Tokyo, pp. 594–598.

Watts, C.W. and Dexter, A.R. (1997) Intensity of tillage of wet soil and the effects on soil structural condition. In: Fotyma, M., Jozefaciuk, A., Malicki, L. and Borowiecki, J. (eds) *Fragmenta Agronomica TOM 2B/97. Proceedings 14th ISTRO Conference, 27 July–1 August 1997, Pulawy, Poland.* Polish Society of Agrotechnical Sciences, Poland, pp. 669–672.

Watts, C.W., Dexter, A.R. and Longstaff, D.J. (1996) An assesment of the vulnerability of soil structure to destabilisation during tillage. Part II. Field trials. *Soil and Tillage Research* 37, 175–190.

WCED (1987) *Our Common Future: the Brundtland Report.* Report from the World Commission on Environment and Development (WCED). Oxford University Press, Oxford.

Wynen, E. (2002) A UN convention on soil health or what are the alternatives? Available at: http:// okologiens-hus.dk/PDFs/Muldrap. doc

Yaalon, D.H. and Arnold, R.W. (2000) Attitudes towards soils and their societal relevance: then and now. *Soil Science* 165, 5–12.

Yang, X.M. and Kay, B.D. (2001) Rotation and tillage effects on soil organic carbon sequestration in a typic Hapludalf in Southern Ontario. *Soil and Tillage Research* 59, 107–114.

Index

Page numbers in *italics* indicate that the reference is to a figure, table or illustration. The index term 'soil quality' is only used to bring together references to very broad discussions on this topic.